Advanced Structured Materials

Volume 70

Series editors

More information about this series at http://www.springer.com/series/8611

Andreas Öchsner · Holm Altenbach
Editors

Mechanical and Materials Engineering of Modern Structure and Component Design

Springer

Editors
Andreas Öchsner
Griffith School of Engineering
Griffith University
Southport Queensland, QLD
Australia

and

The University of Newcastle
Callaghan, QLD
Australia

Holm Altenbach
Fakultät für Maschinenbau, Lehrstuhl für Technische Mechanik
Institut für Mechanik
Otto-von-Guericke-Universität Magdeburg
Magdeburg
Germany

ISSN 1869-8433 ISSN 1869-8441 (electronic)
Advanced Structured Materials
ISBN 978-3-319-19442-4 ISBN 978-3-319-19443-1 (eBook)
DOI 10.1007/978-3-319-19443-1

Library of Congress Control Number: 2015939927

Springer Cham Heidelberg New York Dordrecht London

Printed on acid-free paper

Springer International Publishing AG Switzerland is part of Springer Science+Business Media (www.springer.com)

Preface

The idea of this monograph is to present the latcst results related to mechanical and materials engineering applied to the design of modern engineering materials and components. The contributions cover the classical fields of mechanical, civil, and materials engineering up to bioengineering and advanced materials processing and optimization. The materials and structures covered can be categorized into modern steels, aluminum and titanium alloys, polymers/composite materials, biological and natural materials, material hybrids, and modern nano-based materials. Analytical modeling, numerical simulation, the application of state-of-the-art design tools, and sophisticated experimental techniques are applied to characterize the performance of materials and to design and optimize structures in different fields of engineering applications.

The 8th International Conference on Advanced Computational Engineering and Experimenting, ACE-X 2014, was held in Paris, France, from June 30, 2014 to July 3, 2014 with a strong focus on computational-based and supported engineering. This conference served as an excellent platform for the engineering community to meet with each other and to exchange the latest ideas. This volume contains 35 revised and extended research articles written by experienced researchers participating in the conference. Well-known experts present their research on damage and fracture of material and structures, materials modeling and evaluation up to recent printing, and visualization for advanced analyses and evaluation.

The organizers and editors wish to thank all the authors for their participation and cooperation which made this volume possible. Finally, we would like to thank the team of Springer Publisher, especially Dr. Christoph Baumann, for the excellent cooperation during the preparation of this volume.

April 2015

Andreas Öchsner
Holm Altenbach

Contents

therefore an effective element of duplex microstructures. As an economical development, it is concluded that manganese is a useful replacement element for nickel in duplex alloys, but further work is required before the present alloys, or variations of them, could be commercially viable.

Keywords Free-Nickel duplex stainless steels · Corrosion resistance · Pitting attack · Microstructure development

1 Introduction

The industrial use of duplex stainless steel is rapidly increasing due to the combined advantages of better mechanical and corrosion properties [1–5]. Since the development of first-generation duplex stainless steels in the 1930s, considerable research efforts have been conducted to improve both mechanical and corrosion properties, particularly by controlling alloying elements, such as N, Cr, and Mo [3, 6–16]. Charles [4], for example, reported that the addition of Cr and/or Mo improved the resistance to pitting corrosion and stress corrosion cracking of duplex stainless steels. He further proposed that the addition of such elements needs to be done with caution since they can promote detrimental sigma phases at elevated temperatures [4]. Despite the extensive research works on the mechanical and corrosion behavior of duplex stainless steels, most of the researches have been conducted on the wrought products, and only a limited number of studies are available on the cast products of duplex stainless steels. Furthermore, Mn is often added to duplex stainless steel to increase the solubility of N to maximize the beneficial effect of N [17, 18]. According to the work of Gunn [19] and Kemp et al. [20] however, the effect of Mn on the microstructural evolution, as well as the mechanical and corrosion properties of duplex stainless steels have not been well established. The objective of the present study was therefore to examine the effect of Mn on the tensile and corrosion behaviour of free-nickel cast duplex stainless steels.

2 Experimental Procedures

The two experimentally proposed cast alloys of low and medium low carbon duplex stainless steel have been melted in a 10 kg laboratory induction furnace under nitrogen pressure around 8 bars and then cast in a metallic mould. The commercial duplex steels 2304 and 2205 grades were received in solution-annealed condition. The specimens were cut from plates and prepared for various property evaluations. The contents of alloying elements were determined, as shown in Table 1. The steels

Table 1 Chemical composition of produced DSS

Alloying elements	Low carbon samples		DSS	
	Sample 1	Sample 2	2304	2205
C	0.14	0.232	0.02	0.02
Si	0.27	0.288	0.31	0.32
Mn	6.44	13.45	1.5	1.5
P	0.0056	0.05	0.04	0.04
S	0.031	0.017	0.03	0.03
Cr	24.8	23.33	22.9	21.9
Mo	1.76	1.79	0.4	3.0
Ni	0.19	0.2	4.7	5.7
N	0.21	0.199	–	–

were heat treated by homogenization at 1200 °C for 6 h to decrease chemical segregation by diffusion and to homogenize the overall microstructure.

Test samples were solution annealed at 1050 °C for 15 min and then water quenched to dissolve inter-metallic phases and restore mechanical properties and corrosion resistance to the as-cast duplex stainless steel. In order to reveal the microstructure, the specimens were polished mechanically and then electro etched. The set up for the electro etching was 12 V, 20 °C and the electrolyte was solution of 10 % oxalic acid +2 % nitric acid. In addition, in order to evaluate quantitatively the duplex phases, EDX analysis attached to SEM microscopy was carried out to the solution annealed specimens. Tensile and hardness test samples were prepared and tested to measure the mechanical properties of the different alloys. Eventually, the corrosion resistance of the two samples of free nickel duplex stainless steel was measured by polarization tests and immersion tests, and the results were compared with the commercial duplex stainless steels as DSS 2304 and 2205. All polarization studies were carried out in a single component cell with three electrode configurations. A silver/silver chloride electrode saturated with KCl was used as a reference electrode and a platinum sheet with surface area of 1 cm^2 as a counter electrode. The electrochemical experiments were carried out under computer control using the potentiostat AUTOLAB® PGSTATE 30, where the working electrode was a 1 cm^2 specimen immersed in a 3.5 % NaCl solution. The immersion test where the variation of the corrosion rate for Nitrogen alloyed steel samples was done by measuring the weight loss per unit area with time in 10 wt% absolute HCl solution at room temperature. Prior to the immersion, the samples were mechanically polished using 400, 500 and 600 emery papers and lubricated using distilled water. The samples were then cleansed with distilled water, dried in air, weighed for the original weight (W_o) and then settled in the test solution. The corroded specimens were then removed from the solutions, cleansed with distilled water and dried, then the sample weight will be W_1 and the weight loss is ($W_o - W_1$). The corrosion attack was expressed in terms of weight loss per unit area (mg/cm^2).

3 Results and Discussion

3.1 *Microstructural Evolution*

The microstructural features of cast DSS are significantly influenced by the type and amount of alloying elements and the heat treatment applied. The solidification mode and the morphology of the primary delta ferrite phase as a leading phase, depends mainly on composition. Solidification can begin with primary ferrite or primary austenite.

In general, the microstructure of the as-cast alloys solidifies into the two different morphologies: acicular or globular, according to the amount of interstitial austenitic stabilization elements as Carbon, Manganese and Nitrogen.

According to XRD analysis, the microstructure solidified to a globular structure consisting of approximately 80 % volume fraction of ferrite and consequently, the remaining volume is austenite and some inter-metallic compounds or carbides, as shown in Fig. 1. The effect of homogenization shows a more or less homogenous structure with the disappearance of non-metallic inclusions which was found before

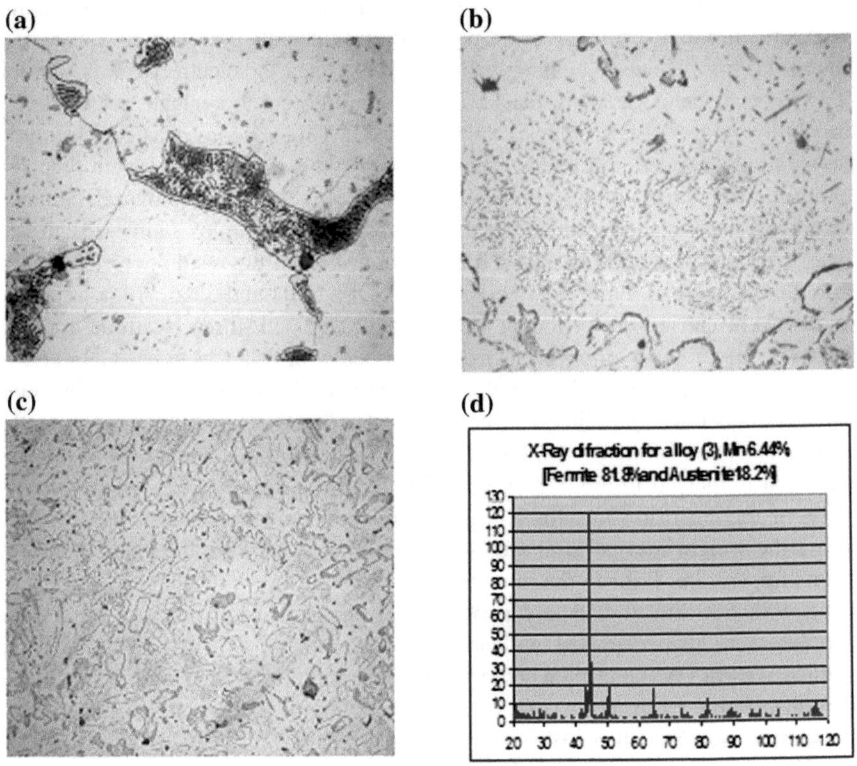

Fig. 1 Microstructure of sample 1 containing 6.44 % Mn and 0.21 % N; **a** as cast, **b** homogenized, **c** solution annealed at 1050 °C, 30 min and **d** XRD analysis

as shown in Fig. 1a and b, respectively. Figure 1c illustrates the effect of the solution annealing treatment in dissolving the carbides and inter-metallic phases. On the other hand, the globular morphology is not modified even after solution annealing at 1050 °C.

As observed from Fig. 2, by increasing the manganese content, the amount of the austenite phase has a moderate increase and the structure had a ferrite volume structure of about 74 %, as investigated from the XRD analysis. Hence, improving in the tensile properties of that alloy was expected due to the increase of the amount of austenite phases compared with the previous commercial steel sample. Unfortunately, that improve in the mechanical properties was associated with a decrease in the corrosion resistance of this steel sample, as it will be discussed and mentioned later in the characterization and evaluation of corrosion properties. The role of homogenization annealing is still to modify the grains size to be more uniform and finer, decreasing the amount of non-metallic inclusion and chemical segregation by diffusion.

Table 2 shows the variation of ferrite volume fractions due to change of manganese contents.

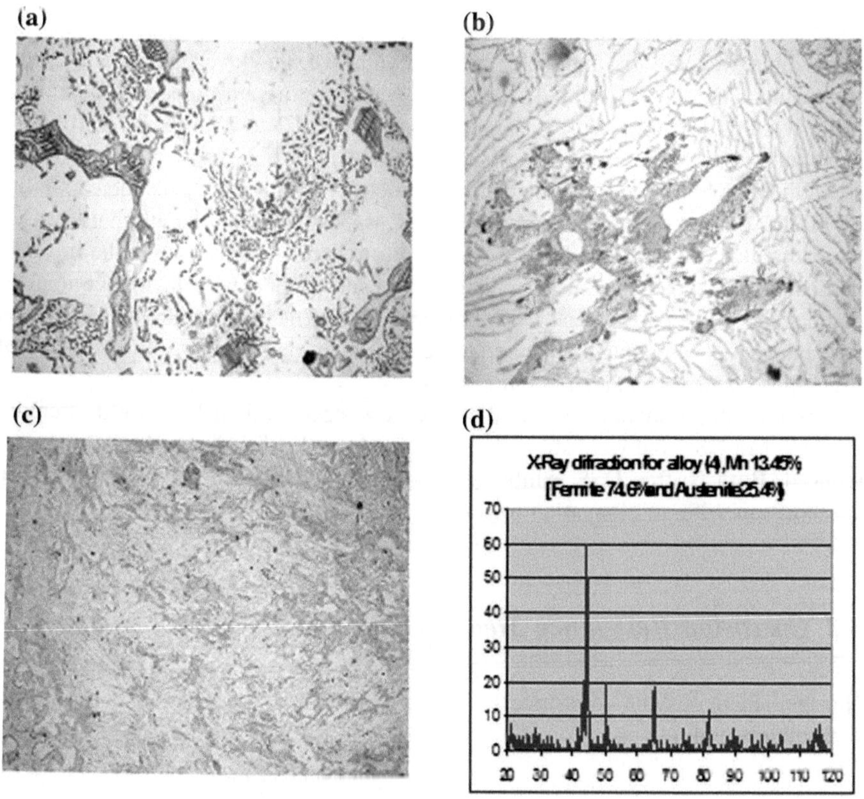

Fig. 2 Microstructure of sample 2 containing 13.45 % Mn and 0.19 % N; **a** as cast, **b** homogenized, **c** solution annealed at 1050 °C, 30 min, and **d** XRD analysis

Table 2 Characterization of phases by X-ray diffraction analysis

Sample no.	1	2
Manganese content	6.44 % Mn	13.45 % Mn
Volume fraction of ferrite	81.8 %	74.6 %

3.2 Mechanical Properties Evaluation

Mn is often added to stainless steels as an austenitic stabilizer for the partitioning effect of N [20]. In the present study, it was evident and demonstrated that different Mn contents significantly affect the tensile behaviour of the cast free nickel duplex stainless steel, as well as the corrosion behaviour. With increasing Mn content, the strength level was greatly increased from 691 to 815 MPa, while the tensile elongation was not notably reduced. Such a complex trend observed in tensile properties with varying Mn contents may be largely related to the microstructural evolution in the present alloy, including the volume fraction of each phase and the shape and size of the austenitic phase, both primary and secondary, along with an intrinsic solid solution hardening effect of Mn. Since, the austenitic phase is a harder phase than the ferritic phase, and the increase in the volume fraction of austenite would improve the strength level of cast duplex stainless steel, the remarkable increase in the yield strength value of the investigated alloy with increasing Mn content from 6.44 to 13.5 % was therefore attributed to the increase in the volume fraction of the austenite phase from 18.2 to 25.4 %, as well as the solid solution hardening effect of Mn as was evident in Fig. 3. The ultimate tensile strength value appeared to be related to the ductility which represents the capability of plastic deformation. The complex trend associated with the ultimate tensile strength values observed in Table 3 was therefore believed to be due to the intrinsic hardening effect of Mn and the change in tensile properties with Mn content in comparison to the commercial DSS 2304 and 2205. The decrease in the elongation with increasing Mn content appeared to be dependent on the change in the shape of the primary and secondary austenitic phase.

The one major advantage of duplex stainless steels is their high yield strength. For high Mn DSS, the yield strength is increased but the ductility and toughness are reduced. Both Ni-free DSS exhibit yield strengths of at least 550 to 750 MPa in combination with an elongation to fracture of at least 20 %.

3.3 Corrosion Resistance Measurement

Nitrogen increases the resistance to pitting corrosion, especially in combination with molybdenum.

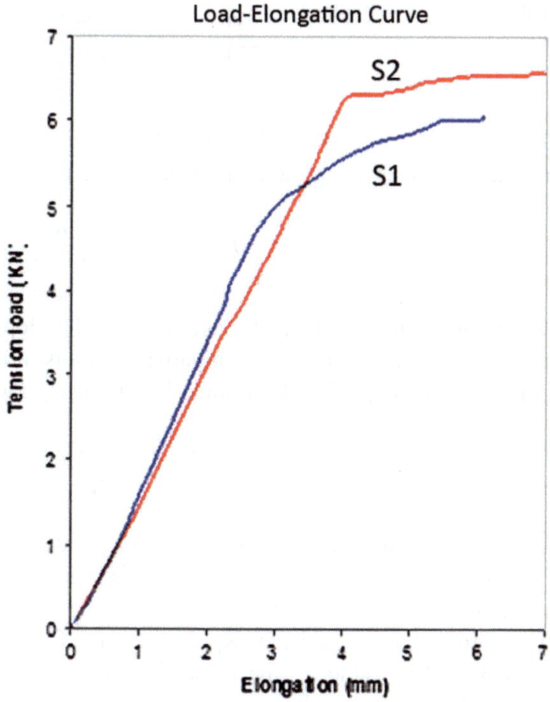

Fig. 3 Load-elongation curve for Sample 1 and 2 with 6.44 and 13.45 wt% Mn, respectively

Table 3 The results of mechanical properties of the experimental samples

Sample no.	U.T.S. (MPa)	Y.S. (MPa)	Elong. (%)	Impact energy j	Brinell hardness (BHN)
1	691	585	19	180	200
2	815	738	21	170	230
DSS 2304	663	480	40	180	190
DSS 2205	758	540	41	170	270

The Pitting Resistance Equivalent (PRE) was calculated according to the following equation [21]:

$$PRE = \%Cr + 3.3 \times \%Mo + 20 \times \%N$$

The pitting resistance results for samples 1 and 2 are illustrated in Table 4.

The polarizations resistance R_p was measured by scanning the potential in a range between ±15 mV around the corrosion potential Ecorr at 0.1 mV/s. From the current density versus potential plot the reciprocal of the slope of the curve dE/dI was determined at the corrosion potential in resistance units (Ohm/cm^2).

Table 4 PRE for samples 1 and 2

Sample no.	Sample 1	Sample 2	Duplex 2304
PRE	33.01	33.17	35

The corrosion current densities in μA/cm^2 were calculated from the polarization resistance values using the Stern–Geary equation, [22]

$$I_{corr} = \frac{B_a \times B_c}{2.303(B_a + B_c)} \tag{1}$$

where the Tafel constants B_a and B_c, are the slopes of tangents drawn on the respective anodic and cathodic polarization plots in mV/decade. Subsequently, the corrosion rate (C_R) was determined in mpy using the relationship:

$$C_R = \frac{0.13 \times I_{corr} \times eq.wt.}{\rho} \tag{2}$$

where ρ the density of the material (g/cm^3) and I_{corr} the current density (μA/cm^2).

The R_p is representative of the degree of protection of the passivation layer of the alloy surface. An increase in values of R_p improves the corrosion resistance of steel. The corrosion characteristics (R_p, I_{corr}, B_a, B_c, C_R) were determined and the results summarized in Table 5. The values of R_p allow a comparison scale to be established. The highest polarization resistances were revealed by S.1 of 2.94 × 10^6 Ohm/cm^2, while S.2 showed lower polarization resistances of 1.21 × 103 Ohm/cm^2. Based on the estimated R_p values, it seems that the general corrosion behaviour of nitrogen–manganese stabilized austenitic steels is better than "conventional" steels of type 2304.

The Tafel slopes B_c, B_a were determined by the fitting of a theoretical polarization curve to the experimental polarization curve plotted in the range of ±150 mV versus Eoc. The corrosion current Icorr is representative of the degree of degradation of the alloy. An alloy with a tendency towards passivity will have a value of B_a greater than B_c, whereas an alloy that corrodes will have a B_a lower than B_c. It can be noticed from Table 5, that regarding to the Tafel slope results, it was found that steels DSS 2304 and S.2 reveal a tendency towards depassivity. For S.1, B_a is higher than B_c indicating to passivation ability.

Table 5 Comparison of the electrochemical quantities measured and calculated for the various steels (37 °C)

Sample	B_a	B_c	R_p, (Ohm)	I_{corr}, (μA/cm^2)	C_R, (mpy)
DSS 2304	0.021	0.038	1.58E4	3.67	0.98
S.1	0.039	0.027	2.94E6	0.15	0.044
S.2	0.007	0.068	1.21E3	22.5	6.0

B_c and B_a tafel slopes, corrosion potential; R_p polarization resistance; I_{corr} corrosion current, C_R corrosion rate

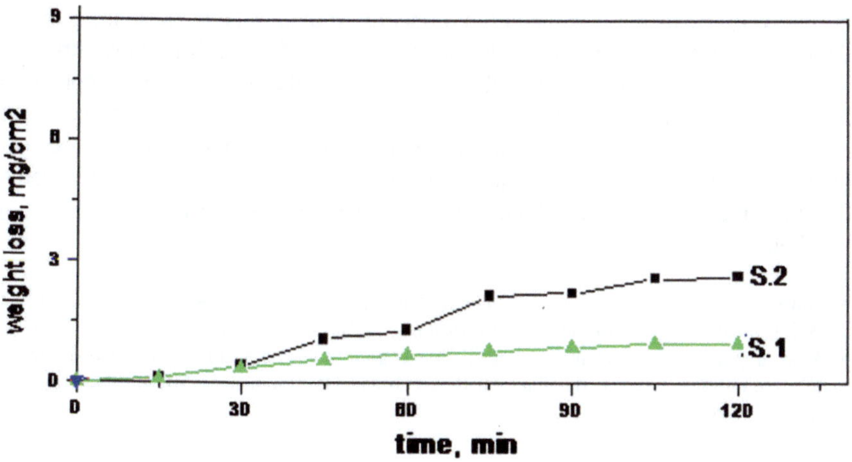

Fig. 4 Corrosion rate versus exposure time in 10 wt% HCl

Generally, the corrosion behavior of nitrogen–manganese stabilized austenitic steels S.1 is better than S.2.

The weight losses of Duplex Stainless Steel alloys in 10 %wt HCl have been determined as a function of the immersion time. The tested sample was immersed at room temperature in 10 %wt HCl for various immersion periods (0–120 min). The variation of the corroded weight (weight loss) with immersion time is shown in Fig. 4. It is obvious that the weight loss W_{loss} (mg/cm^2) increased linearly with immersion time.

The resistance to pitting corrosion is associated with nonmetallic inclusions present in the alloys. Alloys S.1 and S.2 are more pitting resistance due to their low carbon content of 0.14 and 0.232 %, respectively. The data indicated that the annual corrosion rate of S.1 was the lowest due to probably its highest Mn content, i.e. the increase in Mn content of the alloys decreased the resistance to pitting corrosion. It can be concluded that the results of electrochemical polarization and immersion tests are coincident.

4 Conclusion

Two kinds of nickel-free duplex stainless steels have been developed in this scientific research work. These alloys can be characterized by their high strength and toughness, good corrosion resistance and low alloy element cost. In comparison with the commercial DSS with a similar PRE_N, the experimental DSS have a higher yield strength by roughly 100–200 MPa and a lower elongation to fracture of about 20 %. The increase in yield strength and ultimate tensile strength value of the investigated alloy with increasing Mn content from 6.44 to 13.5 % was therefore attributed to the increase in the volume fraction of the austenite phase from 18.2 to 25.4 %, as well as the solid solution hardening effect of Mn. Microstructural investigations showed that such alloys have relatively stable austenite content at high temperatures. Due to the absence of nickel, the experimental DSS exhibit an excellent resistance to SCC in chloride solutions. Finally, these alloys can be cost-efficient because of the total absence of the expensive element nickel and therefore can find applications where high strength and moderate corrosive resistance are required.

Acknowledgments The authors would like to thank the late Prof. Dr.-Ing. Y. Shash the Head of the Mechanical Design and Production Department, Faculty of Engineering, Cairo University for his kind support and wise scientific advices. Thanks goes to Dr. A. Abd Elaal from Department of Metals Technology, Central Metallurgical Research and Development "CMRDI", Helwan-Cairo Egypt for the corrosion tests.

References

1. Huang J, Altstetter C (1995) Cracking of duplex stainless steel due to dissolved hydrogen. Metall Mater Trans A 26A:1079–1085
2. Davis J (1996) ASM specialty handbook stainless steels. ASM International, Materials Park, OH, pp 32
3. Solomon H, Devine T (1983) In: Lula RA (ed) Duplex stainless steels. ASM, Metals Park, OH, pp 693
4. Charles J (1994) In: Proceedings of the fourth international conference on duplex stainless steels, vol 1. Paper KI, Glasgow
5. Nicholls J (1994) In: Proceedings of the fourth international conference on duplex stainless steels, vol 1. Paper KIII, Glasgow
6. Son J, Kim S, Lee J, Choi B (2003) Effect of N addition on tensile and corrosion behaviors of CD4MCU cast duplex stainless steels. Metall Mater Trans A 34A:1617
7. Son J, Kim S, Lee J, Choi B (2002) Slow strain rate tensile behavior of CD4MCU cast duplex stainless steel with different nitrogen contents. J Korean Inst Met Mater 40:723
8. Park Y, Lee Z (2001) The effect of nitrogen and heat treatment on the microstructure and tensile properties of 25Cr–7Ni– 1.5Mo–3W–xN duplex stainless steel castings. Mater Sci Eng A 297:78
9. Son J, Kim S, Lee J, Choi B (2003) Tensile and Corrosion Behaviors of As-solutionized CD4MCU Cast Duplex Stainless Steel with Different Nitrogen Contents. J Korean Inst Met Mater 40:949
10. Sakai J, Matsushima I, Kamemura Y, Tanimura M, Osuka T (1983) In: Lula RA (ed) Duplex stainless steels. ASM, Metals Park, OH, pp 211
11. Kim S, Paik K, Kim Y (1998) Effect of Mo substitution by W on high temperature embrittlement characteristics in duplex stainless steels. Mater Sci Eng A 247:67
12. Poznansky A, Nalbone C, Crawford J (1983) In: Lula RA (ed) Duplex stainless steels. ASM, Metals Park, OH, pp 431
13. Kim J, Park C, Kwon H (1999) Bull Korean Inst Met Mater 12:635
14. Guha P, Clark C (1983) In: Lula RA (ed) Duplex stainless steels. ASM, Metals Park, OH, pp 355
15. Okazaki Y, Miyahara K, Hosoi Y, Tanino M, Komatsu H (1989) Effect of alloying elements of σ phase formation in Fe-Cr-Mn alloys. J Jpn Inst Met 53:512

16. Speidel M (1991) In: Proceedings of the international conference on stainless steels, vol 1. ISIJ, Chiba, p 25
17. Nilsson J (1992) Superduplex stainless steels. Mater Sci Technol 8:685
18. Merello R, Botana F, Botella J, Matres M, Marcos M (2003) Influence of chemical composition on the pitting corrosion resistance of non-standard low-Ni high-Mn-N duplex stainless steels. Corros Sci 45:909
19. Gunn R (1997) Duplex stainless steel. Abington, Cambridge, p 15
20. Kemp AR, Dekker NW, Trinchero P (1995) Differences in inelastic properties of steel and composite beams. J Constr Steel Res 34:187–206
21. Gunn R (ed) (1997) Duplex stainless steels—microstructure, properties, and applications. Abington Publishing, Cambridge
22. Stern M, Geary A (1957) Electrochemical polarization: I. A theoretical analysis of the shape of polarization curves. Electrochem J 56–63

Influence of Al_2O_3 Nano-dispersions on Mechanical and Wear Resistance Properties of Semisolid Cast A356 Al Alloy

Ahmed Y. Shash, Amer E. Amer and Moataz El-Saeed

Abstract The present investigation studies the prospects of using nanoparticles as reinforcement ceramic powders to gain improved performance of A356 Al cast alloy. Alumina nano-powder of 40 nm size was stirred into the A356 matrix with different fraction ratios ranging from (0, 1, 2 and 4 wt%) in a mushy zone (600 °C) using a constant stirring time for one minute. To evaluate the results, the alloys were further characterized by various tribological and mechanical characterization methods. The results showed higher strength values with improved ductility when compared to the monolithic alloy under the same casting conditions. Also, the wear resistance has been positively enhanced as the amount of the Al_2O_3 nano-particles addition increases from 1 to 4 wt% leading to a decrease in the weight loss ranging from 5.5 to 4.0 mg, respectively. The Scanning Electron Microscopy of the fracture surface and the wear surface revealed the presence of nanoparticles at the inter-dendritic space of the fracture surface and was confirmed with an EDX analysis of these particles.

Keywords Nano-metal matrix composites · Al_2O_3 nano-powders · Wear resistance · Mushy zone · Mechanical stirring

---

A.Y. Shash (✉)
Mechanical Design and Production Department, Faculty of Engineering, Cairo University, Giza, Egypt
e-mail: ahmed.shash@cu.edu.eg

A.E. Amer
Mechanical Design and Production Engineering Department, Beni Suef University, Beni Suef, Egypt
e-mail: aeid958@yahoo.com

M. El-Saeed
Department of Mechanical Engineering, Akhbar El-Youm Academy, 6 of October City, Giza, Egypt
e-mail: mo3taz.elsaeed@gmail.com

A. Öchsner and H. Altenbach (eds.), *Mechanical and Materials Engineering of Modern Structure and Component Design*, Advanced Structured Materials 70,
DOI 10.1007/978-3-319-19443-1_2

1 Introduction

History is often marked by the materials and technologies that reflect human capabilities and understanding. Many time scales begins with the stone age, which led to the Bronze, Iron, Steel, Aluminum and Alloy age, as improvements in refining and smelting took place and science made it possible to move towards finding more advanced materials. Composite structures have shown universal savings of at least 20 % over metal counterparts and a lower operational and maintenance cost [1]. Aluminum based alloys and metal matrix composites (MMCs) exhibit attractive tribological and mechanical properties such as a high specific modulus, good strength, long fatigue life, superior wear resistance and improved thermal stability, which allow these alloys to have numerous applications in the aerospace, automobile and military industries. As the data on the service life of composite structures is becoming available, it can be safely said that they are durable, maintain dimensional integrity, resist fatigue loading and are easily maintainable and repairable. Composites will continue to find new applications, but the large scale growth in the marketplace for these materials will require less costly processing methods and the prospect of recycling [2] will have to be solved [3].

Composite materials are emerging chiefly in response to unprecedented demands from technology due to rapidly advancing activities in aircrafts, aerospace and automotive industries. These materials have low specific gravity that makes their properties particularly superior in strength and modulus to many traditional engineering materials such as metals. As a result of intensive studies into the fundamental nature of materials and better understanding of their structure property relationship, it has become possible to develop new composite materials with improved physical and mechanical properties [4–6]. These new materials include high performance composites such as polymer matrix composites, ceramic matrix composites and metal matrix composites etc. Continuous advancements have led to the use of composite materials in more and more diverse applications. The importance of composites as engineering materials is reflected by the fact that out of over 1600 engineering materials available in the market today more than 200 are composites [6].

For most applications, a homogeneous distribution of the particles is desirable in order to maximize the mechanical properties [5]. In order to achieve a good homogeneous distribution of a particle in the matrix, the process parameters related with the stir casting method must be studied [4, 5]. So that it is essential to study the influence of stirring speed and stirring time on the distribution of particles in MMC.

In the Prabu [7] study, the stirring speeds were set at 500, 600 and 700 rpm and the stirring times were set at 5, 10 and 15 min for this study.

Many researchers have claimed enhanced properties for these produced composites relative to those produced by reinforcing with micro-particles.

Therefore the aim of this research work is to improve the mechanical and tribological properties of the A356 aluminum alloy using ceramics Al_2O_3 nano particles.

2 Experimental Procedures

The experimental work carried out through this scientific study consists of the following three stages:

(a) Production of new NMMC alloys.
(b) Identification of the mechanical and wear resistance properties.
(c) Characterization of the new material.

2.1 *Materials Produced*

The hypoeutectic alloy A356 was used as a base metal for the produced material having the chemical composition shown in Table 1. The material used for reinforcement was 1, 2, and 4 % by weight Al_2O_3 ceramic nano-particles with constant particle size of 40 nm, the description of which is given in Table 2.

2.2 *Equipments Used*

2.2.1 Melting Furnace

An electric resistance furnace was designed and constructed for approaching this research work for preparing the NMMCs. It consists of a lift out ceramic crucible of max. 2 kg, a heating system, and is connected to a stirring mechanism with a

Table 1 Chemical composition (in wt%) of A356 cast Al–Si

Alloy	Chemical composition (wt%)							
	Al	Si	Mg	Fe	Cu	Pb	Zn	Mn
A356	Bal.	7.44	0.3	0.27	0.02	0.022	0.01	Nil

Table 2 Properties of Al_2O_3 reinforcement powders

Reinforcement	γ-Al_2O_3
Density (solid) (g/cm^3)	3.95
Crystal structure	FCC
Appearance	White solid
Young's modulus (GPa)	380
Average size (nm)	40
Melting point	2054 °C

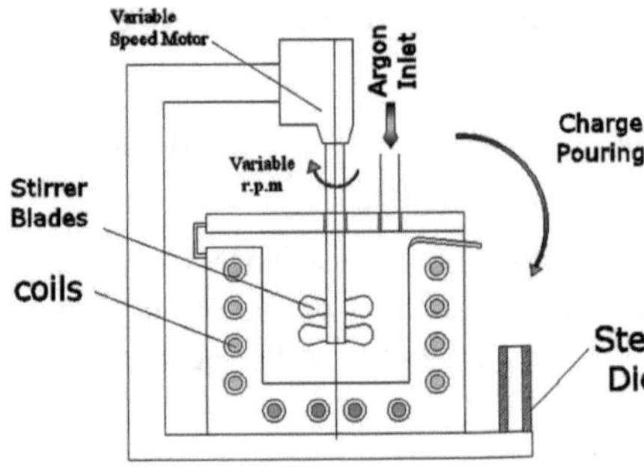

Fig. 1 Schematic apparatus used for preparing the NMMCs

3000 rpm max. rotating speed motor and adjustable height with a control unit of up to 1200 °C connected to a thermocouple for controlling the stirring temperature Fig. 1.

2.2.2 Metallic Mould

The melt was poured into a mild steel mould, in which the casted samples were in a 24 mm diameter as shown in Fig. 2.

2.3 Melting Methodology and Approach

A charge of 0.5 kg of the A356 alloy was introduced to the crucible and heated up to the melting temperature (640 °C). The melt was degassed and shielded with argon before pouring after reaching the liquid state to prevent oxidation of the molten metal. The melt was subsequently brought down to the semi-solid state by around 605 °C and hence the Al_2O_3 nano-powders were preheated to 700 °C and then added to the melt simultaneously with mechanical stirring for 1 min at 1500 rpm. The fabrication conditions of the composites prepared in this investigation are

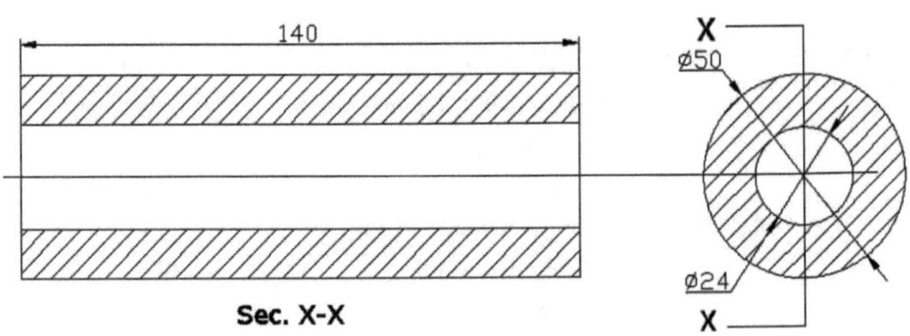

Fig. 2 Schematic drawing for the mould used

Table 3 List of produced alloys and fabrication conditions

Melt no.	Additions	Stirring (rpm)	Pouring temp. (semi-solid)
Melt 1	A356	1500	605 °C
Melt 2	A356 + 1 % Al_2O_3	1500	605 °C
Melt 3	A356 + 2 % Al_2O_3	1500	605 °C
Melt 4	A356 + 4 % Al_2O_3	1500	605 °C

summarized in Table 3. Cast samples were poured into the prepared mould without additions and with additions of the different investigated Al_2O_3 percentages.

2.4 *Mechanical Properties*

Mechanical properties, mainly tensile strength, ductility, hardness and wear resistance, were determined in the as-cast conditions for the investigated NMMC samples.

2.4.1 Tensile Test

The tensile tests were conducted on round tension test specimens of diameter 5.02 mm and gage length 25.2 mm using a universal testing machine according to DIN 50125. The elongation percentage and ultimate tensile strength were calculated. The results were based on the average of three samples taken from each melt.

2.4.2 Hardness Test

The hardness tests were conducted on Rockwell hardness testing machines in the Faculty of Engineering, Cairo University, using a $(\frac{1}{16})''$ diameter hardened steel ball and a 62.5 kg applied load. The reported results are the average of three readings for each case.

2.4.3 Wear Test

A PLINT TE 79 Multi Axis Tribometer Machine was used for measuring friction force; friction coefficient and wear rate for NMMC manufactured materials, as illustrated in Fig. 3, in which a standard specimen with a diameter of 8 and 20 mm length as a computerized pin on disc machine used for friction and wear testing of materials is loaded vertically downwards onto the horizontal disc.

Fig. 3 PLINT TE 79 multi axis tribometer machine

2.5 Material Characterization

2.5.1 Microstructural Evolution

Representative sections from the cast samples were cut into 3 pieces: the 1st from the top, the 2nd from the middle and the 3rd from the bottom. Samples were wet grounded on a rotating disc using silicon carbide abrasive discs of increasing fineness (120, 180, 220, 320, 400, 600, 800, 1000 and 1200 grit). Then they were polished using 10 μm alumina paste.

2.5.2 Optical Microscopy (OM)

The microstructure examination was carried out using an OLYMPUS DP12 optical metallurgical microscope, equipped with a high resolution digital camera for the investigation of the microstructure.

2.5.3 Scanning Electron Microscope (SEM)

The surface topography and the fracture characteristics were studied using SEM to understand the fracture mechanism and also to detect the favorable sites for particle incorporation by using a JSM-5410 Scanning Electron.

The JSM-5410 scanning electron microscope is a high-performance multipurpose SEM with a high-resolution of 3.5 nm, and EDXS (energy dispersive X-ray spectrometer). Its automated features included Auto Focus/Auto Stigmator, and Automatic Contrast and Brightness. The EDS makes the JSM-5410 expandable from morphological observations to multi-purpose high-resolution elemental analysis.

3 Results and Discussions

3.1 Mechanical Properties of the NMMC

Table 4 illustrates the mechanical properties (tensile strength, elongation in %, and hardness) of the produced castings with reinforced Al_2O_3 nanopowder.

As can be seen from Fig. 4, as the wt% fraction of Al_2O_3 nanopowder increases the UTS increases reaching 195 MPa, until a value of 2 wt% of Al_2O_3. Beyond this weight fraction, the UTS decreases as the wt% increases.

As shown in Fig. 5, increasing the weight fraction of Al_2O_3 has no visible effect on ductility until reaching 1 wt%, then with increasing the wt% beyond 1 wt% the ductility increases. The ductility of NMMC increases by about 40 % at 2 wt% of Al_2O_3 nanopowder. At 4 wt% fraction, the ductility reaches to its minimum values; due to the agglomeration of the dispersed particles in the NMMC, while its hardness at this weight fraction increases by about 30 % as shown in Fig. 6. The presence of the ceramic phase increases the hardness of the alloys and hence, reduces the ductility of the composites in comparison with the matrix alloy.

Substantial increases in strength, along with good ductility, have been observed in a number of alloys with multiphase nanoscale microstructures.

Table 4 Mechanical properties (tensile strength, elongation in % and hardness) NMMC using Al_2O_3 Nanopowder

Melt no.	Additions	UTS (MPa)	Elongation %	Hardness RB
Melt 1	A356	155	5	57
Melt 2	A356 + 1 % Al_2O_3	170	5	61
Melt 3	A356 + 2 % Al_2O_3	195	6.8	72
Melt 4	A356 + 4 % Al_2O_3	163	4.2	73

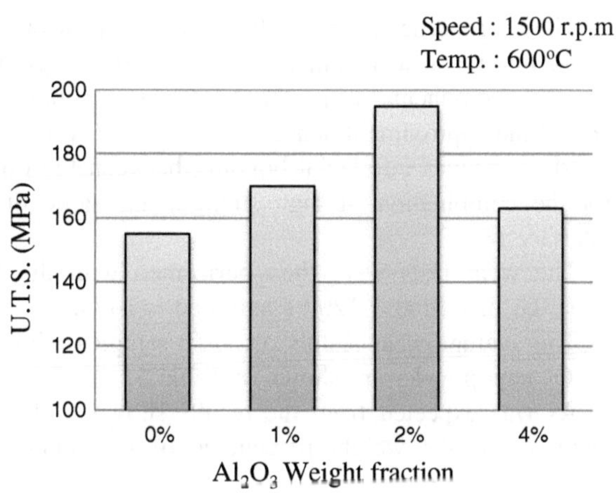

Fig. 4 The effect of wt% fraction of Al_2O_3 nanopowder on the ultimate tensile strength of MMC at 1500 rpm stirring speed at a semi-solid-state (600 °C)

Fig. 5 The effect of wt% fraction of Al_2O_3 nanopowder on elongation % of MMC at 1500 rpm stirring speed when in a semi-solid state (600 °C)

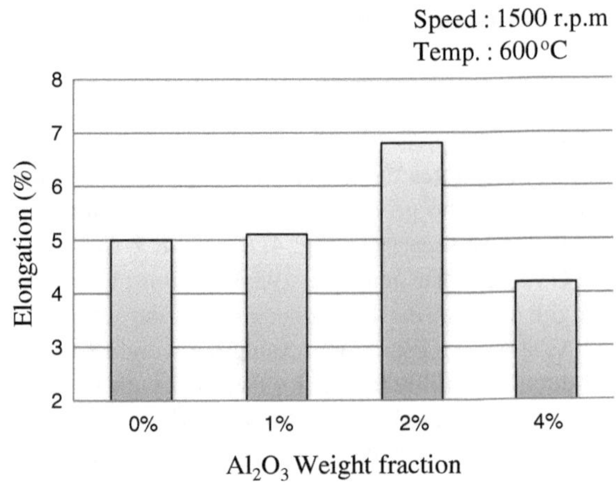

Fig. 6 The effect of wt% fraction of Al_2O_3 nanopowder on the hardness of MMC at 1500 rpm stirring speed when in a semi-solid state (600 °C)

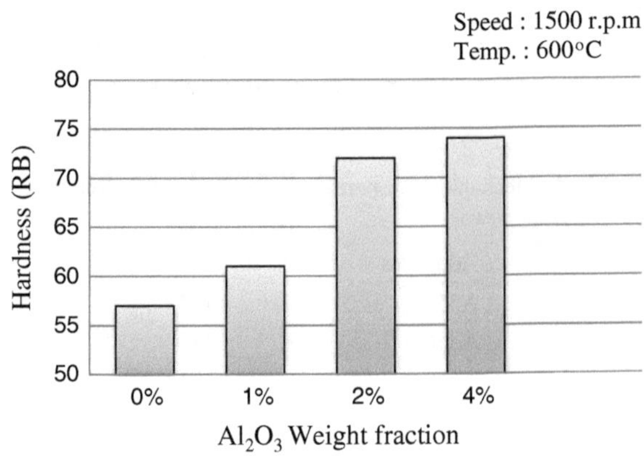

These properties originate from the fine distribution of globular particles in an A356 matrix on a nanometer scale, where the globular particles act as strength bearing components, while the A356 matrix supplies ductility. The existence of a crystalline approximant phase at the interface between the particles and the FCC Al matrix improves interfacial bonding between the different phases, thus important for the combination of high strength and good ductility without failure at the interface [8].

The wear tests were then performed with the following parameters: velocity = 0.8 m/s, time = 1200 s and load = 10 N.

The average wear results of A356 samples reinforced with 0, 1, 2 and 4 wt% Al_2O_3 nanopowder are shown in Table 5.

As was expected from the results shown in Table 5 that the wear resistance increases as the weight percentage of the reinforced nano particles increases.

Table 5 The average wear results of A356 samples reinforced with 0, 1, 2 and 4 wt% Al_2O_3 nanopowder

Sample no.	Additions	Weight loss mg	Friction coefficient
1	A356	3.9	0.4
2	A356 + 1 % Al_2O_3	5.5	0.430
3	A356 + 2 % Al_2O_3	4.5	0.385
4	A356 + 4 % Al_2O_3	4.0	0.361

The wear resistance results were evidently confirmed by increasing the hardness and decreasing the friction coefficient values as the wt% of the nano dispersions increases.

3.2 Microstructure Evolution

Figure 7a, b shows the optical microstructure of the base matrix A356 alloys reinforced with 2 wt% fraction of Al_2O_3 nano-powder. The microstructures of the two castings show that the phases are uniformly distributed. It also clearly show a morphological change in the microstructures. In the base matrix sample, the microstructure is dendritic whereas in the other rheocast samples, the primary dendrites are fragmented due to mechanical stirring.

As the structure contains good amount of eutectic phases it should give a range of mechanical properties when mechanical stirring processed [9]. This was clear in the mechanical properties of alloys reinforced with nano-powder using mechanical stirring.

The SEM illustrated in Fig. 8 shows a typical fracture surface for a specimen reinforced with a 2 wt% fraction of Al_2O_3 using the best conditions of stirring speed

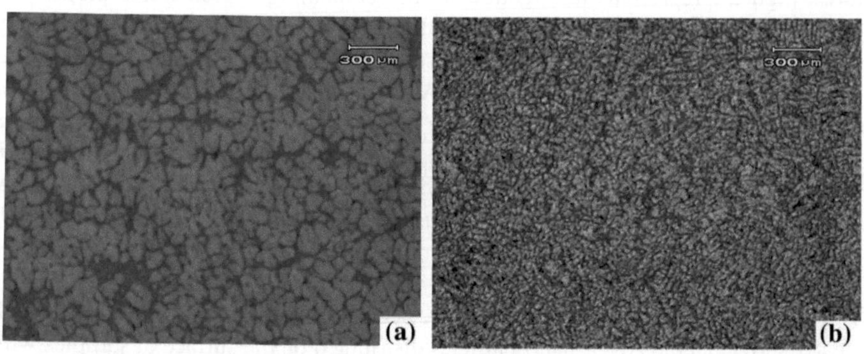

Fig. 7 The optical microstructure of **a** base matrix A356 alloys. **b** reinforced with 2 wt% fraction of Al_2O_3 nanopowder

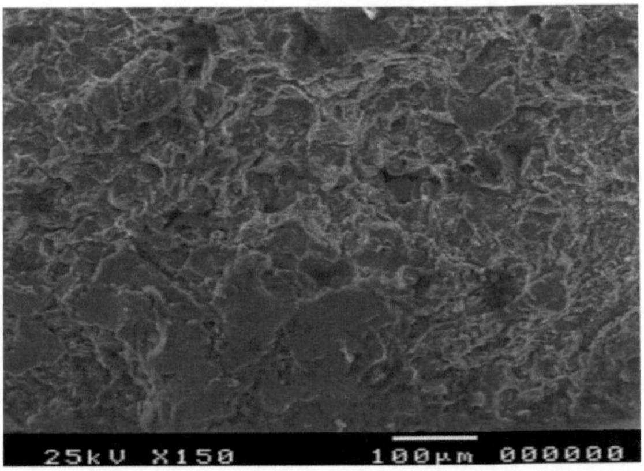

Fig. 8 The SEM of the fracture surface specimen reinforced with 2 wt% fraction of Al_2O_3 nano-powder

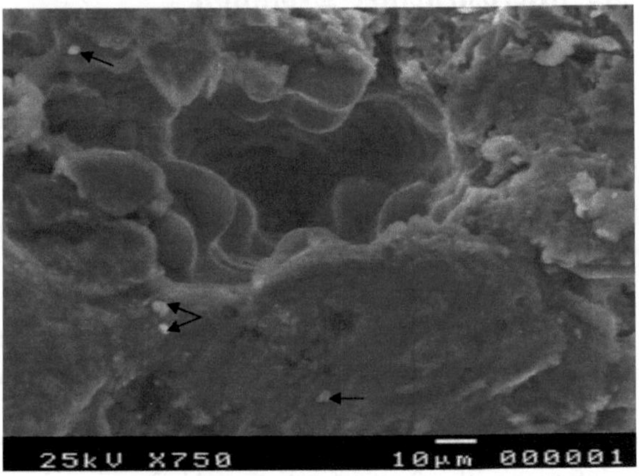

Fig. 9 SEM fracture surface containing agglomerated particles

and temperature. Fracture surface shows mixture of dendrites and globular structures. Figure 9 represents the fracture surface containing agglomerated particles.

The agglomeration could happen by the particles reinforced inside the matrix during the melting stage. Moreover, these agglomerated particles were not homogenously distributed inside the matrix. Al_2O_3 agglomerated particles have size of about 3 μm attached in the interdendritic space and in matrix. The EDX analysis shown in Fig. 10, has evidently confirmed that these are Al_2O_3 particles though a strong reflection from the matrix was inevitable. The specified analysis of the EDX and the percentage of O, Al and Si are illustrated in Table 6. It is clear that the high percentage of the Oxygen and Aluminum, confirms the presence of Al_2O_3 nano-powders in the matrix.

The samples subjected to wear tests have been examined using SEM. Figure 11a, b clearly show that the Al_2O_3 nano-particles distributed on the surface of samples with 1 and 4 wt% Al_2O_3 addition subjected to wear, respectively.

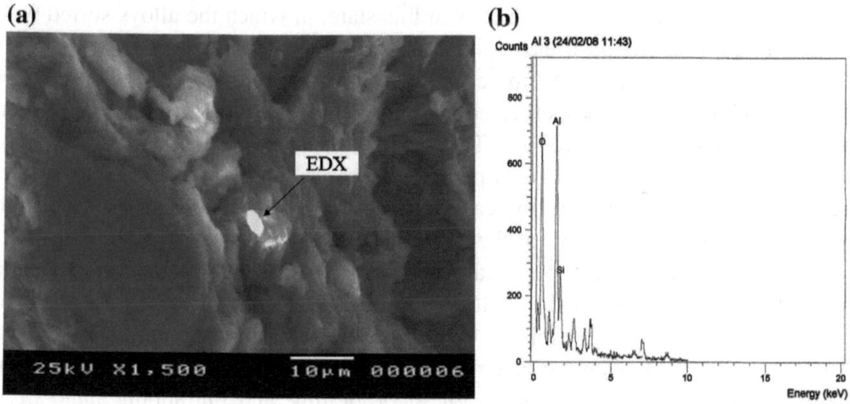

Fig. 10 **a** SEM containing Al_2O_3 agglomerated particles, **b** EDX of Al_2O_3 agglomerated particles

Table 6 Specified analysis of the EDX of Al_2O_3 agglomerated particles

Label	Range (keV)	Gross	Net	Total %
O Ka	0.407–0.668	5459	3149	38.1
Al Ka	1.327–1.628	5942	4222	51.1
Si Ka	1.648–1.888	2347	891	10.8

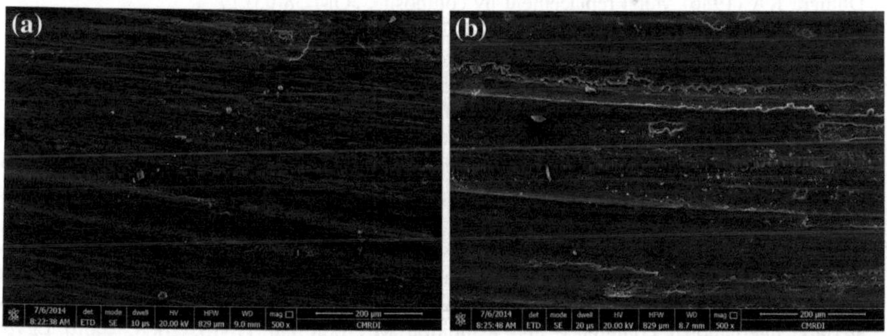

Fig. 11 SEM for wear surface of **a** 1 wt% Al_2O_3 and **b** 2 wt% Al_2O_3 nano-particles

4 Conclusion

The nano-composites manufactured using the semi-solid route exhibited better mechanical properties when compared with those prepared using the liquid metallurgy rout. Also, a high mixing speed is required in order to obtain good distribution of the particles reinforced and introduce it inside the matrix as introducing the reinforced Al_2O_3 particles to the A356 matrix in the semi solid state is difficult due to the higher density of the matrix at this state, in which the alloys stirred with 1500 rpm exhibits the best tensile strength and elongation.

The A356 matrix alloy reinforced with 2 wt% fraction of Al_2O_3 nano-powder has the best mechanical properties at conditions of 1500 rpm stirring speed at semi solid state temperature 600 °C. The wear resistance increases as the weight percentage of the reinforced nano-particles increases. The wear resistance results were evidently confirmed by increasing the hardness and decreasing the friction coefficient values as the wt% of the nano-dispersions increases.

In the base matrix sample without stirring, the microstructure is dendritic whereas in the other rheocast samples, the primary dendrites are fragmented due to mechanical stirring which explains the improvement in the mechanical properties. Analysis using both scanning electron microscope (SEM) and high magnification shows evidence for the possibility of incorporating and entrapping nano-sized particles within the interdendritic interface developing during the solidification of the dispersed alloys.

Acknowledgments The authors would like to thank the late Prof. Dr.-Ing. Y. Shash the head of Mechanical Design and Production Dept., Faculty of Engineering, Cairo University for his kind support and wise scientific advices.

References

1. Dhingra KA (1986) Metal replacement by composite. JOM 38(03):17
2. Mehrabian R, Riek GR, Flemings CM (1974) Preparation and casting of metal-particulate non-metal composites. Metall Trans 5A:1899–1905
3. Eliasson J, Sandstorm R (1995) Applications of aluminium matrix composites Part 1, Newaz GM, Neber-Aeschbacherand H, Wohlbier FH (eds) Trans Tech Publications, Switzerland, pp 3–36
4. El-Mahallawi SI, Rashad MR, Mahmoud ST, Shash YA (2006) Effect of processing parameters on synthesis and mechanical characteristics of A356/(Al_2O_3)p cast metal matrix nano-composites (MMNCs). Sixth arab foundry symposium, (ARABCAST 2006), Sharm El Sheik, Egypt. pp 68–77 Nov 12–15
5. Elmahallawi SI, Egenfeld K, Kouta HF, Hussein A, Mahmoud ST, Rashad MR, Shash YA, Abou-AL-Hassan W (2008) Synthesis and characterization of new cast A356/(Al_2O_3) p metal matrix nano-composites. Proceedings of the 2nd multifunctional nanocomposites and nanomaterials: international conference and exhibition MN2008 Jan 11–13, 2008, Cairo Egypt. Copyright © 2008 by ASME
6. El-Mahallawi SI, Shash Y, Eigenfeld K, Mahmoud T, Rashad R, Shash A, and El-Saeed M (2010) Influence of nano-dispersions on strength ductility properties of semi-solid cast A356 Al Alloy. Mater Sci Technol, online, Oct 2010
7. Prabu BS (2006) Influence of stirring speed and stirring time on distribution of particles in cast MMC. J Mater Process Technol 171:268–273
8. Koch CC (2002) Nanostructured materials processing, properties and potential applications. 1st edn. ISBN-10: 0815514514
9. Dey KA, Poddar P, Singh KK, Sahoo LK (2006) Mechanical and wear properties of rheocast and conventional gravity die cast A356 Alloy. Mater Sci Eng A 435–436:521–529

Evaluation of Mechanical Properties of Natural and Synthetic Rubber Material

Chang-Su Woo and Hyun-Sung Park

Abstract It is critical to investigate the mechanical properties of rubber materials to secure the reliability of rubber components. In this study, we performed mechanical tests of natural and synthetic rubber in various environmental conditions. The hardness, elongation, stress-strain relation, dynamic properties, and the nonlinear material constants that are necessary for a finite element analysis were determined through uniaxial tension, equi-biaxial tension and pure shear tests. The hardness of thermally aged rubber increased in proportion to aging time and temperature while that of elongation decreased. The storage modulus increased in proportion to aging time in dynamic property tests while the loss factor decreased. In mechanical tests according to change in strain, we determined the second Mooney-Rivlin and the third Ogden terms that are necessary for the finite element analysis of rubber components.

Keywords Rubber material · Mechanical test · Aging · Strain energy function

1 Introduction

Most rubber products are composites in which various additives including fillers are mixed with rubber that is selected according to the characteristics of the products. The most important thing in selecting materials is that they meet the standards for mechanical properties that are required for applications and the usage conditions of products [1, 2]. Mechanical properties or dynamical behaviors of rubber differ significantly depending on the kind of vulcanizing agents, the kind and quantity of

---

C.-S. Woo (✉) · H.-S. Park
Korea Institute of Machinery and Materials, 171 Jang-Dong, Yusung-Gu, Daejeon 305-343, Korea
e-mail: cswoo@kimm.re.kr

H.-S. Park
e-mail: hspark@kimm.re.kr

A. Öchsner and H. Altenbach (eds.), *Mechanical and Materials Engineering of Modern Structure and Component Design*, Advanced Structured Materials 70,
DOI 10.1007/978-3-319-19443-1_3

antiozonants or the quantity of modifiers or tenderizers. Data on the mechanical properties of rubber can be determined only through experiments because the components and mechanical properties of rubber compounds demanded by clients vary greatly. To improve the mechanical test methods and estimation skills of rubber material, we obtained data on the mechanical properties for selected natural and synthetic rubbers in this study, through mechanical tests under various conditions such as room temperature, high and low temperatures, and aging in the uniaxial and equi-biaxial tension and pure shear states. Owing to various advantages of rubber, it has been widely used in the following applications: anti-vibration rubber mounts for vehicles and trains, rubber rollers for printing and paper-making, and components in semiconductors, IT, and aerospace [3]. Establishing mechanical test methods and estimation skills in this study can result in savings in both time and cost required to develop rubber components and achieve advances in related technology by promoting localization and strengthening global competitiveness by securing reliability and stability.

2 Experiment

2.1 Rubber Material

Six kinds of rubber materials were selected for material tests: natural rubber (NR), copolymers of acrylonitrile and butadiene (NBR), polychloroprene rubber with chloroprene as a repetitive unit (CR), polymers of ethylene, propylene, and diene (EPDM), chloroethylsulfonyl polyethylene (CSM), and silicon rubber (Si).

2.2 Mechanical Test

The viscoelasticity of rubber has an impact on its strain velocity. The faster the strain velocity is, the greater the stress on material is, and vice versa. The suitable range of the strain velocity to obtain static mechanical properties is 0.007–0.17 m/s. Because there was no remarkable difference in the stress-strain curves of rubber in this range, we conducted tests at the same strain velocity of 0.01 m/s [4]. Mechanical properties of the rubber materials were investigated in environmental tests at room temperature (23 °C), low temperature (−40 °C), and high temperature (85, 100, and 125 °C), through uniaxial tension tests, and obtained hardness, elongation, stress-strain relation, and dynamic properties after aging for 1000 h at 70, 85, 100, and 200 °C to construct databases on mechanical properties. Uniaxial tension tests were conducted as shown in Fig. 1a by mounting load cells of 500 N on material testers, in which laser extensometers were used to measure the strain of specimens. A pure compressive stress-strain relation is difficult to obtain in uniaxial

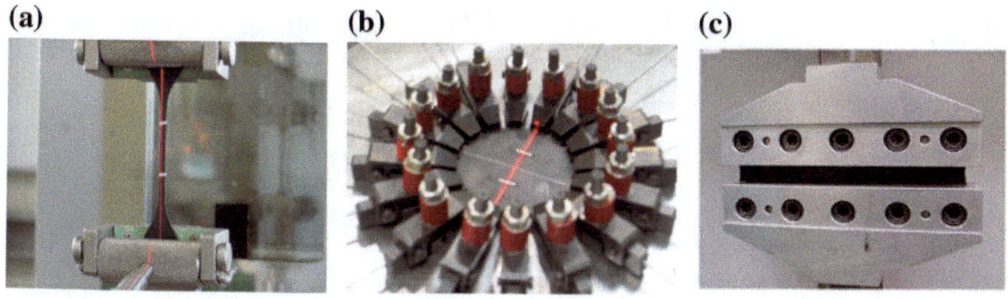

Fig. 1 Mechanical test of rubber materials. **a** Uniaxial tension. **b** Equi-biaxial tension. **c** Pure shear

tests due to friction generated on contact surfaces between rubber specimens and compression plates. To overcome this problem, equi-biaxial tension tests [5] with the same strain mode as in the compression tests were conducted as shown in Fig. 1b to determine the nonlinear material constants according to the range of the strain. Shear strain state is a more important mode of deformation for engineering applications than tension. If the material is incompressible and the width of the specimen is longer than the height, a pure shear state exists in the specimen at 45° angle to the stretching direction. The aspect ratio of the specimen is most significant in pure shear tests because the specimen is perfectly constrained in the horizontal direction. Figure 1c shows the pure shear test by using a non-contacting strain measurement with laser extensometer.

3 Result and Discussions

3.1 Environmental Test

Environmental tests were conducted at various temperatures including room temperature (23 °C), low temperature (−40 °C), and high temperature (85, 100, and 125 °C) to investigate the mechanical properties of the rubber materials depending on temperature conditions. As a result (Fig. 2), the higher the temperature, the lower the strength: changes in strength were significant at low temperature. The strength of rubber including NBR and CR was much greater than the strain below the glass transition temperature (Tg).

3.2 Thermal Aging Test

For most rubber products, the degradation of mechanical properties by environmental effects has impacts on the characteristics and life expectancy of the rubber. In this study, we investigated changes in the mechanical properties of rubber with

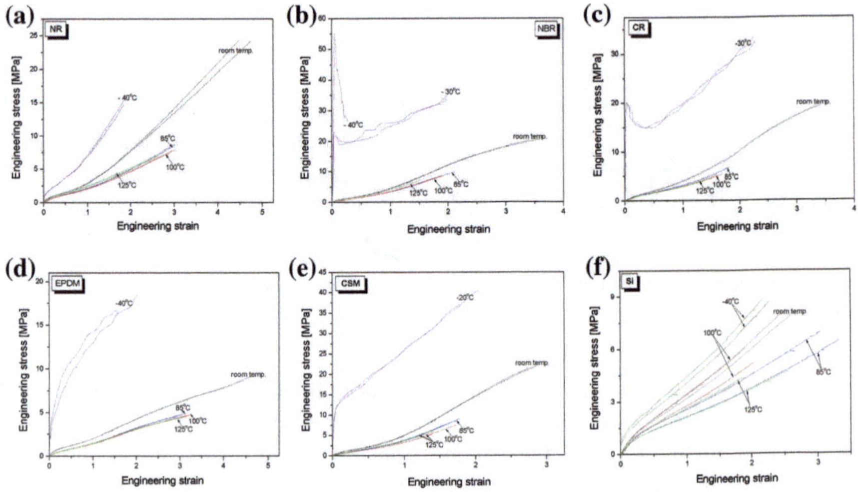

Fig. 2 Stress-strain curves at various temperatures. **a** NR, **b** NBR, **c** CR, **d** EPDM, **e** CSM, **f** Si

accelerated test in which rubber was aged at a higher temperature than rubber parts which are actually used, considering temperature as the most important factor among several degradation factors [6, 7]. Specimens of rubber were aged in the oven for 1000 h at 70, 85, and 100 °C conforming to ASTM D412, and then kept at room temperature for 24 h before measuring hardness, elongation, and stress-strain relations. Figures 3 and 4 show the changes in hardness and elongation depending on the aging temperature: the hardness is proportion to temperatures and aging period and drastically increases at 100 °C; elongation is inversely proportional to

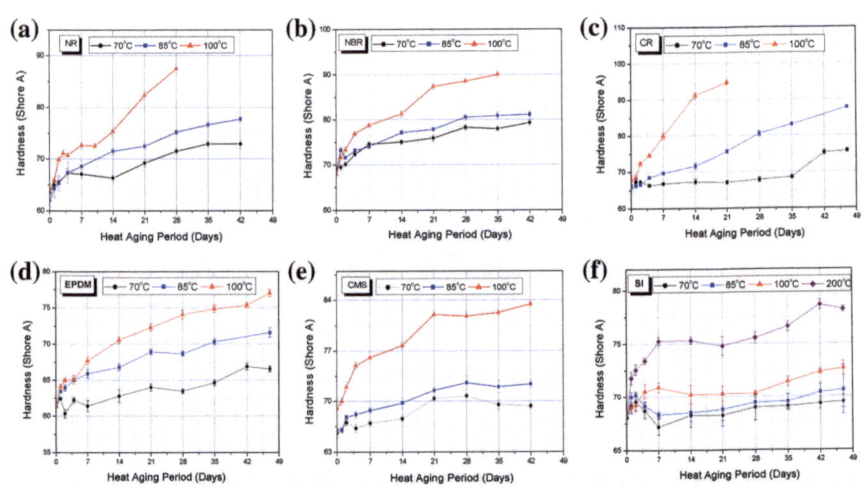

Fig. 3 Change of hardness at thermal aging conditions. **a** NR, **b** NBR, **c** CR, **d** EPDM, **e** CSM, **f** Si

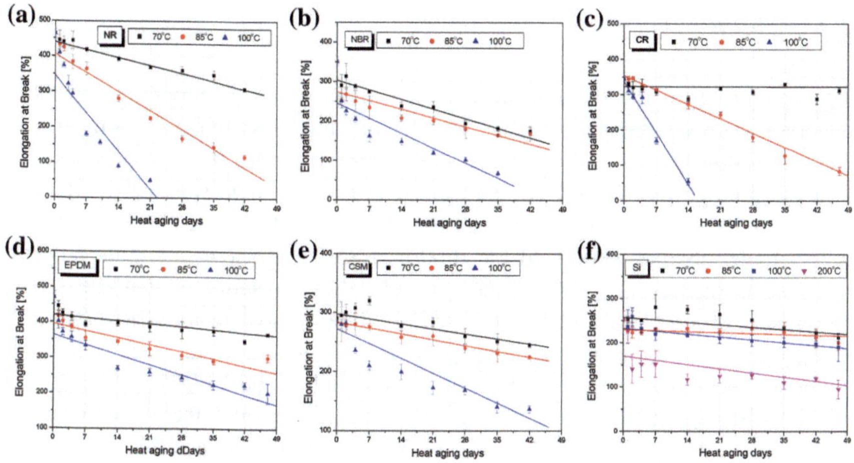

Fig. 4 Change of elongation at thermal aging conditions. **a** NR, **b** NBR, **c** CR, **d** EPDM, **e** CSM, **f** Si

temperature and the aging period decreases drastically at high temperatures. Figure 5 shows the stress-strain relation through tensile strength tests: a remarkable change in strength from the aging property of each rubber material can be seen at 100 °C while changes in properties are not remarkable at 70 °C.

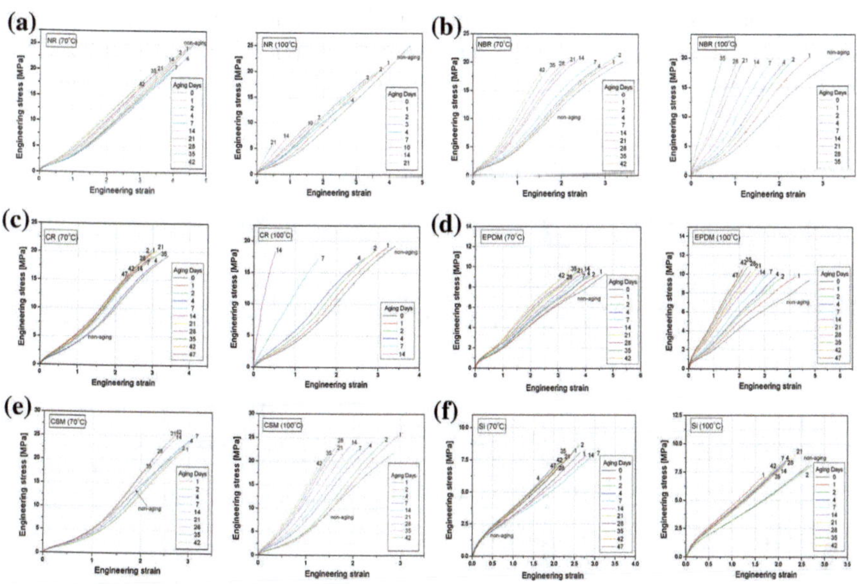

Fig. 5 Stress-strain curves at thermal aging conditions. **a** NR (70 and 100 °C), **b** NBR (70 and 100 °C), **c** CR (70 and 100 °C), **d** EPDM (70 and 100 °C), **e** CSM (70 and 100 °C), **f** Si (70 and 100 °C)

3.3 Dynamic Test

The conditions of the dynamic test to investigate dynamic properties of rubber materials such as storage modulus, loss factor, and glass transition temperatures were as follows: static and dynamic strain were 10 and 1.0 %, respectively, the frequency range was 1–100 Hz, and the temperature range was −80 to 100 °C. Figures 6 and 7 show the changes in the storage modulus and the loss factor of rubber materials that were aged for 1000 h at 85 °C. It shows a similar tendency with the frequency property of the general rubber material in that the storage modulus is proportional to the aging period while the loss factor is inversely proportional to it. Figure 8 shows the glass transition temperature of each rubber material. The transition temperature of natural rubber and EPDM was −45 °C at which the storage modulus and loss factor drastically changed; that of NBR and CR was −20 °C; and that of Si was −80 °C. We, therefore, can see that the physical properties of Si at low temperatures were excellent.

3.4 Stain Energy Function

The material of the rubber component is taken to be an incompressible rubberlike material modeled as a hyper-elastic material. The constitutive behavior of a hyper-elastic material is defined as a total stress–total strain relationship [8, 9]. Hyper-elastic materials are described in terms of their strain energy potential, which defines the strain energy stored in the material per unit of reference volume as a function of the strain at that point in the material. The strain energy functions have

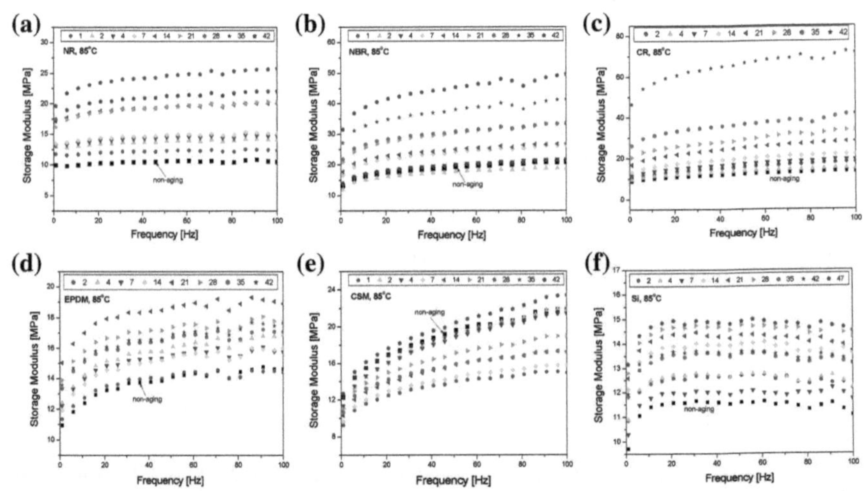

Fig. 6 Storage modulus at thermal aging conditions. **a** NR, **b** NBR, **c** CR, **d** EPDM, **e** CSM, **f** Si

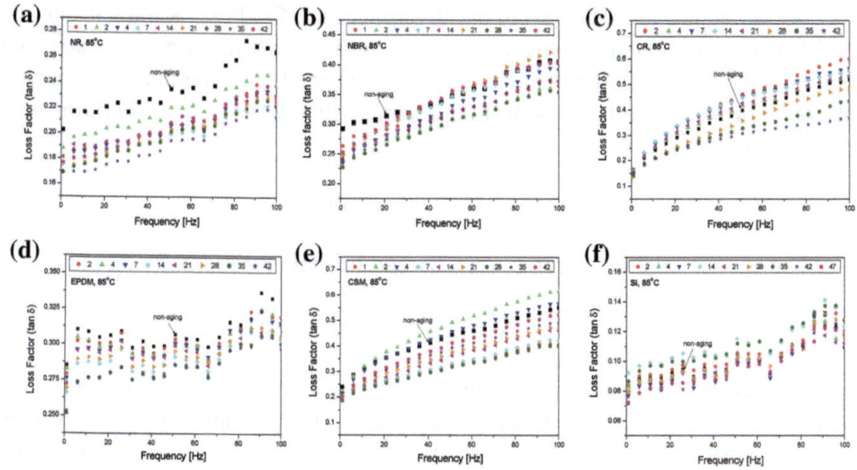

Fig. 7 Loss factor at thermal aging conditions. **a** NR, **b** NBR, **c** CR, **d** EPDM, **e** CSM, **f** Si

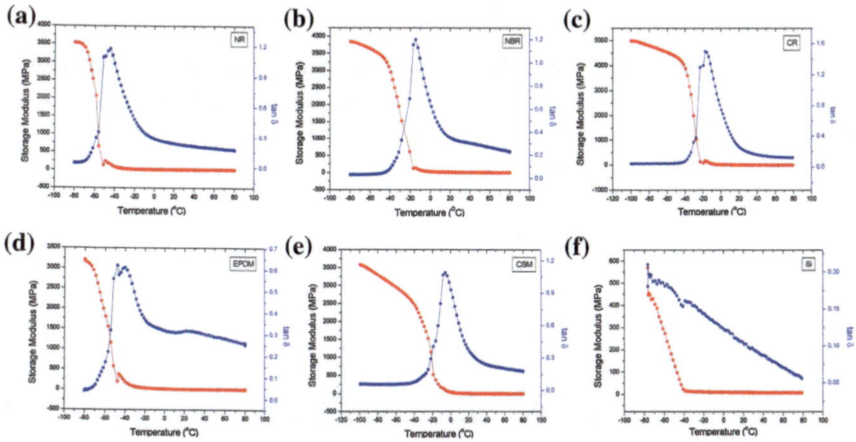

Fig. 8 Glass transition temperatures of rubber materials. **a** NR, **b** NBR, **c** CR, **d** EPDM, **e** CSM, **f** Si

been represented either in term of the strain invariants that are functions of the stretch ratios, or directly in terms of the principal stretch. Successful modeling and design of rubber components relies on both the selection of an appropriate strain energy function and an accurate determination of the material coefficient in the function. Material coefficients in the strain energy functions can be determined from the curve fitting of experimental stress-strain data. There are several different types of experiments, including simple tension, equi-biaxial tension and pure shear tests. In general, a combination of simple tension, equi-biaxial tension and pure shear tests are used to determine the material coefficient. The classical Mooney-Rivlin and Ogden models are an example of a hyper-elastic model that is implemented in the FEA. In order to explain the deformation of the rubber materials, it is assumed that the material has elastic behavior and is isotropic. Then, strain energy function (W) can be written as Eq. (1), with strain invariant functions (I_1, I_2, I_3) and principal stretch functions $(\lambda_1, \lambda_2, \lambda_3)$.

$$W = W(I_1, I_2, I_3), \quad W = W(\lambda_1, \lambda_2, \lambda_3) \tag{1}$$

When the material is isotropic, I_1, I_2, I_3 can be expressed as follows;

$$I_1 = \lambda_1{}^2 + \lambda_2{}^2 + \lambda_3{}^2, \quad I_2 = \lambda_1{}^2\lambda_2{}^2 + \lambda_2{}^2\lambda_3{}^2 + \lambda_3{}^2\lambda_1{}^2, \quad I_3 = \lambda_1{}^2\lambda_2{}^2\lambda_3{}^2 \tag{2}$$

Most rubber materials are incompressible and their bulk modulus is much greater than their shear modulus. Thus, it is widely accepted to presume the materials to be incompressible when they are under less restriction. When the materials are incompressible in Eq. (2), $\lambda_1, \lambda_2, \lambda_3 = 1$ and $I_3 = 1$. Since, Eq. (1) can be rewritten as follows,

$$W = W(I_1, I_2) \tag{3}$$

The strain energy function, which is widely used to analyze deformations of incompressible materials, can be described with Mooney-Rivlin's function and Ogden's function.

Mooney-Rivlin's function:

$$W = \sum_{n=1}^{N} C_{ij}(I_1 - 3)^i(I_2 - 3)^j \tag{4}$$

Ogden's function:

$$W = \sum_{n=1}^{N} \frac{\mu_n}{\alpha_n}(\lambda_1^{\alpha_n} + \lambda_2^{\alpha_n} + \lambda_3^{\alpha_n} - 3) \tag{5}$$

where C_{ij}, μ_n, α_n are the material constants determined experimentally from the stress-strain relationship.

3.5 Material Constant

When rubber is strained, its strength decreases and its damping properties change due to a modification and redistribution of the molecular structure in the initial state. These stress-strain relaxation phenomena are called the Mullins effect [10] that frequently occur in rubber filled with carbon black. That is, the strength of rubber is history dependent. Due to this, the stress-strain curves of rubber are different on loading and unloading. Figure 8 shows the stress-strain curves obtained from the uniaxial tension, equi-biaxial tension and pure shear test in which we applied five repetitive loads in each of the vertical and horizontal direction with 25, 50 and 100 % of the strain range for rubber material. According to Fig. 8, the stress-strain curves during the second repetition showed a greater decline than in the first repetition. The stress-strain curve gradually decreased as the number of repetitions increased, and ultimately stabilized to a fixed stress-strain value.

In order to predict the behavior of the rubber components using the finite element analysis, the rubber material constants must be determined from the stabilized cyclic stress-strain curve. The stress-strain curve varies significantly depending on the cyclic strain levels. A 5th loading cycle was selected as the stabilized stress-strain relationship in this study. But this stabilized relation should be shifted to pass through the origin of the curve, to satisfy the hyper-elastic nature of rubber. Figure 9 shows the stress-strain relation of rubber material. The shift of curve meant that the gage length and initial cross sectional area were changed as shown in Eq. (6).

$$\varepsilon = \frac{\varepsilon' - \varepsilon_p}{1 + \varepsilon_p}, \sigma = \sigma'(1 + \varepsilon_p) \tag{6}$$

Using the stress-strain relation obtained from the strain energy function, the nonlinear material constants of the second Mooney-Rivlin and the third Ogden terms [11, 12] were determined from the results of the stress-strain through uniaxial tension, equi-biaxial tension and pure shear tests, as shown in Fig. 10. The material constants obtained as in Table 1 was be used as material data necessary for finite element analysis of rubber components. It is possible to exactly estimate the properties of rubber components when material tests are conducted in the same level of strain as when components are actually used. We investigated the mechanical properties of selected natural and synthetic rubber through material tests in uniaxial tension, equi-biaxial tension and pure shear states, under various conditions including room temperature, low and high temperature, and aging. Because the mechanical behaviors and dynamic properties of rubber materials even with the

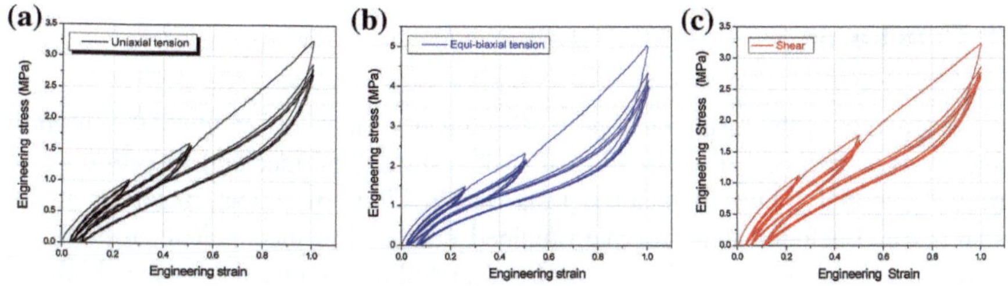

Fig. 9 Stress-strain curves at various loading and strain range. **a** Uniaxial tension, **b** Equi-biaxial tension, **c** Pure shear

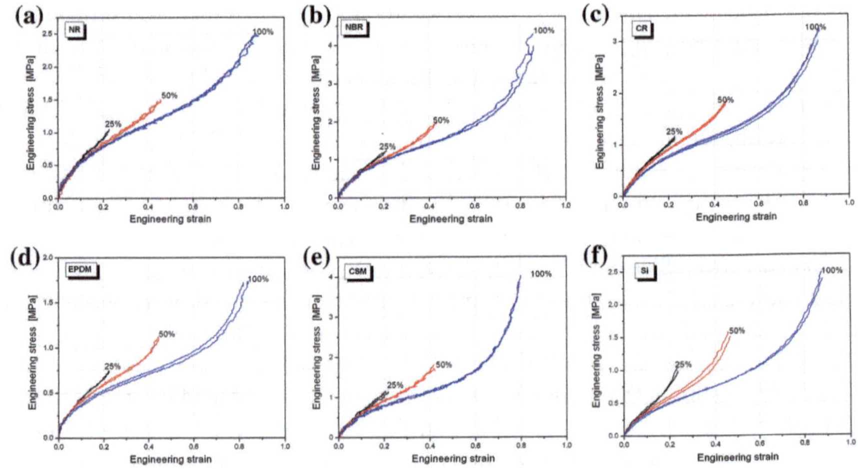

Fig. 10 Stress-strain curves after stabilized. **a** NR, **b** NBR, **c** CR, **d** EPDM, **e** CSM, **f** Si

Table 1 Mooney-Rivlin and Ogden function of rubber materials

Material	Mooney-Rivlin			Ogden						
	C_{10}	C_{01}	G_{mooney}	μ_1	α_1	μ_2	α_2	μ_3	α_3	G_{ogden}
NR	0.587	0.0	1.174	0.53	2.876	0.054	4.473	0.079	2.854	0.995
NBR	0.656	0.0	1.312	0.001	1.012	0.731	3.390	0.001	2.305	1.235
CR	0.170	0.806	1.952	219.2	0.012	129.1	0.012	0.001	34.16	2.106
EPDM	0.369	0.103	0.944	0.004	1.305	0.799	1.214	0.624	1.603	0.987
CSM	0.598	0.317	1.830	0.010	7.550	0.001	13.73	2.340	1.370	1.980
Si	0.257	0.337	1.188	0.001	22.18	0.001	16.97	508.2	0.005	1.290

same raw rubber depend on the kind or quantity of vulcanized agents or antioxidants, the quantity of modifiers or tenderizers, material tests and estimation should be conducted to exactly investigate their mechanical properties.

4 Conclusion

An estimation of the mechanical properties under various environmental conditions and loads is required in order to exactly investigate the mechanical properties of rubber material. Through material tests on six kinds of rubber materials in this study, we concluded that: (1) We obtained data on the mechanical properties of rubber materials under various conditions including room temperature, high and low temperatures, and aging in the uniaxial tension, equi-biaxial tension and pure shear states. (2) As a result of material tests at various temperatures, the strength lowered as temperature increased and its change was remarkable at low temperatures. (3) The hardness of thermally aged rubber materials increased in proportion to temperature and time while the elongation decreased. We found the aging properties of rubber materials that changes in its mechanical properties were not significant over time at 70 °C while a change in the strength was remarkable at 100 °C. (4) The glass transition temperature of rubber materials was obtained through dynamic property tests: the storage modulus increased over aging period while the loss factor decreased. (5) The nonlinear material constants of the second Mooney-Rivlin and the third Ogden terms for the finite element analysis of rubber components were determined through material tests according to changes in the strain. (6) In this study, we constructed databases of the mechanical properties of rubber materials under various environment conditions and by utilizing this, it is expected to greatly contribute to the development of rubber materials and to enhance the quality of components.

Acknowledgments This work was supported by the Industrial Strategic technology development program (10037360, A Multidimensional Design Technology Considering Perceived Quality (BSR) Based on Reliability) funded by the Ministry of Knowledge Economy (MKE, Korea).

References

1. Treloar LRG (1975) The physics of rubber elasticity, 3rd edn. Clarendon press, Oxford
2. Eirich FR (1978) Science and technology of rubber. Academic Press, New York
3. Freakley PK, Payne AR (1978) Theory and practice of engineering with rubber. Applied Science Publishers Ltd, London
4. Brown RP (1996) Physical testing of rubber, 3rd edn. Chapman & Hall, London
5. Kim WD, Woo CS (2003) The study on the equi-biaxial tension test of rubber material. Trans KSAE 11:95–104
6. Brown RP, Burtler T et al (2001) Ageing of rubber-accelerated heat ageing test results, RAPRA Technology, United Kingdom
7. Brown RP, Burtler T et al (2000) Natural ageing of rubber—changes in physical properties over 40 years, RAPRA Technology, United Kingdom
8. Peeters FJH, Kussner M (1999) Material law selection in the finite element formulation of rubber-like materials and its practical application in the industrial design process. Constitutive Model Rubber I 29–36
9. Luo RK, Mortel WJ (1999) Finite element analysis on bolster springs for metro railway vehicles. Constitutive Model Rubber I 275–280
10. Mullins L (1969) Softening of rubber by deformation. Rubber Chem Tech 42:339–362
11. Ogden RW (1984) Non-linear elastic deformations. Ellis Horwood Limited, UK
12. Ogden RW (1972) Large deformation of isotropic elasticity: on the correlation of theory and experiment for incompressible rubberlike solids. Proc Royal Society (A) 326:565–584

Influence of Laser Feeding on Structure and Properties of Cast Aluminium Alloy Surface

K. Labisz

Abstract The modern methods for surface layer engineering in current surface laser treatments, is LSA treatment, where there are small amounts of alloying additions introduced into the surface layer of the matrix material in the form of ceramic particle powders with different properties influencing the surface layer application possibilities. It was possible to produce a layer consisting of the heat affected zone, transition zone and remelted zone, without cracks and defects as well as with a slightly higher hardness value compared to the non remelted material. The laser power range was chosen to be 1.5–2.0 kW and implicated by a process speed rate in the range of 0.25–0.75 m/min. The purpose of this work is to apply High Power Diode Laser (HPDL) for the improvement of aluminum's mechanical properties, especially the surface hardness. This study was conducted to determine the effect of SiC powder addition on the structure and the mechanical properties as well as the structural changes occurring during the rapid solidification process. The main findings were, that the obtained surface layer is without cracks and defects as well as having a comparably higher hardness value when compared to the non-remelted material. The hardness value increases according to the laser power used so that the highest power applied gives the highest hardness value in the remelted layer.Also, the distribution of the SiC particles is good, but the particles are mainly present in the upper part of the surface layer. The hardness value increases in general according to the laser power used so that the highest power applied renders the highest hardness value in the remelted layer. The main goal of this work is to investigate and determine the effect of HPDL remelting and alloying on the cast Al-Si-Cu cast aluminium alloy structure to recognise the possibility for application in real working conditions mainly for light metal constructions as in the many branches of the industry.

Keywords High power diode laser · Surface treatment · Heat treatment · Aluminium cast alloys · Laser alloying · Laser feeding · Ceramic powders

K. Labisz (✉)
Institute of Engineering Materials and Biomaterials, Silesian University of Technology, ul. Konarskiego 18a, 44-100 Gliwice, Poland
e-mail: krzysztof.labisz@polsl.pl

A. Öchsner and H. Altenbach (eds.), *Mechanical and Materials Engineering of Modern Structure and Component Design*, Advanced Structured Materials 70,
DOI 10.1007/978-3-319-19443-1_4

1 Introduction

While analysing the given element in terms of its internal structure as well as possible, future working conditions it should be remembered that the product properties depend mainly on two factors: internal structure of the material, from which it has been made and on the condition of the external surface having both direct mechanical and chemical contact with the environment, as well as during exploitation with the surfaces of other working elements [1–5].

In relation to the current market demand for light and reliable constructions, aluminium alloys, belonging to the group of construction materials, characterised by a number of good mechanical and usable properties, good castability and resistance to corrosion play an important role. Aluminium alloys, constituting of a combination of low density and high strength, which are more and more often used in situations, where reduction of the subassembly element masses is significant, desired and feasible [6–11].

Remelting of cast aluminium alloys with SiC silicon carbide powder was performed using a high power diode laser (HPDL), (Fig. 1). Hardening requires a special form of heat delivery in order to ensure an evenly hardened zone and acceptable quality. The application was performed with a modern high power diode laser. The case studies of these four applications are presented and discussed in scientific papers [12–17].

The analysis of the test results of the mechanical and usable properties as well as tests of the structure of the surface layer of aluminium alloys after thermal and surface processing will allow for specification of the conditions of the laser surface processing of the aluminium alloys, such as the rate of scanning, laser power, used within the scope from 1.0 up to 2.2 kW, type and rate of the powder feeding, in order to produce the best possible surface layers on the surface of the processed alloys. Laser surface treatment is a relatively new generation of techniques used in metal surface technology. Laser treatment is presented and discussed with remelting of Al-Si-Cu cast aluminium alloys with ceramic powders, especially silicon carbide (SiC) particles. The investigation of the structure and the improvement of

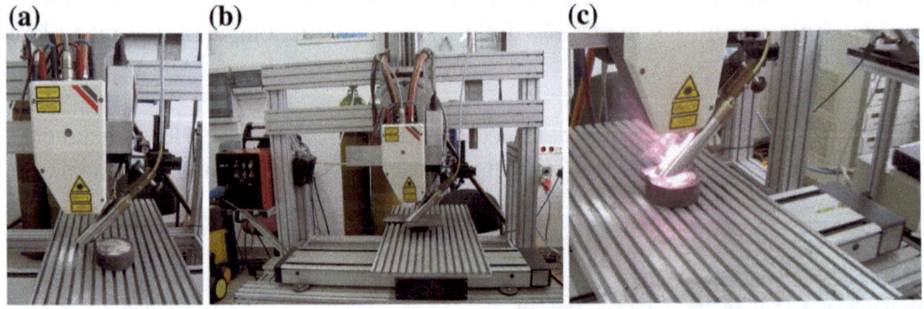

Fig. 1 HPDL laser Rofin DL 020 with: **a** laser head, **b** entire laser stand, **c** sample during working

mechanical properties is the practical aim of this work, as well as improving hardness, a very important property for practical uses. Silicon carbide is a material sometimes used in metal machining because of its high hardness and high resistance to softening at high cutting speeds and at high cutting temperatures [18–21].

The goal of this work is to determine the optimal laser power leading to the surface layer including the remelting zone and the heat influence zone. The standard metallographic investigation using light and an electron microscope was performed, as well as the hardness measurements and microanalysis, for the chemical composition investigation.

2 Experimental Conditions

The material used for investigation was the AlSi9Cu and AlSi9Cu4 aluminium cast alloy. The chemical composition of the investigated aluminium alloys is presented in Table 1. For feeding, the silicon carbide (SiC) powder was used with properties listed in Table 2. It should be mentioned that for this work it is planned to also use other types of powders.

The heat treatment was carried out in the electric resistance furnace U117, for the solution heat treatment process with two holds at 300 and 450 °C performed for 15 min. The cooling of the samples after the heat treatment was performed in the air for the aging process and in water for the solution heat treatment process. The solution heat treatment temperature was 505 °C for 10 h, and then aging was performed at 175 °C for 12 h (Fig. 2).

For remelting the high power diode laser HPDL Rofin DL 020 (Fig. 1) was used. This laser is a device with high power, used in materials science, including for welding purposes. The laser equipment used included: a rotary table which moved in the XY plane, the nozzle of the powder feeder for enrichment or welding,

Table 1 Chemical composition of the investigated aluminium alloys

	Si	Fe	Cu	Mn	Mg	Zn	Ti	Al
AlSi9Cu	9.094	0.1792	1.049	0.3608	0.2682	0.1409	0.0733	rest
AlSi9Cu4	9.268	0.3379	4.64	0.0143	0.2838	0.0478	0.0899	rest

Table 2 Physical and mechanical properties of the ceramic carbide and oxide powders

Properties	WC	TiC	SiC	Al_2O_3
Density, kg/m^3	15.69	4.25	3.44	3.97
Hardness, $HV_{0.05}$	3400	1550	1600	2300
Melting temperature, °C	2870	3140	1900	2047
Thermal expansion coefficient α, 10^{-6}/°C	23.5	8.3	4.0	7.4 ÷ 8.5
Grain size, μm	80	5	100	120

Fig. 2 Morphology of the SiC powder used for feeding, SEM

shielding gas nozzle, laser head, power and cooling system, and the computer system controlling the operation and location of the laser on the working table.

The feeding and remelting was performed with argon, in order to protect the substrate from oxidation. The sample during feeding was subjected to protective gas blowing and covering by mind of two nozzles, one directed axially to the laser-treated sample and the other directed perpendicular to the weld area. The shielding gas supply (Argon 5.0) was 10 l/min.

Based on the preliminary investigations of a high power laser diode HPDL Rofin DL 020 the process rate was v = 0.5 m/min. Other work parameters are presented in Table 3. For laser power values below 1.0 kW there are no remelted areas present at all, due to very high radiation absorption of the aluminium alloy surface.

Structure investigation was performed using the light microscope Leica MEF4A supplied by Zeiss with a magnification range of of 50–500 times. The micrographs of the microstructures were made by means of the KS 300 program a using digital camera. Metallographic investigations were performed also using the scanning electron microscope ZEISS Supra 35 with a magnification of up to 500 times. For the microstructure evaluation the Back Scattered Electrons (BSE) as well as the Secondary Electron (SE) detection method was used, with an accelerating voltage of 20 kV. Chemical composition investigations using EDS microanalysis on the scanning electron microscope Zeiss Supra. The phase composition investigations

Table 3 HPDL laser parameters

Parameter	Value
Laser wave length, nm	940 ± 5
Peak power, W	100 ÷ 2300
Focus length of the laser beam, mm	82/32
Power density range of the laser beam in the focus plane (kW/cm^2)	0.8 ÷ 6.5
Dimensions of the laser beam focus, mm	1.8 × 6.8

were made using a X-ray diffractometer with the filtered copper lamp rays at an acceleration voltage of 45 kV and heater current of 40 mA. The measurements were made with a diffraction angle from 30° to 130° of 2Θ. The X-ray scan was performed parallel to the alloying direction.

The hardness was measured with a Rockwell hardness tester with a load chosen for the HRF scale, with a load of 60 Kgf. Also, the temperature was measured during the remelting process, of which the exact temperature changes are shown in Fig. 15. It can be seen that the temperature remains constant throughout the whole process.

3 Results and Discussion

The structural investigations carried out using the high power diode laser allow us to compare the structure of these alloys used for investigations. On Figs. 3, 4 and 5 the microstructures revealing the phases occurring in the alloy, especially the Si phase, are presented. When using both the laser power of 1.5 and 2.0 kW for both alloys in the remelted zones there was huge grain refinement found, besides the silicon powder presence. The obtained results from the microstructure investigation performed on the scanning electron microscope ZEISS Supra 35 with a magnification up to 5000 times reveals the structure of the aluminium cast alloy used for investigation (Figs. 3, 4 and 5) where the structure refinement can be observed after the proper preformed heat treatment. In general a zone like structure of the obtained surface layer, where a remelted zone was found, and a heat affected zone, were found. The layers achieved by the alloying process are showed in Figs. 6, 7, 8, 9, 10 and 11. The results allow us to state that with increasing laser power the roughness of the remelted metal surface increases.

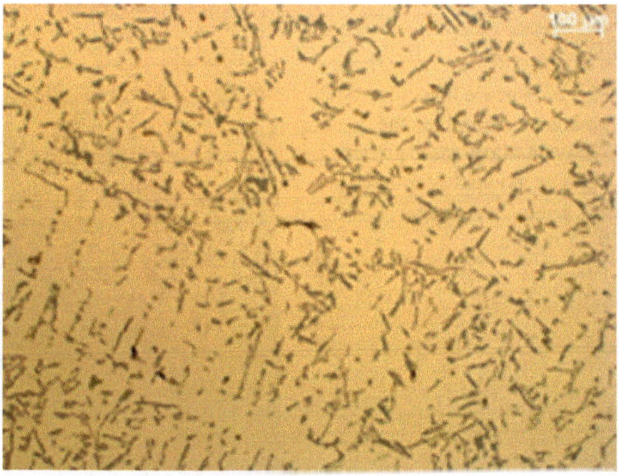

Fig. 3 Microstructure of the cast aluminium AlSi9Cu in its cast state

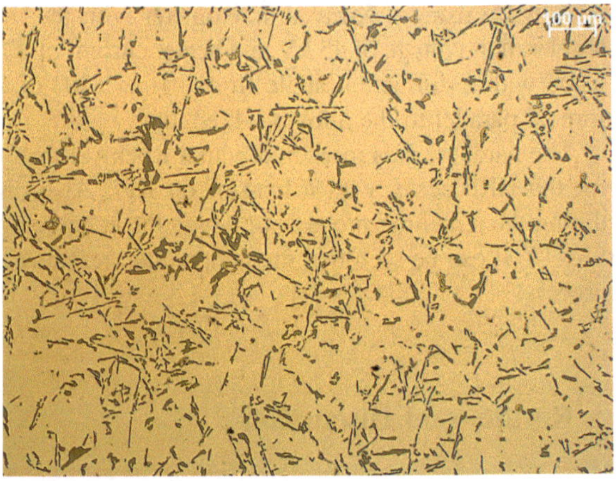

Fig. 4 Microstructure of the cast aluminium AlSi9Cu after aging

Fig. 5 Microstructure of the cast aluminium AlSi9Cu in its cast state

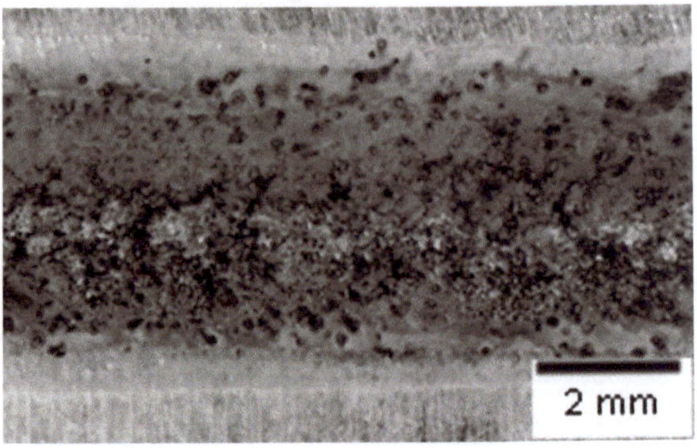

Fig. 6 Surface layer after SiC powder feeding into the AlSi9Cu alloy, laser power 1.5 kW, powder feed rate 1.5 g/min, scan rate 0.25 m/min

Fig. 7 Structure of the surface layer after SiC powder feeding into the AlSi9Cu alloy, laser power 1.5 kW, powder feed rate, 1.5 g/min, scan rate 0.25 m/min

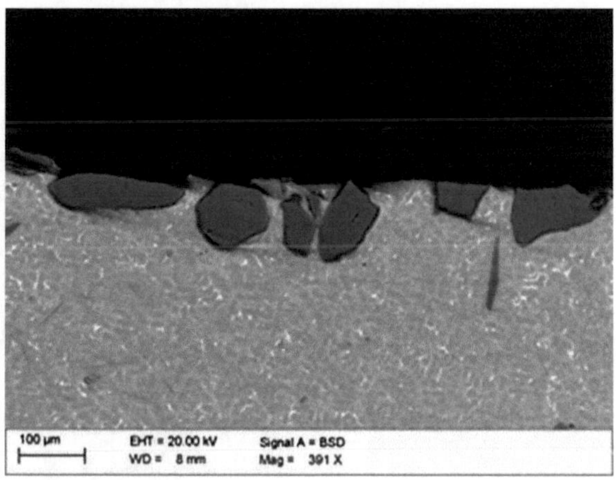

Fig. 8 Structure of the surface layer after SiC powder feeding into the AlSi9Cu4 alloy, laser power 2.0 kW, 1.2 g/min, scan rate 0.25 m/min

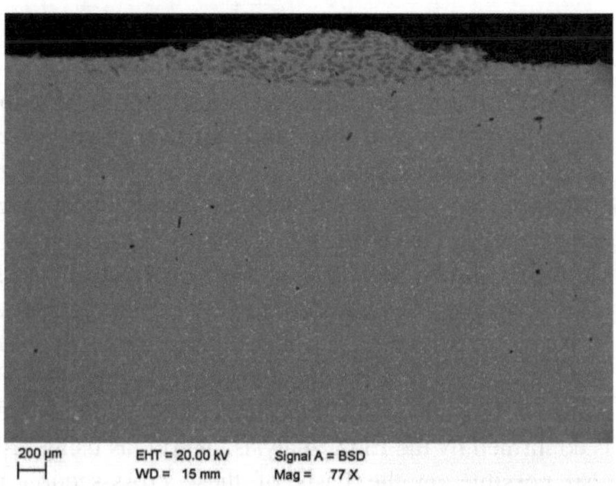

Fig. 9 Structure of the surface layer after SiC powder feeding into the AlSi9Cu4 alloy, laser power 2.0 kW, 1.5 g/min, scan rate 0.25 m/min

Fig. 10 Structure of the surface layer after SiC powder feeding into the AlSi9Cu4 alloy, laser power 2.0 kW, powder feed rate 2.0 g/min, scan rate 0.25 m/min

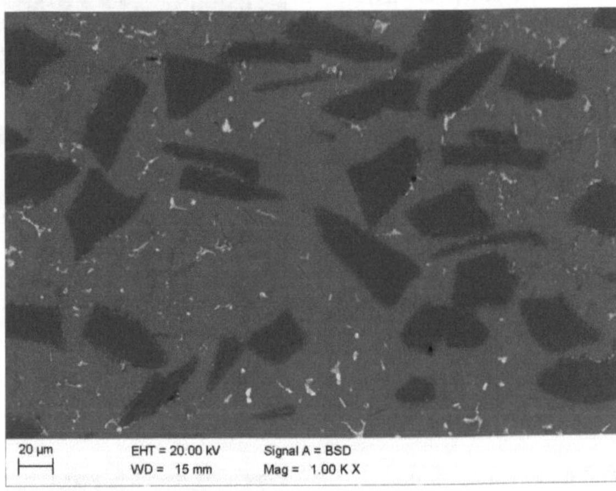

Fig. 11 Structure of the surface layer after SiC powder feeding into the AlSi9Cu4 alloy, laser power 2.0 kW, powder feed rate 1.5 g/min, scan rate 0.25 m/min

Based on these investigation concerning the distribution of the powder particles in the surface layer of the aluminium-silicon-copper cast alloys, it was also found, that in the laser treated surface layer there are no pores or cracks in the produced coating or any defects and failures occurs in this layer, but the fed SiC particles occur mainly in the upper part of the RZ. Occasionally he occurring discontinuity of the layers can be seen as a product of the heat transfer process and may be neutralised by properly adjusted powder quality and powder feed rate.

As it can be seen for all investigated alloys the used SiC powder does not form any, and occurs in form of loose particles relatively evenly distributed in the alloy matrix but mainly in the upper part of the surface layer (Figs. 12, 13 and 14), what is confirmed by the EDX analysis, as well as using X-Ray diffraction (Fig. 15). It is also possible on the basis of these cross-section micrographs to evaluate the

Fig. 12 EDS microanalysis of the surface layer after SiC WC feeding into the AlSi9Cu alloy: SE SEM image, as well mappings of Al and Si

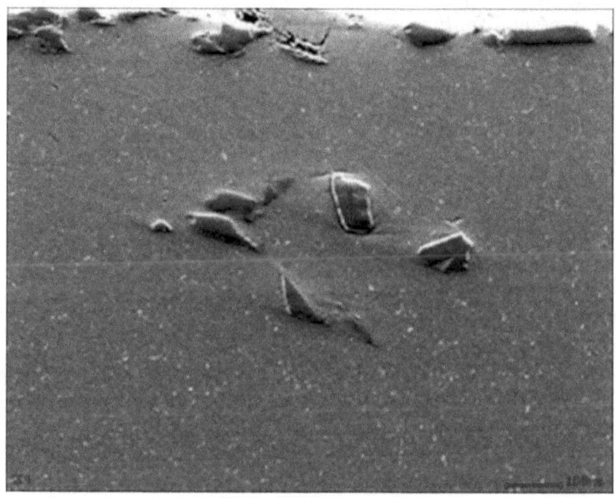

Fig. 13 Surface layer cross section of the AlSi9Cu4 alloy after feeding of SiC particles, laser power 2.0 kW

thickness of the surface layer depth [Remelting Zone (RZ)], which is ca. 0.55 mm (Fig. 16) in the case of the AlSi9Cu4 as well as the AlSi9Cu alloy and the 2.0 kW laser power. For lower laser powers the thickness is ca. half as much as for the highest power for both types of the investigated alloys.

During the investigations also the EDS point analysis was carried out, presented in Figs. 13 and 14, has revealed that the chemical composition of the alloyed area around the SiC particles, so in case of this powder the silicon and carbon is not dissolved in the aluminium matrix, but still in the form of the primary SiC phase. There is no evidence for the dissolving of the ceramic powder Sic grains, because the carbon concentration in the Al matrix will be possible to measure and visible on Fig. 12. The influence of the laser speed on the remelting of the aluminium surface was also investigated. The range was chosen from 0.25 to 0.75 m/min (Figs. 13 and 14), the optimal value was set as 0.25 m/min for both of the investigated aluminium alloys groups. The idea was to find out if the laser power will change the nature of

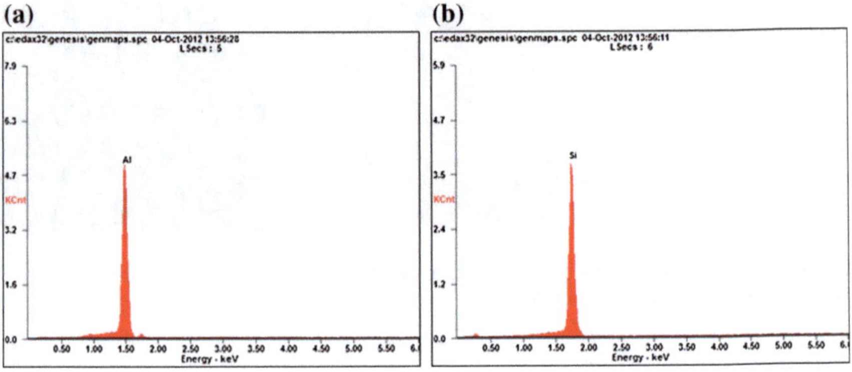

Fig. 14 EDS pointwise analysis of the chemical composition performed: **a** in point 1 on Fig. 12, **b** in point 2 on Fig. 12, AlSi9Cu4 alloy after SiC feeding

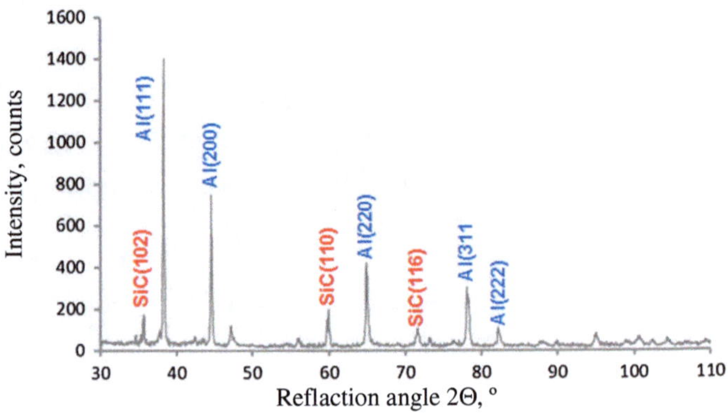

Fig. 15 X-Ray diffraction of the surface layer of the AlSi9Cu4 alloy after feeding of SiC particles obtained using the Bragg-Brentano method

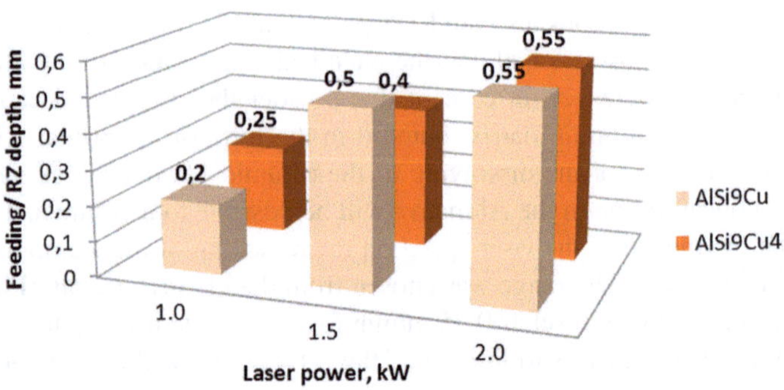

Fig. 16 Thickness of the surface layer (RZ zone) of the aluminium alloys fed with SiC powder

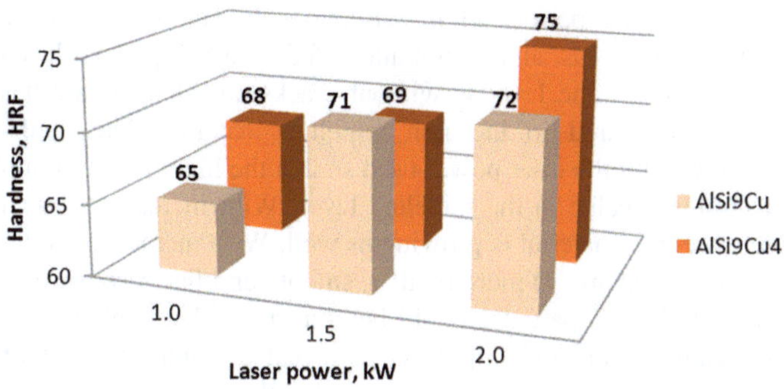

Fig. 17 Hardness of the surface layer of the cast aluminium alloys AlSi9Cu and AlSi9Cu4 fed with SiC powder

the alloying of the SiC powder. But for powers higher than 1.80 kW the SiC powder forms a bulky non-flat surface on the laser treated aluminium.

The measured results, of the surface layer thickness are presented on Fig. 16, which confirm the relationship between the laser power and the remelting zone (RZ) depth, where with increasing laser power the depth of this zone increases and reach even 0.55 mm for 2.0 kW laser power for both of the investigated alloys.

Figure 17 shows the hardness measurement results of the remelted surface for 1.0, 1.5 and 2.0 kW laser power, where it can be seen that the hardness of the remelted zone is smaller than of the non remelted area. It can be found that the hardness value of the surface layer increases together with the laser power and has the highest value for the laser power at 2.0 kW. These results are confirmed for both of the alloy types AlSi9Cu and AlSi9Cu4, where the value reached even 75 HRF for the AlSi9Cu4 alloy. It can be also seen that the hardness of the laser fed samples with the laser power of 1.5 kW is comparable with the value achieved after a conventional heat treatment of this cast aluminium alloys and for 1.0 kW the hardness is even lower—68 HRF.

4 Conclusion

On the basis of the analysis of the test results, it has been pointed out that in the case of the analysed aluminium cast alloys the applied laser surface processing, and the thermal processing preceding it, ensuring occurrence of the mechanisms responsible for material strengthening, enable enhancement of the mechanical and usable properties of the examined alloys. An essential objective is also to indicate the multiple possibilities for continuation of the tests, regarding the light metal alloys aluminium, magnesium and titanium, broadening the current knowledge within the scope of elements and light structures.

On the basis of the performed investigations it can be concluded, that the AlSi9Cu and AlSi9Cu4 cast aluminium alloys fed with SiC powder has achieved a high-quality top layer. The layer is without cracks and defects and has a higher hardness when compared to the non remelted material. The hardness value increases together with the laser power used so that the highest power applied gives the highest hardness value in the remelted layer. With increasing laser power the depth of the remelting material is growing as well. With increasing laser power the surface of the remelted area is more regular, smooth and flat Also the occurrence of the tungsten carbide SiC particles could be confirmed. The metallographic investigations on scanning and light microscope reveal a dendritic structure which is present in the heat affected zone in samples fed with the applied laser powers. The performed investigations of the microstructure evaluation of the Al-Si-Cu alloys, carried out using a light and scanning electron microscope, allowed to confirm the zone-like nature of the surface layer obtained using HPDL laser for alloying of the AlSi9Cu and the AlSi9Cu4 cast aluminium alloys. The remelted zone (RZ), the heat affected zone (HAZ) on top of the substrate material as well as a transition zone were revealed. There were also ceramic powder SiC particles detected in the substrate. Before the alloying with SiC there were performed preliminary tests allowing to find the proper laser scan speed as well as the optimal laser power used for alloying of this type of aluminium alloy substrate. Of course the optimal laser powder and feeding rate will differ slightly compared to the values used for remelting of the aluminium alloy, but the range does not changes significantly.

The conducted tests enabled the fulfilment of the objectives of the this paper, namely that the application of the laser surface processing for improvement of the exploitation properties of surfaces of the aluminium cast alloys allows for formation of the surface layer characterised by better mechanical and tribological properties in comparison as opposed to the core material.

Tests of the usable properties of the surface layers formed by means of laser indicate an increase in the hardness and resistance to abrasion of the surface layer with the fused powders of the WC and SiC carbides and relatively low corrosion resistance. In the case of the aluminium and zirconium oxide powders the surface layer obtained by the laser fusion method has a form of oxide coating, composed by the elements present in the fused ceramic powders. In the case of the surface layers after fusion of the oxide powders occurrence of the melted zone has not been confirmed, and at the same time presence of the particles fused in the matrix of aluminium alloys has also not been confirmed. The WC and SiC powder introduced into the matrix of the selected aluminium alloys during fusion has a form of particles uniformly distributed in the melted zone.

The laser power determination leads to the conclusion, that the optimal power range is ca. 1.5 kW, and that a lower value of ca. 1.0 kW does not achievement of an completely homogenous remelting tray on the sample surface, whereas a higher power of 2.0 kW makes an uneven shape of the remelted surface layer. Particularly it can be also found that: the SiC powder particles are fed successfully into the aluminium alloy matrix during laser feeding. The optimal laser power is in the range of <1.0–1.5 kW, with a laser scan rate determined as 0.5 m/s. It can be summarize that this technique is a great method for the modelling of tools of hot cast aluminium surface layer, which can be deeply recognized in following studies and further investigations. There are also next investigation planned with other ceramic powder particle like: Al_2O_3 or ZrO_2 (see Table 1).

Acknowledgments This research was financed by the National Science Centre (NCN—Narodowe Centrum Nauki) within the framework of the Research Project No. 2011/01/B/ST8/06663 headed by Dr Krzysztof Labisz DSc. Eng.

References

1. Samuel AM, Gauthier J, Samuel FH (1996) Microstructural aspects of the dissolution and melting of Al2Cu phase in AI-Si alloys during solution heat treatment of Al2Cu phase in Al-Si alloys during solution heat treatment. Metall Mater Trans A 27:1785–1798
2. Labisz K (2014) Microstructure and mechanical properties of HPDL laser treated cast aluminium alloys. Mater Sci Eng Technol (Mat.-wiss. u. Werkstofftech) 45:314–324. doi:10.1002/mawe.201400231
3. Tański T, Labisz K, Lukaszkowicz K (2013) Structure and properties of diamond-like carbon coatings deposited on non-ferrous alloys substrate. Mechatron Syst Mater V (Solid State Phenomena) 199:170–176
4. Ozgowicz W, Labisz K (2011) Analysis of the state of the fine-dispersive precipitations in the structure of high strength steel Weldox 1300 by means of electron diffraction. J Iron Steel Res Int 18(1):135–142
5. Dobrzański LA, Krupiński M, Labisz K, Krupińska B, Grajcar A (2010) Phases and structure characteristics of the near eutectic Al-Si-Cu alloy using derivative thermo analysis. Mater Sci Forum 638–642:475–480
6. Kusiński J, Przybyłowicz J, Kąc S, Woldan A (1999) Structure and properties change In case of laser remelting of surface layers and coatings. Hutnik 14–20 (in Polish)
7. Konieczny J, Dobrzański LA, Labisz K, Duszczyk J (2004) The influence of cast method and anodizing parameters on structure and layer thickness of aluminium alloys. J Mater Process Technol 157–158:718–723
8. Dobrzanski LA, Tanski T (2009) Influence of aluminium content on behaviour of magnesium cast alloys in bentonite sand mould. Solid State Phenom 147–149:764–769
9. Kennedy E, Byrne G, Collins DN (2004) Review of the use of high power diode lasers in surface hardening. J Mater Process Tech 155–156:1855–1860
10. Dobrzański LA, Krupiński M, Labisz K, Krupińska B, Grajcar A (2010) Phases and structure characteristics of the near eutectic Al-Si-Cu alloy using derivative thermo analysis. Mater Sci Forum 638–642:475–480
11. Piec M, Dobrzański LA, Labisz K, Jonda E, Klimpel A (2007) Laser alloying with WC ceramic powder in hot work tool steel using a high power diode laser (HPDL). Adv Mater Res 15–17:193–198
12. Horst EF, Mordike BL (2006) Magnesium technology. Metallurgy, design data, application. Springer, Berlin
13. Tanski T (2014) Determining of laser surface treatment parameters used for light metal alloying with ceramic powders. Materialwiss Werkstofftech 45(5):333–343. doi:10.1002/mawe.201400232
14. Dobrzański LA, Tomiczek B, Pawlyta M, Król M (2014) Aluminium AlMg1SiCu matrix composite materials reinforced with halloysite particles. Arch Metall Mater 59(1):335–338

15. Dobrzański LA, Tomiczek B, Pawlyta M, Nuckowski P (2014) TEM and XRD study of nanostructured composite materials reinforced with the halloysite particles. Mater Sci Forum 783:1591–1596
16. Dobrzański LA, Labisz K, Piec M, Lelątko AJ, Klimpel A (2006) Structure and properties of the 32CrMoV12-28 steel alloyed with WC powder using HPDL laser. Mater Sci Forum 530–531:334–339
17. Tanski T, Labisz K (2012) Electron microscope investigation of PVD coated aluminium alloy surface layer. Solid State Phenom 186:192–197. doi:10.4028/www.scientific.net/SSP.186.192
18. Lisiecki A, Klimpel A (2008) Diode laser surface modification of Ti6Al4V alloy to improve erosion wear resistance. Arch Mater Sci Eng 32:5–12
19. Klimpel A, Dobrzański LA, Janicki D, Lisiecki A (2005) Abrasion resistance of GMA metal cored wires surfaced deposits. J Mater Process Technol 164(165):1056–1061
20. Dobrzański LA, Borek W (2012) Thermo-mechanical treatment of Fe–Mn–(Al, Si) TRIP/TWIP steels. Arch Civil Mech Eng 12:299–304
21. Grajcar A, Borek W (2008) Thermo-mechanical processing of high-manganese austenitic TWIP-type steels. Arch Civil Mech Eng 8(4):29–38

Application of the Finite Element Method for Modelling of the Spatial Distribution of Residual Stresses in Hybrid Surface Layers

Tomasz Tański, Krzysztof Labisz, Wojciech Borek, Marcin Staszuk, Zbigniew Brytan and Łukasz Krzemiński

Abstract The presented investigations concern PVD/CVD surface treatment performed on samples of heat treated cast magnesium and aluminium alloys and properties modelling of obtained coatings using the finite element method (FEM). In order to identify the structure and fractures of the analysed surface coatings, investigations using the scanning electron microscope Zeiss Supra 35 were performed. Evaluation of the adhesion of the PVD/CVD coatings was carried out using a scratch test. The obtained coatings—Ti/Ti(C,N)-gradient/CrN; Ti/Ti(C,N)-gradient/(Ti,Al)N; Ti/(Ti,Si)N-gradient/(Ti,Si)N as well coatings: Cr/CrN-gradient/CrN; Cr/CrN-gradient/TiN and Ti/DLC-gradient/DLC are characterized by a clear heterogeneity of the surface associated with the presence of microparticles in the structure in form of droplets broken out of the target during the deposition process, as well immersions occurring in the surface due to the loss of some droplets during solidification. It was also found that the applied coatings are characterized with a mono-, di-, or multi-layer structure according to the applied layer system; the individual layers are applied uniformly and tightly adhere to the substrate and to each other. The obtained results of the numerical FEM analysis, have enabled a full integration of the material engineering knowledge and informatics tools, confirming compliance of the simulation model with the obtained experimental results.

Keywords Light alloys · Surface coatings · Residual stress · FEM · Properties

T. Tański (✉) · K. Labisz · W. Borek · M. Staszuk · Z. Brytan · Ł. Krzemiński
Institute of Engineering Materials and Biomaterials, Silesian University of Technology, Konarskiego St. 18a, 44-100 Gliwice, Poland
e-mail: tomasz.tanski@polsl.pl

A. Öchsner and H. Altenbach (eds.), *Mechanical and Materials Engineering of Modern Structure and Component Design*, Advanced Structured Materials 70,
DOI 10.1007/978-3-319-19443-1_5

1 Introduction

Modern technological development makes it necessary to look for new design solutions to improve efficiency and product quality, to minimize size and to increase dimensional stability in real working conditions. The proper choice of material is always preceded by an analysis of a number of factors including following requirements like: mechanical, physical, design, environmental, cost-related, availability and weight. The general problem of high material weight verifies the application of various groups of materials in order to improve the lightweight properties. Therefore over the past few decades, there is an increase in the demand of materials with a low density and relatively high strength, such as titanium, aluminium and magnesium [1–6].

From the listed materials group special attention should be devoted to magnesium and aluminium alloys, because of their low densities in addition to other benefits, such as good damping capacity, high dimensional stability, good castability, the combination of low density and high strength, and easy recycling procedure. However, the undisputed advantage of magnesium and aluminium is mainly the low density of Mg-1.7 g/cm^3 and Al-2.7 g/cm^3. Currently, the greatest need for magnesium and aluminium cast alloy reveals the automotive industry. In the case of transportation, the general rule is to reduce the weight of vehicles because of fuel savings. The need to reduce the primary weight of vehicles is more and more important, because an increasing number of vehicles is equipped with new accessories (such as air bags, traction and parking control systems) and, increasing its mass, which are aimed not only at improving the safety of driving, but also increases the attractiveness of the utility of these vehicles [1–6].

However, besides the above-mentioned advantages of aluminium and magnesium alloys, there are also some disadvantages, most commonly poor corrosion resistance and low wear resistance of these materials. A promising solution, which is based on many years of research, is laser treatment, particularly laser remelting and the feeding of hard ceramic powder particles in the surface layer as well as the technologies of physical and chemical vapour deposition PVD and CVD, which is an alternative for applied thermal spraying techniques or anodic protection techniques [7–16].

As a result of proper material selection for special elements together with the technologies, which determine the structure and properties of the alloy and its surface layer, providing the required functional properties, it is possible to obtain the best possible properties of the substrate and the surface layer of the produced element, which consequently leads to the design and delivery of a material meeting all demands given by the final consumer [7–16].

One of the goals of this work was to achieve the best combination of hybrid coatings, including a gradient transition layer, with a stable change or one or more of its components in a linear way reaching from the substrate to the surface top. It was also important to achieve an outer coating using the CAEPVD/PACVD method in order to increase the low strength of the substrate material. This article focuses on a very important surface engineering issue, namely the surface treatment of selected light cast alloys using vacuum techniques. These alloys are currently recognized by materials science experts as future materials combining low density and high strength. A very important issue is to develop the production and processing technology of construction materials or surface protection against environmental influences. Such a way could help to find a proper relationship between the modern substrate material and production of coatings of a new generation, reaching even to IV generation [7–21].

The manufacturing processes makes it necessary to carry out the analysis of the impact of process parameters on the substrate material and the interaction between the surface layer and the substrate material in the transition zone. The used computational methods allow us to perform an analysis of complex phenomena occurring during the coating process, as well as the analysis of the phenomena occurring as the end result of the layer deposition process. As a result of the application of these methods it is possible to create a numerical model describing the mechanical and functional properties of the surface layer depending on the process parameters and the type of substrate material and the surface layer [7–21].

The describe vacuum deposition method are not the only one usable for surface engineering, there exists a lot of other surface treatment processes, but the achieved surface quality and relative low costs makes this methods very usable and of wide application range in material engineering [22–24].

2 Methodology

PACVD is a process where a relatively low surface treatment temperature can be obtained to produce a carbon DLC coating at a specified pressure and in an C_2H_2 acetylene atmosphere. The gradient of the resulting coating was achieved by a variable concentration of silicon (Me) in the middle layer. Silicon was supplied to the furnace chamber in the gas phase—Ti/a-C:H-Me/a-C:H. The second method was implemented using the device DREVA ARC400 supplied by Vakuumtechnik using the electric arc cathodic evaporation method. The device is equipped with three independent sources of metal vapours.

Prior to the coating process the substrates were cleaned using a chemical cleaning process and rinsed in ultrasonic cleaners and dried in a stream of hot air. Furthermore, the samples were cleaned with Argon ions at the voltage of 800/200 V for 20 min. For PVD coatings 65 mm diameter discs were used, cooled, cooled with water, containing pure metals (Cr, Ti), Al and TiSi alloys. The coatings were deposited in an inert atmosphere of Ar and a reactive N_2 atmosphere in order to obtain a mixture of nitrides as well as a N_2 and C_2H_2 mixture in order to obtain layers of carbonitrides. The obtained appropriate concentrations of elements, allows control over the properties of the deposited coatings. Gradient concentration change of the chemical composition of the coating cross section was obtained by changing the proportion of reactive gas or by changing the evaporation current on the arc sources. The set coating process conditions are presented in Table 1. During the PVD process the substrates have moved to the vapor sources carrying out rotational movements to achieve uniform thickness of the coatings and preventing the formation of the phenomenon of a so-called shadow coming into existence on the coated surface model, which aims to determine the residual stresses in gradient and single-layer coatings such as: Ti/Ti(C,N)/CrN, Ti/Ti(C,N)/(Ti,Al)N, Cr/CrN/CrN, Cr/CrN/TiN, Ti/(Ti,Si)N/(Ti,Si)N and Ti/DLC/DLC deposited on the surface of cast magnesium and aluminium alloys (Tables 2 and 3) was performed using the finite element method, assuming the real dimensions of the sample. The geometry of the plate with coated gradient and monolayer coatings as well as the calculations were carried out using the ANSYS 12.0 software (see Fig. 1). Due to the expected range of simulations there were elaborated parametric input data files, allowing for a comprehensive analysis.

The calculation model is composed of 12,816 and 11,780 parts of nodes. To avoid an error in the calculation of residual stresses in the coatings, there were variable sizes of the finite elements used, in places where larger stress gradients were expected, because the grid is more dense than in regions where stress should take similar values.

Table 1 Deposition conditions of the investigated layers

Process parameters	Type of the produced coating as well the applied coating deposition technique					
	PVD					PACVD
	Ti/Ti(C,N)-gradient/CrN	Ti/Ti(C,N)-gradient/(Ti,Al)N	Cr/CrN-gradient/CrN	Cr/CrN-gradient/TiN	Ti/(Ti,Si)N-gradient/(Ti,Si)N	Ti/DLC-gradient/DLC
Basic press., Pa	5×10^{-3}	5×10^{-3}	5×10^{-3}	5×10^{-3}	5×10^{-3}	1×10^{-3}
Work press., Pa	0.9/1.1–1.9/2.2	0.9/1.1–1.9/2.8	1.0/1.4–2.3/2.2	1.0/1.4–2.3/2.2	0.89/1.5–2.9/2.9	2
Argon flow, cm^3/min	80[a]	80[a]	80[a]	80[a]	80[a]	80[a]
	10[b]	10[b]	80[b]	80[b]	20[b]	–
	10[c]	10[c]	20[c]	20[c]	20[c]	–
Nitrogen flow, cm^3/min	225 → 0[b]	0 → 225[b]	0 → 250[b]	0 → 250[b]	0 → 300[b]	–
	250[c]	350[c]	250[c]	250[c]		–
Acetylene flow, cm^3/min	0 → 170[b]	140 → 0[b]	–	–		230
Voltage on the substrate, V	70[a]	70[a]	60[a]	60[a]	70[a]	500
	70[b]	70[b]	60[b]	60[b]	100[b]	
	60[c]	70[c]	60[c]	100[c]	100[c]	
Current at the cathode, A	60	60	60	60	60	–
Temperature, °C	<150	<150	<150	<150	<150	<180

[a]During the deposition of the metallic layer
[b]During the deposition of the gradient layer
[c]During the deposition of the ceramic layer

Table 2 Chemical composition of the investigated magnesium alloys

Alloy type	Mass concentration of the chemical alloying elements in the investigated, %						
	Al	Zn	Mn	Si	Fe	Mg	Rest
AZ91	9.09	0.77	0.21	0.037	0.011	89.79	0.092
AZ61	5.92	0.49	0.15	0.037	0.007	93.33	0.066

Table 3 Chemical composition of the investigated aluminium alloys

Alloy type	Mass concentration of the chemical alloying elements in the investigated, %							
	Si	Cu	Fe	Mn	Mg	Zn	Al	Rest
ACAlSi9Cu	9.09	1.05	0.72	0.36	0.27	0.14	88.17	0.15
ACAlSi9Cu4	9.27	4.64	0.17	0.01	0.28	0.05	85.4	0.18

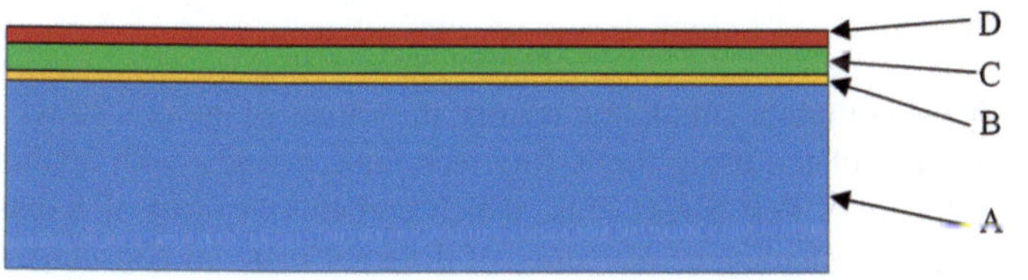

Fig. 1 Schematic diagram of the geometry of the samples of the cast magnesium and aluminium alloys with the deposited coatings onto the surface, where: **a** substrate, **b** intermediate layer, **c** gradient layer, **d** outer layer

Also for the coatings smaller elements were applied smaller elements, more accurately representing the stress gradient, and in the substrate material the applied elements increase with increasing distance from the coatings. Due to the need to calculate the residual stress in the gradient coating with a variable chemical composition in the direction perpendicular to the sample surface, a schematic division of the modelled coatings into zones because of areas with similar chemical composition was carried out. The calculation model with the zone gradient coating division has been designed to be able to determine the average residual stress values in the relevant areas of the coating because of functional properties.

In order to simulate the internal stresses in the analysed coatings on various substrates, the following boundary conditions were set:

- temperature change of the PVD and CVD process reflects the sample cooling process from 150 °C to room temperature of 20 °C,
- the analysed material data form substrate from light magnesium and aluminium alloys are summarized and presented in Table 4.

Table 4 Parameters used in computer simulation of the analysed coatings and substrate material

Coating type	Young modulus [GPa]	Poisson's ratio	Heat expansion coefficient [10^{-6}/K]
DLC	140	0.22	7.0
(Ti,Si)N	450	0.25	13
CrN	360	0.28	2.3
Ti(C,N)	460	0.2	9.4
Cr	140	0.31	6.2
Ti	113	0.34	8.5
(Ti,Al)N	460	0.25	9.35
Substrate material			
AZ61	45	0.35	27.30
AZ91	45	0.35	26
AlSi9Cu	75	0.33	22
AlSi9Cu4	75	0.33	22

To verify the computer simulation results, the values of internal stress in selected coatings were calculated using the X-Ray technique g-sin2ψ with appliance of the computer software X'Pert Stress Plus, which contains in form of a data base the necessary data for calculations of values of different material constants. The stress measurements were performed on the (Ti,Si)N outer layer of the Ti/(Ti,Si)N/(Ti,Si)N as well on the outer TiN layer of the Cr/CrN/TiN coating on the same substrate namely the AZ61cast magnesium alloy substrate. The compliance of the FEM model and its full relevance to the actual measured data were clearly confirmed, because in each of the analysed cases, the calculation results are consistent with the results of the finite element computer analysis and are within the range of error.

Tests of the coatings' adhesion to the substrate material were made using the scratch test on the CSEM Revetest device, by moving the diamond indenter along the examined specimen's surface with a gradually increasing load. The character of the damage caused was evaluated by applying light microscope observations. Microhardness tests of the coatings were made with the Shimadzu DUH 202 ultra-microhardness tester.

3 Investigations Results

3.1 Metallographic Investigations

In order to investigate the structure and the existing relationship between the type of the substrate made from cast magnesium and aluminium alloy, the type and technological conditions of the hybrid coatings production process, metallographic tests were carried out. Coatings produced using the CAE-PVD technique are characterized by a clear heterogeneity connected to a number of microparticles in the form of droplets occurring on the structure (see Figs. 2 and 3).

The highest heterogeneity of the surface area compared to other investigated coatings is characteristic for the coatings of the type: Ti/Ti(C,N)/(Ti,Al)N and Ti/Ti (C,N)/CrN, where a number of solidified droplets of the vaporized metal were identified (see Figs. 2 and 3). The lowest amount of solidified droplets of the vaporized metal was identified in case of the Cr/CrN/TiN and Cr/CrN/CrN coating (see Fig. 4).

The occurrence of these morphological defects is related to the cathodic arc evaporation process itself. Depending on the process conditions, including the kinetic energy of the drops sputtered into the metal substrate and the nature of the metal vapour source, the observed particles are clearly different in terms of shape and size. Moreover, it was also observed that there are present characteristic cavities, which are formed as a result of falling out of droplets after finishing the PVD process. In the case of the DLC coating obtained using the PACVD process, on the surface small droplets are identified, often in a spheroidal form (see Fig. 5). Surface morphology of the obtained DLC coatings differ significantly from the surface morphology obtained in the classical high-temperature CVD processes, where a network of microcracks occurs, wave-like surfaces or surfaces with spherical shapes. A clear transition zone between the substrate and the coating identified was also identified. The obtained coatings reveal a compact structure, without visible delamination and defects, they are uniform and tightly adhere to each other as well as to the substrate (see Figs. 6, 7, 8, 9, 10 and 11). Investigations of fractures

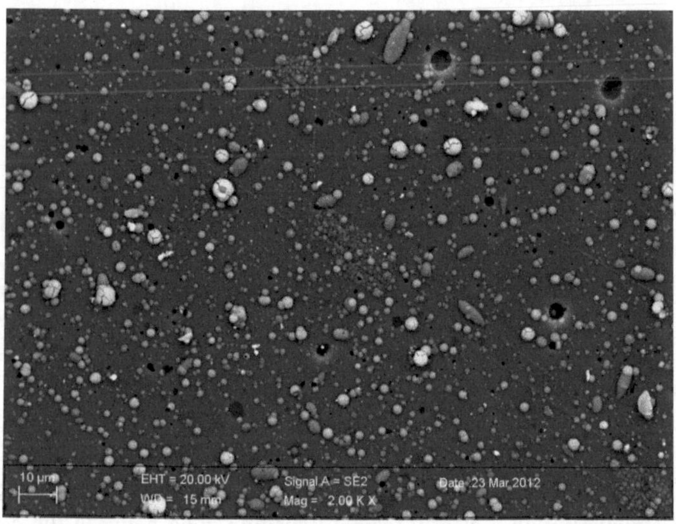

Fig. 2 Surface morphology of the Ti/Ti(C,N)/(Ti,Al)N layer coated on the AZ91 cast magnesium substrate

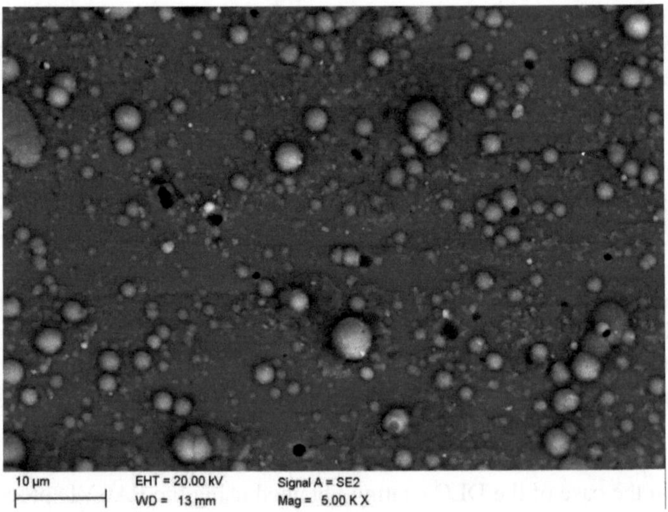

Fig. 3 Surface morphology of the Ti/Ti(C,N)/CrN layer coated on the AlSi9Cu4 cast aluminium substrate

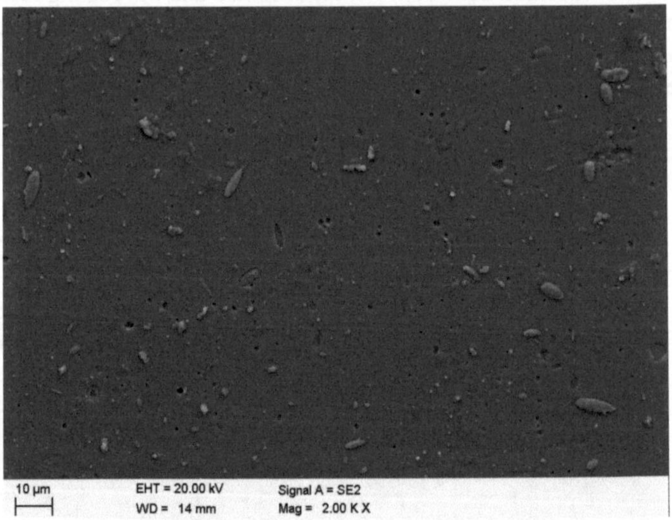

Fig. 4 Surface morphology of the Cr/CrN/TiN layer coated on the AlSi9Cu1 cast aluminium substrate

confirm that the coatings of the type Ti/Ti(C,N)/(Ti,Al)N and Ti/Ti(C,N)/CrN show a layered structure, with a clearly visible transition zone between the coating and the gradient wear resisting coating. (see Figs. 6 and 7).

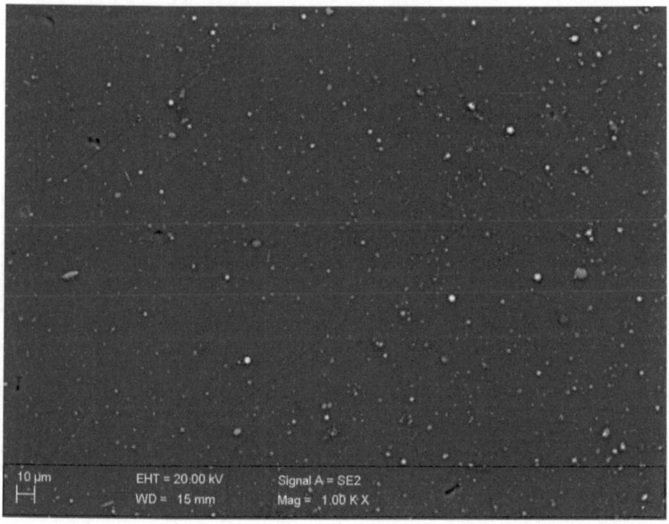

Fig. 5 Surface morphology of the Ti/DLC/DLC layer coated on the AZ61 cast magnesium substrate

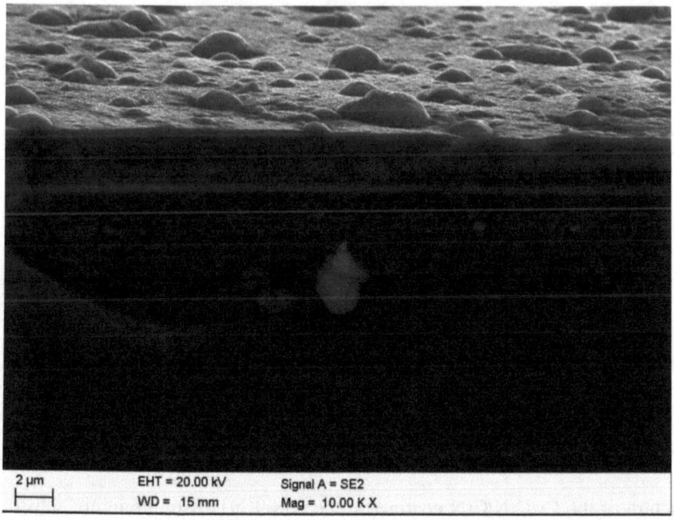

Fig. 6 Fracture of the Ti/Ti(C,N)/(Ti,Al)N coating on the AZ91 cast magnesium substrate

In case of the Cr/CrN/CrN, Ti/(Ti,Si)N/(Ti,Si)N coatings, where the same set of chemical elements in the gradient coating as well as in the wear resistant coating and in the coatings of the Ti/DLC/DLC type was applied, any differences on the cross section has been found. (see Figs. 8, 9 and 10). Moreover, in the case of the

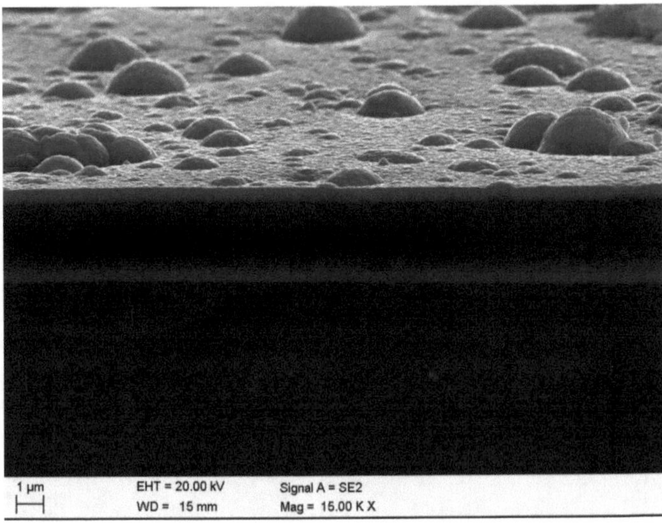

Fig. 7 Fracture of the Ti/Ti(C,N)/CrN coating on the AlSi9Cu4 cast aluminium substrate

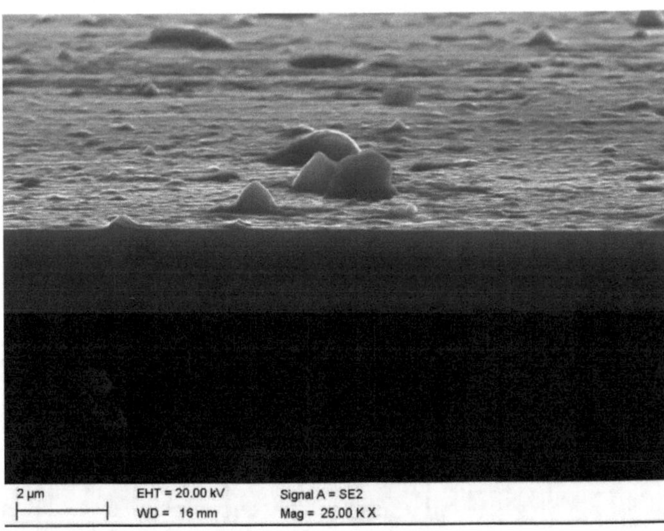

Fig. 8 Fracture of the Cr/CrN/CrN coating on the AlSi9Cu1 cast aluminium substrate

thin adhesive coating, which improves the adhesion of DLC coating to the substrate, there was a characteristic bight continuous titanium layer identified, which was also which was also confirmed using EDS microanalysis. In the case of the layer of the Cr/CrN/TiN type, the characteristic nature similar to a columnar growth

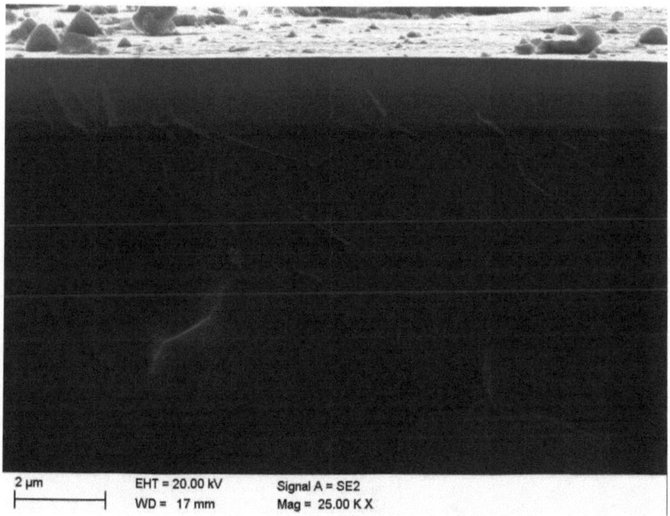

Fig. 9 Fracture of the Ti/(Ti,Si)N/(Ti,Si)N coating on the AlSi9Cu1 cast aluminium substrate

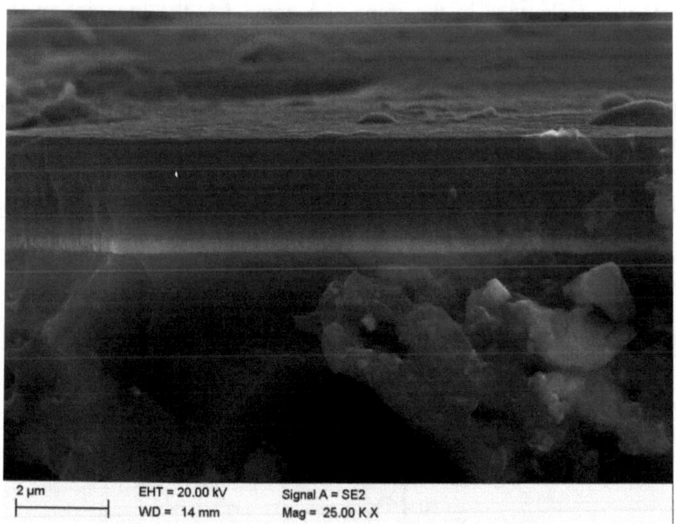

Fig. 10 Fracture of the Ti/DLC/DLC coating on the AZ91 cast magnesium substrate

of crystallites characteristic of coatings was confirmed based on titanium nitride obtained in the cathodic arc evaporation process (see Fig. 11).

Chemical composition investigations performed using the GDOES confirm the presence of the chemical elements included in the analysed coatings layer in a depth, depending on the coating's thickness, from 1.4 to 3.4 μm (Figs. 12 and 13).

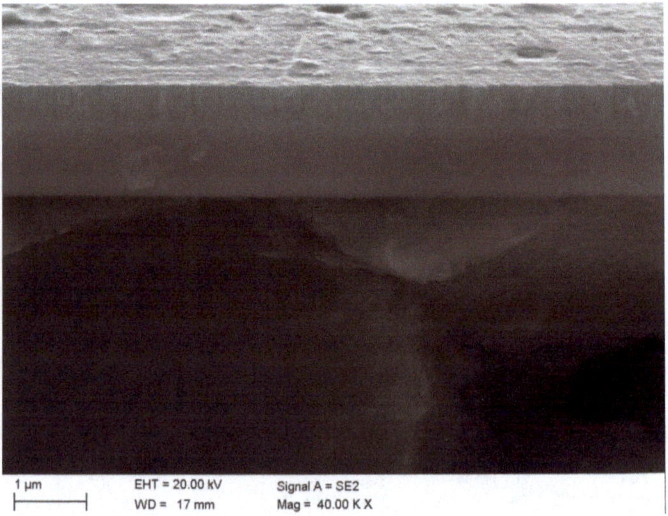

Fig. 11 Fracture of the Cr/CrN/TiN coating on the AZ61 cast magnesium substrate

The nature of the changes in the transition zone, i.e. the increase in the concentration of elements contained in the substrate, while decreasing the concentration of the coating-forming elements, may be indicative of the existence of the diffusion layer of the interface zone between the substrate material and the obtained coating, improving the adhesion of the produced coatings to the magnesium substrate. Furthermore, by using a glow discharge optical spectrometer the presence of a zone of linearly decreasing chemical composition concentration was confirmed, which indicates their gradient-like nature (see Figs. 12 and 13).

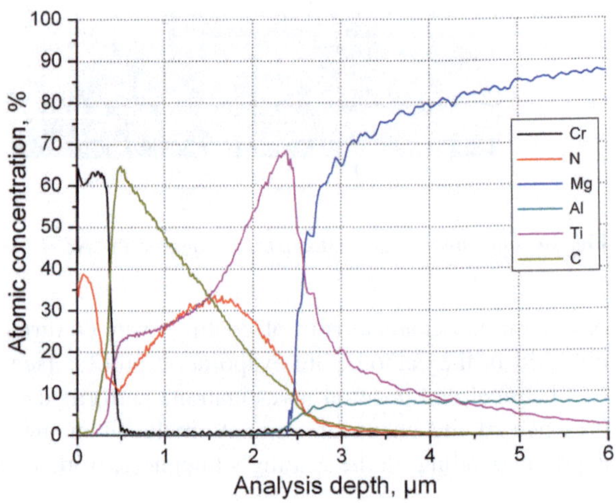

Fig. 12 Change of the concentration of the Ti/Ti(C, N)/CrN coating compound coated on ten substrates of the AZ91cast magnesium alloy

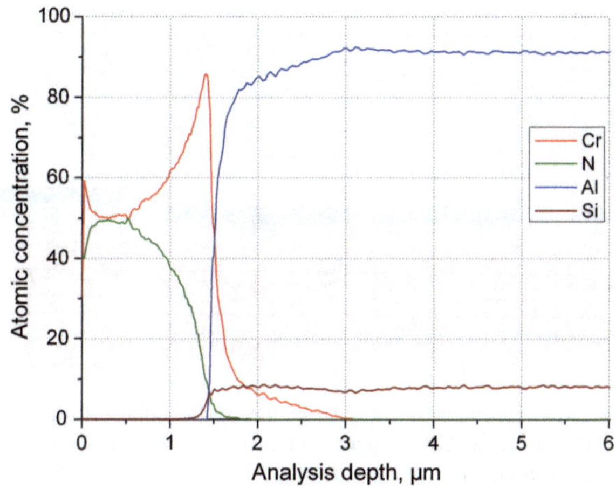

Fig. 13 Concentration change of the Cr/CrN/CrN coating coated on the AlSi9Cu1 substrate

3.2 Mechanical Investigations and Application of the Finite Element Method

The analysis of the obtained simulation results show that in each of the analysed cases the stress has an compressive character, that means with a negative value, causing "compression" of the produced coatings to the substrate surface, with the average absolute value $\sigma = 321 \div 1377$ MPa (see Table 5).

Occurrence of compressive stress in the surface layer (see Figs. 14 and 15), can prevent the formation of cracks under operating conditions, when the element is subjected to stresses resulting from outside forces. However, a too high value of compressive stresses can lead to adhesive wear and to too high tensile stresses in the coating, reducing the fatigue resistance of the element [17–20]. Volvoda [21] noted that the relationship between stress and the hardness of the titanium nitride layer obtained by magnetron scattering, is that the greater the compressive stress, the greater the hardness of the resulting layer. The maximum absolute value of the

Table 5 Reduced stress values occurred in the analysed coatings

Type of substrate	Type of the applied coating					
	Ti/Ti(C,N)-gradient/CrN	Ti/Ti(C,N)-gradient/(Ti,Al)N	Cr/CrN-gradient/CrN	Cr/CrN-gradient/TiN	Ti/(Ti,Si)N-gradient/(Ti,Si)N	Ti/DLC-gradient/DLC
	Values of the achieved stresses, MPa					
AZ61	−1325	−1210	−1377	−1329	−949	−425
AZ91	−1325	−1210	−1377	−1329	−949	−425
AlSi9Cu1	−1060	−871	−1109	−1065	−616	−321
AlSi9Cu4	−1060	−871	−1109	−1065	−616	−321

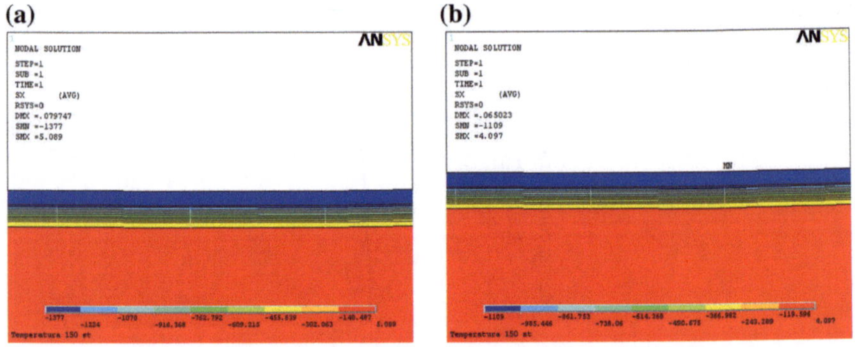

Fig. 14 Examples of the distribution of the simulated stress in the Cr/CrN/CrN coating on the substrate of: **a** AZ91 magnesium cast alloy, **b** AlSi9Cu4 aluminium cast alloy (view from the sample edges)

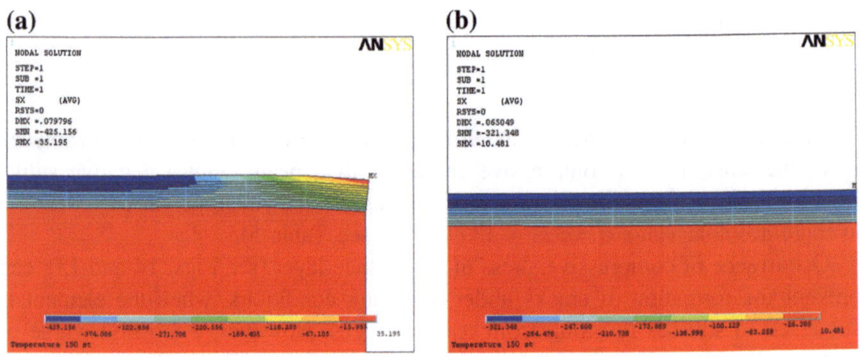

Fig. 15 Examples of the distribution of the simulated stress in the Ti/DLC/DLC coating on the substrate of: **a** AZ61 magnesium cast alloy, **b** AlSi9Cu1 aluminium cast alloy (view from the sample edges)

compressive stress layers were obtained for the coatings of the type: Cr/CrN/CrN, Cr/CrN/TiN, Ti/Ti(C,N)/CrN respectively 1377, 1329, 1325 MPa deposited on magnesium substrate. As a result of the performed simulation it was found, that the use of aluminium substrate as well the application of CVD method for deposition of coating in a hybrid system for lightweight materials results in a significant decrease in the absolute value of the occurring residual stresses (see Table 5).

The hardness of the non-coated magnesium and aluminium alloy substrates are respectively 133 and 108 HV. In the case of coatings produced by the cathodic PVD process of the Cr/CrN/CrN type, the microhardness was found to be 100 % greater than the microhardness of the substrate material (after precipitation hardening). The microhardness of the produced coatings did not exceed the value of 2000 HV in this case. However, for coatings with a gradient carbide-nitride coating

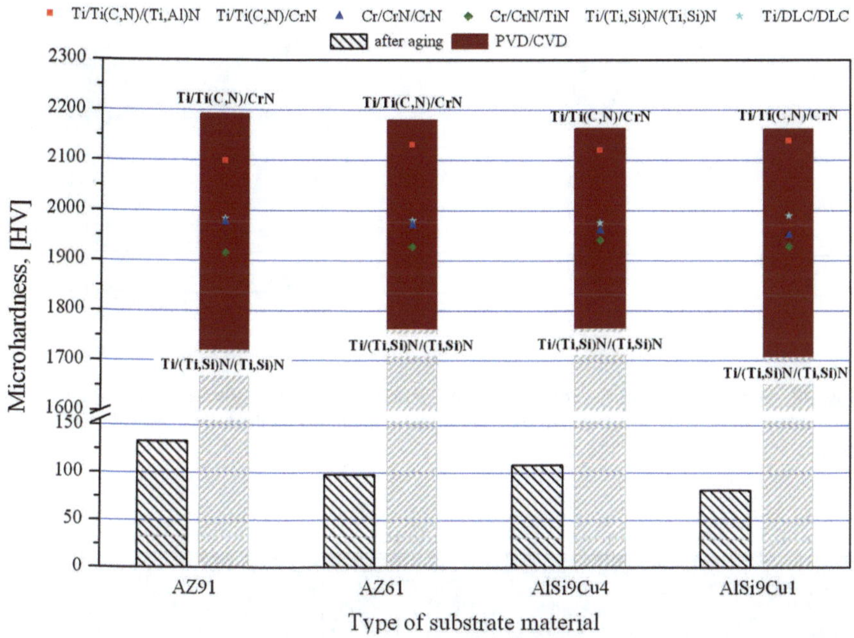

Fig. 16 Microhardness investigation results of the cast magnesium and aluminium alloys Mg–Al–Zn and aluminium Al–Si–Cu after ageing and PVD/CVD treatment

obtained in an atmosphere containing CH_4 and N_2 of the Ti/Ti(C,N)/(Ti,Al)N type, the greatest increase of surface microhardness over 2000 HV was found (see Fig. 16), which also significantly correlates with the wear resistance of these coatings. For the case of DLC coatings produced by the chemical vapour deposition process, the measured microhardness was about 2000 HV (see Fig. 16).

Reduction of residual stress values in the deposited coatings influences also the adhesion to the substrate, which was confirmed using the scratch test, and could be one of the main factors to improve functional properties such as wear resistance (Figs. 17, 18 and 19). The critical load LC1 is recorded on a diagram presenting the interdependences concerning the wear load and acoustic emission to the load, as the first, small increase of the acoustic emission signal. The critical LC2 load refers to the point at which the coating delamination occurs, reaching the substrate material, all this accompanied by an increasing acoustic signal. The highest values of the critical load LC1 and LC2 are respectively 7 and 19 N, and the best adhesion of the coating to the substrate was obtained for the Ti/DLC/DLC coating produced using the CVD process on the AZ91 substrate. Due to the fact that the parameters used for computer simulations, namely the Young's modulus, Poisson's ratio and thermal expansion coefficient of each group of alloys have the same values, except for one case, the difference in the values of the thermal expansion coefficient is relatively

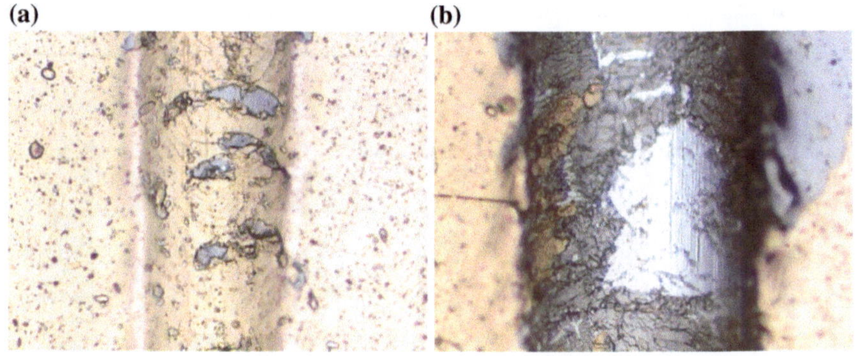

Fig. 17 Scratch on the surface of the Cr/CrN/TiN coating on the substrate of the AlSi9Cu4 cast aluminium alloy using the diamond indenter in the scratch test method, by a critical load: **a** LC1, **b** LC2

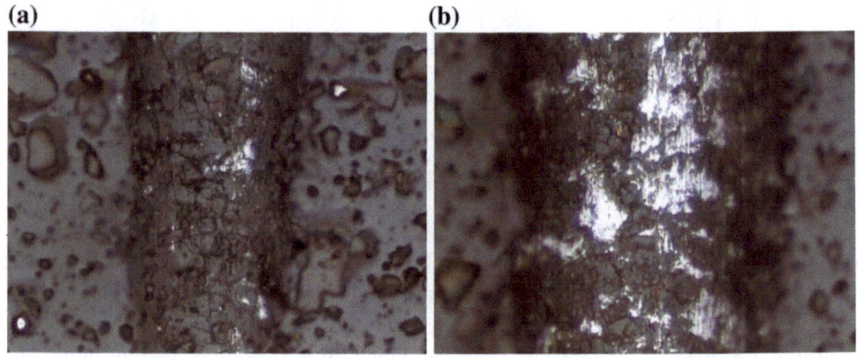

Fig. 18 Scratch on the surface of the Ti/DLC/DLC coating on the substrate of the AZ91 magnesium cast alloy using the diamond indenter in the scratch test method, by a critical load: **a** LC1, **b** LC2

small (see Table 4). The stress values, obtained for two magnesium and two aluminium alloys for various coatings have even identical values (see Table 5). In all considered cases, the maximum stress values were located at the edges of the samples.

4 Summary

As an answer to new demands to eliminate pollution, universal solutions have been developed, combining low-cost, lightweight substrate materials with the best possible properties and properly selected surface treatment technology. Because the

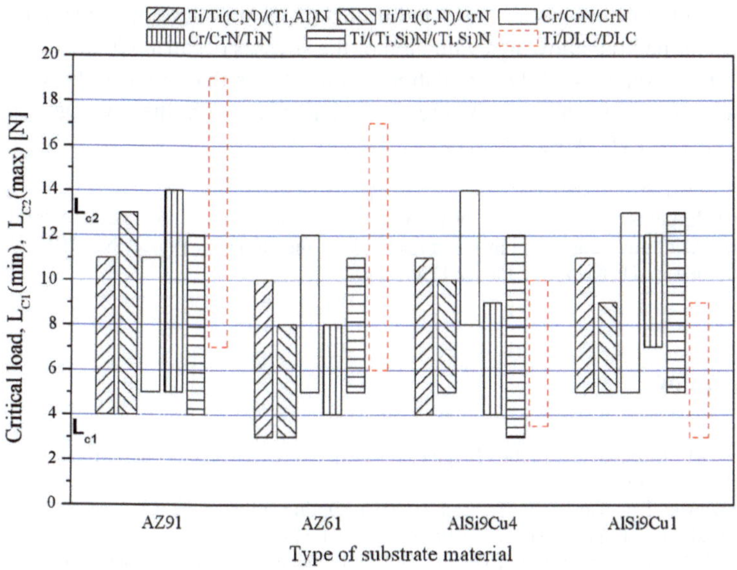

Fig. 19 Critical load LC1 and LC2 of the analysed coatings deposited on the investigated cast magnesium and aluminium alloys

cost of recycling, calculated as the sum of the material cost, labour, energy and general costs, are in the range from a few to tens of percent of a new element value, it can be concluded, that the economic effects achieved after taking into account the gain of the application of a layer or coating with better properties compared to an uncoated material can be much higher. The key issue is also to ensure a simultaneous development of both manufacturing and technology, lightweight construction, in particular of magnesium and aluminium, and the technology of forming and surface protection, which consequently will allow to create a balance between modern substrate material and coating of new generation.

The investigations presented in this paper reveal that it is possible to deposit coatings successfully on light alloys substrate. Particular findings are as follows:

- Investigations of the coatings using the scanning electron microscope reveal a microstructure without any visible delamination, tightly adhering to the substrate. The morphology of the surface of the obtained coatings is characteristic of a significant inhomogeneity connected with the occurrence of multiple drop-shaped particles on the surface, containing Ti as one of the deposited metals.
- Based on the tribological investigations carried out it was also found, that the critical load LC_2 is within the range of 8–19 N and the highest value was achieved for the AZ91alloy coated with the DLC coating. The highest hardness value was obtained for the coatings, which were a compound of the gradient Ti (C,N) layer.

- The mathematical model presented in this paper was developed using the finite element method, allowing evaluation of the stresses occurring in the investigated coatings. Analysis of the simulation results shows that in each of the analysed cases the stress has a compressive character, with the average value of $\sigma = 321 \div 1377$ MPa.

Acknowledgments This research was financed by the National Science Centre (NCN—Narodowe Centrum Nauki) within the framework of the Research Project No. 2011/01/B/ST8/06663 headed by Dr Krzysztof Labisz DSc. Eng.

References

1. Horst EF, Mordike BL (2006) Magnesium technology: metallurgy, design data, application. Springer, Berlin
2. Dobrzański LA, Tański T, Trzaska J (2010) Optimization of heat treatment conditions of magnesium cast alloys. Mater Sci Forum 638–642:1488–1493
3. Tański T, Dobrzański LA, Čížek L (2007) Influence of heat treatment on structure and properties of the cast magnesium alloys. J Advan Mater Res 15–17:491–496
4. Dobrzański LA, Tański T (2009) Influence of aluminium content on behaviour of magnesium cast alloys in bentonite sand mould. Solid State Phenom 147–149:764–769
5. Tański T (2013) Characteristics of hard coatings on AZ61 magnesium alloys. J Mech Eng 59 (3):165–174
6. Tański T, Labisz K, Lukaszkowicz K (2013) Structure and properties of diamond-like carbon coatings deposited on non-ferrous alloys substrate. Solid State Phenom 199:170–175
7. Dobrzański LA, Staszuk M, Gołombek K, Śliwa A, Pancielejko M (2010) Structure and properties PVD and CVD coatings deposited onto edges of sintered cutting tools. Arch Metall Mater 55(1):187–193
8. Musztyfaga-Staszuk M, Dobrzanski LA, Rusz S, Staszuk M (2014) Arch Metall Mater 59 (1):247–252
9. Staszuk M, Dobrzański LA, Tański T, Kwaśny W, Musztyfaga M (2014) The effect of PVD and CVD coating structures on the durability of sintered cutting edges. Arch Metall Mater 59 (1):269–274
10. Dobrzański LA, Żukowska L, Mikuła W, Gołombek K, Pakuła D, Pancielejko M (2008) Structure and mechanical properties of gradient PVD coatings. J Mater Process Technol 201 (1–3):310–314
11. Zhong Ch, Liu F, Wu Y, Le J, Liu L, He M, Zhu J, Hu W (2012) Protective diffusion coatings on magnesium alloys: a review of recent developments. J Alloy Compd 520:11–21
12. Dobrzański LA, Pakuła D, Mikuła J, Gołombek K (2007) Investigation of the structure and properties of coatings deposited on ceramic tool materials. Int J Surf Sci Eng 1(1):111–124
13. Cunha T, Rebouta L, Vaz F, Staszuk M, Malara S, Barbosa J, Carvalho P, Alves E, Le Bourhis E, Goudeau Ph, Riviere JP (2008) Effect of thermal treatments on the structure of MoNxOy thin films. Vacuum 82:1428–1432
14. Martini C, Morri A (2011) Face milling of the EN AB-43300 aluminium alloy by PVD- and CVD-coated cemented carbide inserts. Int J Refract Metal Hard Mater 29:662–673
15. Shanaghi A, Reza A, Rouhaghdam S, Ahangarani S, Chu PK, Farahani TS (2012) Effects of duty cycle on microstructure and corrosion behavior of TiC coatings prepared by DC pulsed plasma CVD. Appl Surf Sci 258:3051–3057

16. Endler I, Höhn M, Herrmann M, Holzschuh H, Pitonak R, Ruppi S, van den Berg H, Westphal H, Wilde L (2010) Aluminium-rich TiAlCN coatings by Low Pressure CVD. Surf Coat Technol 205:1307–1312
17. Algren M, Blomqvist H (2005) Influence of bias variation on residual stress and texture in TiAlN PVD coatings. Surf Coat Technol 200:157–160
18. Perry A, Sue JA, Martin PJ (1996) Practical measurement of the residual stress in coatings. Surf Coat Technol 81:17–28
19. Welzel U, Ligot J, Lamparter P (2005) Stress analysis of polycrystalline thin films and surface regions by X-ray diffraction. Appl Crystallogr 38:1–29
20. Dobrzanski LA, Sliwa A, Kwasny W (2005) Employment of the finite element method for determining stresses in coatings obtained on high-speed steel with the PVD process. J Mater Process Technol 164:1192–1196
21. Volvoda V (1995) Structure of thin films of titanium nitride. J Alloy Compd 219:83–87
22. Labisz K (2014) Microstructure and mechanical properties of HPDL laser treated cast aluminium alloys. Mater Sci Eng Technol (Mat.-wiss. u. Werkstofftech.). doi:10.1002/mawe.201400231
23. Tanski T (2014) Determining of laser surface treatment parameters used for light metal alloying with ceramic powders. Mater Sci Eng Technol (Mat.-wiss. u. Werkstofftech.). doi:10.1002/mawe.201400232
24. Dobrzański LA, Krupiński M, Labisz K, Krupińska B, Grajcar A (2010) Phases and structure characteristics of the near eutectic Al-Si-Cu alloy using derivative thermo analysis. Mater Sci Forum 638–642:475–480

Study of the Utilization of Polyamide Composite with Fiberglass Reinforcement in Automotive Engine Mounts

Leandro Cardoso da Silva, Antonio Augusto Couto, Renato Baldan and Jan Vatavuk

Abstract The aim of this work is to study the replacement of the support of the engine mounts from aluminum to commercial polyamide composite (PA 6.6) reinforced with 30, 35 and 50 % of fiberglass. The purpose of this replacement is to reduce the weight of the component. Flexion and fatigue tests were performed at 120 °C utilizing the Staircase Method. The results have shown the excellent adherence of polyamide in fiberglass. The polyamide composite with 50 % fiberglass presented the lowest deflection. The comparison of the tension distribution utilizing the Finite Element Method between the supports of the bearing engine made of polyamide with fiberglass and Al alloy have shown almost the same results. However, the polyamide composite presented higher values of deflection. Finally, due to the weight reduction of 32 %, it was possible to confirm that the polyamide composite is viable to change the Al alloys in automotive engine bearings.

Keywords Composite · Polyamide · Fiberglass · Fatigue

L.C. da Silva · A.A. Couto (✉) · J. Vatavuk
Mackenzie Presbyterian University, UPM, São Paulo, Brazil
e-mail: acouto@ipen.br

L.C. da Silva
e-mail: leandro.dasilva@mpsa.com

J. Vatavuk
e-mail: janvatavuk@uol.com.br

A.A. Couto · R. Baldan
Nuclear and Energy Research Institute, IPEN-CNEN/SP, São Paulo, Brazil
e-mail: renatobaldan@gmail.com

A. Öchsner and H. Altenbach (eds.), *Mechanical and Materials Engineering of Modern Structure and Component Design*, Advanced Structured Materials 70,
DOI 10.1007/978-3-319-19443-1_6

1 Introduction

Nowadays, most researches in the automotive field are related with the weight reduction of the components in order to increase the performance of the vehicle. In order to achieve this purpose, the replacement of the metallic components by polymers or polymer-based composites is increasing. Traditionally, the supports of the engine mount are made of metals such as aluminum and steel, which have excellent fatigue and degradation resistance. However, they present higher weights than some composites. Based on this, the aim of this work is to study the replacement of the support of the engine bearing from aluminum to commercial polyamide composite (PA 6.6) reinforced with 30, 35 and 50 % of fiberglass. This composite presents good mechanical and fatigue resistance, a low friction coefficient and a high melting point [1–4].

2 Experimental Procedure

The materials utilized in this work were the grains of polyamide composite with 30, 35 and 50 % of fiberglass, designated as PA 6.6 GF30, PA 6.6 GF35 and PA 6.6 GF50, respectively. The fiberglass present in the composite was classified as short (4.5 mm length and 10 μm diameter). The specimens for the bending and fatigue tests were produced in a Romi 65R injection machine with a pressure capacity of 175 bar. The feeding and the mold temperature were 280 and 50 °C, respectively. The bending and fatigue tests were performed in a MTS 810 Shock Absorber Test System with the Test Star II Software with a 15 kN load cell and a stove with a maximum temperature of 200 °C. The tests were performed at 120 °C, temperature at which the mount works. The bending tests were performed at 20 mm/min and the fatigue tests were performed at frequencies of 25 Hz. Figure 1 shows the illustration of the specimen utilized in the bending and fatigue tests. Figure 2a shows the schematic illustration and Fig. 2b shows the photo of the device utilized in the bending and fatigue tests.

The fatigue tests were controlled by the variation of the displacement (deflection). The values of the initial tension utilized in the Staircase Method were

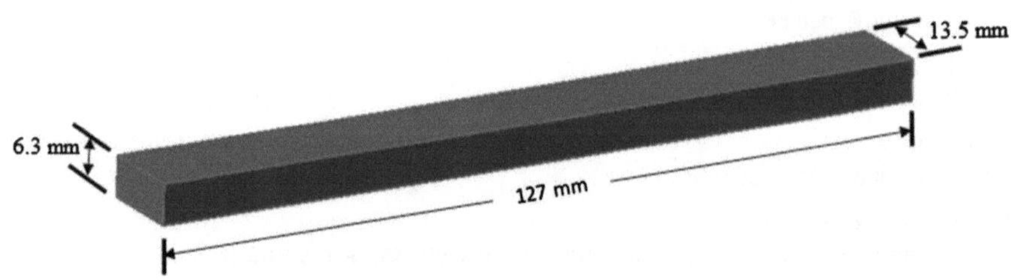

Fig. 1 Specimen utilized in the bending and fatigue tests

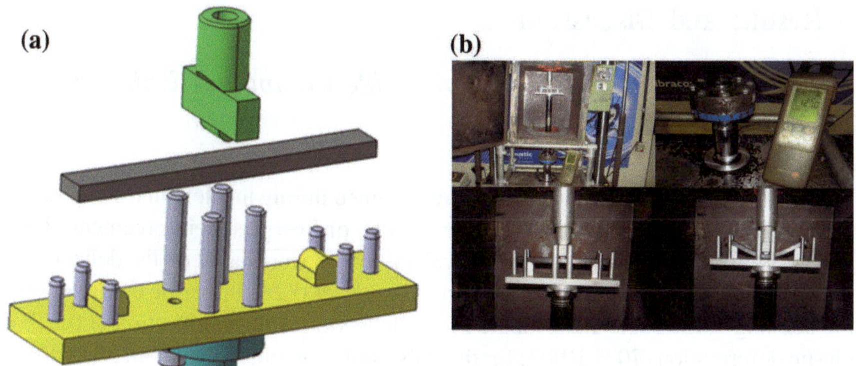

Fig. 2 Schematic illustration (**a**) and photo, (**b**) of the device utilized in the flexion and fatigue tests

determined with the aid of the load-deflection curves obtained by the bending tests. The Staircase Method is an interactive process with the following steps: the first is to submit the specimen (i) to a Fd_i load for 10^6 cycles. If the specimen fails in this condition, a new test is initiated decreasing the load by 10 %. If the specimen resists to 10^6 cycles at this low loading, the same specimen is subjected to the second step, increasing the load by 10 % and submitted again to 10^6 cycles. These steps occur up to the rupture of the specimen. Another specimen (i + 1) is submitted to the first step (10^6 cycles) with the loading $Fd_{i+1} = Fd_i + 10\ \%$. The number of tested specimens is between 6 and 10. The number of the fractured specimens must be the same as the non-fractured specimens in order to guarantee a sufficient approximation of the mean calculated values.

The initial value of the fatigue tests utilizing the Staircase Method was the last deflection value before the fracture utilizing the Locati Method. The load-deflection curves also allowed obtaining the initial values for utilizing in the Locati Method, which corresponds to 70 % of the tension and 0.2 % of plastic deformation obtained in the bending test. The fatigue tests utilizing the Locati Method were performed with the deflection value of 70 % RP02 and 3×10^5 cycles. If the fracture does not occur in this step, the deflection is increased by 10 % and the specimen is subjected to 3×10^5 cycles. This procedure is utilized up to the rupture of the specimen.

The fracture surface of the specimens are tested under fatigue were analysed with the aid of the scanning electron microscope JEOL JSM-6510. The Finite Element Method was also utilized to support this work. The structural calculation was performed with the ABAQUS software. The hexaedric elements were chosen and the Finite Element Method was utilized in order to obtain the fatigue limit and to study the viability of changing the Al alloy with the polyamide composite reinforced with fiberglass.

3 Results and Discussion

3.1 Fatigue Behavior of the Polyamide Composites Utilizing the Staircase Method

Figure 3 shows the load-deflection curves obtained during the flexion tests at 120 °C for pure polyamide and with 30, 35 and 50 % of fiberglass reinforcement. These curves allowed to obtain the deflection values, which were based on the deflection of the specimen when submitted to a tension that causes 0.2 % of plastic deformation in the bending tests. Table 1 shows the initial deflection and the deflection at 0.2 % of plastic deformation (70 % RP02) for the polyamide composites with 30, 35 and 50 % of fiberglass reinforcement. As can be seen, the values of the deflection at 0.2 % of plastic deformation for the polyamide composites with 30, 35 and 50 % of fiberglass reinforcement were 3.5, 3.5 and 3.0, respectively. These values were utilized to initiate the fatigue tests utilizing the Locati Method.

Table 2 shows the results of the initial deflection obtained by the Locati Method for the fatigue tests utilizing the Staircase Method. The polyamide composite with 50 % of fiberglass presented lower deflection values than the polyamide with 30 and

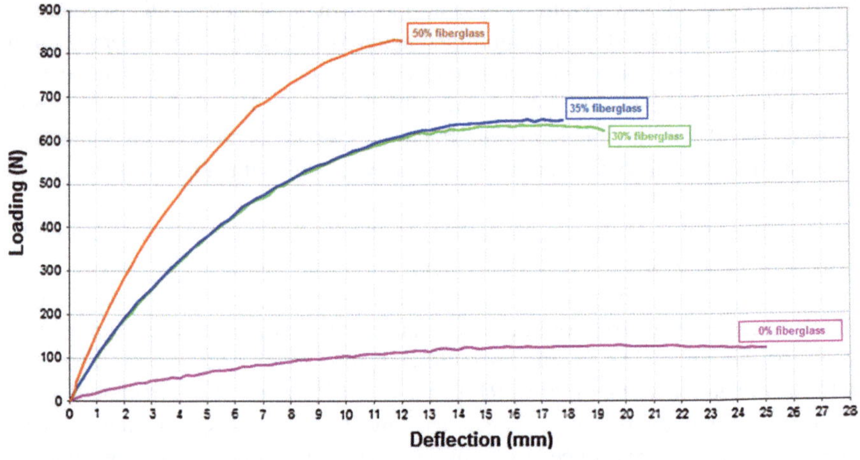

Fig. 3 Load-deflection curves obtained during the bending tests at 120 °C for pure polyamide and with 30, 35 and 50 % of fiberglass reinforcement

Table 1 Initial deflection and deflection at 0.2 % of plastic deformation (70 % RP02) for the polyamide composites with 30, 35 and 50 % of fiberglass reinforcement

Polyamide composite	RP02 (mm)	70 % RP02 (mm)
30 % of fiberglass (PA 6.6 GF30)	5.0	3.5
35 % of fiberglass (PA 6.6 GF35)	5.0	3.5
50 % of fiberglass (PA 6.6 GF50)	4.4	3.0

Table 2 Results of the initial deflection obtained by the Locati Method for the fatigue tests utilizing the Staircase Method

Step	30 % of fiberglass PA 6.6 GF30		35 % of fiberglass PA 6.6 GF35		50 % of fiberglass PA 6.6 GF50	
	D	N	D	N	D	N
1	3.5	300	3.5	300	3.0	300
2	3.85	300	3.85	300	3.3	300
3	4.24	290	4.24	282	3.63	153

D deflection (mm); *N* number of cycles × 1000

35 % of fiberglass. The fatigue tests utilizing the Staircase Method were initiated with the deflection values in the step before the rupture, which is the step 2 showed in Table 2.

Table 3 shows the results of the fatigue tests utilizing the Staircase Method. The "F" letter in the parenthesis indicates that the rupture of the specimen occurred before 10^6 cycles while the "S" letter indicates that the test was interrupted when the specimen reached 10^6 cycles without rupture. It can be seen that PA 6.6 GF30 and PA 6.6 GF35 composites presented 4 overlife/3 fails relationship and 5.12 and 4.66 mm of maximum deflection, respectively. For the PA 6.6 GF50, the results of the six tests presented 3 overlife/3 fails relationship and the maximum deflection was 3.63 mm. This deflection value is 29 % lower than PA 6.6 GF30 and PA 6.6 GF35 due to the higher fiberglass content in the PA 6.6 GF50.

Figure 4 shows the general aspect of the fractured specimens during the fatigue tests utilizing the Staircase Method. Figures 5, 6, 7 and 8 show the fracture surfaces of these specimens. As can be seen in Fig. 5, the fiberglass was broken. Additionally, the cyclic loading promotes the formation of cavities. Figure 6 shows the good adhesion between the fiberglass and the polyamide. Figure 7 shows, in detail, the typical tension and compression regions of the specimen fractured in the fatigue test. The plane faces of the polyamide in the tension region and the overloading feature in the compression region can be seen. Figure 8 shows the rupture of the fiberglass in the tension region evidenced by the fatigue propagation lines. It indicates that the crack initiates and grows up to a critical size in the tension region and the fracture of the compression region occurring due to overloading.

Table 3 Results of the fatigue tests utilizing the Staircase Method for PA 6.6 GF30, PA 6.6 GF35 and PA 6.6 GF50 polyamide composites

Polyamide composite	Deflection in each specimen (mm)						
	1	2	3	4	5	6	7
30 % of fiberglass (PA 6.6 GF30)	3.85 (S)	4.66 (S)	5.12 (F)	4.66 (F)	4.42 (S)	4.66 (S)	5.12 (F)
35 % of fiberglass (PA 6.6 GF35)	3.85 (S)	4.24 (F)	3.85 (S)	4.24 (S)	4.66 (F)	4.24 (S)	4.66 (F)
50 % of fiberglass (PA 6.6 GF50)	3.30 (S)	3.63 (F)	3.30 (F)	2.97 (S)	3.30 (F)	2.97 (S)	

S overlife and *F* failure

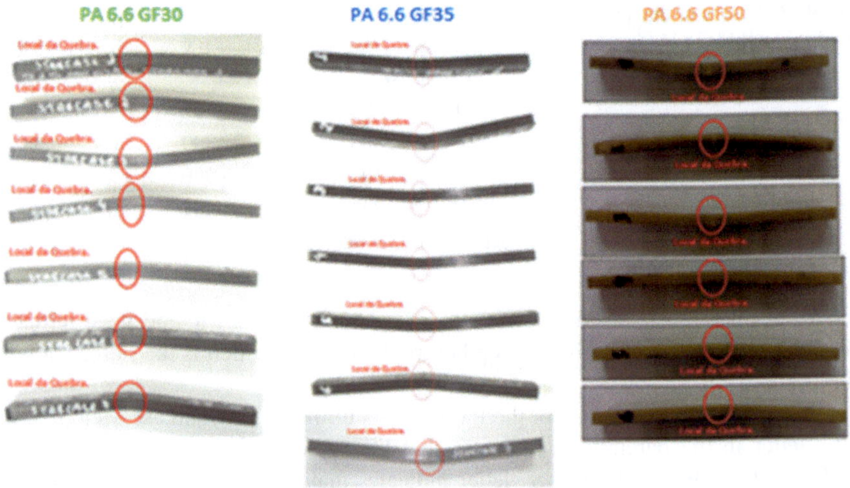

Fig. 4 General aspect of the fractured specimens during the fatigue tests utilizing the Staircase Method

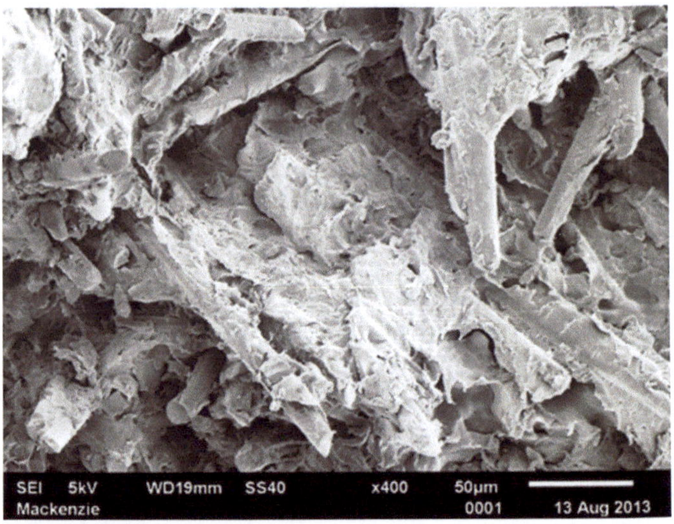

Fig. 5 Fracture surface of the specimen fractured during the fatigue test

3.2 Results Obtained Utilizing the Finite Element Method

The Finite Element Method was performed to simulate the flexion and fatigue tests for the polyamide composites considering the same geometry and adjusting the characteristic curve of the material to obtain the same load-displacement behavior.

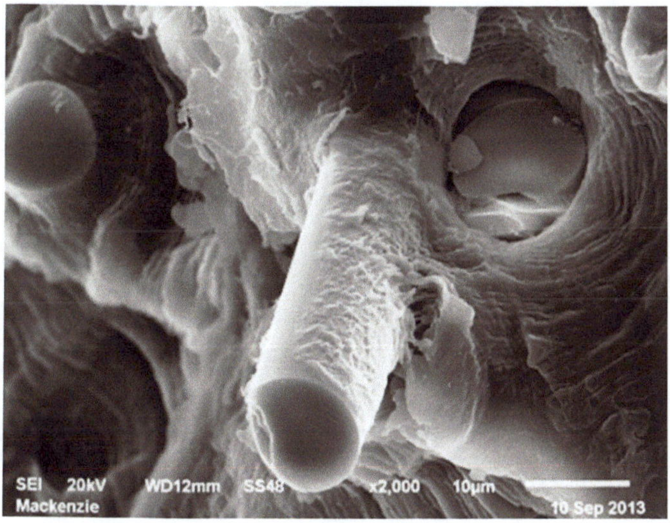

Fig. 6 Good adherence of the polyamide in the fiberglass

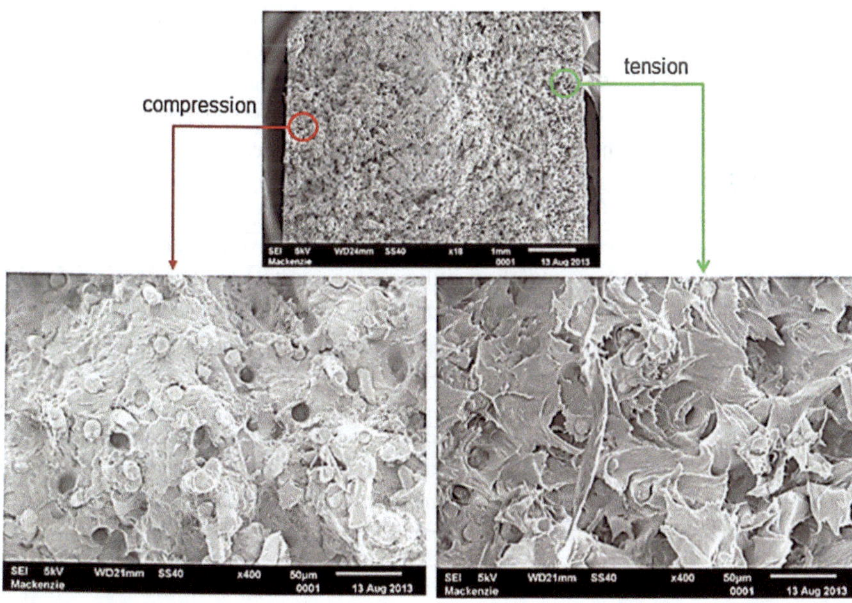

Fig. 7 Fracture surface (tension and compression regions) of the specimen fractured during the fatigue tests

This adequacy procedure is based on an interactive process to equiparate the response of the virtual load-displacement with the real test, which finishes when the error between the real and the virtual tests is less than 5 %.

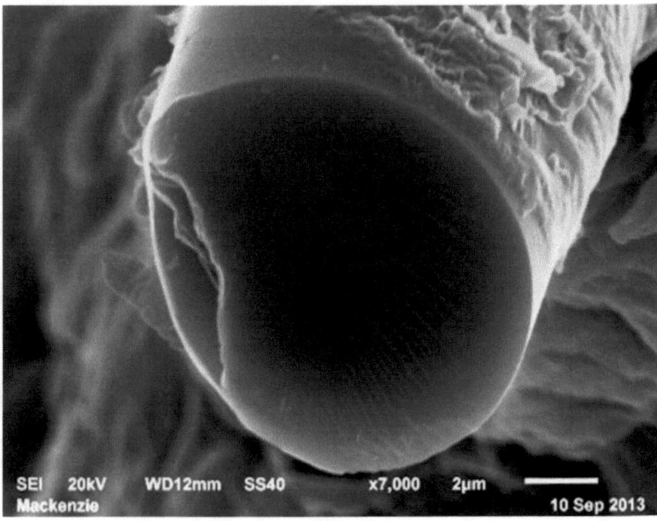

Fig. 8 Rupture of the fiberglass in the tension region evidenced by the fatigue propagation lines

Table 4 shows the average value of the deflection (m*) obtained from the fatigue tests utilizing the Staircase Method for the polyamide composites with 30 (PA 6.6 GF30), 35 (PA 6.6 GF35) and 50 % (PA 6.6 GF50) of fiberglass reinforcement. In order to use these results to develop the component, it is necessary to obtain the fatigue limit values in the region in which the crack appeared on the specimen. It can be calculated by the Finite Element Method utilizing the displacement (m*) shown in Table 4.

Table 5 shows the results of the fatigue limit, elastic limit and tension in the rupture obtained from the simulation utilizing the Finite Element Method for the polyamide composites with 30 (PA 6.6 GF30), 35 (PA 6.6 GF35) and 50 % (PA 6.6 GF50) of fiberglass reinforcement. The maximum limits are related with static loadings which the component must resist without any plastic deformation. The exceptional limits are overloadings that may occur during the useful life of the vehicle and the component must resist to this loading without rupture. Therefore, these values obtained utilizing the Finite Element Method were adopted as the criterion to validate the fatigue tests, considering the elastic and rupture limits as maximum and exceptional tensions, respectively.

Table 4 Average value of the deflection (m*) obtained from the fatigue tests utilizing the Staircase Method

Polyamide composite	Average value of the deflection (m*) (mm)
30 % of fiberglass (PA 6.6 GF30)	4.64
35 % of fiberglass (PA 6.6 GF35)	4.25
50 % of fiberglass (PA 6.6 GF50)	3.25

Table 5 Values of the fatigue, elastic and rupture limits obtained utilizing the Finite Element Method to validate the fatigue tests, considering the elastic and rupture limits as maximum and exceptional tensions, respectively

Polyamide composite	Fatigue limit (MPa)	Elastic limit/Maximum tension (MPa)	Rupture limit/ Exceptional tension (MPa)
30 % of fiberglass (PA 6.6 GF30)	64.8	69.3	130.0
35 % of fiberglass (PA 6.6 GF35)	66.5	70.4	131.0
50 % of fiberglass (PA 6.6 GF50)	76.6	97.3	139.0

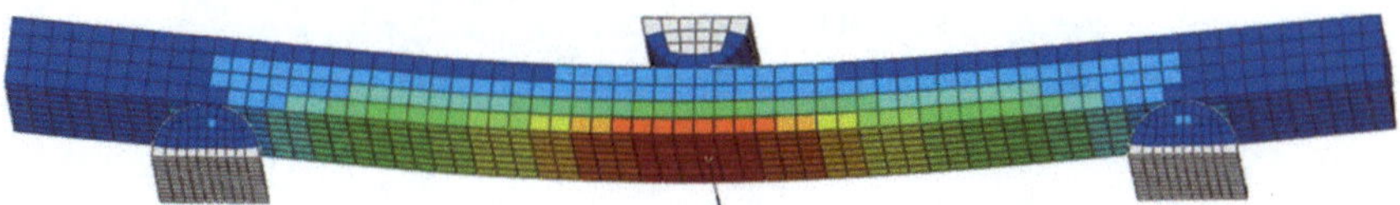

Fig. 9 Results of the load-deflection obtained utilizing the Finite Element Method and obtained from the data of the bending and fatigue tests for the polyamide composite with 50 % of fiberglass (PA 6.6 GF50)

The virtual test utilizing the Finite Element Method represented the maximum condition of all existent variables in the real test. Figure 9 shows the results of the load-deflection utilizing the Finite Element Method obtained from the data of the bending and fatigue tests for the polyamide composite with 50 % of fiberglass (PA 6.6 GF50). The value of the fatigue limit for this composite was 76.54 MPa.

3.3 Comparison of the Engine Mounts Made of Polyamide Composite Reinforced with Fiberglass with Aluminum Alloys Utilizing the Finite Element Method

The project of the structure made of polyamide composite was based on the pre-existent geometry of the engine mount made of Al alloys. Figure 10a shows the geometry defined for the engine mount made of polyamide composites with fiberglass reinforcement. In the regions used for attaching of the polyamide composite were used inserts of the SAE 1020 steel, assembled with the screw M10 × 1.5 with 60 Nm of torque and ±10 % of tolerance. This proposal will be compared with the actual project, which is composed by the structure made of injected SAE 306 aluminum alloys and a plastic base made of polyamide composite with 35 % of fiberglass (PA6.6GF35), as can be seen in Fig. 10b.

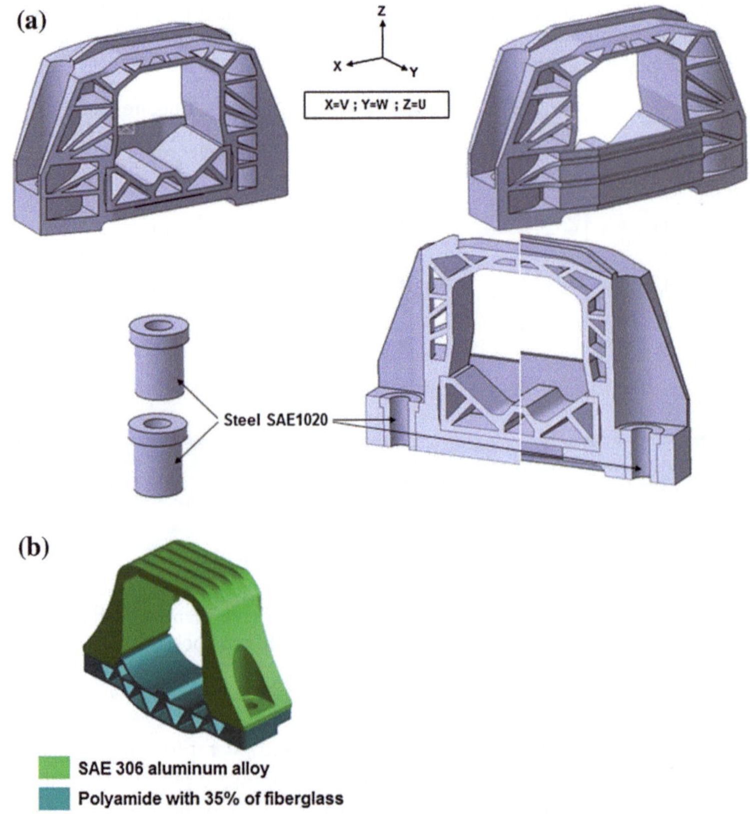

Fig. 10 **a** Geometry defined for the engine mount made of polyamide composite with fiberglass reinforcement. **b** Actual structure of the engine mount made of injected SAE 306 aluminum alloy and a base made of polyamide composite with 35 % of fiberglass (PA 6.6 GF35)

Table 6 shows the directions and the values of the loading applied and utilized for the structural calculation of the polyamide composites utilizing the Finite Element Method. Figure 11 shows, in detail, the directions of the loading applied in the engine bearing. The same analysis utilizing the Finite Element Method performed in the polyamide composites was done in the engine mount currently used (injected SAE 306 aluminum alloy and a plastic base made of polyamide composite with 35 % of fiberglass). Table 7 shows the specific values of the fatigue, elastic and rupture limits utilized as criterion to validate the engine mount currently used. The loadings were performed utilizing the same method and directions utilized in the composite structure, as can be seen in Fig. 11.

The Finite Element Method results have shown that the deflection is lower in the SAE 306 aluminum alloy than the polyamide composites. Additionally, the polyamide composite with 50 % of fiberglass reinforcement can be used in the engine mount, as proposed in the present work. The results of the fatigue and elastic limits,

Table 6 Directions and the values of the loading applied and utilized for the structural calculation of the component utilizing the Finite Element Method

Loading Values (N)			
Direction	Maximum	Exceptional	Fatigue
X+	4.73	16.20	2.16
X−	4.20	16.20	2.16
Y	200	450	0
Z+	6.75	7.00	3.90
Z−	9.45	10.00	3.90

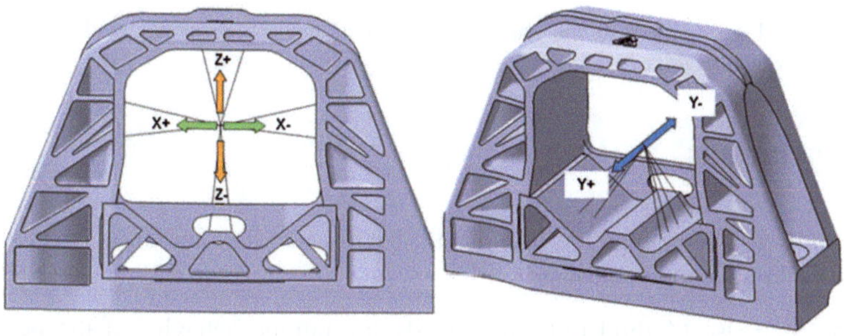

Fig. 11 Directions of the loading applied in the engine mount

Table 7 Criterion of the fatigue limit, maximum and exceptional tensions obtained by Finite Element Method for the engine mount currently used

Material	Fatigue limit (MPa)	Elastic limit/ Maximum tension (MPa)	Rupture limit/ Exceptional tension (MPa)
SAE306 aluminum alloy	70.0	140.0	240.0
Polyamide composite with 35 % of fiberglass (PA 6.6 GF35)	66.5	70.4	131.0

maximum and exceptional tension obtained utilizing the Finite Element Method has shown that the polyamide composites with different amounts of fiberglass were approved. The composites with 30 and 35 % of fiberglass presented higher deflection than the composite with 50 % of fiberglass.

The purpose of the structure made of polyamide composite with fiberglass reinforcement was based on the actual project of the engine mount composed by the structure made of injected SAE 306 aluminum alloy and a base made of polyamide composite with 35 % of fiberglass (PA6.6 GF35). The actual component has 0.625 kg whilst the engine bearing developed in this work (polyamide composite with fiberglass reinforcement) presented 0.427 kg of the total weight. The 0.198 kg of mass reduction represents 32 % of the total weight of the engine mount currently used. Figure 12 shows the final geometry with the mass distribution of the engine mount made of polyamide composites with fiberglass.

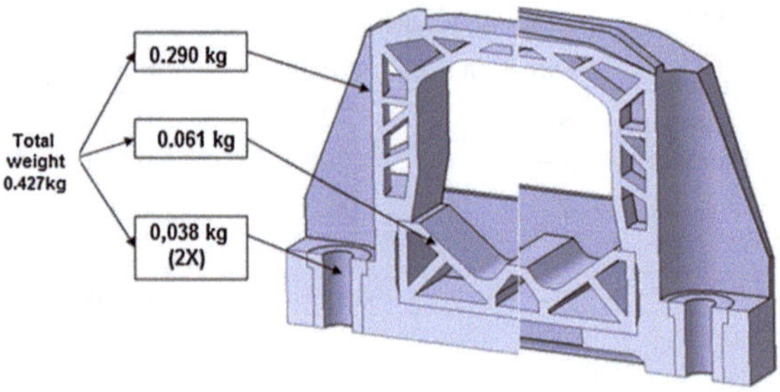

Fig. 12 Final geometry with the mass distribution of the engine mount made of polyamide composite with fiberglass

4 Conclusions

The study of the polyamide composite reinforced with fiberglass in the engine mounts allow to conclude that:

- The Staircase Method utilized to study the fatigue behavior of the polyamide composite with 30, 35 and 50 % of fiberglass reinforcement allowed to define the fatigue limits of the materials. The polyamide composite with 50 % of fiberglass presented the lowest deflection;
- The tension region of the fractured specimens tested under fatigue presented plane faces of the polyamide whilst the compression region presented an overloading feature;
- The results of the Finite Element Method showed that all the polyamide composites with different amount of fiberglass reinforcement were able to be used in the engine mount.
- The results of the tension values obtained for the engine mount made of polyamide composites with fiberglass reinforcement and the aluminum alloy utilizing the Finite Element Method were similar. The deflection values of the polyamide composites were higher than the aluminum alloy;
- The engine mount made of polyamide composite with fiberglass reinforcement presented 32 % of weight reduction when compared with the engine bearing made of aluminum alloy.

References

1. Courteille E, Mortier F, Leotoing L, Ragneau E (2003) Optimisation d'un système de suspension moteur pour une amélioration du confort vibro-acoustique. In: CONGRÈS FRANÇAIS DE MÉCANIQUE, France
2. Chauvet L(2012) Trelleborg weight reduction strategy—utilization of plastic for structural components for power train mounting system. SAE technical paper 2012-28-0017

3. Silva LC (2014) Substituição do Alumínio pelo Compósito de Poliamida no Suporte do Coxim do Motor. Dissertação (mestrado)—Universidade de São Paulo, São Paulo
4. Silva LC, Turra G, Santos WF (2012) Substituição do material do suporte do coxim motor—metal para Poliamida. SAE Brasil, São Paulo, 2012-36-0144, pp 1–6

Shaping of Surface Layer Structure and Mechanical Properties After Laser Treatment of Aluminium Alloys

Tomasz Tański, Wojciech Pakieła, Maciej Wiśniowski and Leszek Adam Dobrzański

Abstract The influence of laser treatment on the structure, mechanical properties and wear resistant casting of aluminium alloys has been studied. The main objective of this investigation was to improve the tribological and mechanical properties of the surface layer of the aluminium alloy AlMg5Si2Mn by remelting and feeding the chromium particles into the melt pool with a rapid solidification. The applied size of the chromium particles have been in the range 50–120 µm. For the remelting of the surface a high power diode laser (HPDL) was used. The applied laser beam power is in the range from 1.8 to 2.2 kW. The linear laser scan rate of the beam was set to 0.5 m/min. The chromium powder has been introduced in the melt pool using a gravity feeder at a constant rate of 2 g/min. The application of the laser surface treatment of aluminum alloys enables us to obtain too much harder and better wear resistance compared to based materials.

Keywords Laser alloying · Aluminium alloy · Tribological test · Ball on plate · Intermetallic phases

---

T. Tański (✉) · W. Pakieła · M. Wiśniowski · L.A. Dobrzański
Faculty of Mechanical Engineering, Institute of Engineering Materials and Biomaterials, Silesian University of Technology, Konarskiego 18a, Gliwice 44-100, Poland
e-mail: tomasz.tanski@polsl.pl

W. Pakieła
e-mail: wojciech.pakiela@polsl.pl

M. Wiśniowski
e-mail: maciej.wisniowski@polsl.pl

L.A. Dobrzański
e-mail: leszek.dobrzanski@polsl.pl

A. Öchsner and H. Altenbach (eds.), *Mechanical and Materials Engineering of Modern Structure and Component Design*, Advanced Structured Materials 70,
DOI 10.1007/978-3-319-19443-1_7

1 Introduction

The dynamic development of the industrial economy makes it necessary to find better and more advanced engineering materials able to meet the new demands [1–26]. Research is being conducted to improve mechanical and functional properties of all groups of engineering materials. There are very interesting possibilities given to materials such as aluminum and magnesium through light alloys [1–8]. The low density of aluminum or magnesium in comparison to steel and the rather simple possibilities to improve mechanical properties and to make it wear resistant is one of the main reasons why these materials are increasingly being used in particular applications where it is important to reduce the mass of elements or where corrosion resistant materials are necessary, such as in the automotive and aerospace industry and air transport. Very significant treatment enhancing properties of metals such as aluminium, magnesium and elements made therefrom are often subject to widely used surface treatment technologies [11–15, 20, 23–26]. A laser beam provides very precise delivery of energy and consequently can better and faster implement technological operations in layer treatments. The layer formed on the metal must be characterized through the high hardness and toughness, high fatigue strength and impact resistance as well as resistance to high and low temperature (creep and fracture toughness), thermal shock and the appropriate thermal conductivity. The properties of the obtained surface layers to a large extent depend on their structure, porous, material discontinuities, uniform chemical composition and phase composition. Laser radiation is also very often used for improvement of mechanical and tribological properties different engineering materials [9–23]. The laser surface treatment is currently often used for forming the structure and properties of the surface layer of not only light metals. Laser is used for reduce porosity and discontinuity in the material on the top surface in order to increase corrosion resistance [18, 19].

Goal of this investigation was to improve the mechanical properties and wear resistance in comparison to the substrate material by remelting the substrate with a small depth (max 1.5 mm) and feeding the chromium particles into the molten pool followed by a rapid solidification.

2 Methodology of Research, Material for Research

As the substrate an aluminum alloy with EN-AC 51500 magnesium was used. The chemical composition of the applied aluminium alloy is presented in Table 1. Chromium particles have been used to improve the mechanical properties and wear resistance of the surface layers. The size and shape of the particles used in the process of laser treatment has been presented in Fig. 1. The gradation of the applied particles of chromium powder was in the range of 50–120 μm.

Table 1 Chemical composition of aluminium alloy ENAC-AlMg5Si2Mn

Fe	Si	Mn	Ti	Cu	Mg	Zn	Others	Al
Max 0.25	1.8–2.6	0.4–0.8	Max 0.25	Max 0.05	4.7–6	Max 0.07	Each 0.05 Total 0.15	Remainder

Fig. 1 Morphology of the chromium powder in the initial state

The heat source was a high power diode laser (HPDL). The high power diode laser was characterized by the very high power density of the laser beam under normal conditions of up to 10^7 W/cm^2. This makes the thermal impact on the detail limited and thus causes only minor thermal stress and strain. The high power diode laser (HPDL) was used to introduce the chromium powder into the aluminum alloy matrix. Because of the limited diffusion of hydrogen, oxygen and nitrogen gas from the atmosphere the process of melting the surface, has been carried out in argon atmosphere. The chromium powder was introduced into the molten pool by a rotary powder feeder with a fixed and predetermined amount of 2 g/m. The parameters of the laser treatment process is shown in Table 2. The shape and distribution of the undissolved Cr particles and precipitates in the aluminum alloy matrix was examined by scanning electron microscopy. The reinforcing phase constituted of undissolved particles of chromium powder and intermetallic phases formed on the

Table 2 Parameters of the laser alloying process

Laser power range	1.8; 2.0; 2.2 kW
Velocity of the laser beam	0.5 m/min
Laser spot size	1.8 × 6.8
Wavelength of the laser radiation	808–940 nm
Reinforcing particles	Cr
Quantity of the powder per min	2 g/min
Gradation of chromium powder	50–120 µm

Table 3 The parameters of the wear test "ball on plate"

Parameter	Value
Load	5 N
Distance	200 m
The length of the test	4 mm
Speed linear motion	2 cm/s

basis of chromium. To verify mechanical and tribological properties of the obtained layers such tests were made:

- hardness of the surface layers,
- microhardness along the cross-section of the solidification molten pool,
- wear resistant test "ball on plate".

The hardness of the surface has been measured steel ball with a steel ball with a diameter of 1/16 in. and a load 60 kgf (HRF scale). The microhardness of the cross-section remelted layers were measured by using Vickers Microhardness testers with an applied load of 100 gf. The wear resistance of the layers was obtained by the laser surface modification was examined using the tribological "ball on plate" test. The surface before the tribological test was grinded using an abrasive paper of grain size 68 μm. The aluminum oxide ball (Al_2O_3) was used as a counter sample in the tribological test. Parameters of wear resistant test is presented in the Table 3. Wear tack and product of wear obtained as a result of the tribological test was observed in the scanning electron microscope and analyzed using the EDS detector. Scanning electron microscopy to determine the shape and placement of the undissolved chromium particles and precipitation in the aluminium matrix also has been used.

3 Results and Discussion

In order to obtain a quasi-composite structure of the surface layer of the aluminum alloy ENAC-AlMg5Si2Mn chromium powder was used. A high power diode laser (HPDL) was used for the melting of the surface. During the laser alloying most of the powder was dissolved in the aluminum alloy matrix. There were also single undissolved chromium particles observed in the remelted zone (Fig. 2c, d).

The greatest amount of metal powder introduced to the molten pool was observed for the smallest applied laser power 1.8 kW (Fig. 3). This phenomenon is due to the moderate the impact of the laser beam (heat and lower pressure produced in melting area) on the dosed chromium powder and a liquid pool, when compared to the higher powered laser at 2.2 kW. The structure of the layers in the AlMg5Si2Mn aluminium alloy obtained by the laser alloying is presented in the Fig. 2a, b.

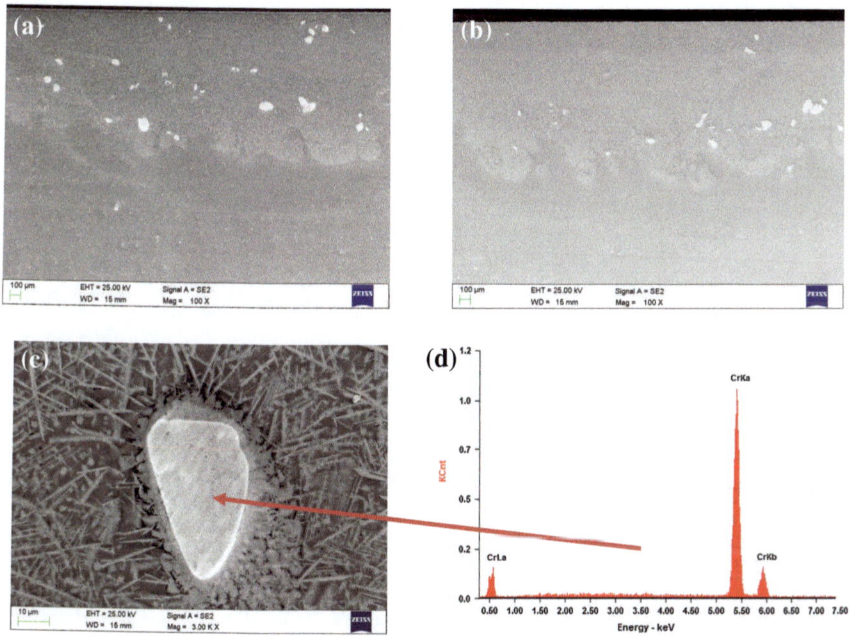

Fig. 2 Structure of the layers obtained during the laser treatment with the power **a** 1.8 kW, **b** 2.2 kW and **c**, **d** chemical analysis of the undissolved particles

Structure observation of the composite layers has shown uniform distribution of chromium powder in the liquid molten pool on a depth of about 0.7–1.7 mm (Fig. 2a, b). Analysis of the structure of the layers showed that the chromium particles are closely associated with the aluminium alloy matrix. No cracks, voids and pores were observed around the embedded particles which may indicate a good wettability of the particles by the matrix material.

Analysis of the cross section of the obtained layer and the surface topography showed that both on the top surface and inside surface layer, there is no demonstrable porosity or discontinuity. In the surface layer obtained during the alloying by the lower power of the laser beam there were observed many more undissolved particles of chromium (Figs. 2a, b and 4). Also the obtained depth was lower at about 0.4 mm compared to the maximum power of the laser. The topography of the layers obtained by the laser treatment are presented in Fig. 4.

The "ball-on-plate" test of the layers and based material AlMg5Si2Mn confirmed the increase of the wear resistance of the surface after laser treatment. It was also observed that the higher power of the laser beam did not affect wear resistance (Fig. 5). This phenomenon is caused by the distribution of the introduced metal particles in a larger volume of the deeper layer obtained during the laser treatment with the maximum power of the laser beam. The smaller power of the laser beam created a shallower layer. In result the same amount of the powder was introduced in the shallower remelting as with the maximum power laser beam. The lower heat

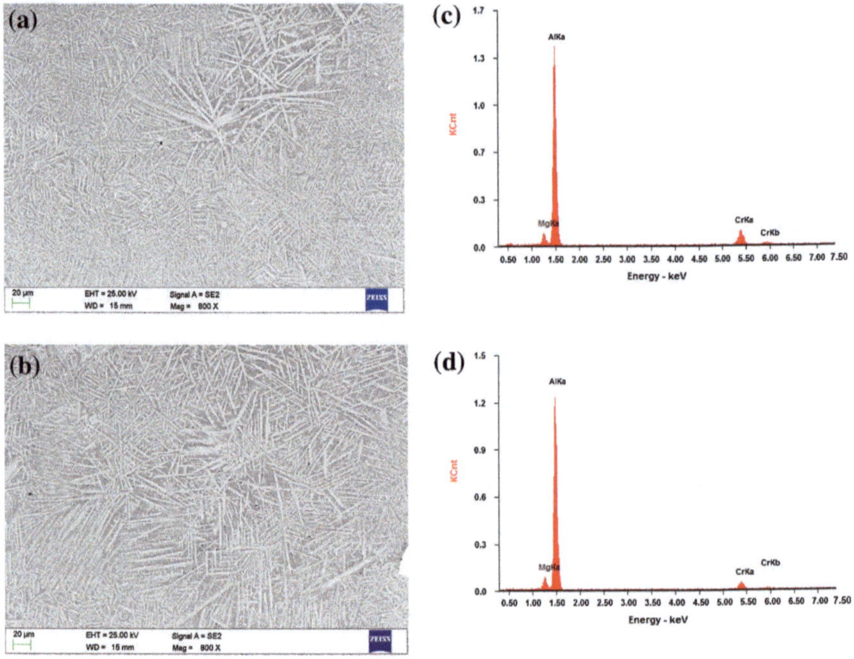

Fig. 3 Structure of the layers obtained during the laser alloying with the power **a** 1.8 kW, **b** 2.2 kW and **c**, **d** chemical analysis of the layers

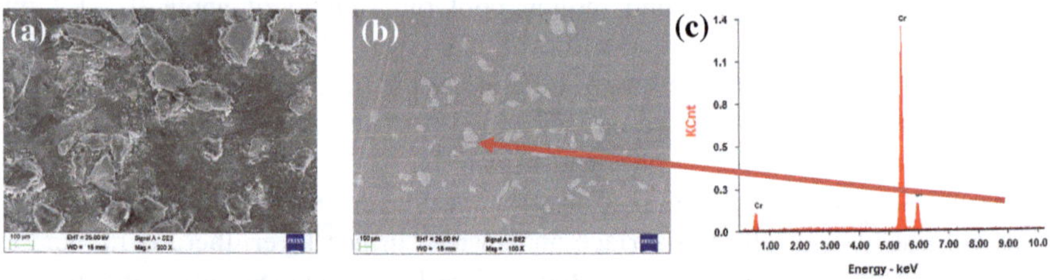

Fig. 4 Topography of the surface **a** after alloying with the 2.0 kW powered laser, **b** alloying and grinding and **c** chemical analysis at the point

input caused faster heat transfer from the volume of the aluminium and faster crystallization of the remelted area after the laser treatment. In all cases the layers enriched with chromium have not been interrupted. The topography of the wear track with the chemical analysis is presented in the Fig. 6.

The analysis of the product (the powder obtained during the scratched and abrasion) of the wear test did not reveal the presence of large particles of the metal powder removed from the surface or hard phases created by the laser alloying. The wear product after the tribological test of the aluminium alloy without the layers

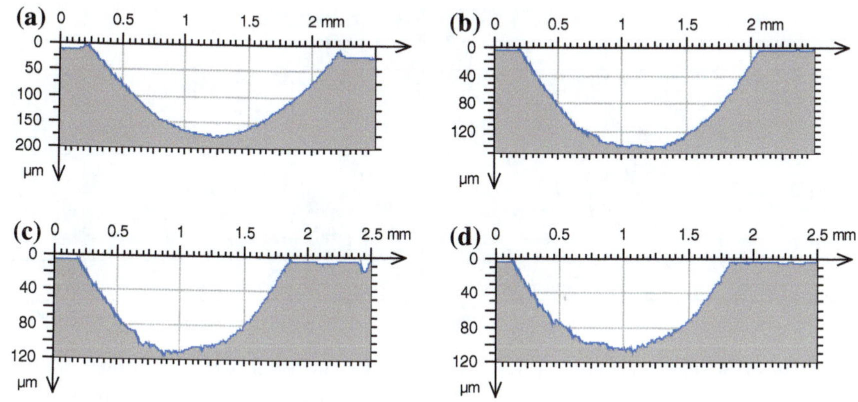

Fig. 5 Wear track after "ball-on-plate test" **a** based aluminium alloy AlMg5Si2Mn, and layers obtained the laser alloying with the power of the beam, **b** 2.2 kW, **c** 2.0 kW and **d** 1.8 kW

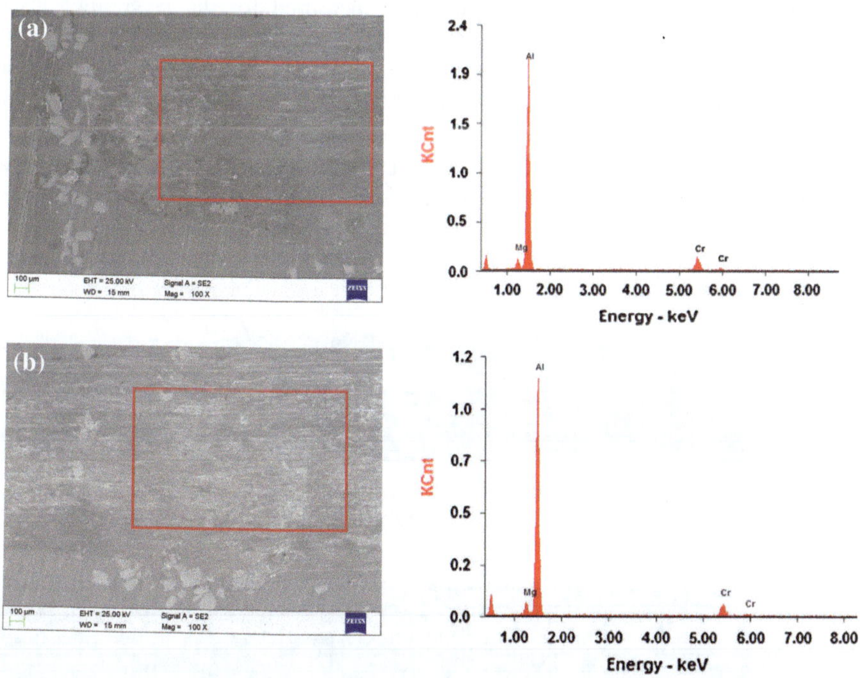

Fig. 6 Chemical analysis of the wear track after tribological test the surface layers obtained during laser the alloying with the beam power **a** 2.2 kW and **b** 1.8 kW

include large particles of aluminum uprooted from the surface particles. This kind of wear product confirms that the dominant mechanism was destructive chipping. The wear product of the layers enriched by the the chromium powder particles for

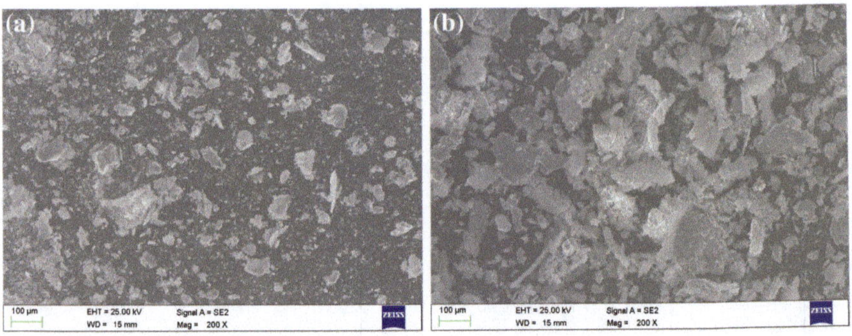

Fig. 7 Wear product after the "ball on plate" test. The samples **a** with layer formed by laser treatment with the power of 2.2 kW and **b** AlMg5Si2Mn without the laser surface treatment

all cases consisted of the fine powder. The product of the wear test is presented in the Fig. 7.

The smallest roughness of the track was measured for the layer after laser alloying by the chromium powder with power of the laser beam 1.8 kW. The highest measured roughness was identified for untreated sample and their surface. It is closely related with the increasing hardness of the tested materials after the laser treatment.

During the wear resistance test the friction coefficient was also measured (Fig. 8). The analysis of the data showed a lower friction coefficient for the samples

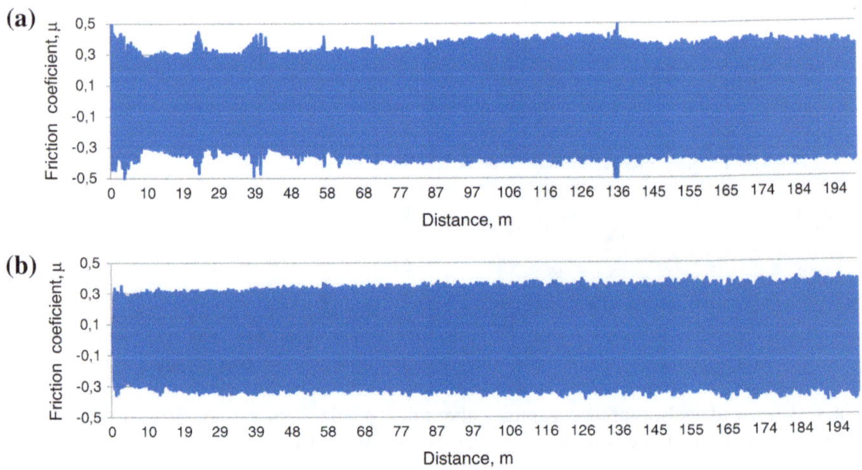

Fig. 8 The friction coefficient as a function of the distance registered during the "ball on plate" test for samples with composite layers obtained by laser treatment at the powers: **a** 1.8 and **b** 2.2 kW

after laser treatment by about 0.2. Futhermore, the friction coefficient curve for samples after laser alloying was smoother when compared to the based material.

The irregular nature of the friction coefficient curve from the AlMg5SiMn2 sample is caused by the removal of particles from the aluminum surface and the adhesive which connects the aluminum with the ceramic counter-specimen. The reason for this phenomenon is the lower tendency of the adhesive connect the counter-specimen (Al_2O_3) with the remelted layer enriched in chromium particles and created during the laser alloying phases with chromium. The presence in the volume of solidification molten pool the phase was created as a result of the laser alloying providing an increase of the wear resistance and as a result the wear track is more smooth. The results of the measurements of size and roughness of the wear track surface i shown in Fig. 5 and Table 4.

Observations of the wear track using scanning electron microscopy and analysis of the chemical composition using X-ray spectrometry confirmed the nature of the wear of the based material and the samples after laser treatment. Analysis of the wear track using scanning electron microscopy and analysis of the chemical composition using X-ray spectrometry confirmed that the layer on the aluminium alloy surface had not been interrupted. On the bottom part of the crater the chromium particles in aluminium matrix was confirmed. The topography of the wear track layers obtained during the laser treatment was much smoother and did not contain particles torn from the surface (Fig. 9b). The topography of the wear track and its the chemical analysis is presented in the Fig. 6a–b. The wear track of the samples without laser treatment have visible traces of losses caused by fissures, wear and ridging of material particles from the substrate (Fig. 9a).

Comparing the hardness of the based aluminum alloy AlMg5Si2Mn and the layers obtained during the surface alloying has shown the significant impact of the laser treatment on the obtained results. The greatest increase of hardness was observed for the sample after laser alloying with a 1.8 kW powered laser beam, which was due to the great amount of powder applied to the top part of the obtained layers. The results of the hardness test are presented in Table 4.

Table 4 Roughness and hardness of the surface before the wear test, friction coefficient registered during the process and dimension of the wear track after "ball-on-plate" test

Power of the laser beam, kW	Roughness of the surface Ra μm	Hardness of the surface HRF	Friction coefficient	Dimension of the wear profile	
				Depth, μm	Width, mm
Aluminium alloy ENAC 51–100 before the laser treatment					
–	0.59	67	0.38	169	2.12
Aluminium alloy ENAC 51–100 after the laser treatment					
1.8	0.57	115	0.33	137	1.88
2.0	0.44	113	0.34	104	1.75
2.2	0.36	109	0.334	109	1.68

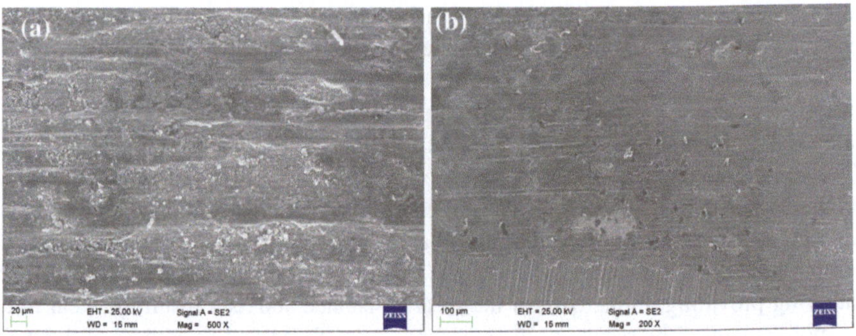

Fig. 9 Wear track after tribological "ball-on-plate" test of the sample: **a** auminium alloy AlMg5Si2Mn without laser treatment and **b** layers obtained during the laser treatment with the power 2.0 kW

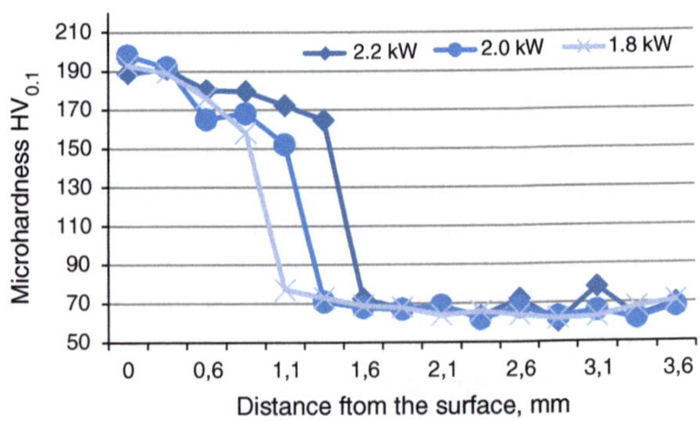

Fig. 10 Microhardness along the cross section of composite layers obtained at different laser powers

For testing the change of hardness on the obtained layers as a correlate of their depth along the cross-section, the Vickers microhardness test has been used. The test shows an increase of the hardness in solidification molten pool in all cases. The greater hardness increase was measured from the layers obtained during the alloying with the smaller power of the laser beam, then layers depth was the smallest. The greater depth of the remelted area was observed from the layers obtained with the highest power of the laser, but the maximum hardness was smaller when compared to those obtained by the lower powered laser (at 1.8 kW). The results of the microhardness test is presented on the Fig. 10.

4 Conclusions

Based on the findings our analysis it can be unambiguously stated that the resulting layer has a greater hardness and better wear resistance compared to the base material. The wear resistance test demonstrated that the best properties of wear resistance belonged to samples which were obtained by using the lowest power laser, at 1.8 kW. An increase in the laser beam power during the alloying of Al Mg5Si2Mn aluminium alloy by the high power diode laser (HPDL) did not cause a growth of the wear resistance composite layers. The analysis of the results of the friction coefficient at a function of distance confirmed that the introduction of chromium particles into the aluminum alloy matrix reduces the friction coefficient by 0.2 compared to the based aluminium alloy. Introducing the chromium particles to the aluminium matrix greatly increases the mechanical properties and and the wear resistance of the top surface layer of aluminum alloy. Increasing the laser beam causes a greater remelting zone but the maximum hardness of this layers is lower compared to the layers obtained with the power of the laser beam of 1.8 kW.

Acknowledgments Wojciech Pakieła is a holder of a scholarship from project "DoktoRIS—Scholarship Program for Innovative Silesia", co-financed by the European Union under the European Social Fund.

This research was financed partially within the framework of the Scientific Research Project No. 2011/01/B/ST8/06663 headed by Dr Krzysztof Labisz.

References

1. ASM International (1993) ASM Handbook, aluminum and aluminum alloys. ASM International, Ohio
2. Bäckerud L, Chai G, Tamminen J (1990) Solidification characteristics of aluminum alloys. AFS/ Skanaluminium, Illinois
3. Dobrzański L, Maniara R, Sokolowski J et al (2007) Applications of the artificial intelligence methods for modeling of the ACAlSi7Cu alloy crystallization process. J Mater Process Technol 192–193:582–587. doi:10.1016/j.jmatprotec.2007.04.022
4. Dobrzański L, Sitek W, Krupiński M, Dobrzański J (2004) Computer aided method for evaluation of failure class of materials working in creep conditions. J Mater Process Technol 157–158:102–106. doi:10.1016/j.jmatprotec.2004.09.020
5. Dobrzański L, Tomiczek B, Pawlyta M, Król M (2014) Aluminium AlMg1SiCu matrix composite materials reinforced with halloysite particles. Arch Metall Mater. doi:10.2478/amm-2014-0055
6. Dobrzański L, Tański T, Trzaska J (2010) Optimization of heat treatment conditions of magnesium cast alloys. MSF 638–642:1488–1493. doi:10.4028/www.scientific.net/msf.638-642.1488
7. Tański T, Dobrzański L, Čížek L (2007) Influence of heat treatment on structure and properties of the cast magnesium alloys. AMR 15–17:491–496. doi:10.4028/www.scientific.net/amr.15-17.491
8. Dobrzański L, Tański T (2009) Influence of aluminium content on behaviour of magnesium cast alloys in bentonite sand mould. SSP 147–149:764–769. doi:10.4028/www.scientific.net/ssp.147-149.764
9. Partes K, Sepold G (2008) Modulation of power density distribution in time and space for high speed laser cladding. J Mater Process Technol 195:27–33. doi:10.1016/j.jmatprotec.2007.05.052
10. Kennedy E, Byrne G, Collins D (2004) A review of the use of high power diode lasers in surface hardening. J Mater Process Technol 155–156:1855–1860. doi:10.1016/j.jmatprotec.2004.04.276

11. Hashim J, Looney L, Hashmi M (2001) The wettability of SiC particles by molten aluminium alloy. J Mater Process Technol 119:324–328. doi:10.1016/s0924-0136(01)00975-x
12. Rana RS, Purohit R, Das S (2012) Reviews on the influences of alloying elements on the microstructure and mechanical properties of aluminum alloys and aluminum alloy composites. Int J Sci Res Publ 2(6):2250–3153
13. Das S, Mondal D, Sawla S, Ramakrishnan N (2008) Synergic effect of reinforcement and heat treatment on the two body abrasive wear of an Al–Si alloy under varying loads and abrasive sizes. Wear 264:47–59. doi:10.1016/j.wear.2007.01.039
14. Liu Z, Zu G, Luo H et al (2010) Influence of Mg addition on graphite particle distribution in the Al alloy matrix composites. J Mater Sci Technol 26:244–250. doi:10.1016/s1005-0302(10) 60041-2
15. Ren S, He X, Qu X et al (2007) Effect of Mg and Si in the aluminum on the thermo-mechanical properties of pressureless infiltrated SiCp/Al composites. Compos Sci Technol 67:2103–2113. doi:10.1016/j.compscitech.2006.11.006
16. Bonek M, Dobrzański L (2010) Characterization performance of laser melted commercial tool steels. MSF 654–656:1848–1851. doi:10.4028/www.scientific.net/msf.654-656.1848
17. Dobrzański L, Bonek M, Piec M, Jonda E (2006) Diode laser modification of surface gradient layer properties of a hot-work tool steel. MSF 532–533:657–660. doi:10.4028/www.scientific.net/msf.532-533.657
18. Brytan Z, Bonek M, Dobrzański L, Pakieła W (2011) Surface layer properties of sintered ferritic stainless steel remelted and alloyed with FeNi and Ni by HPDL laser. AMR 291–294:1425–1428. doi:10.4028/www.scientific.net/amr.291-294.1425
19. Brytan Z, Grande M, Rosso M et al (2011) Stainless steels sintered form the mixture of prealloyed stainless steel and alloying element powders. MSF 672:165–170. doi:10.4028/www.scientific.net/msf.672.165
20. Tański T (2014) Determining of laser surface treatment parameters used for light metal alloying with ceramic powders. Materialwissenschaft und Werkstofftechnik 45:n/a-n/a. doi:10.1002/mawe.201400232
21. Dobrzański L, Labisz K, Jonda E, Klimpel A (2007) Comparison of the surface alloying of the 32CrMoV12-28 tool steel using TiC and WC powder. J Mater Process Technol 191:321–325. doi:10.1016/j.jmatprotec.2007.03.091
22. Dobrzański L, Labisz K, Klimpel A (2006) Comparison of mechanical properties of the 32CrMoV12-28 hot work tool steels alloyed with WC, VC and TaC powder using HPDL laser. KEM 324–325:1233–1236. doi:10.4028/www.scientific.net/kem.324-325.1233
23. Labisz K (2014) Microstructure and mechanical properties of high power diode laser (HPDL) treated cast aluminium alloys. Materialwiss Werkstofftech 45:314–324. doi:10.1002/mawe.201400231
24. Tański T (2013) Characteristics of hard coatings on AZ61 magnesium alloys. Strojniški vestnik. J Mech Eng 59:165–174. doi:10.5545/sv-jme.2012.522
25. Tański T, Labisz K, Lukaszkowicz K (2013) Structure and properties of diamond-like carbon coatings deposited on non-ferrous alloys substrate. SSP 199:170–175. doi:10.4028/www.scientific.net/ssp.199.170
26. Tański T, Labisz K (2012) Electron microscope investigation of PVD coated aluminium alloy surface layer. SSP 186:192–197. doi:10.4028/www.scientific.net/ssp.186.192

On Shearography Testing of Tires Separations

Helena Hajská, Pavel Koštial, Oldřich Kodym, Zora Jančíková, Jiří David, Roman Meca and Vladimír Rusnák

Abstract The paper deals with the systematic experimental study of tires wear accompanied by separation created in tires of different construction exposed to dynamic loading. 12 pieces of tires with the dimensions of 225/75 R 16 C with the same tread pattern were manufactured for the test in order to examine the impact on the material and design changes on the spread of separations in a tire. The atlas of separation is presented. The results of separation detection obtained by contact-less shearography are compared with those of optical. Graphic relevant analysis of tire wearing process is effectuated.

Keywords Tires · Separations · Shearography · Defects

P. Koštial · O. Kodym · Z. Jančíková (✉) · J. David · R. Meca · V. Rusnák
VŠB – TU Ostrava, 17. listopadu 15, 708 33 Ostrava, Poruba, Czech Republic
e-mail: zora.jancikova@vsb.cz

P. Koštial
e-mail: pavel.kostial@vsb.cz

O. Kodym
e-mail: oldrich.kodym@vsb.cz

J. David
e-mail: jiri.david@vsb.cz

R. Meca
e-mail: roman.meca@vsb.cz

V. Rusnák
e-mail: vladimir.rusnak@form-composite.sk

H. Hajská
Vipotest s.r.o, T. Vansovej 1054/45, 02001 Púchov, Slovak Republic
e-mail: helena.hajska@continental.sk

A. Öchsner and H. Altenbach (eds.), *Mechanical and Materials Engineering of Modern Structure and Component Design*, Advanced Structured Materials 70,
DOI 10.1007/978-3-319-19443-1_8

1 Introduction

Security of the road transport depends on the quality of basic and applied research concerning materials and internal construction of tires. The design shapes and material properties characterized by low hysteretic losses as well as a construction have an influence on the driving comfort, adhesion, wear resistance and fatigue resistance [1]. Thick fibre reinforced composites are used extensively in rubber products such as tires and conveyer belts. Generally, the reinforced parts of rubber products on a sub macroscopic level are highly heterogeneous and anisotropic because they are composed of rubber compounds, and textile and steel cords [2]. Rubber compounds consist of natural or synthetic rubber, carbon black, curing agents, cure accelerators, plasticizers, protective agents and other ingredients [3]. The relatively complicated composite structure of real tire so can be a source of defects.

Defects in tires can have a different origin. Some defects occur during manufacturing and other ones during tire operating. In the frame of our interest air pockets in the tire carcass can be included under manufacturing defects. It can be a source of later separations and decreasing reliability. Some other manufacturing defects are closely concerned with irregular or asymmetric construction of tire casings and these defects cause lower wear resistance. Operating defects can occur during mounting on vehicle and they can occur after non-professional service or after wrong operating of service devise. Other defect sources are due to incorrect operating procedures of cars, such as the overloading of the vehicle, incorrect tire pressure or even the wrong driving technique [4].

Non-destructive experimental methods used for controlling of new and retreated tires and rubber products that allow us to discover defects and flaws occurring in production, or as a result of strain during application, can be divided into the following groups:

Ultrasonic flaw detection, X-Ray analysis, computed tomography (X-Ray + PC), microscopic methods (raster tunnel microscopes), laser holography, laser interferometry (shearography)—a higher stage of holography.

The method of non-destructive analysis of tires based on laser shearography was patented in 1987 by Laser Technology, Inc. A special camera composed of special optic member, lens and CCD camcorder visualizes the structural defects manifested by sub-microscopic surface separations in the whole tire profile from bead to bead. The scope of different imaging technologies is subsequently described in recent works of Kumar et al.

In the work [5] a three beam illumination television holographic method is described for the measurement of the surface shape of three-dimensional objects. The arrangement combines an in-plane sensitive arrangement with an out-of-plane sensitive arrangement. Four sets of speckle frames each with five phase shifted sequentially recorded patterns are used to determine the surface shape. In this three beam illumination arrangement the rotation given to the object during the experiment can also be determined from the recorded data, making an independent measurement of the rotation unnecessary.

In laser based interferometry, the unambiguous measurement range is limited to half a wavelength. Multiple methods based on multiple wavelengths or white light interferometers wavelength or white light interferometer is used to overcome this difficulty. In this paper the application of a white light interferometer recorded with a colour CCD camera is discussed. Authors accessed interference intensity information from the three channels of the colour CCD simulating three-wavelength measurements. This makes the data acquisition as simple as in single wavelength interferometry [6].

Quantitative phase information from a single interferogram can be obtained using the Hilbert transform (HT). Authors have applied the HT method for the quantitative evaluation of Bessel fringes obtained in time average TV holography. The method only requires one fringe pattern for the extraction of vibration amplitude and reduces the complexity in quantifying the data experienced in the time average reference bias modulation method, which uses multiple fringe frames. The technique is demonstrated for the measurement of out-of-plane vibration amplitudes on a small scale specimen using a time average microscopic TV holography system [7].

Paper [8] describes a microscopic TV holographic arrangement to study the static and vibrating microsystems. In the optical setup, the object beam and the reference beam arms are provided with a phase shifting mirror and a bias phase modulation mirror to carry out the measurement of the out-of-plane deformation and the vibration amplitude fields, respectively.

The multiple frame digital fringe projection technique is widely used for measuring the 3-D surface shape. In dynamic situations single frame analysis techniques are desirable. In this paper authors discuss the Hilbert transform based single-frame analysis. The Hilbert transformation method requires only one fringe pattern for the extraction of the phase therefore reducing the calculation time. The method is easy to implement, and it is capable of conducting automated measurements at video frame rate. The application of the proposed method for curved surfaces is emphasized [9].

Shearography in contrast to holography measures not the deformation but the deformation gradient (output plane of derivation shift). Shearography allows discovering defective places as holography does, but it is not so sensitive to a shock. The last development stage of this method is the exclusion of film usage and the full implementation of electronic features allowing real time evaluation. The history of the method's development, its basic theoretical principles, instrumentation and applications for evaluation of tires can be found in [10].

Laser holography, shearography and digital shearography along with X-ray technology are powerful tools for the evaluation of tire endurance. Today, digital shearography presents the most effective tool regarding simplicity and costs of laser technology [11].

In this paper we present the study of a tire wear caused by separation generation. We applied shearography as a tool for experimental investigation of such processes. The advantage of this method is that no other special preparation of tires is necessary and no damage occurs during their analysis.

2 Experimental Part

The creation of the shearographic image will be explained in the following.

An extended laser beam illuminates the tested object. A laser beam reflected from the object surface is focused on the image plane of a CCD camera. The peculiarity of this camera lies in the fact that the wedge prism, bi-prism or Michelson's interferometric arrangement, used in non-destructive analyzer, is integrated in the front of camera's lens of the shearing apparatus. The shearing elements cause two minor dislocated images of the object in the image plane.

Finally, the interferometric intersection of both minor dislocated images gives us the interferogram, thus creating the so-called speckle interferogram. That is why this technique is referred to as shearography. The mentioned settings make it obvious that shearography does not need an additional reference beam for an production of interferogram, in contrast to holography and speckle pattern interferometry, but it needs the so-called self-reference light beam because of the shearing effect. Shearography thus uses its own optical reference system (Fig. 1).

In the case of analogous to holographic interferometry, there are also two speckle interferograms stored during the shearographic examination induced via two deformation states: first non-deformed, the second at load. A visible strip pattern occurs also with the contribution of the intensity distribution intersection of double speckle interferograms, simple digital subtraction of intensity distribution of two speckle interferograms and with the help of signal sensors of the computer. This visible strip pattern is the shearogram, that is, in comparison to hologram, not interpreted as a contour line of deformation but it is marked by a deformation gradient in the direction of dislocation [9] (Fig. 2).

The non-destructive analyzer used in our experiments allows a quick and simple discovery of structure defects (closed separations) in tires. A tire test is performed

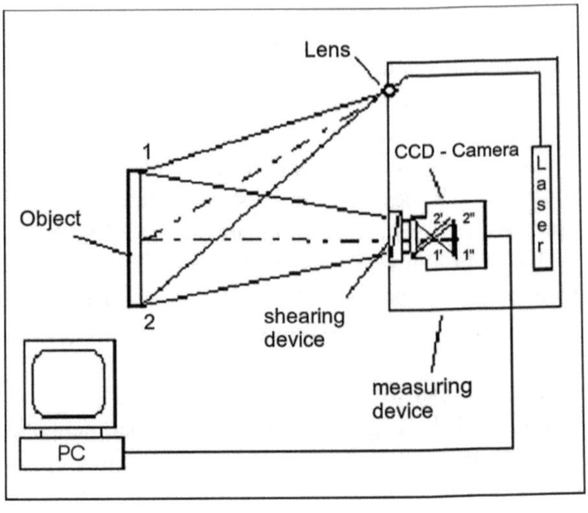

Fig. 1 Schema of the shearographic apparatus

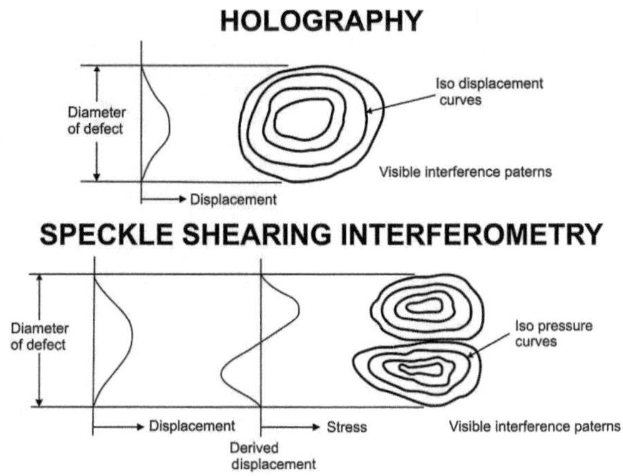

Fig. 2 Difference between holography (*up*) and shearography (*down*)

by developing a small external vacuum on a tire causing the swelling of externally invisible separations (Fig. 3a, b). The higher the vacuum the higher the defect visibility. These structure defects are recorded by means of the interferometric measuring method, speckle shearing interferometry also called TV-shearography. The tire surface is illuminated by laser light and monitored by a CCD camera (Fig. 4). Two images of the shift phases—before and after decreasing the surrounding pressure—they are superposed and pre-processed. The interferential image of the tire surface at normal pressure is stored in computer memory in order to display characteristic separations. The second image, measured in the deformed state at decreased pressure (vacuum application) is subtracted from the first stored image. The difference between these two images is visualized on a computer screen. Characteristic separations are shown as speckles on the computer screen. A phase image and a video image are available for each sector.

12 pieces of tires with the dimensions of 225/75 R 16 C with the same tread pattern were manufactured for the first test in order to examine the impact of material and design changes on the spread of separations in a tire. These tires were included in the survey in this work because great attention was devoted to their critical area of tread shoulder, even if these tires don't belong to the group of passenger car tires, but they belong to a group of van tires. The attention during the production and testing of these tires was focused mainly on the area of the tread shoulders, where the influence of the used material and various construction changes in the tire aimed at reducing the separations in the tread shoulders were examined. We have manufactured the individual options of tires together with the developers.

The individual tires were marked as samples S01–S12. These samples used two kinds of mixtures, but also a variety of construction changes in the area of tread shoulders, and one of the options also included a tire from regular production (Table 1).

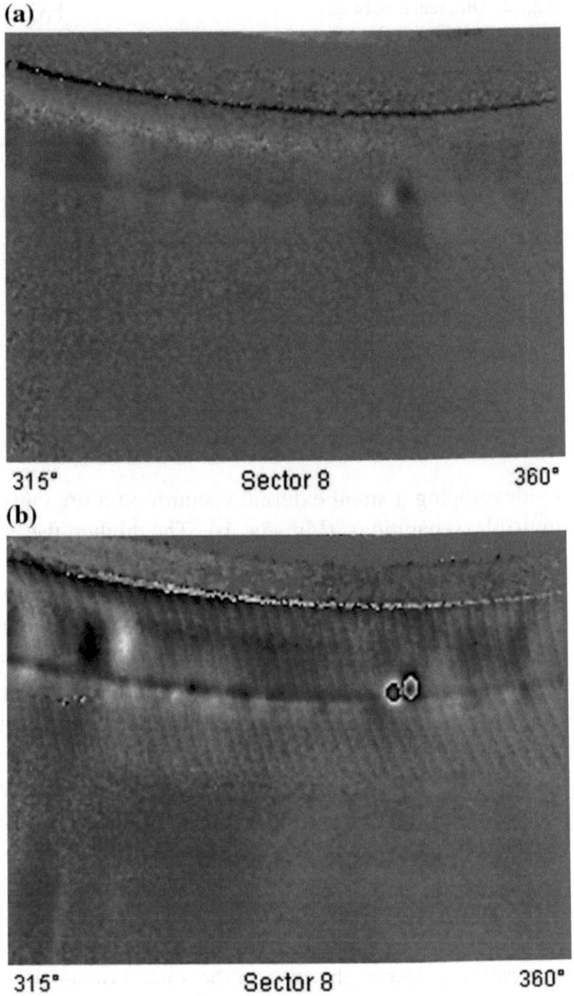

Fig. 3 **a** Visibility of tire structure under vacuum 1 kPa. **b** Visibility of tire structure under vacuum 7 kPa

The tires were tested using the test machine 75 kN-G and a non-destructive analyzer ITT-1.

The following procedure was used to monitor the spread of separations:

- Performing non-destructive analysis of new tires
- Dynamic tests of tires using a fatigue test on a drum testing machine
- Continuous testing of tires on a non-destructive analyser
- Destructive analysis of tires—cross-sections
- Evaluation

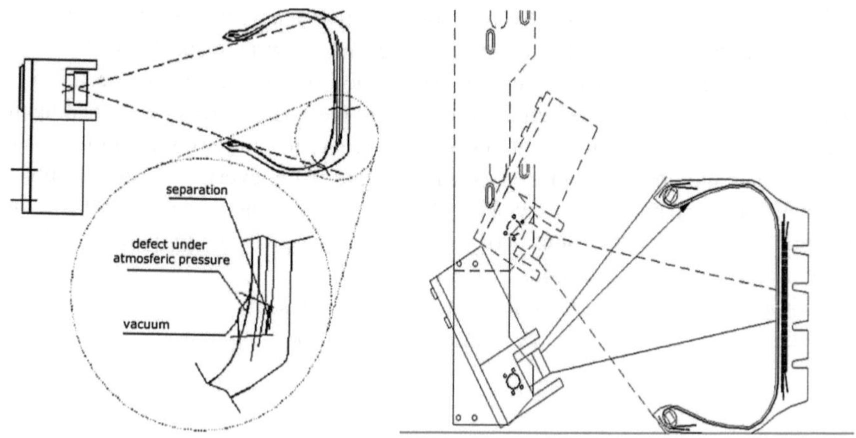

Fig. 4 CCD camera setting of non-destructive analyzer at tire casing test

Table 1 Overview of the individual tire options

Sample number	Material	Construction
S01	SC—A, TC—*A*, mixture A	**N1—A, N2—B**
S02	SC—A, *TC*—*A*, mixture A	**N1—A, N2—B, N3—C**
S03	SC—A, *TC*—*A*, mixture A	**N1—A, N2—B, N3—D**
S04	SC—B, *TC*—*B*, mixture B	**N1—E, N2—F, N3—G**
S05	SC—A, *TC*—*A*, mixture A	**N1—A1, N2—B, N3—H**
S06	SC—A, *TC*—*A*, mixture A	**N1—A2, N2—B, N3—H**
S07	SC—A, *TC*—*A*, mixture A	**N1—A3, N2—B1, N3—H**
S08	SC—A, *TC*—*A*, mixture A	**N1—A4, N2—B2, N3—H**
S09	SC—A, *TC*—*A*, mixture A	**N1—I, N2—J, N3—H**
S10	SC—A, *TC*—*A*, mixture A	**N1—A2, N2—B3, N3—H**
S11	SC—B, *TC*—*B*, mixture B	**N1—A, N2—B, N3—H**
S12	SC—A, *TC*—*A*, mixture A	**Regular production**

Note: *SC*—steel cord, *TC*—thread count types (*A*, *B*) (number of textile or steel cords per 1 m), **N1**, **N2**, **N3**—the individual breakers order, **A**, **B**—type of breakers

All the tires were put to a fatigue test under the following conditions:

Test load:	22,760 N
Inflation during test:	575 kPa
Test speed:	40 km/h
Duration:	20 h
Conditioning time including the analyser testing:	app. 4 h

The tires were first tested by a non-destructive analyzer, aimed at detecting eventual hidden internal defects. Tires No. 05 and 06 were suspected of having a local minor tread shoulder separation on the non-serial side of the tire (NSST). Other tires were without defects. A non-destructive analysis was performed again after 20 h, which represents about 800 km run, where local minor separations in the area of tread shoulder or bead were already visible. After every 800 km, the spread of small local separations was visible. This procedure was repeated until the tire was put aside by a contact sensor or until it was damaged. The individual surfaces showing the defects (separations) were calculated.

Although there were 12 samples in total, we present only one from a non-destructive analyser in this work. Other results will be in graphic output.

3 Results and Discussion

During the initial testing stage there was no damage detected in the tires, which was confirmed by a test on a non-destructive analyzer (see Fig. 5).

The tire was further put to a fatigue test on a drum testing machine with a load of 75 kN. The first small local separations in the tire tread shoulders began appearing after 20 h.

After another 20 h, when the tire had run for the equivalent of 1600 km, there were separations above the bead and small local separations in the tread shoulders,

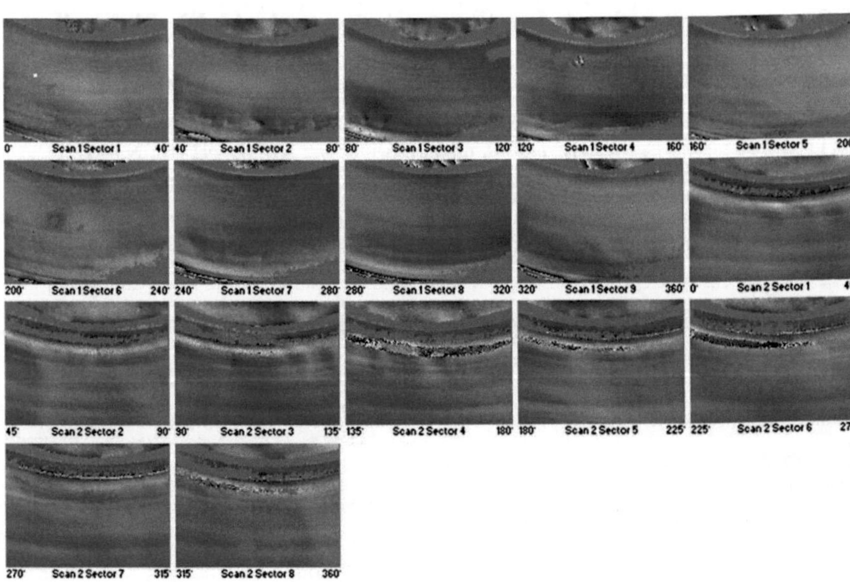

Fig. 5 Interferometry of sample S01 before testing

which gradually began to join along the circumference, as one can see on the scans of the non-destructive analyser (see Fig. 6).

After 80 h representing about the equivalent of about 3200 km, the test was finished as the tire was put aside by the contact sensor due to a large separation above the bead. The separations above the bead were expanding and joining along the circumference. They were expanding more and more from the tread shoulder into the tire crown and sidewall (see Fig. 7).

Prior to the fatigue test, footprints of the individual tire options were taken on a pressure sensitive film, where the pressure distribution in the tire under static load is visible. In Fig. 8 you can see the footprint prior to the dynamic test.

The tire load was 14,225 N and the tire was inflated to 575 kPa. The average measured pressure in the tire was 353 kPa.

As we have observed from the pressure distribution in the footprints of the tire prior to the test, the tires had lower pressure in the crown and tread shoulders, and the highest pressure was in the outer tread pattern towards the tire crown. After the equivalent to 800 km, the pressure distribution had changed; the pressure had increased not only in the crown area, but especially in the outer tread pattern. The same behaviour was observed in all tires, except for samples S08 and S09, where the pressure was higher prior to the test than after the test. This pressure distribution

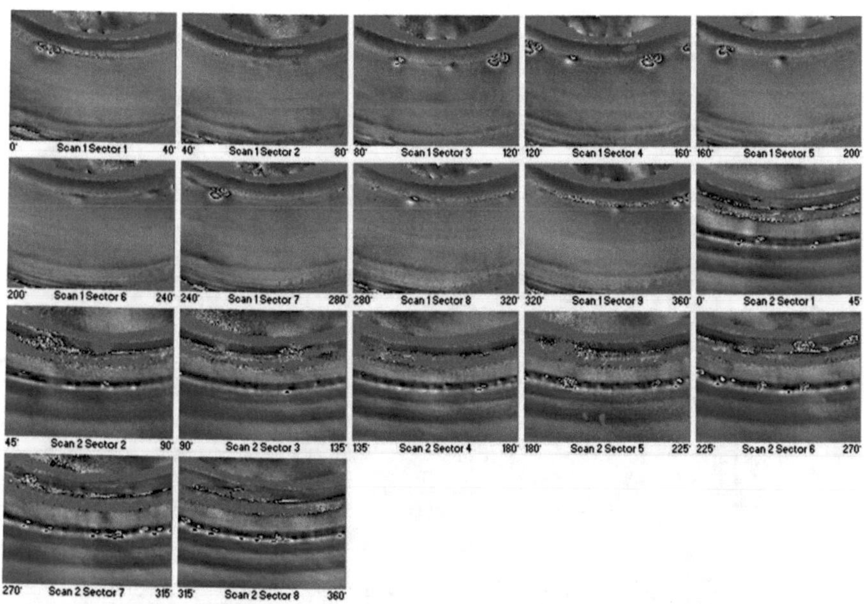

Fig. 6 The third measurement of tire S01 after running for 40 h or 1600 km

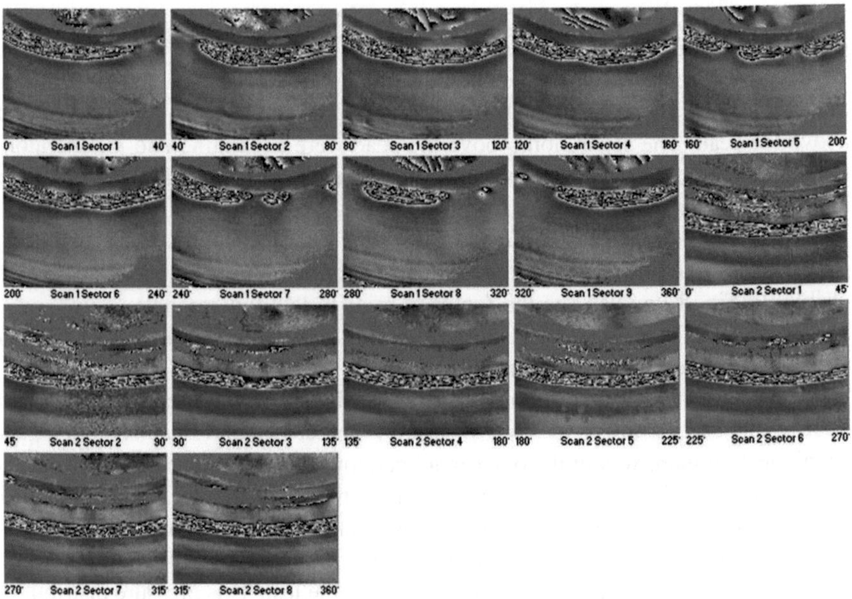

Fig. 7 The fifth measurement of tire S01 after 80 h or 3200 km, the test was terminated

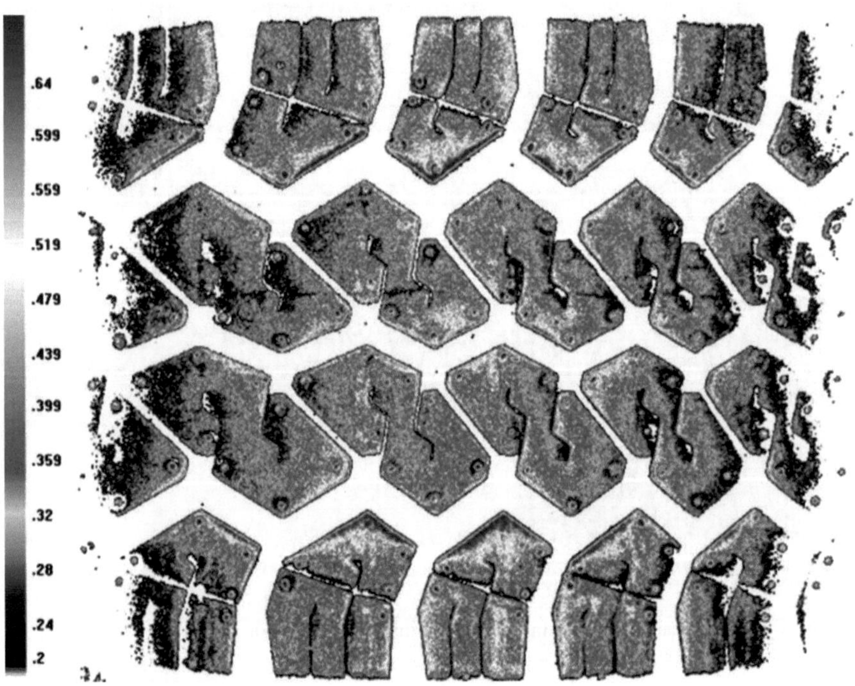

Fig. 8 Foot print of tire S01 prior to the dynamic test

shows that there will be defects in the tread shoulder area and the tire tread pattern will be most worn-out in the outer treads.

After completion of the test, cross-section cuts were made in places of defects, where separations had been detected by a non-destructive analyzer. The individual cross-sections really show the separations in the tread shoulder and above the bead. The separations of the tread shoulder were created and spread between the first and second breakers at their ends. In the bead area, they arose above the bead wire between the tire body and the bead filler. The real image of defects in chosen tire sectors 1 and 6 are presented in Figs. 9 and 10.

Other types of observed separation in investigated samples in shearography view (left) and real view (right) are in Fig. 11a–n.

Exactly the same procedure was used to examine all other options of this tire size. The tire performances ranged from 40 to 100 h. The performances of the individual tire options are shown in Table 2.

Figures 12, 13 and 14 present charts comparing the separations of the individual options in the area of bead tread shoulder and the whole tire casing after 40 h, i.e. after 1600 km. The comparison was made after a 40 h run, which was the lowest performance of tires S04 and S11. The highest performance was achieved by tires S10 and S12.

Comparison of the individual tire options with regards to their hour performance and % of defect area (separations) in the whole tire is presented in Fig. 15.

The charts presented above show that:

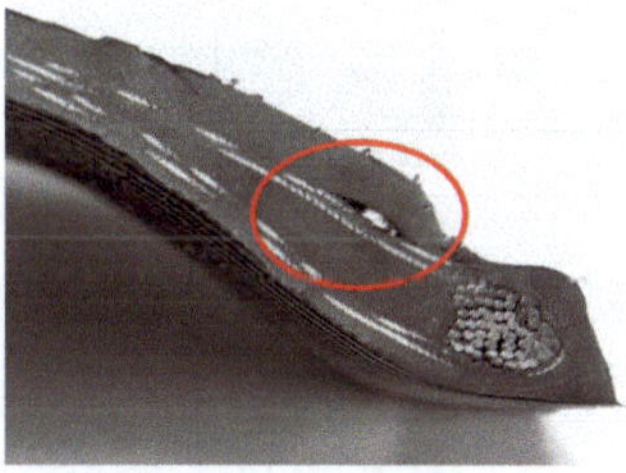

Fig. 9 Cross-sections of tire S01 in Scane1 Sector 1 separation above bead

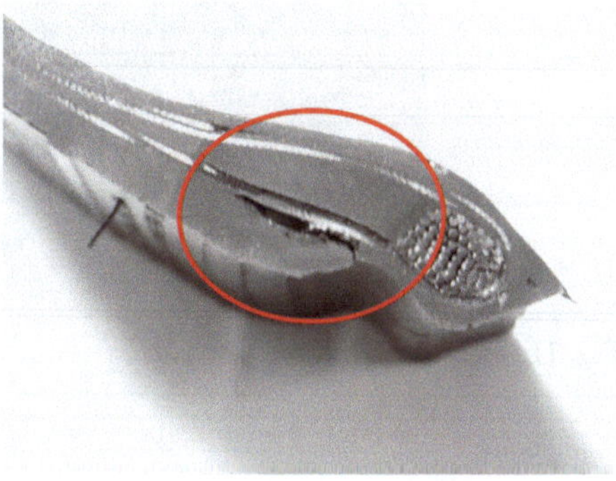

Fig. 10 Cross-sections of tire S01 in Scane1 Sector 6 separation above bead

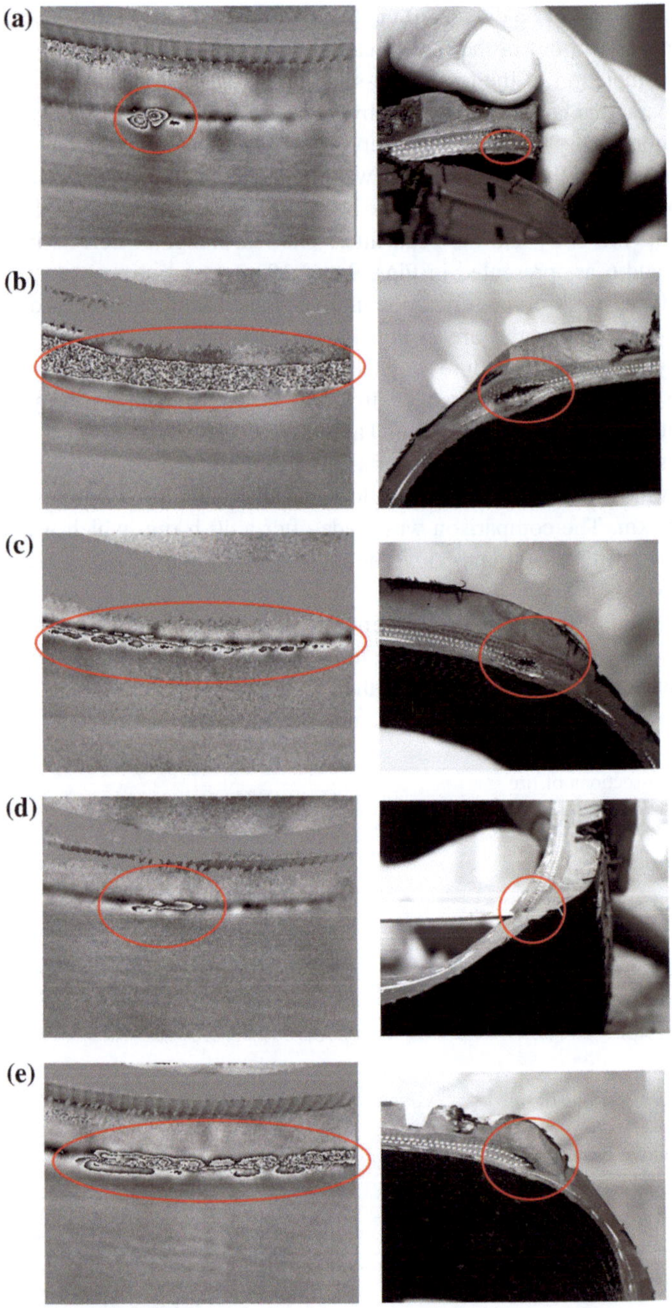

Fig. 11 (continued)

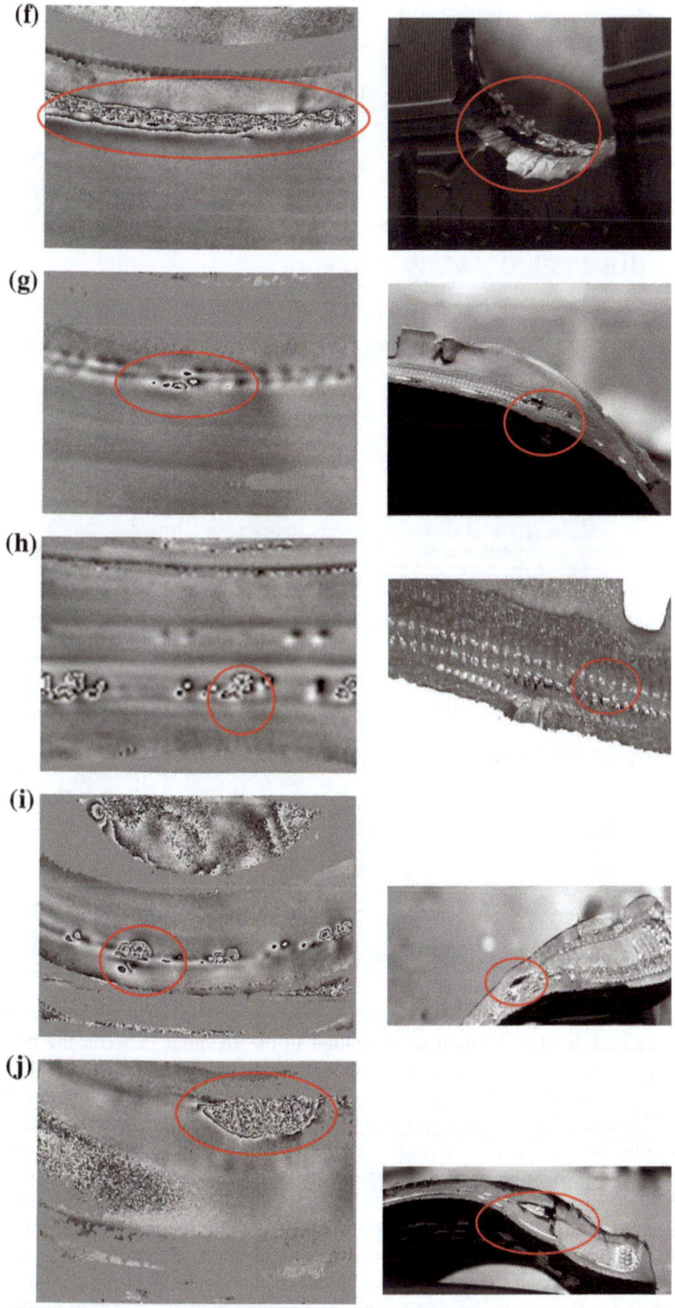

Fig. 11 (continued)

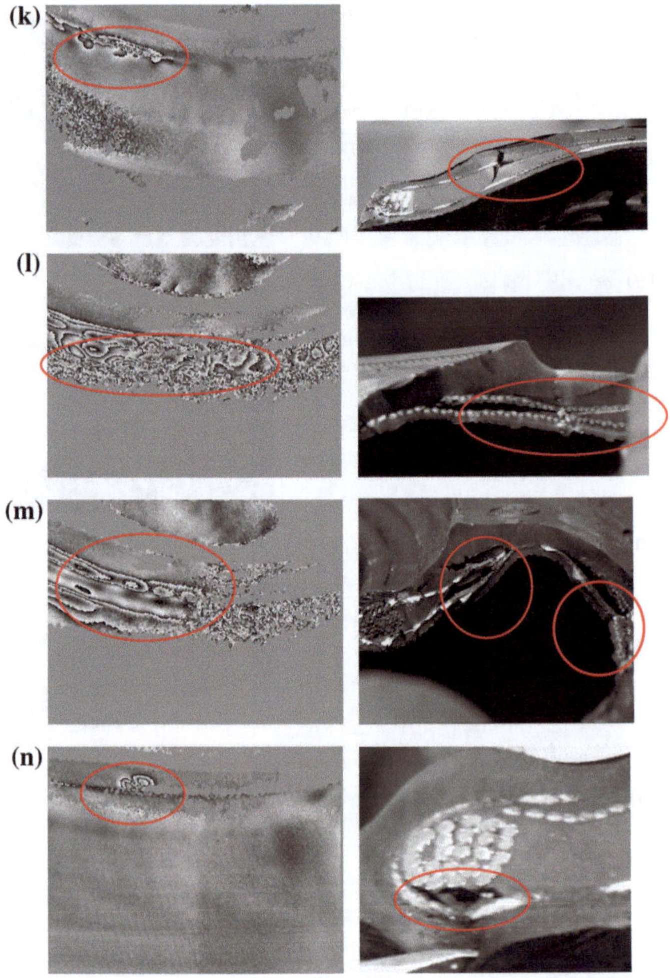

Fig. 11 The atlas of separations with their description. **a** The local separation at the second bumper lining end. **b** The longitudinal separation in the shoulder between the first and second bumper lining. **c** The chain separation in the shoulder between the first and second bumper lining. **d** The local separation in the shoulder at the first bumper lining end. **e** The longitudinal separation in the shoulder where the first and second bumper lining ends and spreads to the crown. **f** The separation in the shoulder and its spreading to the side directions. **g** The local separations in the shoulder in the site where the second bumper lining ends. **h** The local separations in the crown between the first bumper lining and the cord carcass. **i** The local separation in the sidewall between the end of carcass lining and bead bundle. **j** The local separation in the sidewall-burst bead rubber and bundle, separation is between the end of carcass lining and bead bundle. **k** The local separation of the tire casing-burst bead rubber and bead bundle. **l** The longitudinal separation in the sidewall of the tire casing-at the carcass lining bending. **m** The longitudinal separation in the shoulder and sidewall of the tire casing between the bent carcass lining and filling linings of bead. **n** The local separation in the bead between bead plies and bent-burst bead rubber and bead bundle carcass lining

Table 2 Overview of performances and defects of tested tires

Sample	Interferometry prior to test	Perf. in h.	Specific pressure in foot print prior to test in kPa	Specific pressure in foot print after 20 h of test in kPa	Defect after test
S01	No defects	80	353	–	Separation in tread shoulder and above bead
S02	No defects	80	365	–	Separation in tread shoulder and above bead
S03	No defects	60	343	390	Separation in tread shoulder and above bead
S04	No defects	40	355	392	Separation in tread shoulder and above bead
S05	Suspicion of separation in tread shoulder	60	360	373	Separation in tread shoulder and above bead
S06	Suspicion of separation in tread shoulder	60	341	366	Separation in tread shoulder and above bead
S07	No defects	60	332	397	Separation in tread shoulder and above bead
S08	No defects	60	397	379	Separation in tread shoulder and above bead
S09	No defects	60	390	374	Separation in tread shoulder and above bead
S10	No defects	100	339	387	Separation in tread shoulder and above bead
S11	No defects	40	343	373	Separation in tread shoulder
S12	No defects	100	369	361	Separation in tread shoulder

- the smallest defect area in tread shoulder after 40 h was in tire S10
- the smallest defect area in bead after 40 h was in tire S12

Tires S04 and S11 options marked had the lowest performance of only 1600 km. These two tires were made from mixture B. The results show that the material labelled as B is not suitable to be used with components in the area of tire shoulder, with high dynamic property requirements. The material labelled A (see Table 1)

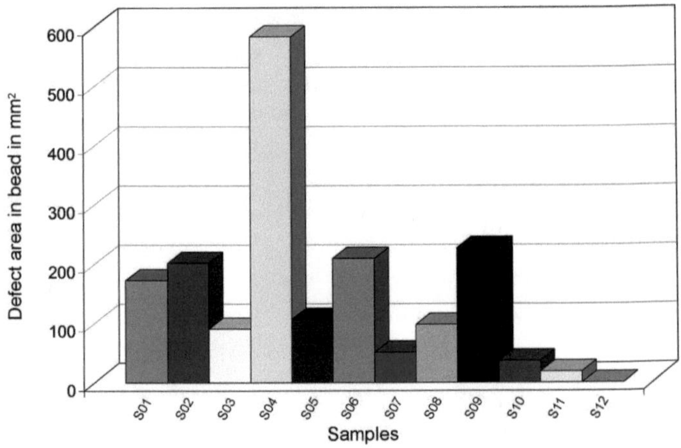

Fig. 12 Illustration of separation areas above the bead after 40 h run or app. 1600 km

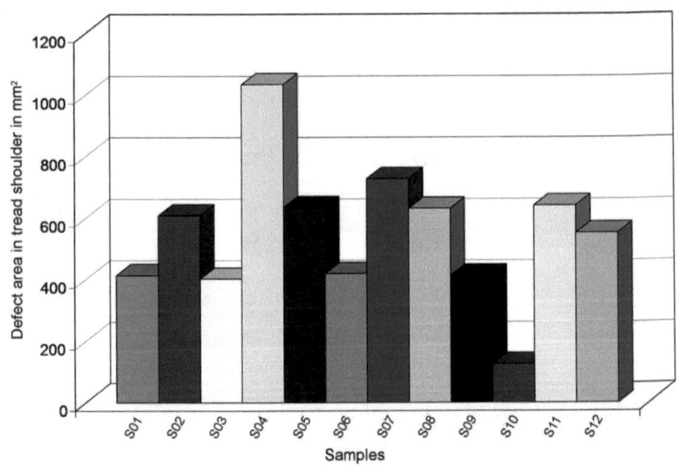

Fig. 13 Illustration of separation areas in the tread shoulder after 40 h run or app. 1600 km

was used in the other tires. The performance of these tires ranged from 2400 to 4000 km. It shows that the lifetime of tires is affected not only by the material, but it is also significantly affected by the design of the relevant tire areas.

The highest performance and the slowest spreading of separations were monitored in tires no. 10 and no. 12 according to the percentage of defects in the whole tire casing; the best results were monitored in tires S10. The bead area of tire S12 is proper for application in production of tire casings due to the lowest number of defects in this area.

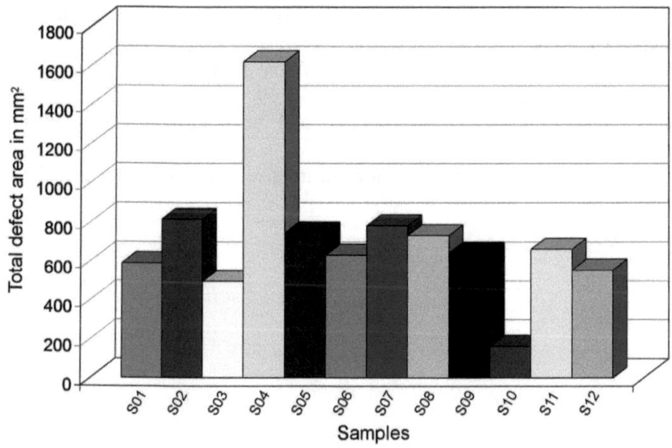

Fig. 14 Illustration of separation areas in the whole tire after 40 h run or app. 1600 km

Fig. 15 Comparison of the individual tire options with regards to their hour performance and % of defect area (separations) in the whole tire

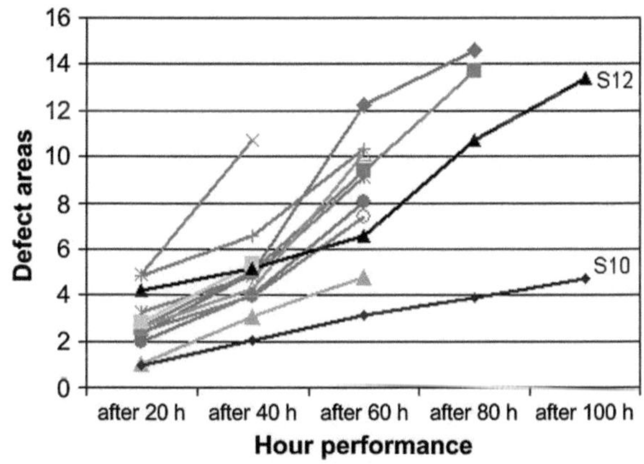

References

1. Koštial P, Ružiak I, Krmela J, Frydrýšek K (2012) The chosen aspects of materials and construction influence on the tire safety. In: Hu N (ed) Composites and their properties, InTech
2. Marcín J, Zítek P (1985) Rubber products tires 1. SNTL, Praha
3. Prekop Š, Várkoly L (2003) Rubber technology II. GCTech, Trenčín
4. Symens R (2003) Sidewall bulge and depression testing. The annual review pp 112–114
5. Kumar P, Pal S et al (2011) Television (TV) holography for three-dimensional surface shape measurement. Lasers Eng 21:231–240
6. Kumar P, Wang H et al (2012) White light interferometry for surface profiling with a colour CCD. Elsevier, Amsterdam, pp 1084–1088
7. Kumar P, Mohan K et al (2010) Time average vibration fringe analysis using Hilbert transformation. Appl Opt 49:5777–5786

8. Kumar P, Mohan K et al (2011) Measurement of static and vibrating microsystems using microscopic TV holography. Optik 122:49–54
9. Kumar P, Somasundaram U et al (2013) Single frame digital fringe projection profilometry for 3-D surface shape measurement. Optik 124:166–169
10. Yeager W (2004) Principles of shearography and its application in non-destructive testing. Tire Technol Int 106–110
11. Yeager B (2002) Standard testing labs-non-destructive testing of radial tires. Tire Technol Int 110–113

Compression Behaviour of Finite Dimensional Cellular Metals by Generalization of Cell Buckling Effects

Renato V. Linn and Branca F. Oliveira

Abstract Metal foams are materials of recent developments and applications that show interesting combinations of physical and mechanical properties. Foams are commonly used as of passive safety components due to their high capacity of energy absorption under impact conditions. In this work the foam is represented as a cellular material with a regular structure and specimens of a cellular metal are used to study the foam behaviour. Considering that compression is the dominant loading in impact situations, the deformation behaviour of finite dimensional cellular metal specimens under compression is investigated. The specimen deformation configuration is determined by means of fundamental buckling effects on cells walls evaluated from simple representative volume elements. Damage effects under a finite strain context are included together with self-contact considerations. The overall behaviour of a finite specimen is derived from an analytical and numerical framework based on the boundary conditions present on the foam. The main advantage of this method is the capability of determine the full behaviour of a complex foam configuration with only simple case analyses, with low computational cost.

Keywords Cellular metals · Metallic foams · Damage · Gurson models · Finite elements

R.V. Linn · B.F. Oliveira (✉)
Virtual Design Research Group—ViD, Federal University of Rio Grande do Sul, Av. Osvaldo Aranha, 99, s.408, 90035-190 Porto Alegre, Brazil
e-mail: branca@ufrgs.br

R.V. Linn
e-mail: renatolinn@gmail.com

A. Öchsner and H. Altenbach (eds.), *Mechanical and Materials Engineering of Modern Structure and Component Design*, Advanced Structured Materials 70,
DOI 10.1007/978-3-319-19443-1_9

1 Introduction

Metallic foams are materials composed of a metallic matrix with internal voids (Fig. 1). Their use has increased by the automotive industry, particularly as a recourse of passive safety. In that case, they act to absorb impact energy.

Metallic foams show mechanical behavior and physical properties that strongly differ from those of solid materials and show interesting combinations of properties. For example, high stiffness combined with low specific weight, or permeability for gas flow combined with high thermal conductivity, offering possibilities of use in the industry. The schematic stress-strain curve for a metallic foam in compression (see Fig. 2), shows a large area in the plateau region corresponding to high energy absorption at constant stress.

The characterization of the stress-strain curve for a metallic foam under compression involves several complexities, like the material disorder (aleatority foam distribution), the influence of the finite dimensions of the specimen and geometric and physical nonlinearities involved.

Several approaches have been proposed and analysed by this characterization. Some works of Shulmeister [2] and Hallström and Ribeiro-Ayeh [3] investigated

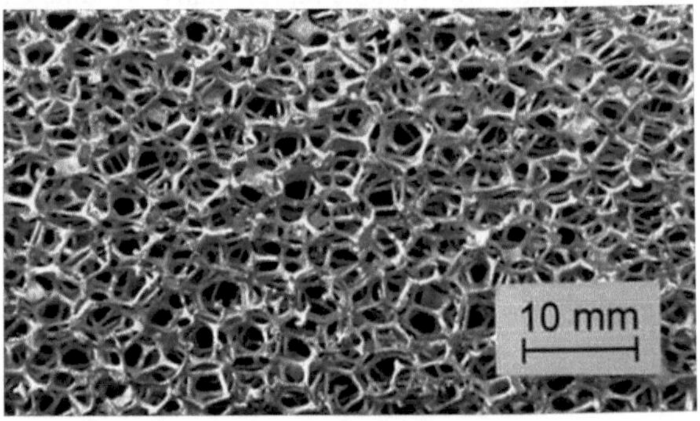

Fig. 1 Aluminum foam with open cell (DUOCEL®) [1]

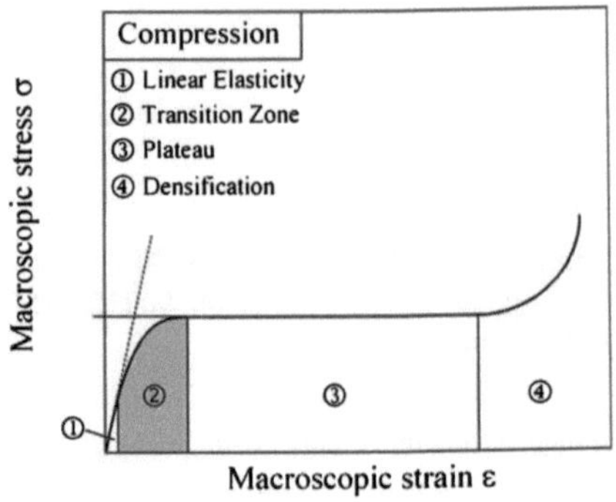

Fig. 2 Stress-strain curve for a foam showing large capacity of energy absorption at constant stress [7]

the influence of the representation of the material disorder for the characterization of the foam behavior under compression and tension. Periodic ordered lattice structures for foam simulations were examined in ordered and in randomised versions by Grenestedt and Tanaka [4] and Shulmeister [2]. Works of Duvaut [5] and Xia et al. [6] analysed the characterization of metallic foams with the use of periodically infinite boundary conditions to simulate the average behavior of a foam.

In this work, a numerical and analytical framework is proposed, which determines the overall behavior of a finite dimensional specimen of metallic foam through the generalization of the buckling effects that can occur on an homogenized representative volume model. The model is simulated using the finite element method with damage considerations in a large strain and displacement context. The main advantages of such a methodology are the generalization for finite dimensional foam cases instead of an average infinite cell approach for the characterization of the behavior of the foam and very low computational cost required.

The characterization of finite dimensional foams is important for applications where the foam shows a predominant dimension under loading condition. For example in sandwich foam structures, where the foam usually has a much smaller dimension in the direction of applied load. For such a case, the behavior of the structure under compression or tension is not fully characterized through the use of an infinite dimensional model because there is a loss of stiffness in the loading direction in the true finite dimensional structure.

2 Modeling Considerations for Metallic Foams

Metal foam properties can change along the production process, particularly according to the cell being open or closed and the relative density of the foam in relation to the base material. Moreover, foams show (see Fig. 1) a fairly random structure. It is possible to determine the exact geometry using a CT-scan (computed tomography) and modeling it with a dense mesh of finite elements, as seen in Fig. 3a. A different approach towards the understanding and modeling of these materials, adopted in this work, is an idealization as a cellular material with regular structure, as seen in Fig. 3b. In this case, modeling a RVE (representative volume element, seen in Fig. 3c) and still considering the existent symmetries, good results can be obtained with fewer elements. Then, specimens with regular structure may be tested to validate results and determine material parameters. These homogenization procedures have been used with success in other areas of continuum mechanics. A specimen of a cellular metal made to represent a foam is shown in Fig. 4.

The main purpose of the present framework is to determine the behavior of a finite dimensional specimen (like the one shown in Fig. 3b) based only on RVE models (see Fig. 3c) with different boundary condition considerations. The simulation of a full specimen as seen in Fig. 3b is computationally costly, and the results depend on the number and distribution of the cells analysed. The proposed method

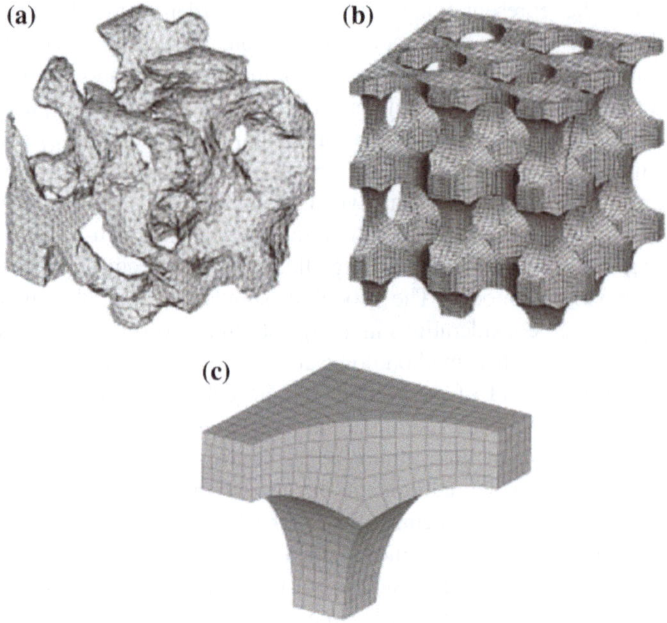

Fig. 3 Meshes used to represent a cellular metal [1]

Fig. 4 Cube of cellular metal

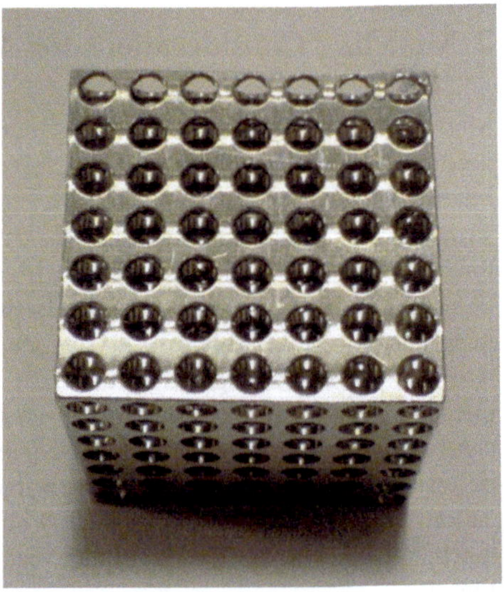

is generalized to any number of cells to be considered on the foam and any proportion of the distribution of the cells along the finite dimensions. The applied boundary conditions, the contact hypothesis, the damage model and the numerical and experimental simulation are first detailed.

2.1 Gurson Damage Model

The Gurson damage model was developed to describe the mechanical effect of high plastic deformations in ductile metals. The loss of resistance is governed by the porosity level. The (isotropic) damage variable employed is the volumetric void fraction, represented by f and defined by $f = V_v/V$, where V_v is the volume of voids in a representative small volume V, corrected for effects as stress concentration; f is defined at each point of the continuum. The presence of voids alters the elasto-plastic constitutive relations. The equations usually employed in computational damage analyses, the Gurson-Tvergaard model [8, 9], considers a yield surface defined by:

$$\Phi = \sqrt{\frac{3}{2}S_{ij}S_{ij}} - \overline{\omega}\sigma_y = 0 \tag{1}$$

where

$$\overline{\omega} = \left[1 - 2\alpha_1 f \cosh\left(\frac{\alpha_2 3p}{2\sigma_y}\right) + \alpha_3 f^2\right]^{1/2} \tag{2}$$

$$S_{ij} = \sigma_{ij} - p\delta_{ij} \quad p = {}^1/_3\sigma_{ij}\delta_{ij} \tag{3}$$

$$\alpha_1 = \frac{1}{f_U} = 1.5 \quad \alpha_2 = 1.0 \quad \alpha_3 = \alpha_1^2 \tag{4}$$

and σ_{ij} are the Cauchy stresses, σ_y indicates the yield stress in simple tension. α_i are material parameters. The parameter $f_U = 1/\alpha_1$ is the maximum volumetric void fraction admissible before rupture in absence of pressure. Another possible interpretation for the α_1 and α_2 parameters is that they work as multipliers acting on porosity f and pressure p, respectively.

In Fig. 5, yield surfaces for different levels of void content are shown, in a plot of normalized deviatoric stress versus normalized pressure.

It can be seen that the plastic domain depends on the hydrostatic pressure. When the volumetric void fraction f decreases, it decreases the influence of pressure, leading to a larger elastic domain. For $f = 0$, the model reduces to the von Mises model, which is independent of hydrostatic pressure.

The basic mechanisms of damage evolution are nucleation, growth and coalescence of voids. Nucleation occurs mainly due to material defects, in the presence of tension. Growth occurs when the voids (preexistent or nucleated) change their size according to the volume change in the continuum.

Coalescence is related to the fast rupture process that occurs after that the volumetric void fraction reaches a limit, indicated by f_C. Coalescence consists in the union of neighbor voids due to the rupture of a ligament.

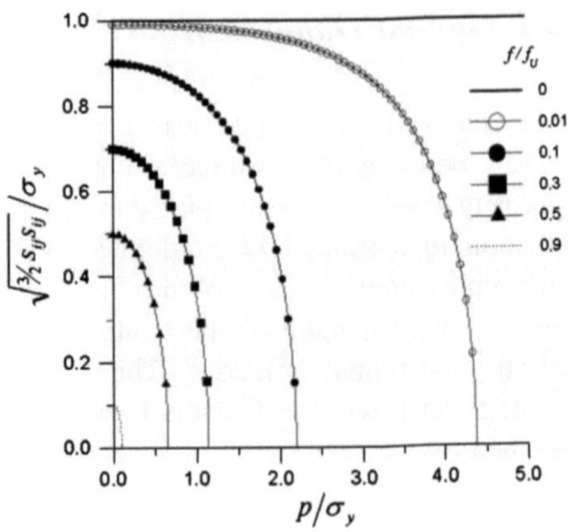

Fig. 5 Yield surface for a porous material: influence of volumetric void fraction

The equations that govern damage evolution are modeled in a simplified form as follows. First, it is assumed that total void rate is given by:

$$\dot{f} = \begin{cases} \dot{f}_n + \dot{f}_g & f \leq f_C \\ \dot{f}_C & f > f_C \end{cases} \tag{5}$$

where $\dot{f}_n$ is the void nucleation rate, $\dot{f}_g$ is the void growth rate and $\dot{f}_C$ is the void coalescence rate. Thus, as long as f is smaller than a characteristic value f_C only nucleation and growth develop. If it is above f_C, then only coalescence takes place.

Or in a alternative way [9, 10] employing a corrected volumetric void fraction given by:

$$f^* = \begin{cases} f \\ f_C(1.0f_C)(f - f_C)/f_F f_C \end{cases} \tag{6}$$

In this case, only nucleation and growth are considered in Eq. (5).

The commercial finite element code ABAQUS is used in this work and the Gurson damage model is one of the available constitutive relationships.

2.2 Application of Boundary Conditions

One important aspect to take into account when modeling cells as those described above are the boundary conditions. For modelling the average behavior of the foam with RVE, the displacement of some of the boundaries should be constrained in order to represent an infinite (periodically) cell foam. This behavior can be computationally imposed using multiple point constraints (MPC). The used FE code (ABAQUS) offers the possibility to realize such a boundary condition where all nodes on a certain surface have the same x-displacement: $u_{xi} = \cdots = u_{xj}$. The effect of differing boundary conditions on the deformation is shown in Fig. 6.

If there are no free boundaries (MPC applied to the three planes of the RVE as shown in Fig. 6a) the walls of the cell remain vertical for tension and no lateral buckling occurs at compression, while the distance between them may change. This is the idealization of a full enclosed cell inside of the foam and is referred to as then MPC-0 boundary condition case in the present work.

Another possibility of the applied boundary condition of the RVE is when MPC is applied on only one plane of the RVE, as shown in Fig. 6b (here, MPC is applied to a plane parallel to x–z plane). This case represents a situation where the cell has a free displacement at one boundary, as the case when it is located on the boundary of the finite dimensional foam, and is referred as the MPC-1 boundary condition here. In this situation, as illustrated in Fig. 6b, the cell shows displacement of one of its walls along a direction orthogonal to the applied load: unidimensional flatness under tension and unidimensional buckling under compression.

The last possibility of an applied boundary condition for the RVE is the three-dimensional extension of the MPC-1 case, as shown in Fig. 6c. In this situation, no boundary condition is applied. This way, the walls are free to show displacement in both orthogonal directions with the applied load or displacement. This is the case

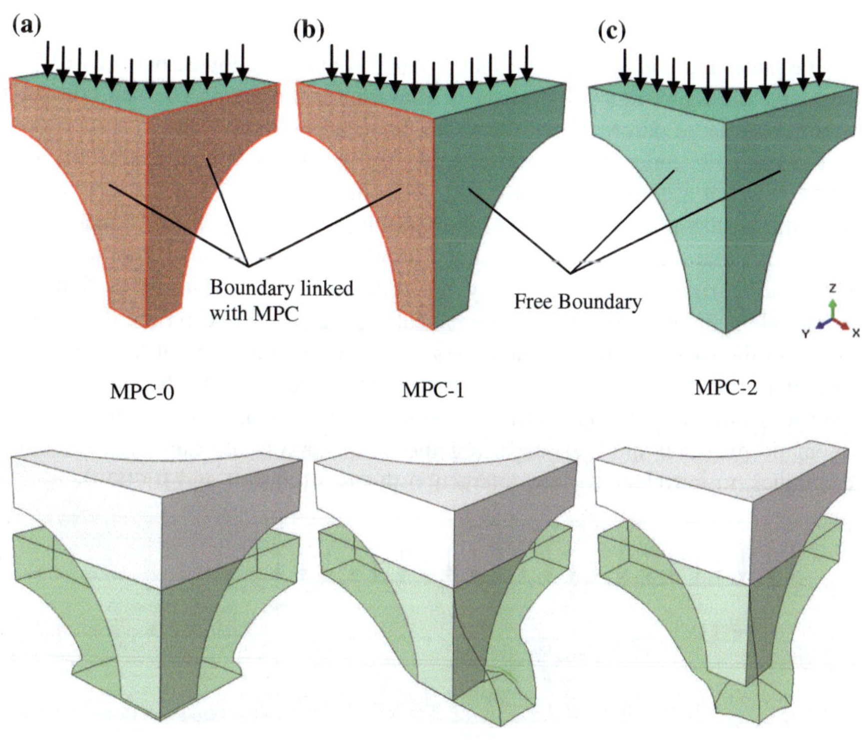

Fig. 6 Results of different applied boundary conditions

for some cells on the boundary of the finite dimensional foam. The behavior of the RVE in this case is two-dimensional flatness under tension and two-dimensional buckling under compression and is referred as the MPC-2 boundary condition case in the present work.

To determine the full behavior of a generic finite dimensional foam under compression using only a single RVE, we assume that each of the cells of the foam can develop only one of the three boundary conditions cases previously determined: MPC-0, MPC-1 or MPC-2. This condition is true for a simple compression case with absence of shear or imperfections on the foam. Furthermore, we assume that there are $m \times n \times p$ cells on the foam oriented with the x, y and z axes, respectively. The compression direction is oriented following the z axes.

2.3 Contact Hypothesis

Considering the contact between each part of the cellular metal perfect, the implementation of the self-contact can be imposed considering the symmetry of the problem. The compression is simulated using three different bodies: the foam and two rigid plates (see Fig. 7). The first plate stands static and an imposed displacement is applied on the other, compressing the foam between them, simulating an experimental compression test.

The plates are considered rigid and a 'hard' contact is allowed between each of them and the foam, and self-contact is allowed for the foam. The self-contact is important at large deformations, when the foam becomes crushed.

The used FE code (ABAQUS) provides a formulation for modeling the interaction between a deformable body and an arbitrarily shaped rigid body that may move during the analysis. A finite-sliding formulation is used where separation and sliding of finite amplitude and arbitrary rotation of the surfaces may arise. The finite-sliding rigid contact capability is implemented by means of a family of contact elements that ABAQUS automatically generates based on the data associated with the user-specified contact pairs. At each integration point these elements construct a measure of overclosure (penetration of the point on the surface of the deforming body into the rigid surface) and measures of relative shear sliding. These kinematic measures are then used, together with appropriate Lagrange multiplier techniques, to introduce surface interaction theories (contact and friction).

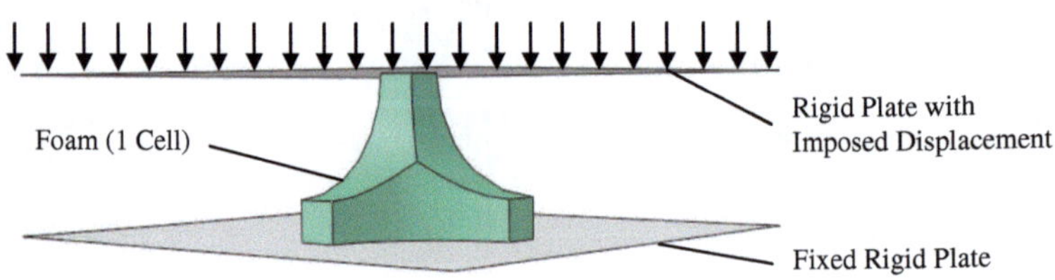

Fig. 7 Bodies considered for the compression simulation

3 Generalization of the Buckling Effects on a Finite Dimensional Cellular Metal

The behavior of a finite dimensional cellular metals for a pure compression case is determined by means of simple RVE compression cases here. For this, a generic foam with $m \times n \times p$ cells is considered as detailed in Sect. 2.2. Each of the cells on the foam can have only 3 different types of buckling effects. The cells can have a total buckling effect, like the behavior observed with the application of the MPC-2 boundary condition, a partial buckling effect, like that one obtained with the application of the MPC-1 boundary condition on the RVE or no buckling effects like that obtained with the MPC-0 boundary condition (see Fig. 6). The most outward cells in the corners of the structure have stronger buckling effects in two directions because there are no restrictions that prevent the deformations (MPC-2). The outward cells between the corners of the structure have a unidirectional buckling effect because there is one degree of freedom restricted in this configuration (MPC-1). The internal cells have all degrees of freedom restricted, and no buckling effects occur (MPC-0).

The stress-strain curve for a finite dimensional foam must lie between the two curves obtained for the extreme cases (MPC-0 and MPC-2). It is an intermediary case representing a specimen with a certain number of cells. Accepting that only the 3 cases proposed govern the overall behavior of the cellular metal, the number of cells and their respective type of buckling effect are needed to characterize the overall behavior. Assuming that the full behavior can be obtained as a linear combination of the different buckling effects that can be developed on the foam, then, the stress $\sigma_{m\times n}(\varepsilon)$ for a $m \times n \times p$ foam as function of the strain can be expressed as:

$$\sigma_{m\times n}(\varepsilon) = \frac{n_2\sigma_2(\varepsilon) + n_1\sigma_1(\varepsilon) + n_0\sigma_0(\varepsilon)}{n_2 + n_1 + n_0} \tag{7}$$

where n_2, n_1 and n_0 are the number of cells on the finite dimensional foam that display the same behavior observed for the MPC-2, MPC-1 and MPC-0 boundary condition cases applied to the RVE, respectively and $\sigma_2(\varepsilon)$, $\sigma_1(\varepsilon)$ and $\sigma_0(\varepsilon)$ are the stress value for a given strain ε obtained for the MPC-2, MPC-2 and MPC-0 analysis cases, respectively. The sum in the denominator is also the number of cells in the structure.

To simplify the application of Eq. (7), the concept of Representative Unit Cells, (RUC) is used. A group of unit cells is chosen to represent the symmetry of the cellular structure, wherein all the other unit cells in the structure may be generated by translations of the representative unit cell integral distances along each of its edges. Thus, the RUC is the basic structural unit or building block of the cellular structure and defines its behavior by virtue of its geometry and proportion of the buckling effects cases.

The determination of the RUC is made for a cellular metal whose applied load follow the direction in which the p cells of the foam are aligned (see Fig. 8). As it is considered that the foam is compressed by a plane on the top face and limited by another plane in the bottom face (with contact conditions for both planes and the foam and the foam itself), then, each $m \times n$ cells for a given p position show exactly the same buckling behavior of other $m \times n$ cells for any $p + 1$ or $p - 1$ position. As a result, the compression behavior (RUC) is characterized by a single $m \times n$ cell group. This is in accordance with Eq. (7) because adding any number of $m \times n$ cells with the same proportion of different buckling effects implies in the multiplication of the numerator and denominator by the same factor.

Figure 8 shows the parameters used to determine the RUC. The number of cells with two free boundary faces is 4 (purple cells on Fig. 8), so $n_2 = 4$, with only one is $2(m-2+n-2)$ (green cells on Fig. 8), so $n_1 = 2m + 2n - 8$ and there are $(n-2)(m-2)$ cells with no free boundaries (orange cells on Fig. 8) and then $n_0 = nm - 2n - 2m + 4$, with $m, n \geq 2$ as even numbers. The total number of cells analysed with the RUC is $n_o + n_1 + n_2 = mn$. These terms can be substituted in Eq. (7), resulting in:

$$\sigma_{m\times n}(\varepsilon) = \left[\frac{4}{mn}\right]\sigma_2(\varepsilon) + \left[\frac{2}{n} + \frac{2}{m} - \frac{8}{mn}\right]\sigma_1(\varepsilon) + \left[1 - \frac{2}{m} - \frac{2}{n} + \frac{4}{mn}\right]\sigma_0(\varepsilon) \quad (8)$$

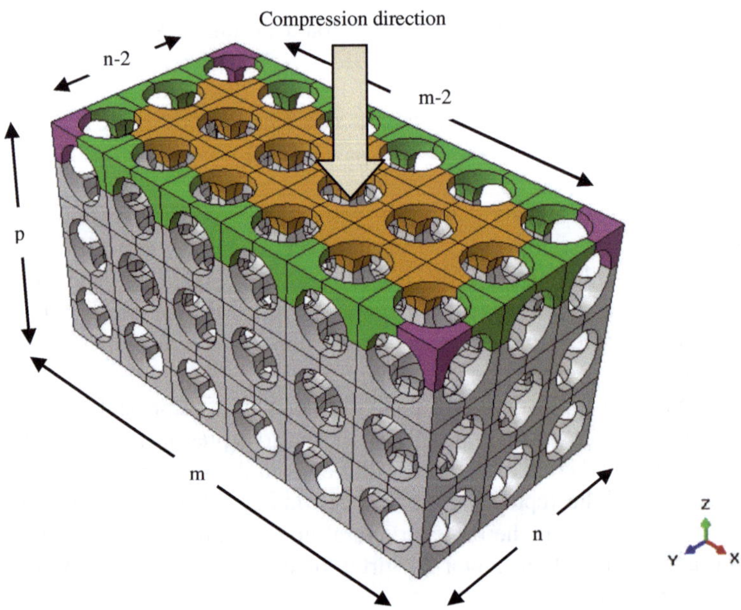

Fig. 8 Parameters used to determine the RUC configuration

This equation expresses the stress for a given strain that a finite dimensional foam exhibits in terms of the stress developed by the three RVE considered at the same strain level. The limit cases:

$$\lim_{m,n\to\infty} \sigma_{m\times n}(\varepsilon) = \sigma_0(\varepsilon) \tag{9}$$

$$\lim_{m,n\to 1} \sigma_{m\times n}(\varepsilon) = \sigma_2(\varepsilon) \tag{10}$$

represent the behavior of two RVE cases chosen as the limit cases, MPC-0 (infinite cells behavior) and MPC-2 (one single cell behavior), and are in agreement with the initial proposal (the second limit case can be obtained by taking $m, n \to 2$ without loss of generality). The integers m and n are considered both even numbers here, with the arrangement shown in Fig. 8, but an odd or mixed version of Eq. (8) can also be formulated resulting in a slight change of n_2 and n_1 proportions. Other important observations that can be done with Eq. (8) are the cases where only one dimension is predominant:

$$\lim_{m\to\infty} \sigma_{m\times n}(\varepsilon) = \left[\frac{2}{n}\right]\sigma_1(\varepsilon) + \left[1 - \frac{2}{n}\right]\sigma_0(\varepsilon) \neq \sigma_0(\varepsilon) \tag{11}$$

$$\lim_{n\to\infty} \sigma_{m\times n}(\varepsilon) = \left[\frac{2}{m}\right]\sigma_1(\varepsilon) + \left[1 - \frac{2}{m}\right]\sigma_0(\varepsilon) \neq \sigma_0(\varepsilon) \tag{12}$$

which shows explicitly that the behavior of a finite dimensional cell can not be correctly described in terms of an infinite cell case when only one dimension is truly large.

In order to validate the proposed formulation, the next section shows the simulation of the three RVU cases used and compares the results with those obtained with full finite element simulation and experimental results.

4 Simulation of a Compression and Model Validation

For the simulation, the geometry of the cells are obtained from a symmetrical double partition of a cube that contains 3 mm diameter holes with distance of 4 mm among them. The presence of holes simulates a cellular metal with a relative density of 0.2712. The considered material properties are: Elastic Modulus E = 72.7 GPa, Poisson's ratio $\nu = 0.34$, initial yield stress value of $\sigma_y^0 = 250\text{MPa}$ and final yield stress value $\sigma_y^\infty = 410\text{MPa}$. Hardening is taken into account by the relation $\sigma_y = \sigma_y^0 + (\sigma_y^\infty - \sigma_y^0)[1 - \exp(-k\varepsilon^p)]$, with k = 25. The Gurson model of the ABAQUS code is used and the parameters considered are 5 % of initial porosity, $\alpha_1 = 1.5$ and $\alpha_2 = 1.5$. The material parameters are obtained from experimental results [1, 11–15].

Load is applied as a vertical displacement on the top rigid surface in contact with the cell and the bottom surface is kept with no displacement. Contact between the surfaces and the cell and self-contact of the cell is considered.

Firstly, the three fundamental cases with one cell are detailed. Then, the obtained results of those simulations are used to predict the behavior of a finite dimensional cell and compared with the results obtained with full finite element simulation. After, an experimental simulation is compared with the results obtained with the proposed framework.

4.1 Fundamental Buckling Cases

The single cell model of Fig. 3c is studied for the three MPC boundary conditions. Figure 9 shows a comparison among macroscopic stress and strain plots for the three cases. The macroscopic stress is the relation between the resultant forces and the original solid surface area. The macroscopic strain is the ratio between change of length and original length. Figure 10 shows the von Mises stress distribution for the MPC-0, Fig. 11 for the MPC-1 and Fig. 12 for the MPC-2 case for an applied macroscopic strain of 25 and 50 %.

It can be seen that the first case MPC-0 possibly overestimates the stiffness of a finite dimensional model. For this case, due to the nature of the restriction (see Sect. 2.2), both of the side walls remain vertical. This effect is more significant after 2.5 % macroscopic strain, when buckling occurs in the walls between the holes (see Fig. 10).

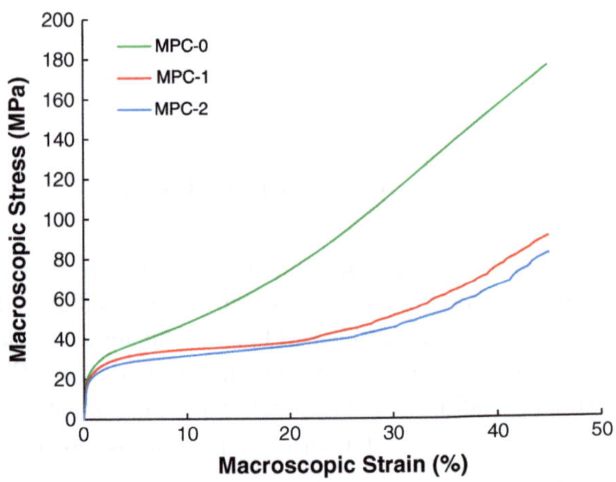

Fig. 9 Stress strain plots for the three MPC boundary condition cases

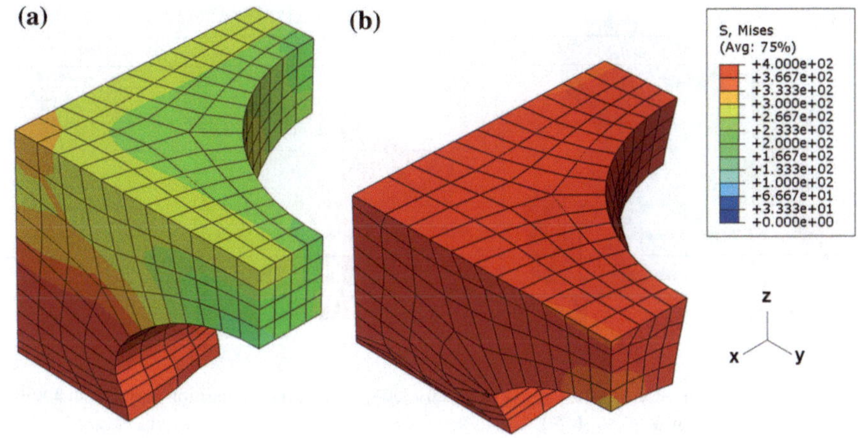

Fig. 10 Distribution of von Mises stress (MPa) for MPC-0 boundary condition case for an applied macroscopic strain of **a** 25 %, **b** 50 %

The MPC-1 case is shown in Fig. 11. This one has less stiffness in compression due to the fact that only one wall suffers strong buckling effects and the other remains vertical.

The last type of applied boundary condition, the MPC-2 case (see Fig. 12) possibly underestimates the stiffness of a finite dimensional foam because in this case all the walls suffer strong buckling effects in two different directions, which is not the predominant behavior of the real specimens.

A finite dimensional case can be determined by using Eq. (8) in combination with the stress-strain curves of Fig. 9. Some results are plotted in Fig. 13. The stiffness increases as the number of cells increases in one or two directions (it is assumed to be independent of the third direction because load is applied along it,

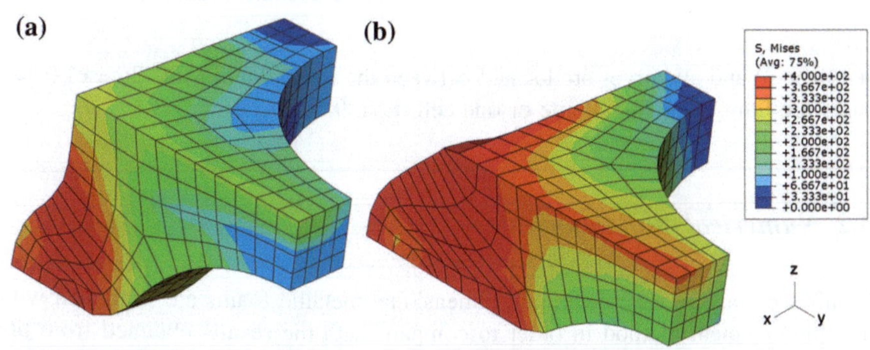

Fig. 11 Distribution of von Mises stress (MPa) for MPC-1 boundary condition case for an applied macroscopic strain of **a** 25 %, **b** 50 %

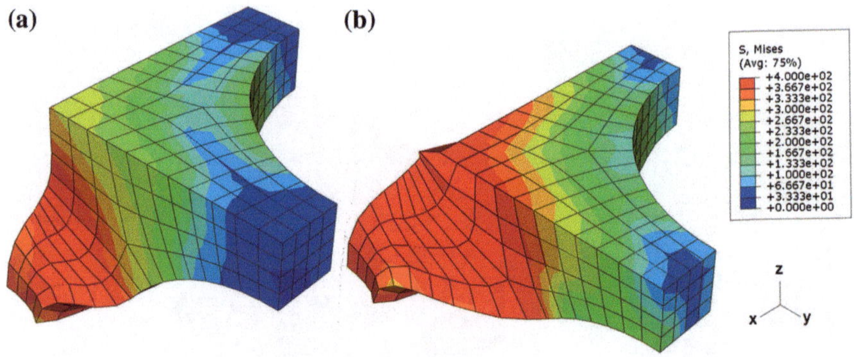

Fig. 12 Distribution of von Mises stress (MPa) for MPC-2 boundary condition case for an applied macroscopic strain of **a** 25 %, **b** 50 %

Fig. 13 Stress strain plots for some finite dimensional foam cases using the present analytical and numerical framework, Eq. (8)

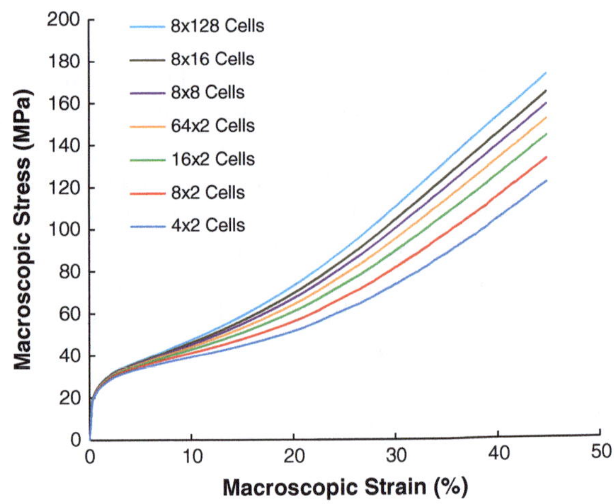

see Sect. 3) and all curves are located between the MPC-0 and MPC-2 cases (such cases are limit cases for infinite or one cell, Eqs. 9, 10).

4.2 Numerical Validation

In this section, two complete finite dimensional metallic foams are simulated with the finite element method in order to compare with the results obtained from the analytical and numerical framework proposed. The first case is a 4×6 foam, and the results obtained are shown in Fig. 14 for the von Mises stress distribution for an applied macroscopic strain of 25 and 50 %.

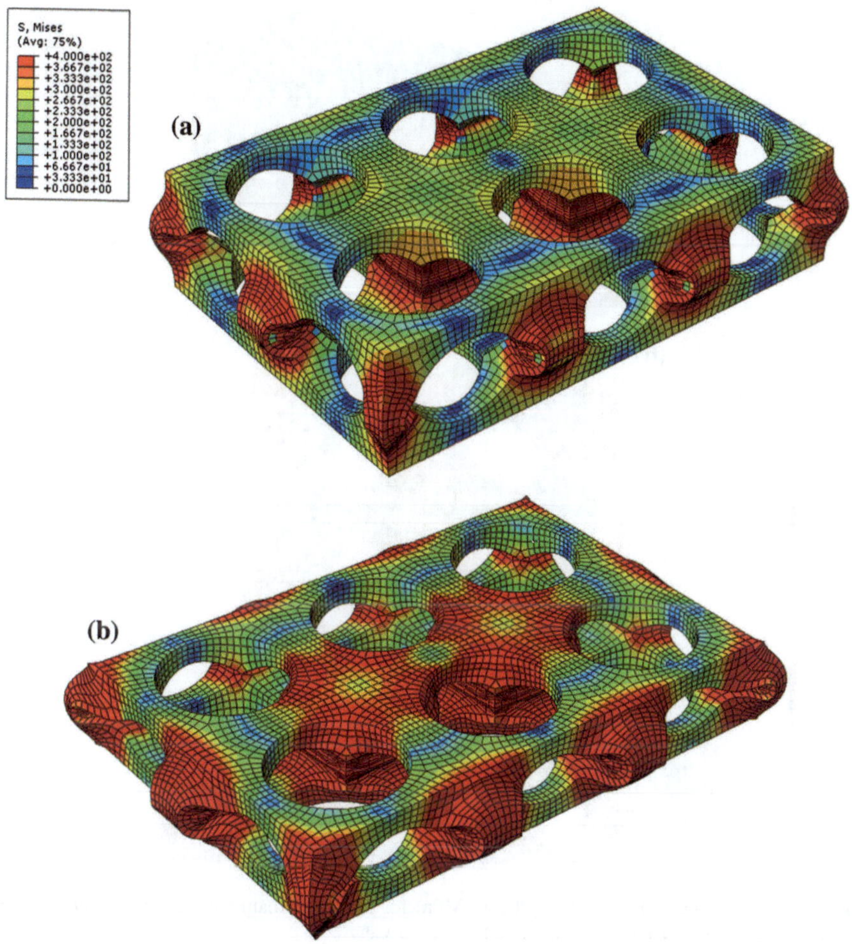

Fig. 14 Distribution of von Mises stress (MPa) for a 4 × 6 foam for an applied macroscopic strain of **a** 25 %, **b** 50 %

The results for the second case, a 4 × 12 foam, are shown in Fig. 15 for the von Mises stress distribution for an applied macroscopic strain of 25 % in a perspective, a frontal view and a lateral view. It can be observed that all the three types of buckling cases are fully characterized in the foam structure.

A comparison among macroscopic stress and strain plots for those two simulations and the results obtained using the present framework are shown in Fig. 16. The results obtained by the full finite element simulation are in accordance with those evaluated by the present methodology. However, it is important to highlight the strong difference in computational resources required to achieve these results in each case. While the present framework uses data obtained from simulations that take only a few minutes to be performed, the full finite element simulation takes

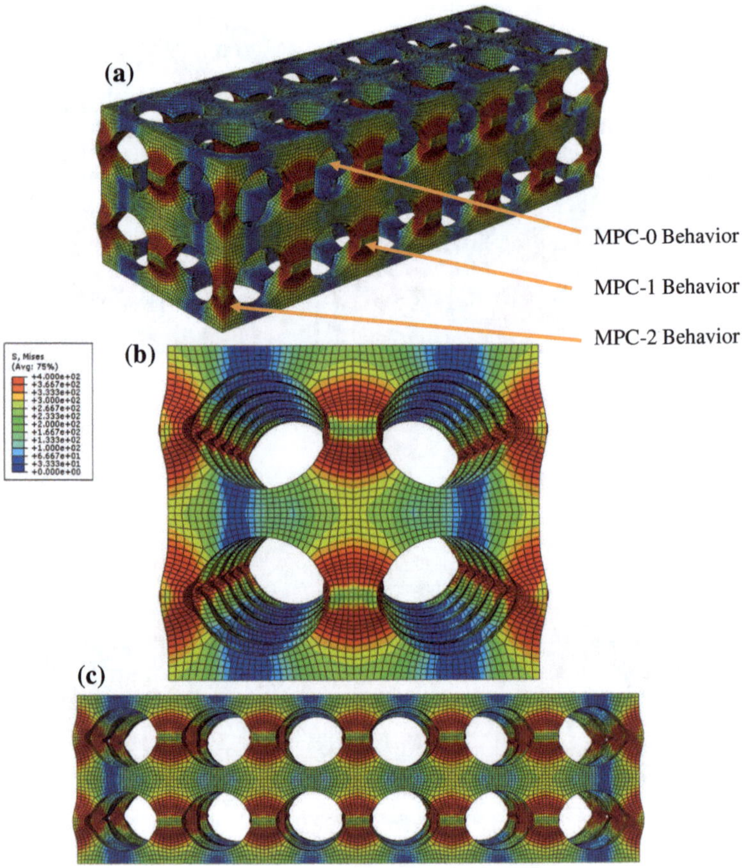

Fig. 15 Distribution of von Mises stress (MPa) for a 4 × 12 foam for an applied macroscopic strain of 25 % in **a** perspective, **b** frontal, **c** lateral view

several hours for the 4 × 6 case and days for the 4 × 12 case. Cases involving a larger number of cells are computationally prohibitive for full finite element simulation. The damage considerations and the contact formulation are the most costly of the simulation, and these costs grow rapidly as the number of cells grows. Finally, as the results obtained are close, the present framework is a more attractive toll for determining the compression behavior of metallic foams instead of full simulation without loss of precision.

4.3 Experimental Investigation

This last section contains an investigation through experimental analysis. A real 14 × 14 × 14 cell foam as that shown in Fig. 4 is used here, with the same material

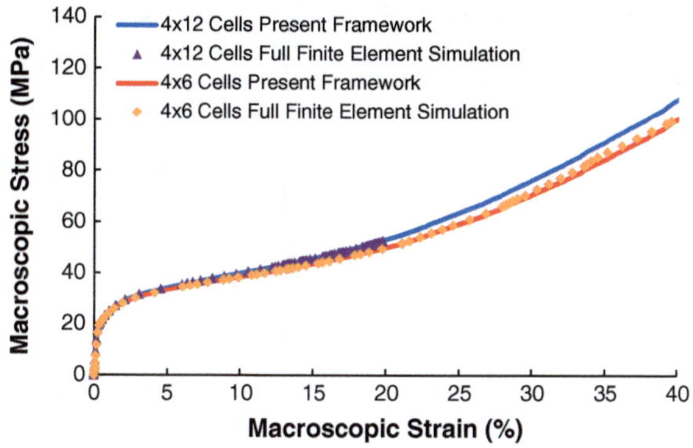

Fig. 16 Stress strain plots for some finite dimensional foam cases using the present analytical and numerical framework, Eq. (8), and full finite element simulation

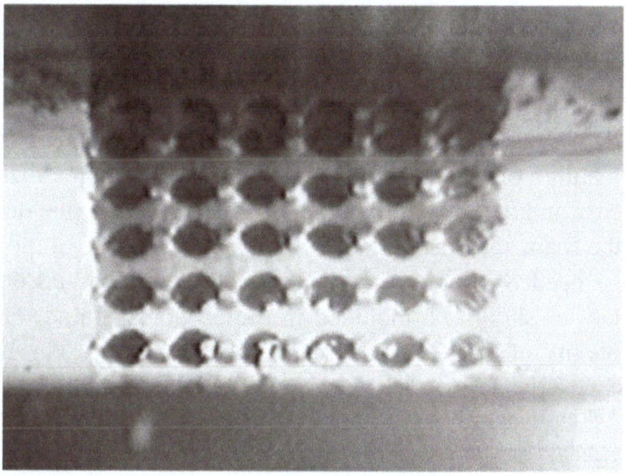

Fig. 17 Picture showing buckling process at the walls between holes

properties as that described in Sect. 4. A similar compressed specimen is shown in Fig. 17. A strong correlation can be observed with the simulation presented in Fig. 17 where the same pattern of buckling is observed.

However, the experiment also presented a cracking effect, which can be observed in Fig. 18 that shows a comparison among macroscopic stress and strain plots for the experiment and the results obtained using the present framework. The loss of stiffness which comes with a decrease of stress for the experimental data is due to the propagation of cracking through a set of cells. This effect can be generated due to imperfections and more experimental investigation is needed for a fully understanding of this observed difference.

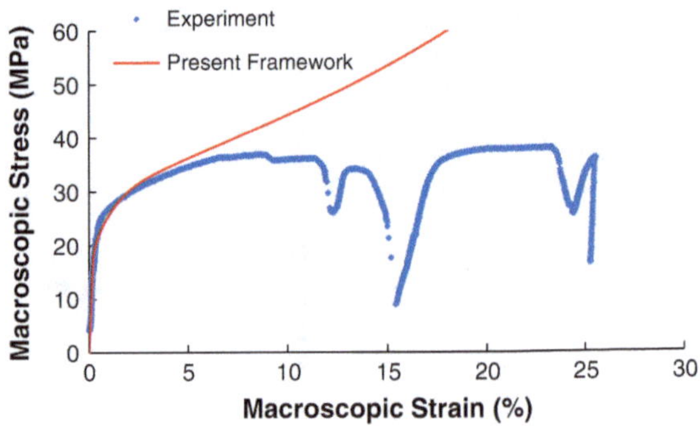

Fig. 18 Stress strain plots for the experiment analysis and the results obtained from the present framework

5 Final Comments

The analytical and numerical framework proposed in the present work based on the study of the internal buckling effects acting in the structure to determine the full behavior of a finite dimensional foam under compression showed agreement with numerical investigation through a full finite element simulation. The more attractive features of the present formulation are the generalization for a finite dimensional case and the low computational cost required to the determination of the behavior of the foam specimen under compression. The use of more experimental results should be needed to confirm the numerical approach and other nonlinearities and irregularities should be included in the proposed framework in order to refine the complexity of the simulation capability.

Acknowledgments We thank FAPERGS, CNPq, CAPES and PROPESQ-UFRGS for continuous support of our research projects.

References

1. Öchsner A, Lamprecht K (2003) On the uniaxial compression behavior of regular shaped cellular metals. Mech Res Commun 30:573–579
2. Shulmeister V (1998) Modelling of the mechanical properties of low-density foams. Delft University of Technology, Proefshrift
3. Hallström S, Ribeiro-Ayeh S (2005) Stochastic Finite Element Models of foam materials. In: Thomsen OT, Bozhevolnaya E, Lyckegaard A (eds) Sandwich structures 7: advancing with sandwich structures and materials, 1st edn. Springer Aalborg, Aalborg
4. Grenestedt JL, Tanaka K (1999) Influence of cell shape variations on elastic stiffness of closed cell cellular solids. Scripta Mater 40(1):71–77
5. Duvaut G (1984) Homogeneisation et materiaux composites. In: Ciarlet PG, Roseau M (eds) Trends and applications of pure mathematics to mechanics (Proceedings of Lecture notes in physics, vol 195), 1st edn. Springer, Berlin
6. Xia Z, Zhang Y, Ellyin F (2003) A unified periodical boundary conditions for representative volume elements of composites and applications. Int J Solids Struct 40:1907–1921

7. Öchsner A, Winter W, Kuhn G (2003) On an elastic-plastic transition zone in cellular metals. Arch Appl Mech 73:261–269
8. Gurson AL (1977) Continuum theory of ductile rupture by void nucleation and growth: part I-yield criteria and flow rules for porous ductile media. J Eng Mater T ASME 99:2–15
9. Tvergaard V (1981) Influence of voids on shear band instabilities under plane strain conditions. Int J Fract 17:389–407
10. ABAQUS (1992) Theory manual v. 5.2. Hibbitt, Karlsson & Sorensen Inc, Providence
11. Munoz-Rojas PA, Fiedler T, Cunda LAB, Öchsner A, Creus GJ (2007) Parameter identification to simulate a traction test applying Gurson damage model. In: Proceedings of the CMNE 2007—Congress on Numerical Methods in Engineering/CILAMCE 2007—XXVIII Latin-American Congress on Computational Methods in Engineering
12. Oliveira BF, Cunda LAB, Öchsner A, Creus GJ (2006) Gurson damage model: applications to case studies. In: Proceedings of the CILAMCE 2006—XXVII Iberian-Latin American Congress on Computational Methods in Engineering
13. Linn RV, Oliveira BF (2008) Finite element simulation of compression behavior of cellular. In: Proceedings of the CILAMCE 2008—XXIX Iberian-Latin American Congress on Computational Methods in Engineering
14. Oliveira BF, Cunda LAB, Öchsner A, Creus GJ (2008) Comparison between RVE and full mesh approaches for the simulation of compression tests on cellular metals. Materialwiss Werkst 39(2):133–138
15. Oliveira BF, Cunda LAB, Öchsner A, Creus GJ (2007) Finite element simulation of compression tests on cellular metals. In: Proceedings of the CILAMCE 2007—XXVIII Iberian-Latin American Congress on Computational Methods in Engineering

Modelling of the Surface Morphology by Means of 2D Numerical Filters

Andrzej Golabczak, Andrzej Konstantynowicz and Marcin Golabczak

Abstract When examining a surface machined by means of a repeatingly acting tool, especially a cyclically acting, as in the case of rotating machines, a characteristic patter can be observed when analyzing the surface morphology. Such a patter can be recognized as a characteristic symptom of the cooperation between the tool and the machined element, being usually in the tight contact. Obtaining any quantitative results evaluating this cooperation, demands establishing of the well suited mathematical model. In this paper a relationship is presented based on the numerical two-dimensional model filters of the quadrant type, acting on a square lattice representing surface. This appears to be an efficient numerical tool, providing a sufficient number of degrees of freedom and a fast execution, even though special measures have to be taken to assure their stability. The filters have been excited with simple stochastic processes simulating randomness of the abrasive machining at its basic level. The obtained results have been examined mainly by to methods: by using classical statistical analysis and by using the surface autocorrelation idea. This second method has been proven to be a good tool for the quantitative evaluation of the surface interaction during the machining process.

Keywords Surface texture · Stochastic processes · DSP · Spatial filtering

A. Golabczak (✉) · A. Konstantynowicz
Department of Production Engineering, Lodz University of Technology,
Stefanowskiego 1/15 Street, 90-924 Lodz, Poland
e-mail: andrzej.golabczak@p.lodz.pl

A. Konstantynowicz
e-mail: andrzej.konst@gmail.com

M. Golabczak
Institute of Machine Tools and Production Engineering, Lodz University of Technology,
Stefanowskiego 1/15 Street, 90-924 Lodz, Poland
e-mail: marcin.golabczak@p.lodz.pl

A. Öchsner and H. Altenbach (eds.), *Mechanical and Materials Engineering of Modern Structure and Component Design*, Advanced Structured Materials 70,
DOI 10.1007/978-3-319-19443-1_10

1 Stochastic Processes Basics

The term “stochastic process” term, often referred to as “random process” is a set of random variables (random varieties) representing the development of some system state over the independent variable *X*, very often representing time. In our case it represents the spatial variable. The exact mathematical definition is of less value for non-mathematicians and therefore we present in Fig. 1 a graphical representation of all the necessary elements and mutual dependencies. Two elements play the main role [7, 8]:

- the stochastic (random) mechanism generating the exact process value which is in our case simply the height of the roughness profile at the given X variable value which is in our case simply the point in physical space—depicted in green,
- assumed to be non-stochastic, a mechanism of the internal dependency between subsequent elementary events which in our case determine the similitude of roughness height in the juxtaposed spatial points—depicted in olive.

The first element is described by the probability density distribution, which is usually chosen from the small, widely used set. In this paper we will make primary use of the normal (Gaussian) distribution:

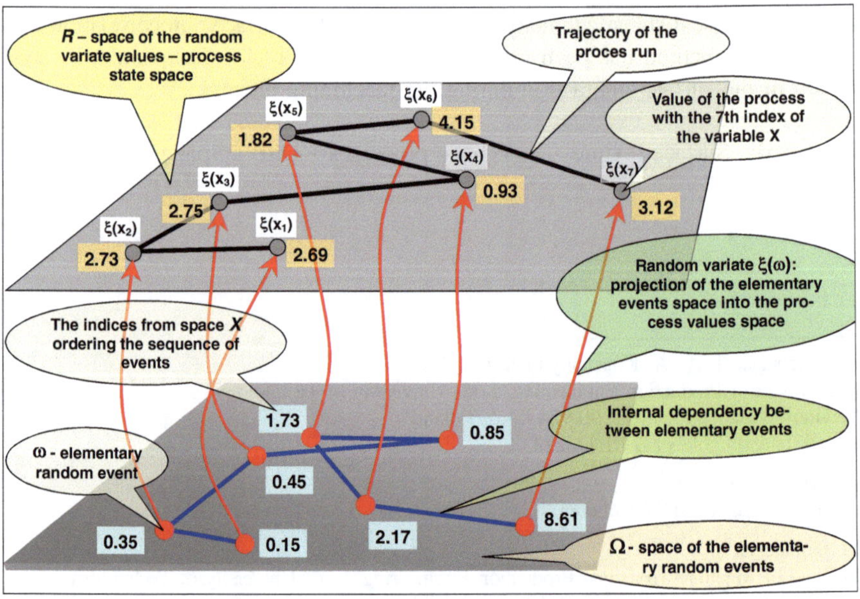

Fig. 1 The structure of mutual dependencies among the basic constituents of the “stochastic process”, as it will be used in the present paper

$$p(h) = \frac{1}{\sqrt{2\pi}\sigma} \exp\left(\frac{(h-\mu)^2}{2\sigma^2} \right) \tag{1}$$

where μ—mean value of the variable h, and σ^2—variance of the variable h.

The second distribution we will take into account is the logarithmic-normal distribution, called shortly "log-normal", with the probability density:

$$p(h) = \frac{1}{\sqrt{2\pi} \cdot \sigma \cdot h} \exp\left(\frac{(\ln(h)-\mu)^2}{2\sigma^2} \right) \tag{2}$$

where the description of variables is as previously stated. The relationship between the variable Y of the log-normal distribution and the N_0 variable of the normal distribution with $\mu = 0$, $\sigma^2 = 1$, is given by:

$$Y = \exp\left(k_\mu + k_\sigma \cdot N_0\right) \tag{3}$$

where

$$k_\mu = \ln\left(\frac{\mu_{\ln}^2}{\sqrt{\sigma_{\ln}^2 + \mu_{\ln}^2}} \right) \tag{4}$$

$$k_\sigma = \sqrt{\ln\left(1 + \frac{\sigma_{\ln}^2}{\mu_{\ln}^2}\right)} \tag{5}$$

For modelling purposes we have to be equipped with an efficient tool to generate variables with prescribed probability densities. The basic tool is the random number generator with uniform probability density, usually provided as embedded in the numerical software. Making use of simple mathematical rules related to the transformation of random variables we can use the following equation to get the first random variables N1 with a normal probability distribution from random variables U_i with uniform probability distribution:

$$\begin{aligned} N_1 &= \sqrt{-2\ln(U_1)} \cdot \cos(2\pi \cdot U_2) \\ N_2 &= \sqrt{-2\ln(U_1)} \cdot \sin(2\pi \cdot U_2) \end{aligned} \tag{6}$$

This "twin generation" is very often used to obtain two "orthogonal" random variables, which, for our purpose, are the independent random variables.

2 Surface Modeling Using Digital Filters

In the classical division of the surface texture into three major components: shape, waviness and roughness [5], the application of digital filters for modelling allows us to successfully link two of these components: waviness and roughness. The way this is achieved is by using the aforementioned stochastic process theory and the application of digital filters to model the "internal dependency among spatial events". This provides researchers with a tool to model the waviness, as seen in the olive mark in Fig. 1 [1, 2, 4]. Also the use of random signal generators to introduce the "unpredictability" element in modelling the roughness, as seen in Fig. 1 in light green [1, 7, 8], is also used to achieve this link. In this paper we have aimed at applying digital filters [2, 4, 6, 8], referring reader, for example, to the classical textbook, in the case [3] tools to accomplish it.

3 Digital Filtering Basics

So called digital filter are given by the numerical algorithm operating on samples of the signal, expressed as numbers. Usually one train of samples is referred to as the input—independent signal, and the second train of pulses is referred to as the output—dependent signal. These signals could be ordered by integer indices. If there is one index—the signal is unidimensional, if two indices are necessary—the signal is two-dimensional. The object of our interest is a special type of two-dimensional digital filter described in terms of the recursive filter equation [2, 6], in the "spatial" domain:

$$\begin{aligned} h(i,j) = {} & A_{w0} \cdot x_w(i) + A_{w1} \cdot h(i-1,j) + A_{w2} \cdot h(i-2,j) \\ & + A_{d0} \cdot x_d(j) + A_{d1} \cdot h(i,j-1) + A_{d2} \cdot h(i,j-2) \end{aligned} \tag{7}$$

where

A_{w0}, A_{d0}	input amplitude coefficients: width and depth respectively,
$x_w(i), x_d(j)$	input excitation series: width and depth respectively,
A_{w1}, A_{w2}	width spatial memory coefficients for retarded samples,
A_{d1}, A_{d2}	depth spatial memory coefficients for retarded samples,
$h(i,j)$	output surface shape value for the point (*i*, *j*).

The filter structure is depicted in Fig. 2. Input signals are marked with light green, initial conditions' cells are marked with green and the active cells are marked with light yellow. In the active cells the filter equation is inscribed to show changes in indices. The dark green marked cells are involved during the calculation of the exemplary cell (4, 4) value. The whole filter is organized on the 100 × 100 grid, which is not big enough for the targeted purpose, but is more than enough to show the filter behavior.

	X	0	1	2	3	4
Y	Xinput, Yinput	x(0)	x(1)	x(2)	x(3)	x(4)
0	y(0)	h(0,0)	h(1,0)	h(2,0)	h(3,0)	h(4,0)
1	y(1)	h(0,1)	h(1,1)	h(2,1)	h(3,1)	h(4,1)
2	y(2)	h(0,2)	h(1,2)	h(2,2)=Aw0*x(2)+Aw1*h(1,2)+Aw2*h(0,2)+ Ad0*y(2)+Ad1*h(2,1)+Ad2*h(2,0)	h(3,2)=Aw0*x(3)+Aw1*h(2,2)+Aw2*h(1,2)+ Ad0*y(2)+Ad1*h(3,1)+Ad2*h(3,0)	h(4,2)=Aw0*x(4)+Aw1*h(3,2)+Aw2*h(2,2)+ Ad0*y(2)+Ad1*h(4,1)+Ad2*h(4,0)
3	y(3)	h(0,3)	h(1,3)	h(2,3)=Aw0*x(2)+Aw1*h(1,3)+Aw2*h(0,3)+ Ad0*y(3)+Ad1*h(2,2)+Ad2*h(2,1)	h(3,3)=Aw0*x(3)+Aw1*h(2,3)+Aw2*h(1,3)+ Ad0*y(3)+Ad1*h(3,2)+Ad2*h(3,1)	h(4,3)=Aw0*x(4)+Aw1*h(3,3)+Aw2*h(2,3)+ Ad0*y(3)+Ad1*h(4,2)+Ad2*h(4,1)
4	y(4)	h(0,4)	h(1,4)	h(2,4)=Aw0*x(2)+Aw1*h(1,4)+Aw2*h(0,4)+ Ad0*y(4)+Ad1*h(2,3)+Ad2*h(2,2)	h(3,4)=Aw0*x(3)+Aw1*h(2,4)+Aw2*h(1,4)+ Ad0*y(4)+Ad1*h(3,3)+Ad2*h(3,2)	h(4,4)=Aw0*x(4)+Aw1*h(3,4)+Aw2*h(2,4)+ Ad0*y(4)+Ad1*h(4,3)+Ad2*h(4,2)

Fig. 2 Proposed filter structure with excitation and initial conditions setting. Cells marked with *dark green color* are involved during the calculating the exemplary cell value

For any digital filter it is essential to determine the conditions of stable work. The appropriate theoretical considerations could be found in positions [2, 6]. For the unidimensional filters they are relatively simple, but for two-dimensional filters it becomes more complicated. Even in the case of relatively simple filters proposed in this work, the exact numerical evaluation is complicated. Two theorems for the two-dimensional filter stability have been developed:

Theorem I *The two-dimensional filter with indefinite impulse responses (recursive), with the characteristics described by the rational polynomial complex function:* $H(z_1, z_2) = \frac{N(z_1, z_2)}{D(z_1, z_2)}$ *is stable when and only when* $D(z_1, z_2) \neq 0$ *for any* z_1, z_2 *for which* $|z_1| \geq 1, |z_2| \geq 1$. *This situation is illustrated in* Fig. 3 *for the exemplary filter we deal with further in this paper.*

Theorem II *The two-dimensional filter with rational polynomial characteristics, as in Theorem I, is stable when and only when the projection of the plane* $|z_1| = 1$ *onto the plane* Z_2, *is equivalent the equation* $D(z_1, z_2) = 0$, *and is located entirely*

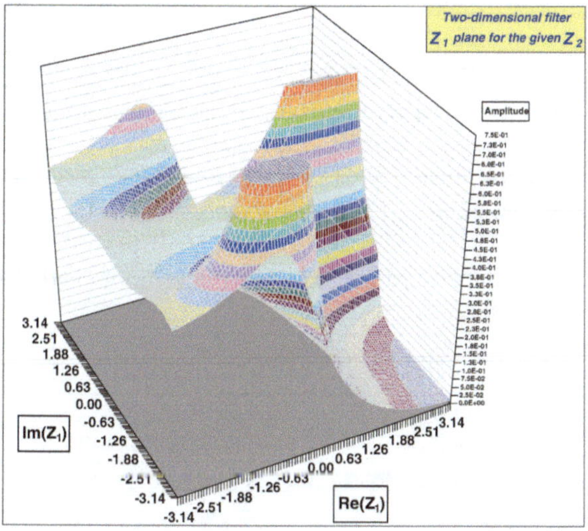

Fig. 3 The exemplary checking of the stability criterion I of the two-dimensional filter

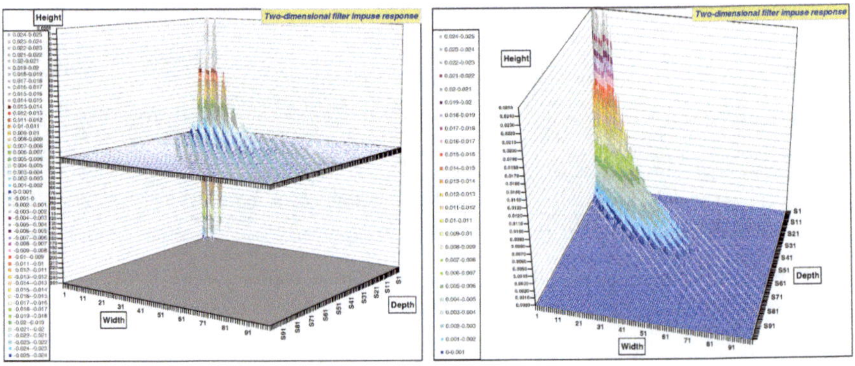

Fig. 4 The exemplary impulse response of the 2D filter (7) for the given coefficients: two different views

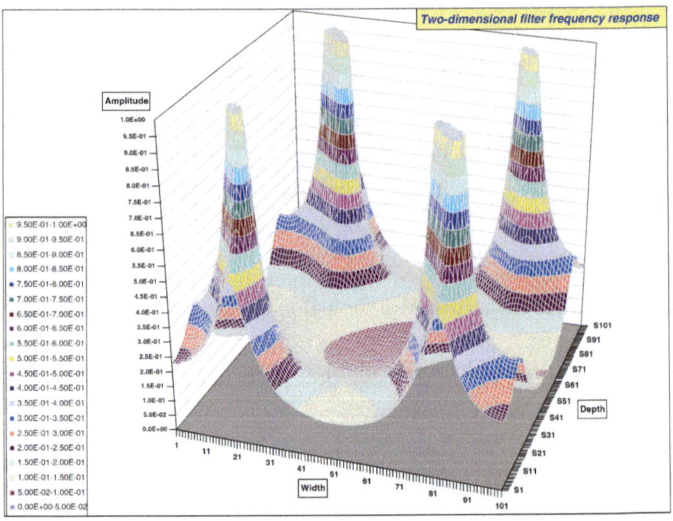

Fig. 5 The exemplary filter impulse response in the frequency domain

inside the plane $|z_2| \geq 1$, *and, the projection* $D(z_1, z_2) = 0$ *does not reflect any point from the plane* $|z_1| \geq 1$ *onto the point* $|z_2| = 0$.

The impulse response at Fig. 4 has been charted for the exemplary filter coefficients (Fig. 5):

$A_{w0} = 0.100000$	$A_{w1} = 0.375000$	$A_{w2} = -0.425000$
$A_{d0} = 0.100000$	$A_{d1} = 0.575000$	$A_{d2} = -0.395000$
Exemplary grid step	$\Delta w = 1\ \mu m$, $\Delta d = 1\ \mu m$	
Dimension of the grid	$N \times N = 100 \times 100$	

The frequency response of the filter (7) is given as the Z-transformation of the filter impulse response taken at the unitary circle at the complex z-plane, i.e. for $z_w = \exp(j\omega_w \Delta w), z_d = \exp(j\omega_d \Delta d)$:

$$F(\omega_w, \omega_d) = \frac{A_{w0} + A_{d0}}{1 - A_{w1} \cdot e^{-j\omega_w \Delta w} - A_{w2} \cdot e^{-2j\omega_w \Delta w} - A_{d1} \cdot e^{-j\omega_d \Delta d} - A_{d2} \cdot e^{-2j\omega_d \Delta d}} \tag{8}$$

where

$\Delta w, \Delta d$ spatial sampling step of the filter grid: width and depth respectively,
ω_w, ω_d spatial frequencies: width and depth respectively, other coefficients as i Eq. (7).

4 Numerical Results

The numerical simulations we have performed, have aimed at demonstrating the usability and functionality of the proposed filter in generation surfaces widely occurring in surface machining at different stages of mechanical accuracy, performed with using very different tools ranged from raw milling to abrasive cloth polishing. The exemplary excitation source for surfaces depicted in Figs. 6 and 7

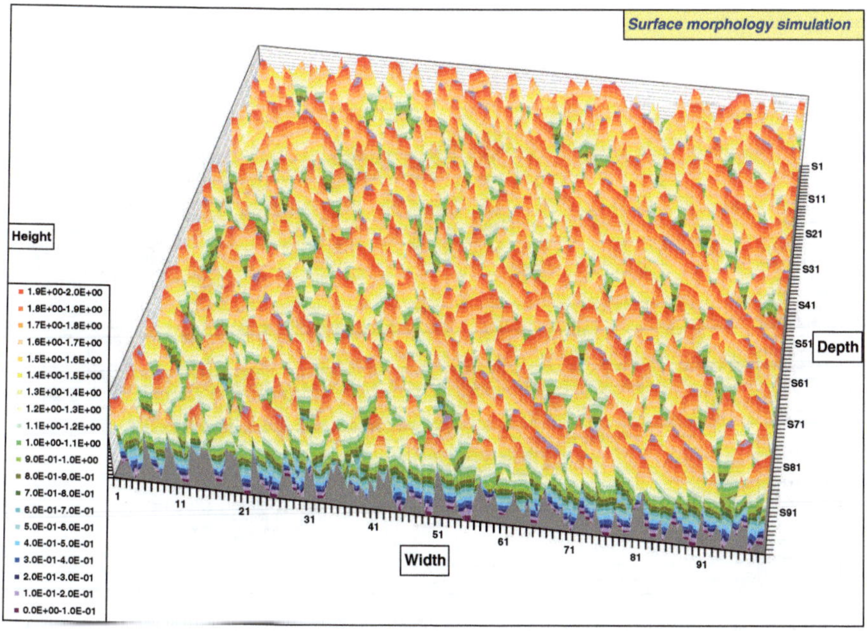

Fig. 6 Surface texture generated by using the filter of the impulse response depicted in Fig. 4

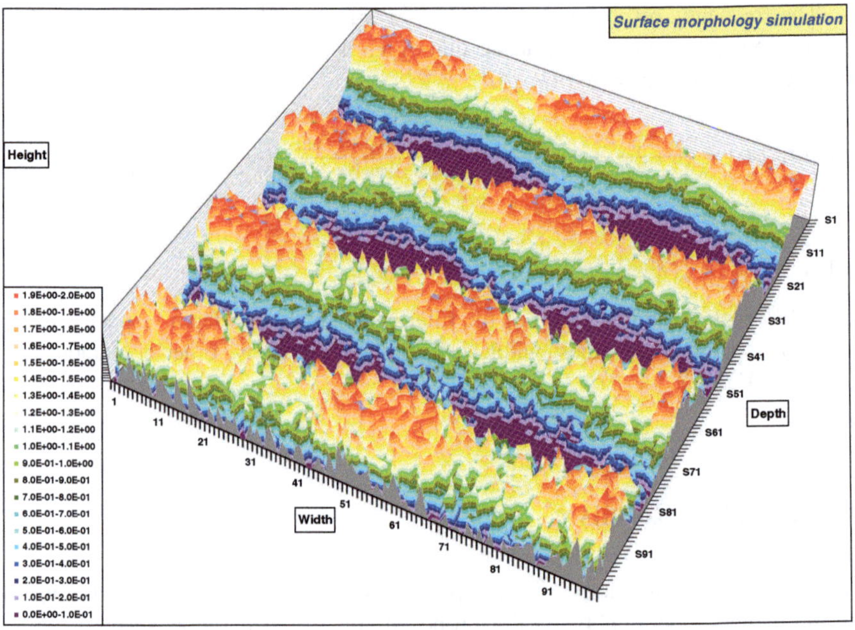

Fig. 7 Surface texture generated by using the filter of another impulse response, more suitable for modelling of the rotating machine rough machining

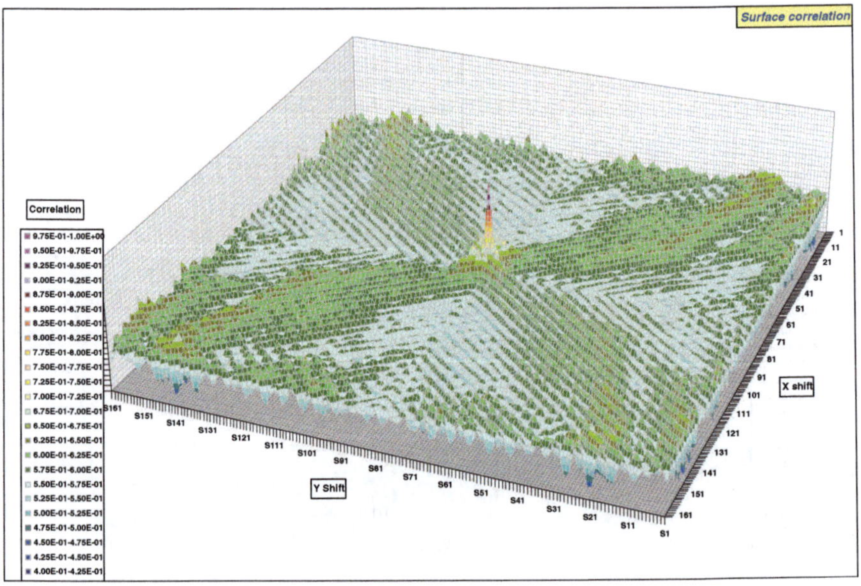

Fig. 8 Two-dimensional correlation function for the surface from Fig. 6

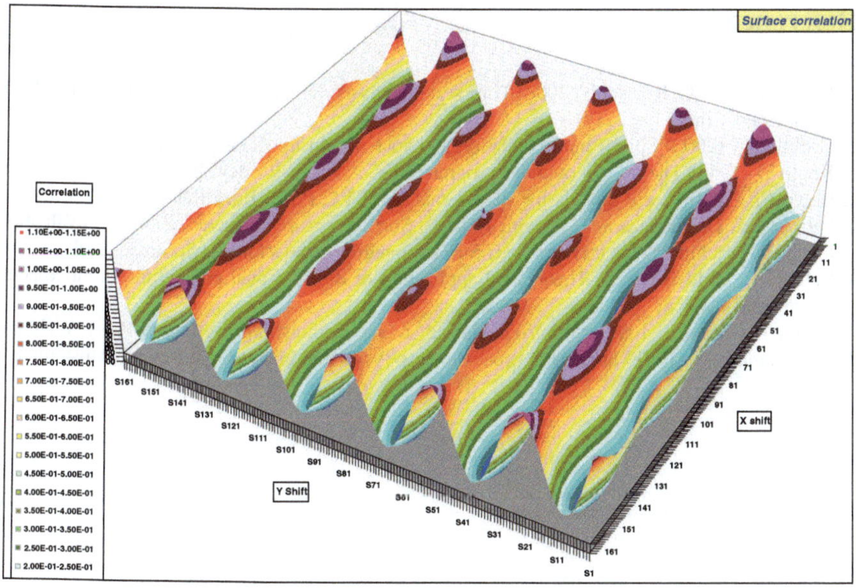

Fig. 9 Two-dimensional correlation function for the surface from Fig. 7

were Gaussian random number generators with variance parameters: Aw = 1.00, Ad = 0.01 respectively. The referenced surfaces exemplify abrasive cloth polishing and raw milling respectively.

For to bring out the proposed filter span of use, the two-dimensional autocorrelation function:

$$R(m,n) = \frac{1}{(N-m)\cdot(N-n)}\sum_{i=1}^{N-m}\sum_{j=1}^{N-n} h(i,j)\cdot h(m+i,n+j) \tag{9}$$

has also been calculated for the surfaces from Figs. 6 and 7 and depicted in Figs. 8 and 9.

5 Conclusions

The two-dimensional filter presented in this work, although relatively simple in its class, proved to be an efficient tool in generating surface textures from the broad range of textures encountered in the surface machining to different grades of finishing and performed with different tools. In our further works we'll tend to apply it to modelling of the machined surfaces we currently deal with: AZ31 magnesium alloy, Ti6Al4V titanium alloy and X38CrMoV5-1 steel.

References

1. Box G, Jenkins G, Reinsel G (2008) Time series analysis: forecasting and control. Prentice Hall, Englewood Cliffs
2. Griffith DA (2003) Spatial autocorrelation and spatial filtering. Springer, Berlin
3. Knuth D (1968) The art of computer programming. Addison-Wesley, Boston
4. Lloyd Ch (2010) Local models for spatial analysis. CRC Press, Boca Raton
5. Petropoulos G, Pandazaras C, Davim J (2010) Surface texture characterization and evaluation related to machining. In: Davim J (ed) Surface integrity in machining. Springer, Berlin
6. Rabiner L, Gold B (1975) Theory and application of digital signal processing. Prentice-Hall, Englewood Cliffs
7. Taylor H, Karlin S (1998) An introduction to stochastic modeling. Academic Press, Waltham
8. Van Kampen N (1992) Stochastic processes in physics and chemistry. Elsevier, Amsterdam

Modelling of the Roughness Profile by Means of the Autoregressive Type Stochastic Processes

Andrzej Golabczak, Andrzej Konstantynowicz and Marcin Golabczak

Abstract The 2D roughness profile resulting from the standard measurement using a mechanical profilometer is usually the basic examination of a machined surface. Obtained results, usually in the form of the statistical parameters' set are used for the surface machining evaluation as well as the forecast of the tribological behavior of the surface. The second mentioned purpose demands a particularly well suited mathematical model to accomplish a quantitative evaluation of the tribological parameters. In this paper a specific method is presented for this modelling based on the stochastic processes. In these processes the amplitude distribution has been modelled with the application of different probabilities densities and the spatial behavior has been modelled with application of the autoregressive process idea. The autoregressive capabilities of the model have also been proved by means of spectral analysis. The obtained results show that some probability densities of the used processes are highly related with the statistical roughness parameters, especially skewness and kurtosis. This in turn gives a good basis to forecast the tribological properties of the examined surface, including its directional characteristics. The numerical results have been compared with the experimental surfaces roughness measurements, showing good compatibility with the forecasted tribological parameters.

Keywords Surface roughness · Mathematical modeling · Autoregression analysis

A. Golabczak (✉) · A. Konstantynowicz
Department of Production Engineering, Lodz University of Technology,
Stefanowskiego 1/15 Street, 90-924 Lodz, Poland
e-mail: andrzej.golabczak@p.lodz.pl

A. Konstantynowicz
e-mail: andrzej.konst@gmail.com

M. Golabczak
Institute of Machine Tools and Production Engineering, Lodz University of Technology,
Stefanowskiego 1/15 Street, 90-924 Lodz, Poland
e-mail: marcin.golabczak@p.lodz.pl

A. Öchsner and H. Altenbach (eds.), *Mechanical and Materials Engineering of Modern Structure and Component Design*, Advanced Structured Materials 70,
DOI 10.1007/978-3-319-19443-1_11

1 Introduction

The “stochastic process” term, also called “random process” is a set of random variables (random varieties) representing the development of some system state over the independent variable *X*, often representing time, which in our case represents the spatial variable. The exact mathematical definition tells little to people outside the mathematics world, therefore we collected all of the necessary elements and mutual dependencies in Fig. 1. Two elements play the main role [1, 2]:

- stochastic (random) mechanism generating the exact process value which is in our case simply the height of the roughness profile at the given X variable value which is in our case simply the point of physical space—depicted in olive,
- assumed to be non-stochastic, a mechanism of the internal dependency between subsequent elementary events which in our case determine the similitude of roughness height in the juxtaposed spatial points—depicted in greenish.

The first element is described by the probability density distribution, which is usually chosen from the small, widely used set. In this paper we have made primarily use of the normal (Gaussian) distribution [1, 3]:

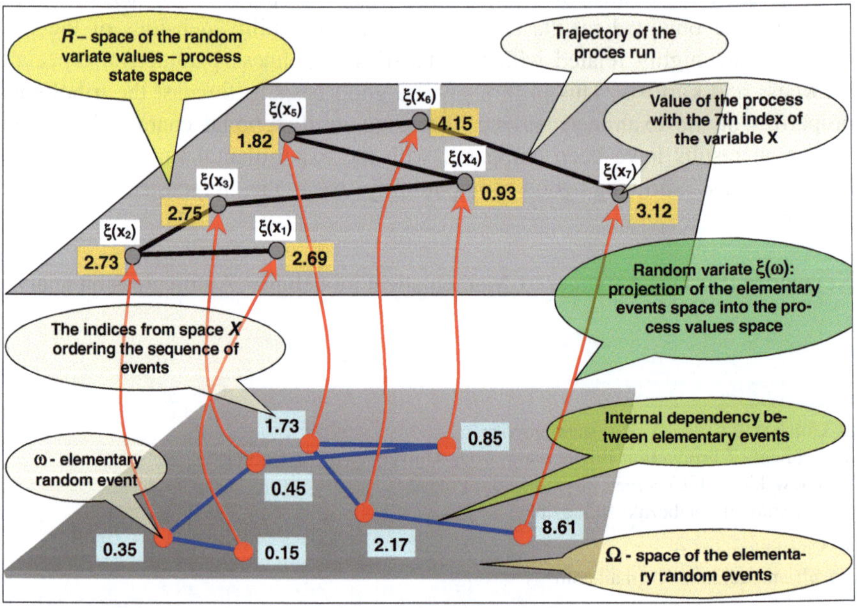

Fig. 1 The structure of mutual dependencies among the basic constituents of the “stochastic process” term, as they have been used in the presented paper

$$p(h) = \frac{1}{\sqrt{2\pi}\sigma} \exp\left(\frac{(h-\mu)^2}{2\sigma^2}\right) \tag{1}$$

where μ—mean value of the variable h, and σ^2—variance of the variable h.

The second distribution we have taken into account is the logarithmic-normal distribution, called shortly “log-normal”, with the probability density [1, 3]:

$$p(h) = \frac{1}{\sqrt{2\pi} \cdot \sigma \cdot h} \exp\left(\frac{(\ln(h)-\mu)^2}{2\sigma^2}\right) \tag{2}$$

where the description of variables is as previously mentioned. The existing relationship between the Y variable of the log-normal distribution and the N_O variable of the normal distribution with $\mu = 0$, $\sigma^2 = 1$, given by:

$$Y = \exp(k_\mu + k_\sigma \cdot N_0) \tag{3}$$

where:

$$k_\mu = \ln\left(\frac{\mu_{\ln}^2}{\sqrt{\sigma_{\ln}^2 + \mu_{\ln}^2}}\right) \tag{4}$$

$$k_\sigma = \sqrt{\ln\left(1 + \frac{\sigma_{\ln}^2}{\mu_{\ln}^2}\right)} \tag{5}$$

For modelling purposes we have to be equipped with an efficient tool to generate variables with prescribed probability densities. The basic tool is the random number generator with uniform probability density, usually provided as embedded in the numerical software. Making use of simple mathematical rules related with transformation of random variables we can use the following equation for to get at first random variables N_{i} with normal probability distribution from random variables U_i with the uniform probability distribution [3, 4]:

$$\begin{aligned} N_1 &= \sqrt{-2\ln(U_1)} \cdot \cos(2\pi \cdot U_2) \\ N_2 &= \sqrt{-2\ln(U_1)} \cdot \sin(2\pi \cdot U_2) \end{aligned} \tag{6}$$

This “twin generation” is very often used to obtain two “orthogonal” random variables, which, for our purpose, we can recognize as the independent random variables.

2 Autoregressive Model of the Stochastic Process

The "autoregressive model" of the stochastic process denotes the process with linear dependency between the current process value and a set of previous values. The term "previous" in our case is related with spatial, not time, sequence of the process values, because is related with the course of roughness profile obtained from profilometer surface examination. This could be formulated as [5–7]:

$$H(x_n) = c + \sum_{i=1}^{M} \phi_i \cdot H(x_{n-i}) + E_n \tag{7}$$

where

$H(x_n)$—the current "nth" value of the roughness height treated as a stochastic process taken at the point n of the profile, denoted further as H_n,
C—constant, eq. describing arbitrary relative level of measurement,
ϕ_i—constant parameter (coefficient) to be estimated (identified) from the experimental data H_E,
M—total number of the autoregressive model coefficients, sometimes called the "memory length" of the model,
$H(x_{n-i})$—previous values of the roughness height, denoted further as $H_{n\text{-}i}$,
E_n—the current "nth" value of the random excitation process, being a realization of so-called "white noise" process, i.e. the process with statistically independent subsequent values.

The basic task of the autoregressive model building is to determine the set of ϕ_i coefficients as well as the nature and parameters of the E process—colored green at the Fig. 1.

At first it is necessary to calculate the autocovariance function $R(n)$ of the H_{Ex} set of data from experiment, which is the autocorrelation function when the mean value of H_{Ex} set is zero, eq. after initial normalization [5, 7]:

$$R(n) = \frac{1}{\sigma_{Ex}^2} \frac{1}{N-n} \sum_{i=1}^{N-n} H_{Ex}(i) H_{Ex}(i+n) \tag{8}$$

where

$H_{Ex}(i)$—samples of the roughness profile,
N—total number of the experimental data samples,
σ_{Ex}^2—estimated variance of the experimental data.

Although the calculations performed according to Eq. (8) could be extended to $n = N-1$ coefficients of the autocorrelation function, it is not reasonable to exceed about the half of experimental data number because the lack of statistical reliability for n tending to N. We have applied limit of the 4096 autocorrelation points for the 9602 data points.

Besides of some objections related with the near-harmonic processes [8], the Yule-Walker method is commonly used to estimate the ϕ_i coefficients set [5, 7, 9]. It is based on the observation relating the Eqs. (7) and (8) giving the following linear equation:

$$\phi = P^{-1} \cdot R \tag{9}$$

where

$$\phi = \begin{bmatrix} \phi_1 \\ \phi_1 \\ \dots \\ \phi_M \end{bmatrix} \quad R = \begin{bmatrix} R_1 \\ R_2 \\ \dots \\ R_M \end{bmatrix} \quad P = \begin{bmatrix} 1 & R_1 & R_2 & \dots & R_{M-1} \\ R_1 & 1 & R_1 & \dots & R_{M-2} \\ \dots & \dots & \dots & \dots & \dots \\ R_{M-1} & R_{M-2} & R_{M-3} & \dots & 1 \end{bmatrix} \tag{10}$$

Solving the Eq. (9) for the relatively big number M might involve some troubles related with the numerical accuracy [4]. In our investigations we adopted the most primary and reliable method, i.e. the direct inversion of the P matrix. This method is implemented in the *Excel* spreadsheet we are using as the basic numerical tool.

3 Experimental Results

The forementioned concept of mathematical modelling has been applied to the experimental data derived from the PGM-1C IOS type mechanical profiler examination of polished X38CrMoV5-1 steel [8]. Samples have been prepared according to multi-phase process including AEDG grinding, initial polishing and final polishing. The X38CrMoV5-1 steel polishing has been carried out using the Phoenix Beta 2 grinding-polishing machine produced by Wirtz Buehler, Germany, equipped with the semi-automatic polishing head Vector Power Head, allowing a pressing force adjustment in the range of 5–200 N. The technological process characterization is presented in Table 1.

The roughness investigations results have been depicted in Fig. 2, including the extraction of the polynomial-type shape line from the raw data. The numerical technique used for the polynomial shape estimation is out of the scope of this work [10, 11]. The resultant data—marked with red, have been applied for building up the autoregressive model, with the roughness profile mean level brought down to zero.

The autocovariance function of the profile has been built accordingly into an Eq. (8), and is presented in Fig. 3, with the initial portion enlarged in the window. The close relationship between autocovariance of the signal and its normalized power spectrum exist—the Fourier Transform [7], which we have used as the additional tool to extract data from an experiment and allow further comparison with the auto regression-generated model curve of the roughness profile—see Fig. 4.

Table 1 The technological conditions of the polishing process of the X38CrMoV5-1 steel sample

Process stages	Polishing surface cover	The polishing material type	Lubricating medium	Process time [min]	Pressing force [N/cm^2]	Polishing speed v [m/s]
Initial polishing	Polishing cloth Buehler Nylon	Polycrystalline diamond (grits 9 μm)	Polysrystalline diamond suspension Buehler MetaDi Supreme	5	3	3
		Polycrystalline diamond (grits 3 μm)	Polysrystalline diamond suspension Buehler MetaDi Supreme	3	3	3
Finishing polishing	Polishing cloth Buehler TriDent	Cerium oxide suspension Buehler MiroMet (grits 1 μm)	–	3	2	3

Ultrasound washing in the solution (formula in weight %): H_2SO_4 (96 %)—12.5 %, oxalic acid—14.5 %, H_2O—73 %

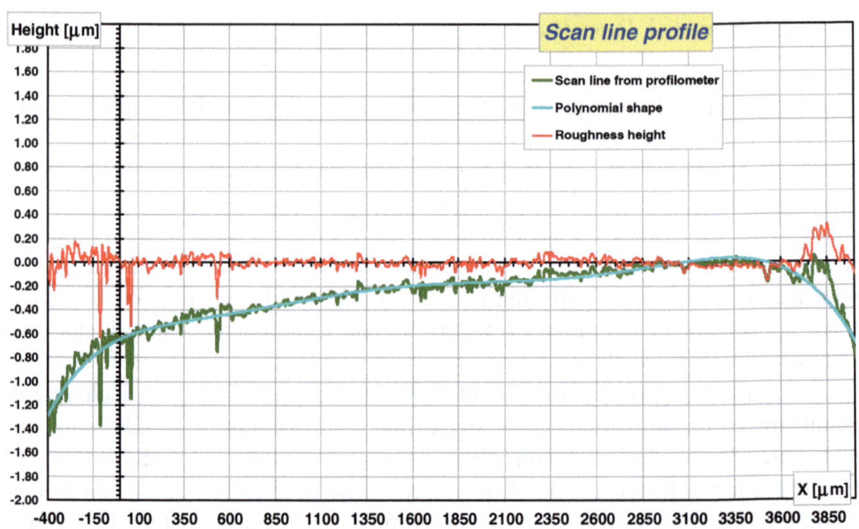

Fig. 2 Scan line from the mechanical profilometer with extracted polynomial shape, normalized to zero level (*red*)

Because of the expected lack of harmonic components in the roughness profile line [6, 9, 12], the Yule-Walker method has been applied for the estimation of the autoregressive model constant coefficients, depicted as a whole set in Fig. 5. The

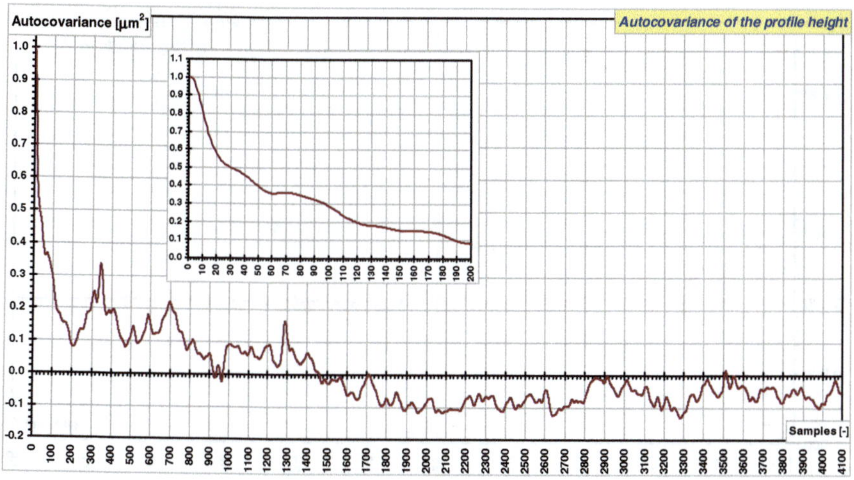

Fig. 3 Autocovariance function of the roughness profile depicted in Fig. 2. Enlarged is the initial portion, which the autoregressive coefficients have been estimated

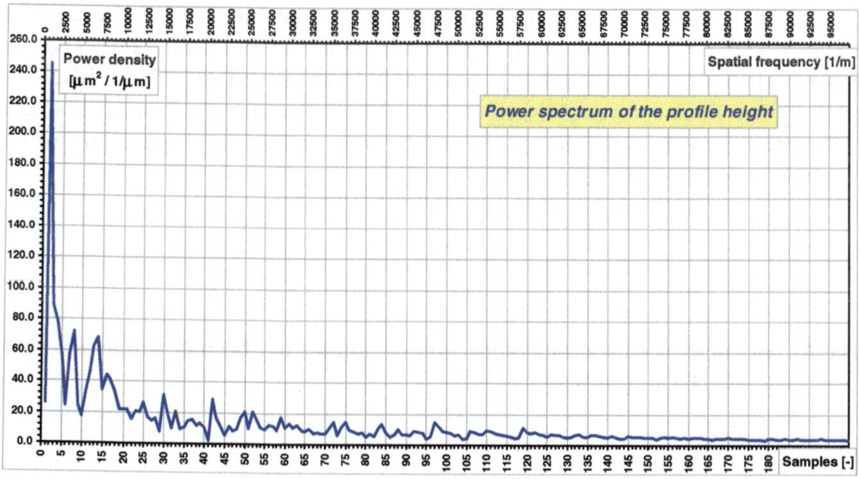

Fig. 4 Normalized spatial power spectrum (FFT transform of the autocovariance function (8) of the roughness profile depicted in Fig. 2. Further coefficients are neglectable

relatively big number of coefficients estimated—50, has been used for to detect spatial dependencies in the roughness profile of the relatively wide span—50 × 0.5 μm = 25.0 μm in comparison with the roughness profile mean—R_a = 0.0445 μm, to extract possible long-distance remains of the earlier stages of surface machining remove "for" [12].

The autoregressive model generated roughness profile has been depicted in Fig. 6. along with the experimental results brought to the same scale for to allow

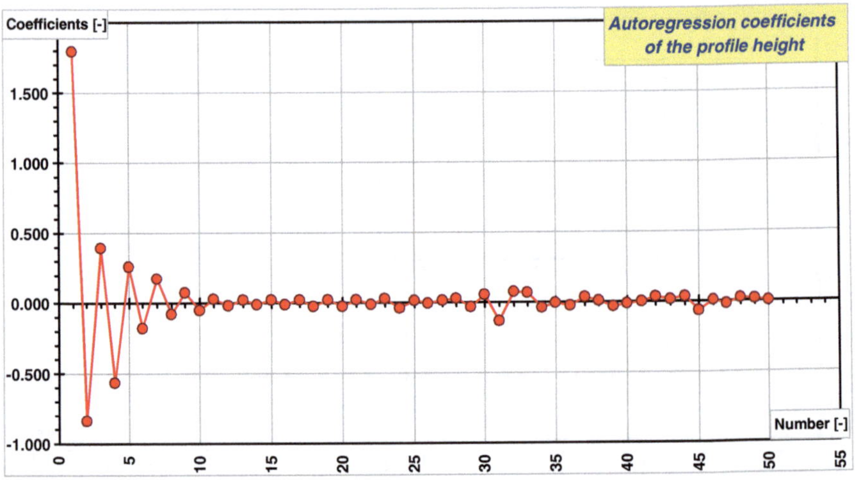

Fig. 5 Distribution and values of the autoregression coefficients for the autocovariance function depicted in Fig. 3

precise comparison. The very characteristic course of the roughness profile has been obtained with application of the initial random generator E—see the Eq. (7) of the log-normal type with the parameters: $\mu = 0.025$ µm, $\sigma^2 = 0.0075$ µm^2. Then the arbitrary level has been established at $c = 0.25$ µm, and the final model has been reflected (which does not change variance!) for to render the "topography"of the roughness profile in the best possible way:

$$H(x_n) = c - \left(\sum_{i=1}^{M} \phi_i \cdot H(x_{n-i}) + E_n \right) \tag{11}$$

The R_q parameter used for the comparison of profiles has been at almost the same level of $R_q \cong 0.07$ µm in both cases.

4 Conclusions

The autoregressive model presented and applied in this work to the experimental data has turned out to be well applicable to the non-harmonic surface roughness obtained after high-grade polishing. The possible use of the autoregressive model along with the spectral analysis gives the well-fitted tool for to analyze effects of the rough-grade remains of surface machining in multi-phase surface machining.

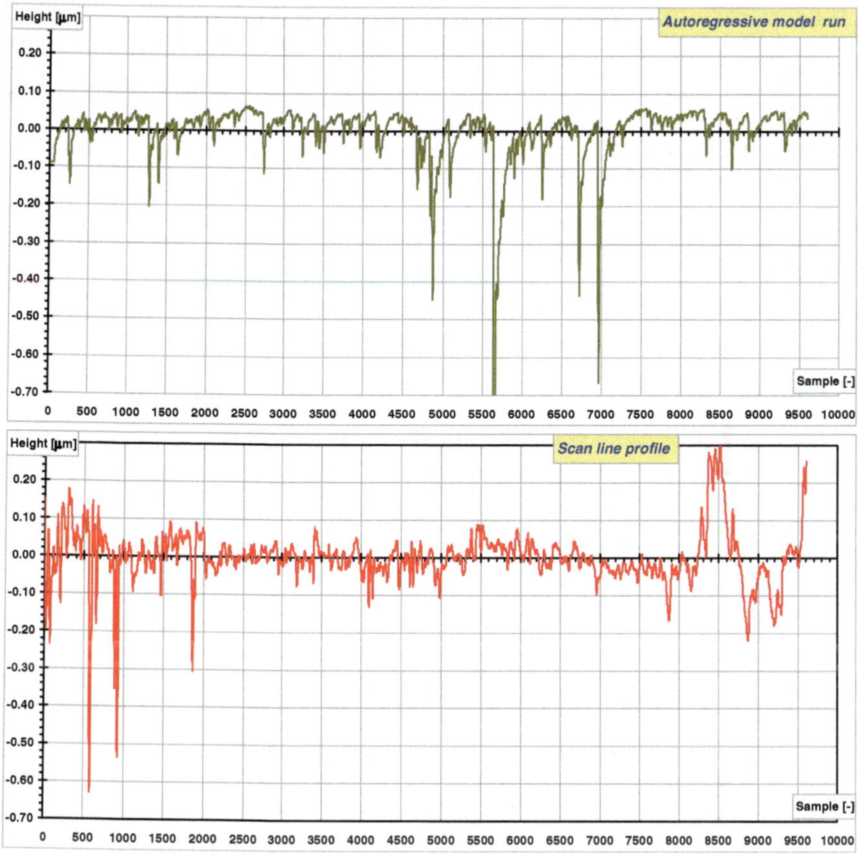

Fig. 6 The exemplary run from the autoregressive model compared with the experimental scan line from profilometer. Both curves have almost the same RMS value

References

1. Feller W (1971) An introduction to probability theory and its applications. Wiley, New York
2. Gentle J (2003) Random number generation and monte carlo methods. Springer, New York
3. Knuth D (1968) The art of computer programming. In: Random numbers. Addison-Wesley, Reading
4. van Kampen NG (1992) Stochastic processes in physics and chemistry. North-Holland, Amsterdam
5. Box G, Jenkins G, Reinsel G (2008) Time series analysis: forecasting and control. Prentice Hall, Engelwood Cliffs
6. de Hoon MJL (1996) Why Yule-Walker should not be used for autoregressive modeling. Ann Nucl Energy 23:1219–1228
7. Stoica P, Moses R (2005) Spectral analysis of signals. Pearson Prentice Hall, Englewood Cliffs
8. Gołąbczak M, Pawlak, Szymański W, Jacquet P, Fliti R (2012) Properties of PVD coatings manufactured on X38CrMoV5-1 steel for plastic moulding applications. J Mach Eng 12:37–45

9. Eshel G (2003) The yule-walker equations for the AR coefficients. Citeulike-article-id: 763363
10. Taylor HM, Karlin S (1998) An introduction to stochastic modelling. Academic Press, New York
11. Wieczorowski M, Ehmann KF, Cellary A (1995) Parametric modelling of 3-D surfaces. ASME Mechanical Engineering Conference
12. DeVor RE, Wu SM (1971) Surface profile characterization by autoregressive moving average models. J Manu Sci Eng 94:825–832

Drainage Concrete Based on Cement Composite and Industrial Waste

Lukáš Gola, Vojtěch Václavík, Jan Valíček, Marta Harničárová, Milena Kušnerová and Tomáš Dvorský

Abstract The ongoing development of urbanization of our landscape has resulted in continuous demand for building materials, which are even nowadays produced mainly from primary natural resources. The continuous reconstructions and modernizations of already built-up areas are the cause of the production of construction

L. Gola · J. Valíček (✉) · M. Harničárová · M. Kušnerová
Faculty of Mining and Geology, Institute of Physics, VŠB—Technical University of Ostrava, 17. listopadu, 708 33 Ostrava, Czech Republic
e-mail: jan.valicek@vsb.cz

L. Gola
e-mail: lukas.gola@vsb.cz

M. Harničárová
e-mail: marta.harnicarova@vsb.cz

M. Kušnerová
e-mail: milena.kusnerova@vsb.cz

V. Václavík · T. Dvorský
Faculty of Mining and Geology, Institute of Environmental Engineering, VŠB—Technical University of Ostrava, 17. listopadu 15, 708 33 Ostrava, Czech Republic
e-mail: vojtech.vaclavik@vsb.cz

T. Dvorský
e-mail: tomas.dvorsky@vsb.cz

V. Václavík · J. Valíček
Institute of Clean Technologies for Mining and Utilization of Raw Materials for Energy Use, VŠB—Technical University of Ostrava, 17. listopadu 15, 708 33 Ostrava, Czech Republic

J. Valíček · M. Kušnerová
Faculty of Metallurgy and Materials Engineering, RMTVC, VŠB—Technical University of Ostrava, 17. listopadu 15, 708 33 Ostrava, Czech Republic

M. Harničárová
Nanotechnology Centre, VŠB—Technical University of Ostrava, 17. listopadu 15, 708 33 Ostrava, Czech Republic

A. Öchsner and H. Altenbach (eds.), *Mechanical and Materials Engineering of Modern Structure and Component Design*, Advanced Structured Materials 70,
DOI 10.1007/978-3-319-19443-1_12

waste which, for example, in Europe represents ¼ of the volume of all waste materials. Such a trend is inconsistent with sustainable development and considerate impact on the environment. The contemporary society is aware of these adverse impacts and it actively participates in the integration of construction waste back into production. Thanks to the systems of recycling, construction waste can return to the building industry as a fully valuable building material. The production of shaped pieces from grey cellular concrete after the autoclave process results in the creation of residual material in the form of waste blocks (rubble). This waste material is stored in dumps. The presence of these dumps has an adverse effect on the surrounding environment. This article presents the first results of a basic research dealing with the treatment process of waste cellular concrete rubble by means of a crushing process and its subsequent use as filler in the production of new porous concretes. The article presents 3 basic recipes of porous concrete, where 100 % of the filler was replaced with crushed porous concrete rubble with the fraction of 0/6 mm. The proposed recipes have been tested in regards to: density of fresh concrete mixture, concrete mixture consistency, strength, and thermal conductivity coefficient.

Keywords Porous concrete · Cellular concrete rubble · Strength · Thermal conductivity coefficient

1 Introduction

Recycling and use of industrial waste in different areas minimize the production of waste, its disposal costs and they protect the environment. This topic is very relevant in terms of research and development of new materials. The issue is especially the use of industrial waste as a secondary raw material in the segment of building materials, particularly concrete. There are known results of the use of fly ash in the production of concrete, as a partial replacement of cement [1, 2], and the production of copolymers [3]. The properties of concrete based on blast furnace slag, as a partial replacement of Portland cement and concrete based on steel slag and as a partial replacement of natural aggregates, are described in [4, 5]. The properties of porous concretes based on natural aggregate with the fractions of 13/20, 5/13, 2.5/5 mm are described in [6]. There are also porous concretes based on latex binder in combination with coarse aggregate and river sand [7]. The use of waste concrete rubble as a complete or partial replacement of the filler in concrete mixtures can be presented as an example of an effective treatment of construction waste. This recycling method turns waste rubble into synthetic aggregate, which is an alternative to natural aggregate. This leads to a balanced utilization of natural resources and helps to tackle the issue of waste management. A specific example of the possible use of recycled waste shaped pieces from grey cellular concrete is porous concrete. It is a lightweight concrete (density after drying at 105 °C reaches the values of 800–2000 kg·m^{-3}), which is specific due to its porosity. Porous concrete typically achieves lower strength classes than plain concrete. The production makes use of porous or dense aggregate. The individual grains are bonded by a binder film on each grain and there is a large volume of air gaps among them [8]. The concrete may consist of one type of fraction or more fractions of small and coarse aggregate, cement, water and optional additives. One can also come across the name "drainage concrete" [9]. In the world, porous concrete is used as a top water-permeable layer on pedestrian foot-paths or roads, where rain water is absorbed in the soil. This leads to a more moderate and more continuous outflow from urbanized areas during tidal rains, improvement of groundwater level and relief of the flow in sewerage systems [10–12]. Another possible application of porous concrete is in partition walls, non-bearing walls, shaped pieces, grass tiles, decorative elements for walls and fences. It is used in places requiring thermal insulation and acoustic insulation.

The aggregate (filler) used in porous concrete can be:

(a) From natural sources:

- Keramzite—ceramic porous aggregate (produced by burning and expansion of natural clay) used in lightweight porous concrete. It is also known under the brand name Liapor.
- Expandit—lightweight porous material (produced by expansion of slate) used as artificial aggregate in concrete.

(b) From industrial waste materials:

- Agloporit (produced by burning fly ash from power plants). It is also know under the brand name Lytag.
- Foamed slag (sudden cooling of hot liquid slag by water).
- Cinder (waste of incineration of solid fuels in grate furnaces).
- Brick rubble (waste from brick production, recycling of brick rubble).

2 Materials and Methods

2.1 *Artificial Filler to Concrete from Industrial Waste Materials*

Generally, the production of grey cellular concrete follows the rule that the siliceous materials (fly ash from coal combustion) together with burnt lime and cement, or other additives are broken down in special mixers with water, border sludge and a gassy agent (aluminium powder) into a liquid slurry. It is then poured into moulds, where the actual loosening will take place—proofing followed by hardening. De-moulding of the moulds is the next process and the hardened material is cut into the required shapes. Curing of grey cellular concrete takes place in autoclaves at higher pressure and temperature. This environment facilitates an effective bonding of the individual components of the concrete mixture. The final product of the production process is a porous shaped piece for very precise walling. The material is safe, it has low density, it is a good thermal insulator, and it is soundproof and permeable. The compressive strength of grey porous concrete is 3.2 MPa.

Nonconforming shaped pieces occur during the production of grey cellular concrete—these are rejects that become the waste material, which is stored in dumps in the production plant area (see Fig. 1).

Figure 2 shows a scheme of treatment of rubble from autoclaved cellular concrete from Fig. 1.

Figure 2 clearly shows that the first stage of mechanical treatment of waste autoclaved cellular concrete is a jaw crusher, where the output is rubble with a maximum grain size of 50 mm. The crushed rubble subsequently goes through a swing-hammer crusher, where the output is pulp with a maximum grain size of 6 mm. Thanks to the absence of foreign substances, we can eliminate the sorting process. The resulting cellular concrete pulp with the fraction of 0/6 mm is used as new filler in the developed porous concrete.

Fig. 1 Waste blocks (rubble) from the production of autoclaved cellular concrete

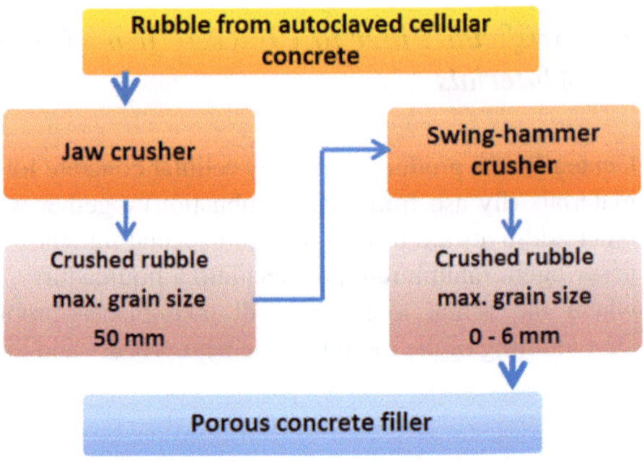

Fig. 2 Scheme of the treatment of waste rubble from autoclaved cellular concrete

2.2 Porous Concrete Components

Crushed rubble from autoclaved grey cellular concrete with the fraction of 0/6 mm was used as the new filler in porous concretes based on industrial waste. Portland cement CEM I 42.5 R from Cement Hranice, a.s. was used as the binding component. Water from the water supply network, i.e. drinking water, was used as the mixture water.

2.3 Preparation of Porous Concrete

The preparation of the experimental mixtures according to the proposed recipes was performed in M80 forced circulation mixer from FILAMOS s.r.o. company. The mixing is carried out by several arms that also ensure that the mixture is scraped from the side and the entire bottom of the mixing tank. The mixture filling is carried out through a sieve in the mixer lid, which is equipped with a shredding comb for bagged mixtures. The mixed material is discharged by turning the sliding segment at the bottom of the tank. The technical parameters of the mixer are shown in Table 1.

2.4 Methods Used to Determine the Physical Properties

The crushed rubble of grey autoclaved cellular concrete was tested for powder density and porosity according to CSN EN 1097-3 [13]. The powder density of freely poured aggregate is determined by weighing the volume of a 5 l standard cylindrical container filled with porous concrete rubble. The pouring of the material into the container is carried out from a minimum height to avoid the compaction of the sample. The resulting value is the average of three measurements.

The density of autoclaved grey cellular concrete was determined according to CSN 72 1171 [14]. The grain density is calculated from the ratio of weight and sample volume. The weight is determined by weighing the water-saturated and surface-dried testing sample backfill and, again, by weighing after drying in a drying plant. The volume is determined from the weight of the water displaced during the application of the pycnometric method. The absorbing power is based on

Table 1 Technical parameters of M80 mixer

Tank volume [l]	111
Max. used volume [l]	69
Electromotor output power [kW]	2.2
Voltage [V]	400
Mixer rotations [rot/min]	47
Max. material grain size [mm]	10
Weight [kg]	137

the increase in weight of the aggregate sample dried in a drying plant, as a result of water penetration into the cavities accessible to water.

The porosity of autoclaved cellular concrete pulp was determined according to CSN EN 1097-3 [13]. It was calculated from the powder density of freely poured aggregate and the grain density.

The geometric properties of cellular concrete pulp were determined by a sieve analysis according to CSN EN 933-1 [15]. The test is based on sorting and separating the material by means of a set of sieves into several grain size parts with decreasing particle size. The mesh size of the sieves and the number of sieves are selected according to the type of sample and the required accuracy. The standard basic set of sieves consists of sieves with square holes with the sizes of: 0.063, 0.125, 0.25, 0.5, 1, 2, 4, 8, 16, 32, 64 and 125 mm.

The consistency of the porous concrete mixture based on autoclaved cellular concrete rubble was examined by a slump test according to CSN EN 12350-2 [16]. Fresh concrete was compacted in a mould of blunted cone shape (diameter of the bottom base was 200 mm, the upper base was 100 mm, the height was 300 mm). The concrete consistency is indicated by the slump distance of concrete after lifting up the blunted cone.

The strength of the newly developed porous concrete was tested on test specimens in the shape of a cube, with the dimension of 150 mm, according to CSN EN 12390-3 [17] after 3, 7, 14 and 28 days.

The thermal conductivity coefficient λ of porous concrete was determined by a measuring device ISOMET 2114 from Applied Precision company. It is a device designed for a direct measurement of the thermal conductivity coefficient of solid, loose or liquid materials. The measuring method is non-stationary and is based on the analysis of the course of the time dependence of the thermal response to a pulse of heat flux of the examined material. The heat flux is generated by a device probe by means of diffused electrical power of resistor in the material. The condition for correct results is a conductive connection between the probe and the measured material. The measured value of the thermal conductivity coefficient can be read directly from the instrument in [$W \cdot m^{-1} \cdot K^{-1}$]. The determination of the thermal conductivity coefficient was performed on 150 mm cubes. After 28 days of curing, the sample was placed in a drying oven, where it was dried at 105 °C ± 1 °C for 48 h. The dried sample was then placed in an exicator, and after cooling to a room temperature, it was measured with a surface probe with the measuring range from 0.04 to 2.00 $W \cdot m^{-1} \cdot K^{-1}$ located in the middle of the test specimen surface.

3 Results and Discussion

Table 2 presents the determined physical properties of cellular concrete pulp, which was obtained by treatment according to the scheme presented in Fig. 2. The pulp is used as a complete replacement of natural aggregate (filler) in the production of porous concrete.

Table 2 Physical properties of crushed pulp of autoclaved cellular concrete with the fraction of 0/6 mm

Parameter	Measured value
Powder density of freely poured aggregate	543 $kg{\cdot}m^{-3}$
Density	844 $kg{\cdot}m^{-3}$
Absorbing power	44.4 %
Porosity	35.6 %

The results of the sieve analysis of cellular concrete pulp are presented in Figs. 3 and 4. A set of sieves with square holes with the dimensions of: 0.063, 0.125, 0.25, 0.5, 1, 2, 2.8, 4, 5.6 and 8 mm was chosen for the actual analysis. Cellular concrete pulp consists mainly of about 40 % of grain size of 1–2 mm, 18 % of grain size of 2–2.8 mm, 13 % of grain size of 0.5–1 mm, 10 % of grain size of 0.063–0.125 mm.

Three basic recipes, with different doses of cement and water-cement ratio, were designed on the basis of the verification of the possibility of the use of waste

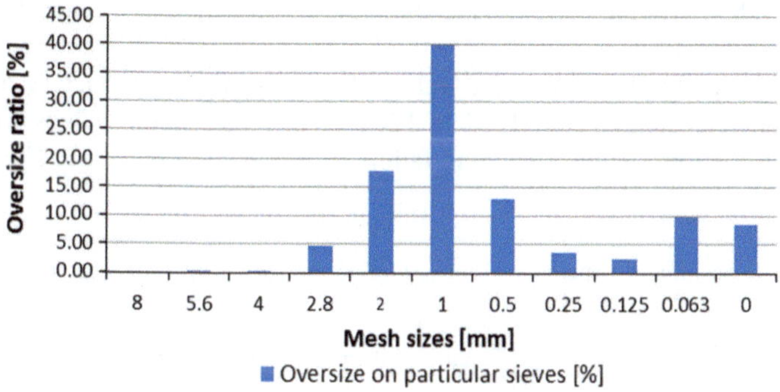

Fig. 3 Share of oversize cellular concrete pulp with the fraction of 0/6 mm on the individual sieves

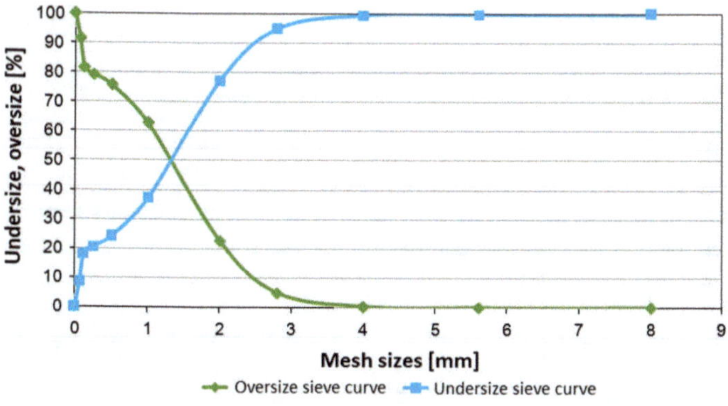

Fig. 4 Cumulative curves of grain fitness of cellular concrete pulp fraction 0/6 mm

cellular concrete pulp as a full replacement of natural aggregate in the production of porous concrete. The composition of the experimental recipes is shown in Table 3.

The purpose of the selected recipes is to determine the sufficient minimum amount of cement, while maintaining the porous structure, strength and good workability of the fresh concrete mixture.

18 test specimens were prepared from each recipe (150 mm cubes). The development of compressive strength of porous concrete was monitored after 3, 7, 14, 21, and 28 days. The results are presented graphically in Fig. 5. It is evident that the strength of porous concrete increases with the amount of cement, which was dosed at the amounts of 200, 300 and 400 kg per m^3. The highest values of compressive strength of 4.1 MPa porous concrete based on waste cellular pulp were achieved after 28 days in recipe no. 3.

In addition to the compressive strength of concrete, we have also monitored other properties of porous concrete. They were: the density of fresh concrete mixture, the consistency of fresh concrete mixture immediately after mixing (slump test), and the thermal conductivity coefficient λ. The results of these tested properties, together with the compressive strength of porous concrete after 28 days, are shown in Table 4.

Table 4 clearly shows that the value of density of fresh concrete mixture of porous concrete ranged from 1172 to 1241 $kg \cdot m^{-3}$, the slump values were in the interval of 2–4 mm, while a shearing failure occurred in recipe 1, see Fig. 6. It was

Table 3 Composition of experimental recipes on 55 dm^3 and 1 m^3 of final porous concrete

Components	Unit	Recipe 1		Recipe 2		Recipe 3	
		55 dm^3	1 m^3	55 dm^3	1 m^3	55 dm^3	1 m^3
Cellular concrete pulp	kg	29.8	543	29.8	543	29.8	543
Cement CEM I 42,5 R	kg	11.0	200	16.5	300	22.0	400
Water	kg	12.1	220	19.0	345	26.4	480
Water-cement ratio w	–	1.10		1.15		1.20	

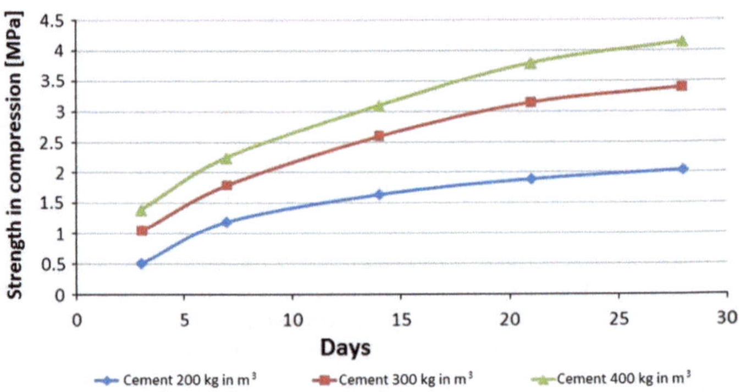

Fig. 5 Strength of porous concrete after 3, 7, 14, 21 and 28 days

Table 4 Test results of physical and mechanical properties of experimental recipes

Marking	Density of fresh concrete mixture [kg·m^{-3}]	Slump [mm]	Compressive strength after 28 days [MPa]	Thermal conductivity coefficient λ [W·m^{-1}·K^{-1}]
Recipe 1	1172	–	2.0	0.2020
Recipe 2	1203	4	3.4	0.2146
Recipe 3	1241	2	4.2	0.2269

Fig. 6 Shearing failure of cone during the slump test of recipe 1

caused by a small dose of cement, where the grains of cellular concrete pulp were not completely coated with cement sealant.

The thermal conductivity coefficient λ ranged from 0.2020 to 0.2269 W·m^{-1}·K^{-1}. It is evident that the increasing amount of cement in the concrete mixture goes hand in hand with the increasing value of the thermal conductivity coefficient.

4 Conclusion

Based on the achieved results, it can be stated that crushed waste rubble from autoclaved cellular concrete with a maximum grain size of 6 mm can be used as a 100 % replacement of natural aggregates in the production of porous concrete. The prepared concrete reaches density values of around 1200 kg·m^{-3}. The compressive strengths are as high as 4.2 MPa after 28 days, the slump is 2–4 mm, and the thermal conductivity coefficient λ reaches the values of 0.2020–0.2269 W·m^{-1}·K^{-1}.

The new feature of the presented solution lies mainly in the use of industrial waste created in the form of grey scrap blocks during the production of grey cellular concrete as new filler for porous concretes. This waste is stored in dumps in the production plant area or it is kept in municipal waste landfills, which is not in compliance with environmental protection and sustainable development.

Further addition of additives and admixtures into the recipes can lead to a significant improvement of the physical and mechanical properties of porous concrete. The future research will also be focused on monitoring frost resistance, shapes and sizes of pores of the newly developed porous concrete and the possibility of studying the topography of its surface.

Acknowledgments This paper has been elaborated in the framework of the projects:

SGS project reg. no. SP2014/114.

Institute of clean technologies for mining and utilization of raw materials for energy use, reg. no. LO1406 supported by Research and Development for Innovations Operational Programme financed by Structural Founds of Europe Union and from the means of state budget of the Czech Republic.

IT4Innovations Centre of Excellence project, reg. no. CZ.1.05/1.1.00/02.0070 supported by Operational Programme Research and Development for Innovations funded by Structural Funds of the European Union and state budget of the Czech Republic.

RMTVC No. LO1203.

References

1. Ondova M, Stevulova N (2011) Benefits of coal fly ash utilization in the area of a pavement building. In: Cygas D, Froehner KD (eds) ICEE-2011: Proceedings in road and railways. 8th International conference on environmental engineering, Vilnius, May 2011, Gediminas Technical University press Technika, Vilnius, vol 1–3, pp 1156–1159
2. Junak J, Stevulova N (2011) Potential of selected industrial wastes in civil eng appl. In: SGEM-2011: 11th international multidisciplinary scientific geoconference, Bulgaria, Sofia, pp 20–25
3. Svarla J, Sisol M, Botula J, Kolesarova M, Krinicka I (2011) The potential use of fly ash with a high content of unburned carbon in geopolymers. Acta Geodyn Geomater 7(2):123–132. ISSN 1214-9705
4. Vaclavik V, Dirner V, Dvorsky T, Daxner J (2012) The use of blast furnace slag. Metallurgy 51(4):461–464. ISSN 0543-5846
5. Stevulova N, Vaclavik V, Junak J, Grul R, Bacikova M (2008) Utilization possibilities of selected waste kinds in building materials preparing. In: SGEM-2008: 11th international multidisciplinary scientific geoconference, Bulgaria, Sofia, pp 16–20
6. Bhutta M, Tsuruta K, Mirza J (2012) Evaluation of high-performance porous concrete properties. Const Build Mater 31:67–73. ISSN 0950-0618
7. Huang B, Wu H, Schu X, Burdette E (2010) Laboratory evaluation of permeability and strength of polymer-modified pervious concrete. Constr build Mater 24(5):818–823. ISSN 0950-0618
8. Drochytka R (1993) Lehké stavební látky. VUT, Brno
9. CSN 73 6124-2 (2008) Road building—hydraulically bound mixture courses—part 2: concrete drainage layers. Czech Normalization Institute, Prague
10. Lian C, Zhuge Y (2010) Optimum mix design of enhanced permeable concrete—an experimental investigation. Constr Build Mater. doi:10.1016/j.conbuildmat.2010.04.057
11. Putman BJ, Neptune AI (2011) Comparison of test specimen preparation techniques for pervious concrete pavements. Constr Build Mater. doi:10.1016/j.conbuildmat.2011.03.039

12. Yang J, Jiang G (2003) Experimental study on properties of pervious concrete pavement materials. Cem Concr Res. doi:10.1016/S0008-8846(02)00966-3
13. CSN EN 1097-3 (1999) Tests for mechanical and physical properties of aggregates—part 3: determination of loose bulk density and voids. Czech Normalization Institute, Prague
14. CSN 72 1113 (1968) Determination of mass, porosity and voids ratio of aggregates. Czech Normalization Institute, Prague
15. CSN EN 933-1 (2012) Tests for geometrical properties of aggregates—part 1: determination of particle size distribution—Sieving method. Czech Normalization Institute, Prague
16. CSN EN 12350-2 (2009) Testing fresh concrete—part 2: slump-test. Czech Normalization Institute, Prague
17. CSN EN 12390-3 (2009) Testing hardened concrete—part 3: compressive strength of test specimens. Czech Normalization Institute, Prague

Numerical Analysis of Impact Behavior of Rotary Centrifuge Guarded Body

Weizhou Zhong, Xicheng Huang, Chengang Luo, Gang Chen and Zhifang Deng

Abstract Numerical simulation of dynamic mechanical responses of a rotary centrifuge at different impact velocities and angles is performed. The impact velocities are 25 m/s, 50 m/s and 270 m/s and the impact attitudes are 0° and 45° respectively. Stress fields and failure modes of the guarded body, cavity wall, cover and rotary components are obtained. It indicates that the cavity wall can withstand impact action. The deformation of the cavity wall and cover is elastic when the turntable velocity is lower than 25 m/s. Centrifuge guarded body will be broken when turntable velocity is over 50 m/s. This analysis can guide design and safety assessments of rotary centrifuge.

Keywords Centrifuge · Impact resistance · Numerical simulation · Safety assessment

1 Introduction

The energy absorption of a cushion material is an important property for the impact safety of a product. Wide range plateau stress is necessary for an ideal cushion material. According to the actual usage of cushion materials, they can be classified

W. Zhong (✉) · X. Huang · C. Luo · G. Chen · Z. Deng
Institute of Systems Engineering, China Academy of Engineering Physics, Mianyang 621999, China
e-mail: zhongwz@caep.cn

X. Huang
e-mail: huangxc@caep.cn

C. Luo
e-mail: luocg@caep.cn

G. Chen
e-mail: chengang@caep.cn

Z. Deng
e-mail: dengzf@caep.cn

A. Öchsner and H. Altenbach (eds.), *Mechanical and Materials Engineering of Modern Structure and Component Design*, Advanced Structured Materials 70,
DOI 10.1007/978-3-319-19443-1_13

into energy dissipation and energy storage. Energy dissipation materials such as foam metals are widely used. For the energy storage material, kinetic energy converts into elastic potential energy under impact conditions, such as rubber materials and springs. So cushion material and their structure mechanics behavior are important for impact safe protection. High strength steels and porous materials have recently been found to be effective for resisting and reducing the force of impact. Examples of recent uses of cushion materials and structures can be found in product packaging buffers and highway guardrails. Many researchers have performed studies in field of structure impact resistance. For instance, Guillow [1] experimentally investigated the axial compression of thin-walled circular tubes, a classical problem studied for several decades. Experimental results show both axi-symmetric and non-symmetric modes lie on a single curve. Li [2] established a close-celled aluminum foam model under low velocity impact condition, and validated with drop hammer test and systematically explored the influence of impact mass ratio, porosity and geometrical dimensions of foam protection on the critical velocity and acceleration. Jeenager [3] produced metal foams and thermally treated them to enhance their properties and examined the changes in the microstructure and thermal treatment. Fracture test affirms the role of the microstructure for property enhancement. Alavi [4] analyzed the energy absorption capacity of simple and multi-cell thin-walled tubes with triangular, square, hexagonal and octagonal sections. The results showed that the energy absorption capacity of multi-cell sections is greater than that of simple sections. Further more, hexagonal and octagonal sections in a multi-cell configuration absorbed the greatest amounts of energy per unit of mass. Ajdari [5] investigated in-plane dynamic crushing of two dimensional honeycombs with both regular hexagonal and irregular arrangements by using detailed finite element models. Numerical simulations showed three distinct crushing modes for honeycombs with a constant relative density: quasi-static, transitional and dynamic. Kumar [6] experimentally investigated the effect of stiffening the syntactic foam core with a resin impregnated paper honeycomb structure on compression behavior and energy absorption capacity of sandwich composites under flatwise and edgewise loading configurations. Lee [7] reported on the mechanical behavior of an interpenetrating carbon/epoxy periodic submicrometer-scale bicontinuous composite material fabricated following the design principles deduced from biological composites. Using microscopic uniaxial compressive tests, the specific energy absorption is quantitatively evaluated and compared with the epoxy/air and carbon/air precursors. Karasek [8] applied dropped weight impact testing to evaluate the influence of temperature and moisture on the impact resistance of unmodified and modified epoxy/graphite fiber composites. The results indicated moisture was found to have little effect on the damage initiation energy or subsequent energy absorption at ambient and low temperatures. Much study on the energy absorption of materials and structures has been and is being investigated by researchers [9–12].

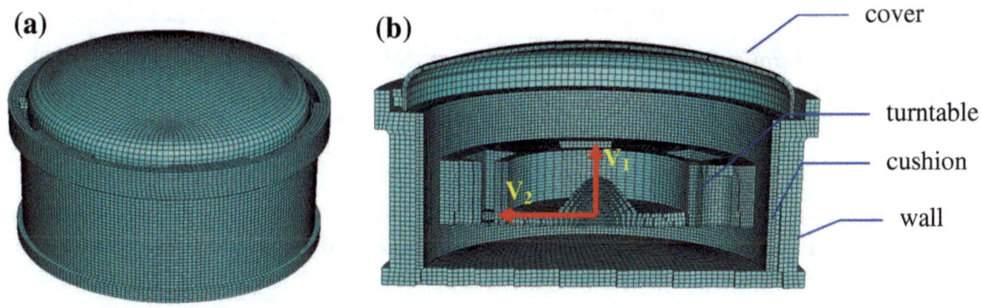

Fig. 1 Finite element structure

The rotary centrifuge is made up of cavity wall, cover, safety gate and steel box. Maximum rotating speed is about 220 rad/s. In order to ensure a safe operation of the centrifuge at high rotating speeds, the impact response of a centrifuge guarded body should be studied. In the present work, numerical analysis on the impact behavior of the rotary centrifuge guarded body is carried out to investigate the centrifuge's protective capacity in an accident environment.

2 Rotary Centrifuge Structure Model

A rotary centrifuge is high speed rotation equipment. The system is made up of spindle bearing, driver motor and centrifuge body. The impact resistance of centrifuge body should be taken into account in the rotary centrifuge safe design stage. Numerical simulation is a feasible way to predict the structure impact response. The finite element model of the centrifuge body is created by the commercial ABAQUS software, as shown in Fig. 1. There are 158,277 hexahedral elements in the numerical model. The junction between the cover and the centrifuge is structural constraint, being similar to a pressure cooker. The bottom of the centrifuge body is fixed in the numerical analysis.

3 Material Properties

In the rotary centrifuge structure, the centrifuge wall and cover are made of Q235 steel. The turntable is made of LD12 alloy metal. The elastic-plastic constitutive model is adopted describe mechanical properties of Q235 and LD12, as shown in Table 1. Rubber is taken as the cushion material on the centrifuge cavity wall. The stress versus strain relation of rubber is described as the potential energy of strain.

Table 1 Material mechanical properties

Material	ρ/kg/m^3	E/GPa	ν	σ_s/MPa	σ_b/MPa	Failure strain
Q235	7820	201	0.33	235	635	0.4
LD2	2700	71	0.32	294	436	0.4

The polynomial model expresses the rubber potential energy in simulation. The mathematical relation is shown in expression (1).

$$U = \sum_{i+j=1}^{N} C_{ij}(i_1 - 3)^i (i_2 - 3)^j + \sum_{i=1}^{N} \frac{1}{D_j} (j_{el} - 1)^{2i} \tag{1}$$

U is the potential energy of strain. Symbol j_{el} is the elastic volume ratio. D_i and C_{ij} designate the material compression and Rinvlin coefficient. As rubber is an incompressible medium, C_{01}, C_{10} and D are equal to 0.36, 0.09 and 0 respectively in this work.

4 Simulation Results

The dynamical response of the rotary centrifuge subjected to four impact cases is simulated with the commercial ABAQUS/Explicit software. The impact cases are: velocities of 25 m/s and 50 m/s normal to the impact cover, 270 m/s normal to the impact cover wall and 270 m/s inclined to the impact cavity wall, respectively. Stress and strain distributions of the cover, centrifuge wall and turntable are recommended in the following.

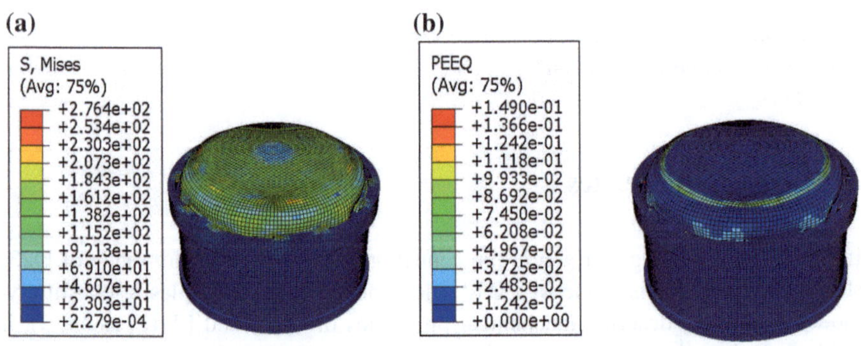

Fig. 2 Stress field and deformation of centrifuge

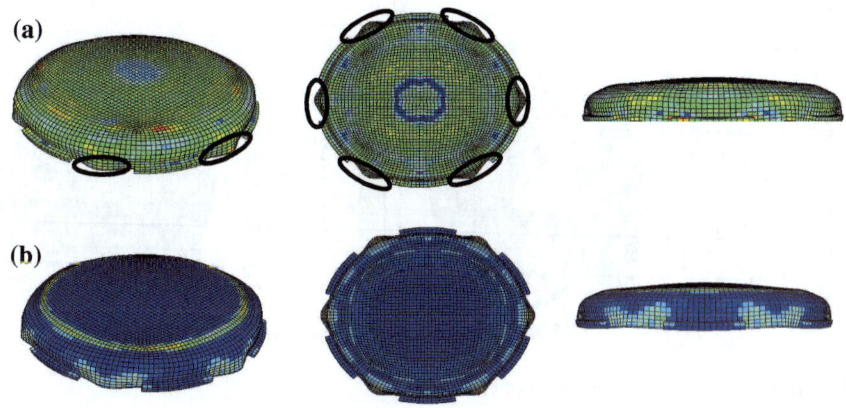

Fig. 3 Stress field and deformation of cover

4.1 Velocity of 25 m/s Normal to the Impact Cover

The dynamic response of the turntable impact normal to the cover is simulated. The turntable impact velocity is 25 m/s. The cover does not separate from the centrifuge body during impact process. The von Mises stress distribution is shown in Fig. 2a. It indicates that the cover brim is under high stress. The equivalent plastic strain of the centrifuge is shown in Fig. 2b. Plastic deformation occurs in the brim of cover's ears. Detailed stress and plastic deformation distribution of the cover is shown in Fig. 3. The elliptical outline means large plastic deformation area in the figure. Therefore we can conclude that the rotary centrifuge can endure turntable impact of 25 m/s. There is no potential safety hazard under this impact condition.

4.2 Velocity of 50 m/s Normal to the Impact Cover

At a normal impact speed of 25 m/s the cover of the rotary centrifuge is safe, therefore the turntable velocity is increased to 50 m/s. The dynamic behavior of the centrifuge under an impact of 50 m/s is simulated. The cover separates from the centrifuge body during the impact process. The von Mises stress distribution is shown in Fig. 4a. It indicates that the cover brim is under high stress. The equivalent plastic strain of the centrifuge is shown in Fig. 4b. Large plastic deformation occurs in the brim of the cover's ears.

Detailed stress and plastic deformation distribution of the cover is shown in Fig. 5. The cover's ears distort to the center during the impact process. The junction between cover and centrifuge is broken. It shows that the rotary centrifuge can not endure a 50 m/s turntable impact. It will bring potential safety hazards under the impact condition.

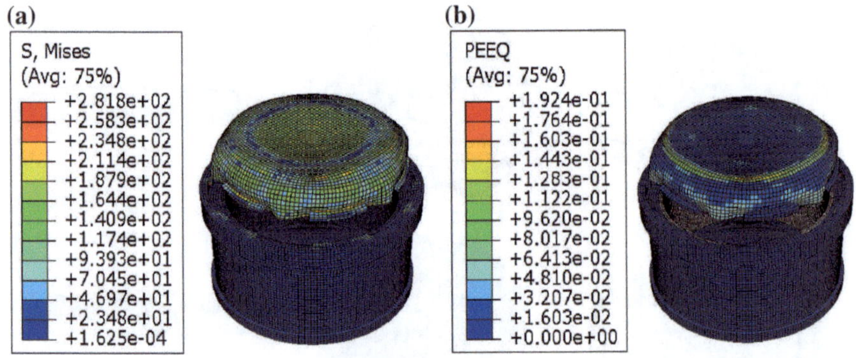

Fig. 4 Stress field and deformation of centrifuge

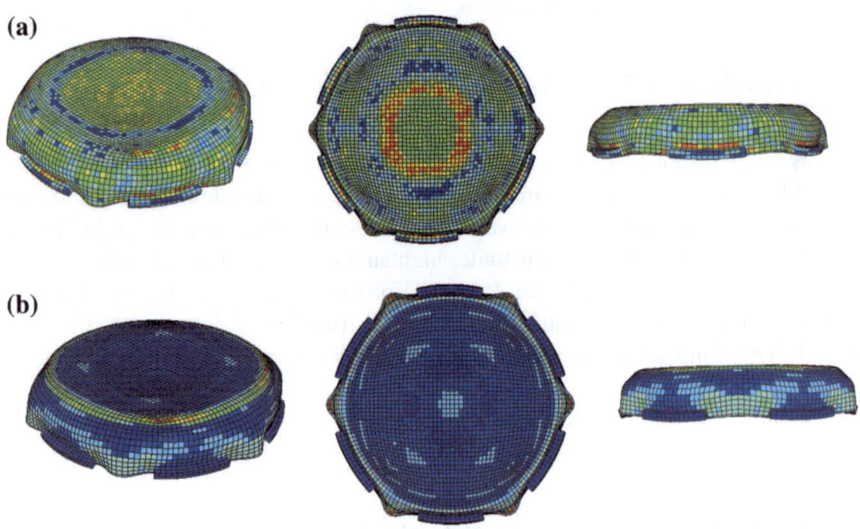

Fig. 5 Plastic deformation of cover

4.3 Velocity of 270 m/s on the Normal Impact Cavity Wall

In order to obtain the centrifuge wall impact resistance, the dynamical behavior of the turntable impacting cavity wall is analyzed. The speed of the turntable is 270 m/s. Stress and equivalent plastic strain distribution of the centrifuge wall is shown in Fig. 6. It becomes drum shape during the impact process. The wall undergoes large plastic strain, but without breaking.

The cover is well connected with the centrifuge body. The deformation of the cover and turntable is shown in Fig. 7. The cover almost undergoes elastic deformation in the impact process, as shown in Fig. 7a. The turntable breaks into pieces,

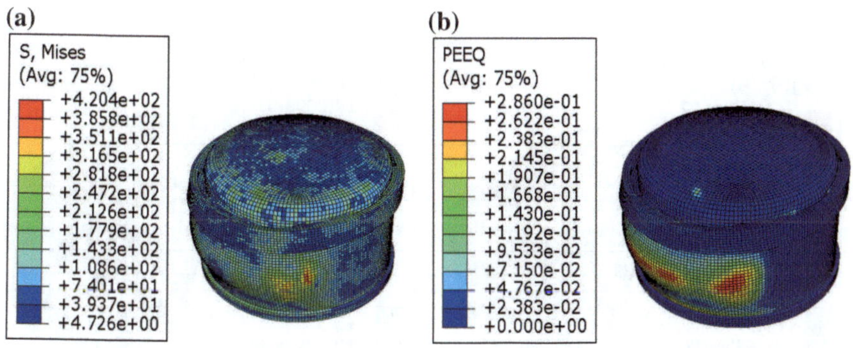

Fig. 6 Stress field and deformation of centrifuge

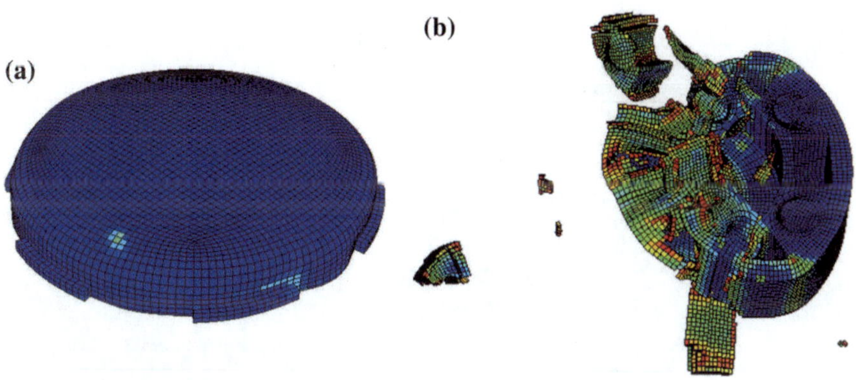

Fig. 7 Deformation of cover and turntable

as shown in Fig. 7b. It indicated that the centrifuge wall is strong enough to resist a turntable impact of 270 m/s.

4.4 Velocity of 270 m/s Impact Slanted to the Cavity Wall

The slanted impact should be considered in the centrifuge safety evaluation. The dynamic behavior of the slanted turntable impacting cavity wall is analyzed. The speed of the turntable is 270 m/s. The stress and equivalent plastic strain distribution of the model is shown in Fig. 8. The centrifuge wall becomes drum shape, without breaking, while the Cover breaks and separates from the centrifuge body.

The deformation of cover and turntable is shown in Fig. 9. The cover suffers large plastic deformation, as shown in Fig. 9a. The turntable is compressed up to

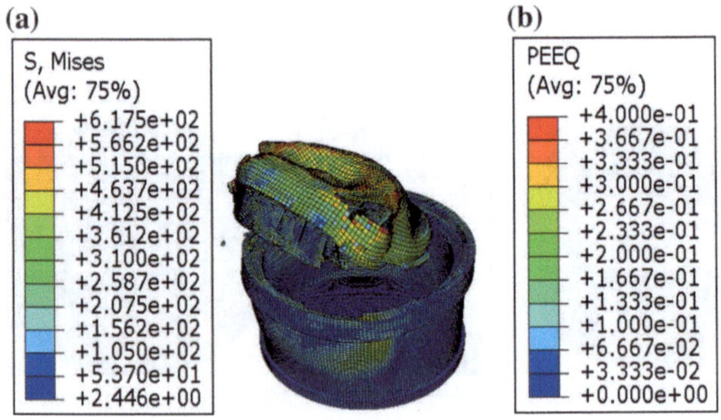

Fig. 8 Stress field and deformation of centrifuge

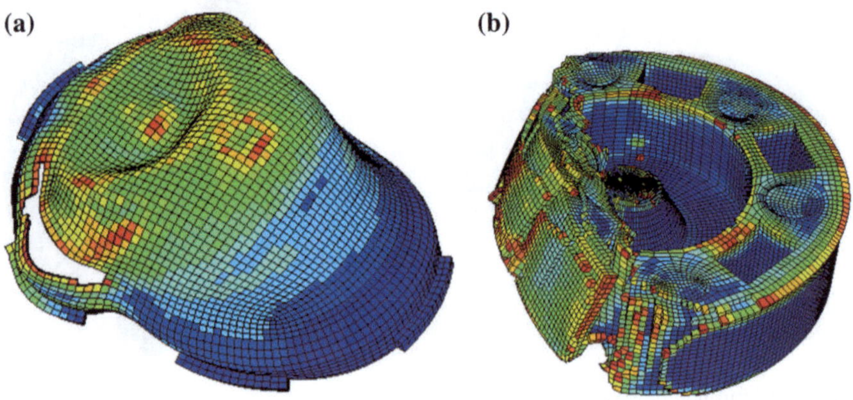

Fig. 9 Deformation of cover and turntable

two-thirds, as shown in Fig. 9b. It can be concluded that it is dangerous in the high speed oblique impact condition.

5 Discussion

For a conventional impact cushion structure, the stress versus strain curve of an ideal cushion material is with wide range plastic plateau phase. The stress value is almost a constant in the compression process. Cushion materials are usually classified into energy dissipation and energy storage. The rotary centrifuge guarded body is taken as energy dissipation type in this work. It shows that the cover separates from the centrifuge body under normal impact cavity wall conditions, as

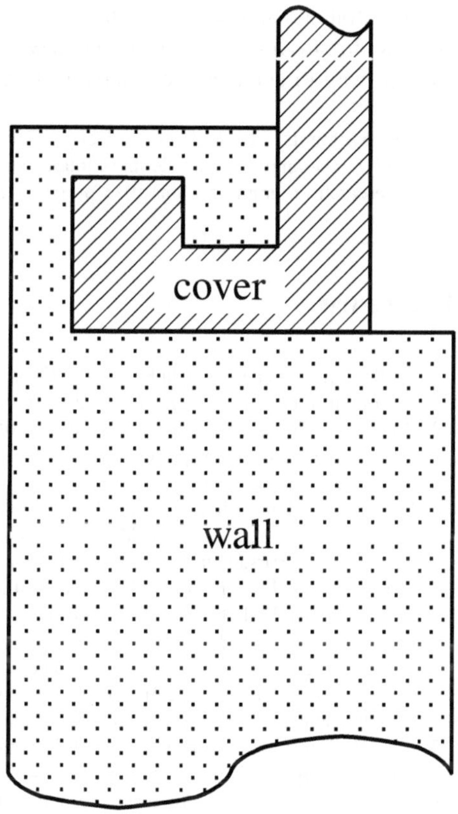

Fig. 10 Schematic illustration of cover and cavity wall

shown in Figs. 4 and 8. The cover is with large plastic deformation, but without breaking. An optimization constraint design between the cover and the cavity wall is provided to enhance the rotary centrifuge structure safety, as shown in Fig. 10. The improved constraint structure is like a high pressure pot, which can restrict axial and radial displacements of the cover. It will induce deformation mode of cavity wall and cover under high speed impact condition. Much more energy dissipation is realized for the improved structure. It is benefit to enhance rotary centrifuge impact guarded properties.

6 Conclusions

According to the simulation results of the rotary centrifuge under different impact conditions, a number of conclusions are obtained. The rotary centrifuge wall is strong enough to endure a 270 m/s impact in a 45° angle. The rotary centrifuge cover can endure 25 m/s impact at normal conditions. The cover suffers from large deformation and separates from the centrifuge body when the impact speed is over 50 m/s. Improving the connecting between the centrifuge cover and body is a feasible way to enhance the mechanical shock resistance of the rotary centrifuge.

Acknowledgments The authors gratefully acknowledge the funding by the National Natural Science Foundation of China under the contract No. 11302211, 51174173, 11390361 and National Basic Research Program of China (973 program) under the contract No. 2010CB832700.

References

1. Guillow SR, Lu G, Grzebieta RH (2001) Quasi-static axial compression of thin-walled circular aluminium tubes. Int J Mech Sci 43(9):2103–2123
2. Binchao L, Guiping Z, Tianjian L (2011) Critical conditions and optimal design of closed-celled aluminum foam protection under low velocity impact. Chin J Solid Mech 32 (4):325–338 (in Chinese)
3. Jeenager VK, Pancholi V (2014) Influence of cell wall microstructure on the energy absorption capability of aluminium foam. Mater Des 56:454–459
4. Alavi Nia A, Parsapour M (2014) Comparative analysis of energy absorption capacity of simple and multi-cell thin-walled tubes with triangular, square, hexagonal and octagonal sections. Thin-Walled Struct 74:155–165
5. Ajdari A, Nayeb-Hashemi H, Vaziri A (2011) Dynamic crushing and energy absorption of regular, irregular and functionally graded cellular structures. Int J Solids Struct 48(3):506–516
6. Kumar SJA, Ahmed KS (2013) Compression behavior and energy absorption capacity of stiffened syntactic foam core sandwich composites. J Reinf Plast Compos 32(18):1370–1379
7. Lee JH, Wang L, Boyce MC et al (2012) Periodic bicontinuous composites for high specific energy absorption. Nano Lett 12(8):4392–4396
8. Karasek ML, Strait LH, Amateau MF et al (1995) Effect of temperature and moisture on the impact behavior of graphite/epoxy composites. I: Impact energy absorption. J Compos Tech Res 17(1):3–10
9. Hanssen AG, Langseth M, Hopperstad OS (2001) Optimum design for energy absorption of square aluminium columns with aluminium foam filler. Int J Mech Sci 43(1):153–176
10. Surani FB, Kong X, Panchal DB et al (2005) Energy absorption of a nanoporous system subjected to dynamic loadings. Appl Phys Lett 87(16):163111
11. Jang BZ, Chen LC, Wang CZ et al (1989) Impact resistance and energy absorption mechanisms in hybrid composites. Compos Sci Technol 34(4):305–335
12. Duan Y, Keefe M, Bogetti TA et al (2006) A numerical investigation of the influence of friction on energy absorption by a high-strength fabric subjected to ballistic impact. Int J Impact Eng 32(8):1299–1312

production of capillary active insulations, thus facilitating an increase of the thermal resistance of buildings with damaged waterproofing or a high degree of moistening. The aim of the development of a suitable recipe was to produce a material with strength in the range of 0.3–0.6 MPa, without the use of an autoclave, with the thermal conductivity coefficient within the range of 0.06–0.08 W/m·K. The monitored properties of these materials include the thermal conductivity coefficient, the diffusion resistance coefficient and the capillary absorption coefficient. The function of the developed recipes for the production of non-autoclaved diffusion-open board insulations is presented on calculation simulations performed in the Delphin software [6], which allows you to take into account the real moisture processes taking place in the construction. Another view of the preparation of lightweight cement composites using hemp shive is presented in works [7–9]. The best results in these works are achieved by lightweight cement composites using binders based on MgO cements, which we also want to deal with in the next stages of development in similar fashion, i.e. replacing the primary binding matrix with fly ash from power plants.

2 Methodology

2.1 *Theory of Moisture Transport in Materials*

According to the way in which moisture can get into the building structure, we distinguish [1]:

- building moisture (initial—built-in during the production): disappears from the structures after a certain period of operation of the building;
- subsoil moisture (action of capillary forces of the surrounding soil): is considered only in the case of missing or non-functioning waterproofing;
- rain moisture (the effect of atmospheric precipitation);
- operating moisture (from the internal environment): depends on the method of operation of the building;
- sorption moisture (from the internal and from the external environment, as a result of the hygroscopic properties of the material): at steady temperature and humidity, there is a balance between the material moisture and the moisture of air that surrounds it—the so-called equilibrium moisture (sorption and operating moisture can be the same under certain conditions);
- condensed moisture (water vapour condensation on the surface of or inside the structure).

If the moisture in the structure after some time of operation of the building (2–4 years) is affected only by rain, sorption and condensed moisture (initial building moisture has already vaporized and the waterproofing is fully functional), this moisture is referred to as stable or practical.

In conventional building materials (porous), moisture spreads in two ways [2]:

- diffusion of water vapours (moisture movement in the vapour phase): it takes place when there is a gradient of the partial pressures of water vapour between the external and internal environment, which separates the building structures—the condition is that the pores and capillaries in the material of the structure must have a larger diameter than the diameter of a water molecule (generally $>10^{-7}$ m);
- moisture conductivity (also: capillary conductivity—the ability of a material to convey moisture in the liquid phase towards its surface, where it evaporates or diffuses): it takes place when there is a temperature gradient (the temperature affects the viscidity of water and its surface tension), and when there is a moisture gradient in the material (the conveyance of a water "film" on the surface of the pores depends on the difference of relative humidity on the opposite sides of the pores)—the condition is a continuous network of pores and capillaries.

2.2 Description of the Simulation Tool Delphin

The calculation performed in the Delphin simulation software tries to achieve the most accurate result of hygrothermal processes in structures approaching reality. The calculation uses the actual climatic conditions in the region of Ostrava.

The dynamic processes taken into account in the simulation are:

- conduction of heat—thermal conductivity depending on moisture (not taking into account latent heat);
- accumulation of heat—accumulation of heat depending on moisture;
- diffusion of water vapour—diffusion of water vapour depending on the changing moisture content of the material;
- transformations of phases—balancing of the evaporation and condensation processes with regard to the evaporating cooling;
- capillary transfer of liquid water—transfer of liquid water depending on the moisture;
- accumulation of moisture—from the difference of moisture flows to and from the space (diffusion of water vapour + capillary transfer); hygroscopic charge according to the measured function of the accumulation of moisture;
- air flow—calculation of the air pressure profile, convective air flow due to pressure gradients.

The possible phenomena that have been excluded from the calculation can include:

- thermo-diffusion, diffusion thermics (Dufour and Soret phenomenon) and the production of internal energy due to compression and friction;
- the transport properties are isotropic, with no directional dependence.
- the effects of electric fields (gravity acts as a single volume force);

- turbulent flow;
- the hysteresis of the function of moisture accumulation is not taken into account,
- the material properties are homogeneous in each discretized volume element,
- time changes of the baric field during the pre-definable time step (in the order of 1–10 min).

Numerical simulation of the critical detail of the construction in the Delphin software The presented critical detail of construction, see Fig. 1, is numerically evaluated from the point of view of the course of the weight of moisture in the foundation structure after gluing thermal insulation calcium silicate boards.

The examined detail was evaluated in unsteady weather conditions, with constant added moisture representing damaged waterproofing, and increased initial weight of moisture representing an object unused for a long time. These boundary conditions were used to model a structure in seven versions of redevelopment treatments and a reference state with lime-cement plaster only. The measured thermal and technical parameters of the developed recipes (materials) were incorporated into the Delphin software, which was used to prepare a 2D moisture model of the peripheral structure of the house at the point of connection to the foundation structure of the house. The moisture simulation takes into account both the formation of condensates in the place of application of calcium silicate boards and the moisture added due to the penetration of moisture through the damaged waterproofing. The area in question was evaluated using six options, see Fig. 3, by applying the materials developed with a thickness of 50 mm. The simulations also included the currently used autoclaved board materials. The best boundary value, which the technical measure is attempting to get close to, is the simulation of 30 mm lime-cement plaster on the exterior of the building representing the original operating condition. The boundary conditions of the calculation are presented in Table 1.

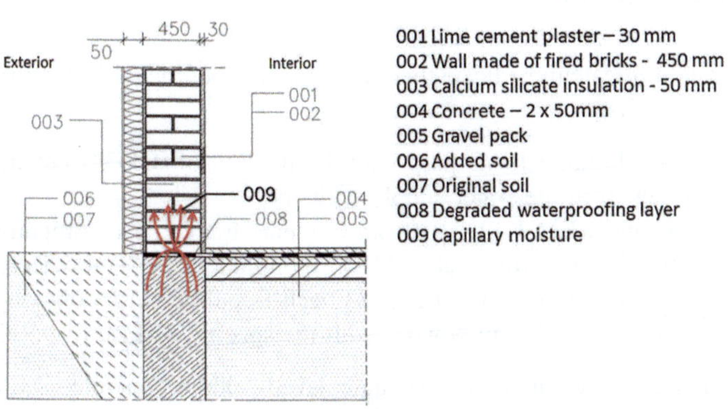

Fig. 1 Detail of the most common method of the construction of a peripheral building foundation with damaged waterproofing evaluated by the Delphin simulation software

Table 1 The initial boundary conditions of the calculation if the moisture behaviour of the building in the Delphin software

[a]The boundary conditions calculation		
Thermal resistance to heat transfer in the interior	Calculation of condensation on the surface	[b]Rsi: 0.25 $m^2 \cdot K/W$
Thermal resistance to heat transfer in the exterior	Calculation of condensation on the surface	[b]Rse: 0.04 $m^2 \cdot K/W$
The design outdoor temperature	Long-term average temperature measured by [c]CHMI	
The design temperature of indoor air	[e]Sinusoidal variable temperature in the range of 19.5–20.5 °C	
The design relative humidity of outside air	Long-term average moisture content measured by [c]CHMI	
The design relative humidity of indoor air	[d]Sinusoidal variable moisture content in the range of 44–55 %	
Direct sun radiation (short wave component)	[c]Reflection coefficient of the surrounding ground: 0.4 [−]	
Diffuse sun radiation (short wave component)	[c]Absorption coefficient of the building surface: 0.6 [−]	
Atmospheric counter radiation	[c]Emission coefficient of the building surface 0.9 [−]	

[a]All the used input data is based on hourly measurements of an average year
[b]CSN EN ISO 13788 [10] determines a standard value of the heat resistance during the transfer on the inner side of the structure with the value of Rsi = 0.25 $m^2 \cdot K/W$ and on the outer side with the value of Rse = 0.04 $m^2 \cdot K/W$, for the calculations of water vapour condensation
[c]The calculation uses long-term average hourly values for the area of Ostrava CZ, taken from the Czech Hydro-meteorological Institute (CHMI)
[d]The design temperatures of indoor air are based on the recommendations of CSN EN 12831 [11], which have been adjusted by taking into account the hysteresis of the thermostat of indoor environment ±0.5 °C
[e]The dispersion of the design relative humidity of indoor air is based on long-term measurements of relative humidity of indoor air in the reference building in order to more accurately describe the hygroscopic load of the structure

2.3 Input Material

The production of calcium silicate thermal insulation boards takes advantage of the following materials:

- Aggregate—Aluminosilicate pulp with the fraction of 0.063/0.125 mm;
- Water—water from the water supply network;
- Binder—cement CEM I 42, 5R (from Cement Hranice, a.s. company);
- Binder—fly ash K12 (from Dalkia Česká republika, a.s.—Třebovice company);
- lime Cl90 (6–9) min (from Carmeuse Czech Republic s.r.o. company);
- foaming agent—aluminium powder with the specific surface of 9500 cm^2/g.

A scheme of the treatment of waste autoclaved cellular concrete used as filler in calcium silicate insulation is presented in Fig. 2.

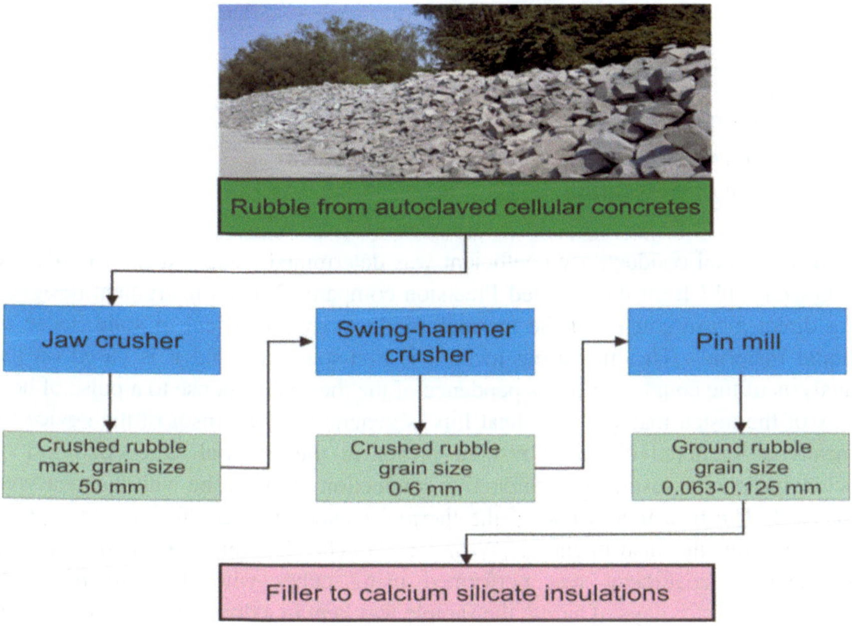

Fig. 2 The treatment process of waste autoclaved cellular concrete to be used as filler in calcium silicate insulations

2.4 Tested Properties

The newly developed material of thermal insulation calcium silicate boards based on waste autoclaved cellular concrete with a maximum grain size of 0.125 mm prepared according to the scheduled recipes has been tested to: density according to CSN EN 1015-10 [12], compressive strength according to CSN EN 12390-3 [13], capillary absorption coefficient according to CSN EN 1015-18 [14], diffusion resistance coefficient, thermal conductivity coefficient λ and porosity.

The determination of the diffusion resistance coefficient was carried out using the wet bowl method. The test specimen was placed between two environments of the same temperature and of different relative humidities. A bowl of water was placed in the environment with a higher relative humidity of 95 % and it was weighed at the beginning of the test. There is a diffusion of water vapour through the material from the environment with the higher relative humidity into the environment of relative humidity below 50 %. The bowl with water was weighed again, including the increase of weight in the measured body, and Eq. (1) was used to calculate the equivalent diffusion thickness.

$$s_{dw} = \frac{\delta_o \cdot A \cdot \Delta t \cdot \Delta p}{\Delta m} \tag{1}$$

where

δ_0 water vapor permeability of air ($kg \cdot Pa^{-1} \cdot s^{-1} \cdot m^{-1}$);
A specimen area (m^2);
Δt time difference (s);
Δp difference of partial pressures of water vapours (Pa);
Δm change of weight (kg);

The thermal conductivity coefficient was determined using a measuring device ISOMET 2114 from the Applied Precision company. It is an instrument designed for direct measurements of the thermal conductivity coefficient of solid, loose or liquid materials. The measuring method is non-stationary and it is based on the analysis of the course of time dependence of the thermal response to a pulse of heat flux of the tested material. The heat flux is generated by a sensor of the device by means of scattered electrical resistor power in the material. The condition of achieving correct results is a conductive connection of the probe with the measured material. The measured value of the thermal conductivity coefficient can be read directly from the instrument in ($W \cdot m^{-1} \cdot K^{-1}$). The determination of the thermal conductivity coefficient was performed using cubes with the dimensions of 100 mm. After 28 days of aging, the sample was kept in a drying oven, where it was dried at 105 ± 1 °C for 48 h. The dried sample was then placed in an exicator and, after cooling, it was inserted into a propylene bag. The measurements were performed using a surface probe with the range of 0.04–2.00 $W \cdot m^{-1} \cdot K^{-1}$ placed in the middle of the test specimen area. The determination of the porosity was conducted on a dried test specimen with the dimensions of 100 × 100 × 100 mm, which was crushed and ground to a grain size of 0.06 mm. The resulting pulp of the sample was poured into a measuring cylinder and the value of the volume was subsequently read from the measuring cylinder scale. The difference between the above presented volumes provides an approximate total porosity in the material.

3 Results and Discussion

3.1 Proposed Experimental Recipes

5 experimental recipes have been prepared in order to verify the incorporation of finely ground waste autoclaved cellular concrete with the fraction of 0.063/0.125 mm as a source filler of calcium silicate cement composite, and their compositions are shown in Table 2.

Only Portland cement CEM I 42.5R was used as the binding agent in recipe 1. In order to save cement, this binding agent was replaced with fly ash K12 from Trebovic with amounts of 10 % of the weight (recipe 2), 20 % of the weight

Table 2 Composition of experimental recipes of capillary active insulation

Mixture components	Measuring unit	Recipe 1	Recipe 2	Recipe 3	Recipe 4	Recipe 5
Aluminosilicate pulp (crushed 0.063–0.125 mm)	g	261.5	261.5	261.5	261.5	261.5
Water	g	351	351	351	351	351
Cement CEM I 42.5R	g	60.0	54.0	48.0	42.0	36.0
Lime Cl90 (6–9) min	g	16.2	16.2	16.2	16.2	16.2
Al powder 9500 cm^2/g	g	1.9	1.9	1.9	1.9	1.9
Fly ash K12 Třebovice	g	–	6.0	12.0	18.0	24.0

(recipe 3), 30 % of the weight (recipe 4) and 40 % of the weight (recipe 5). The mixtures for the production of lightweight concrete boards were prepared in laboratory conditions at the Faculty of Mining and Geology, VŠB—Technical University of Ostrava. The dosing of the individual components was performed according to weight. The mixing time of a mixture was 5 min. to obtain a compact (homogeneous) mixture, which was further applied in moulds with the volume of 8 dm^3.

3.2 Properties of Experimental Recipes

The results of the tested properties of experimental recipes are presented in Table 3.

Table 3 clearly shows that the compressive strength of the experimental recipes ranged from 0.38 to 0.82 $N\cdot mm^{-2}$, the density of hardened lightweight concrete in the dry state ranged from 347 to 369 $kg\cdot m^{-3}$, the capillary absorption coefficient was within the interval of 0.095–0.149 $kg\cdot m^{-2}\cdot s^{-0.5}$, the thermal conductivity coefficient λ was within the range from 0.070 to 0.081 $W\cdot m^{-1}\cdot K^{-1}$ and the porosity was within the range from 78.27 to 81.01 %.

Table 3 Results of the tests of physical and mechanical properties of experimental recipes

Marking	Compressive strength ($N\cdot mm^2$)	Density ($kg\cdot m^{-3}$)	Coefficient of capillary absorption Aw ($kg\cdot m^{-2}.\cdot s^{-0.5}$)	Thermal conductiv. coefficient λ ($W\cdot m^{-1}.\cdot K^{-1}$)	Total porosity (%)
Recipe 1	0.82	347	0.095	0.070	81.01
Recipe 2	0.75	357	0.128	0.073	78.31
Recipe 3	0.64	357	0.121	0.073	78.27
Recipe 4	0.52	357	0.149	0.081	78.39
Recipe 5	0.38	369	0.145	0.079	75.23

3.3 The Course of Moisture Contained in the Evaluated Structure After the Installation of Calcium Silicate Boards During the Following 5 Years

Figure 3 presents curves characterizing the course of moisture in the substructure made from solid bricks that was redeveloped using the developed insulator with an indication of the specific recipes according to Table 2.

The curve in the lowest part of the graph describes the course of moisture in the original structure protected only by lime-cement plaster. The best results have been obtained by applying the board insulation with the thickness of 5 cm manufactured according to the experimental recipe no. 5, where the weight of moisture in the substructure increases from the initial upper limit value of 4.31 % to the value of 4.65 %. According to [15], the structure with the weight of moisture in the range of 3–5 % is included in the moisture category of "low", which is why we rate the effect of this technical measure with very low impact on the increase of the built-in weight of moisture as functional. The difference in the weights of moisture corresponds to approximately 2 kg of water in 1 m^3 of construction. The purpose of this technical measure is to improve the transmission heat loss coefficient U = 1.38 $W/m^2{\cdot}K$ to U = 0.74 $W/m^2{\cdot}K$, i.e., by 46 %. On the other hand, if the same structure is insulated using an insulator based on expanded polystyrene with the thickness of 2.5 cm, thus obtaining the same heat resistance as in the case of recipe no. 5 with the thickness of 5 cm, 10 % of the weight of moisture in the structure is exceeded as early as during the third year. The amount of moisture is classified as very high and leads to moisture effects in the interior of the building. The lower part of the graph makes it possible to compare the course of moisture in the developed recipes with

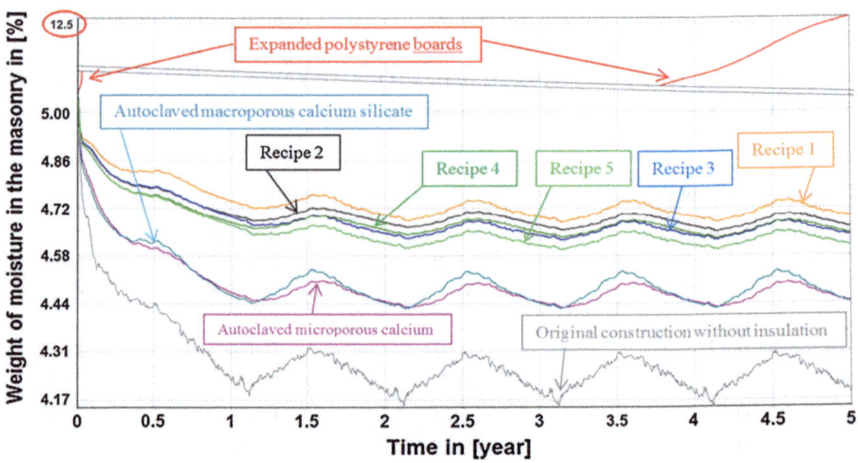

Fig. 3 Output of a numerical simulation of the course of the weight of moisture in masonry substrates after application the insulating material

autoclaved materials, which are available on the market. These materials differ from each other in their pore structure, where one material is of macro-porous and other one of micro-porous character.

4 Conclusion

The presented results have demonstrated that modified rubble from autoclaved cellular concrete with the fraction of 0.063–0.125 mm is suitable as a new type of filler for the production of thermal insulation calcium silicate boards for the segment of energy redevelopment. It is also possible to say that all the developed recipes appear to be suitable for use in the area of energy redevelopment of buildings with damaged waterproofing. The new feature of the research lies in the determination of the recipes for the production of thermal insulation boards based on waste cellular concrete pulp, omitting the autoclave process. The resulting board retains the good properties approaching autoclaved materials, i.e., it can naturally remove moisture from the structure and increase the thermal resistance of the building envelope, while improving the quality of the waste materials.

Acknowledgments Article has been done in connection with projects:

SGS reg. no. SP2014/9 Utilization of industrial waste containing aluminosilicate for material structures with defined surface.

Institute of clean technologies for mining and utilization of raw materials for energy use—Sustainability program, reg. no. LO1406 supported by Research and Development for Innovations Operational Programme financed by Structural Founds of Europe Union and from the means of state budget of the Czech Republic.

IT4Innovations Centre of Excellence project, reg. no. CZ.1.05/1.1.00/02.0070 supported by Operational Programme Research and Development for Innovations funded by Structural Funds of the European Union and state budget of the Czech Republic.

RMTVC No. LO1203.

References

1. Grunewald J, Ruisinger U, Häupl P (2006) The Rijksmuseum Amsterdam—hygrothermal analysis and dimensioning of thermal insulation. In: 3rd international building physics conference—research in building physics and building engineering, Canada
2. Ruochong Y, Jipeng Z, Zhishen W, Zhiren W, Min L, Changhai P (2011) Thermal insulation and strength of autoclaved light concrete. J Wuh Uni Tech Mater Sci Edn. doi:10.1007/s11595-011-0184-6
3. Ruisinger U, Grunewald J (2009) Title of preprint. http://tu-dresden.de/die_tu_dresden/fakultaeten/fakultaet_architektur/ibk/research/researchprojects/2006_Feuchteatlas/Feuchteatlas.pdf. Accessed 22 April 2009
4. Václavik V, Dvorský T, Dirner V, Daxner J, Šťastný M (2012) Polyurethane foam as aggregate for thermal insulating mortars and lightweight concrete. Tehnicki Vjesnik 19 (3):665–672

5. Václavik V, Valíček J, Dvorský T, Hryniewicz T, Rokosz K, Harničárová M, Kušnerová M, Daxner J (2012) A method of utilization of polyurethane after the end of its life cycle. Rocz Ochrona Srodowiska 14:96–106
6. Scheffler G, Grunewald J (2003) Material development and optimisation supported by numerical simulation for a capillary-active inside insulation materiál. In: 2nd international conference on building physics, Antwerp
7. Stevulova N, Kidalova L, Cigasova J, Junak J, Sicakova A, Terpakova E (2013) Lightweight composites containing hemp hurds. In: 6th international conference on concrete and concrete structures 2013, Zilina
8. Cigasova J, Stevuvola N, Sicakova A, Junak J (2013) Some aspects of lightweight composites durability. Chem Eng Trans 32:1615–1620. doi:10.3303/CET1332270
9. Junak J, Stevulova N (2011) Potential of selected industrial wastes in civil engineering applications. In: 1st international multidisciplinary scientific geoconference and EXPO, Varna, 20–25 June 2011
10. CSN EN ISO 13788 (2013) Hygrothermal performance of building components and building elements—internal surface temperature to avoid critical surface humidity and interstitial condensation—calculation methods, (Czech normalization institute, Prague, 2013)
11. CSN EN 12831 (2005) Heating systems in buildings—method for calculation of the design heat load, (Czech normalization institute, Prague, 2005)
12. CSN EN 1015-10 (2000) Methods of test for mortar for masonry—part 10: determination of dry bulk density of hardened mortar, (Czech normalization institute, Prague, 2000)
13. CSN EN 12390-3 (2009) Testing hardened concrete—part 3: compressive strength of test specimens, (Czech normalization institute, Prague, 2009)
14. CSN EN 1015-18 (2003) Methods of test for mortar for masonry—part 18: determination of water absorption coefficient due to capillarity action of hardened mortar, (Czech normalization institute, Prague, 2003)
15. CSN P 73 0610 (2000) Waterproofing of buildings—the rehabilitation of damp masonry and additional protection of buildings against ground moisture and against atmospheric water—basic provision, (Czech normalization institute, Prague, 2000)

Comparison of Some Structural and Stainless Steels Based on the Mechanical Properties and Resistance to Creep

Josip Brnic, Goran Vukelic and Sanjin Krscanski

Abstract Knowledge of the properties of materials and their behavior in certain environmental conditions is one of the most important factors in the procedure of materials selection. In accordance with this fact, this paper presents and analyzes the experimental results relating to two structural (1.0044, 1.7228) and two stainless steel (1.4305, 1.4122) materials. Stress-strain diagrams as well as creep curves related to short-time creep are presented. According to the mentioned diagrams, the ultimate tensile strength, yield strength and modulus of elasticity are determined. On the other hand, based on material creep curves, some conclusions regarding to creep resistance may be given. Also some data related to Charpy impact energy is shown as well as fracture toughness assessment based on impact energy is made. Based on experimental results it can be said that all of the investigated materials have quite high tensile strength and yield strength. Also, these materials may be treated as creep resistant at temperature of 400 °C if the stress level does not exceed 50 % of the yield strength at this temperature.

Keywords Structural and stainless steels · Mechanical properties · Creep · Impact energy

J. Brnic (✉) · S. Krscanski
Faculty of Engineering, University of Rijeka, Vukovarska 58, Rijeka, Croatia
e-mail: brnic@riteh.hr

S. Krscanski
e-mail: sanjin.krscanski@riteh.hr

G. Vukelic
Faculty of Maritime Studies, University of Rijeka, Studentska ulica 2, Rijeka, Croatia
e-mail: gvukelic@pfri.hr

A. Öchsner and H. Altenbach (eds.), *Mechanical and Materials Engineering of Modern Structure and Component Design*, Advanced Structured Materials 70,
DOI 10.1007/978-3-319-19443-1_15

1 Introduction

The material for the design of structure is usually selected in accordance with purpose of the structure. The structure can, for example, be designed as a repository for gas under pressure, for movement of a load, for working at elevated temperatures, etc. However, the material properties of the designed structure must meet the conditions in which the structure will operate. Properties of the material are linked to the material chemical composition, processing path and material microstructure. Properties that depend on microstructure are called structure-sensitive properties. Among these properties can be counted mechanical properties like yield strength of the material, hardness, toughness, fatigue resistance, ductility. Processing is a way to develop and control microstructure, for example, hot rolling or something similar. A material should meet some of the requirements such as: high tensile strength, high creep resistance, fatigue strength, ductility, high temperature strength, heat resistance, resistance to high temperature corrosion, etc. In that way, the designer of the structure must be familiar with the knowledge of the material properties, e.g. he has to assess the material behavior under certain environmental conditions. Design philosophy includes material selection as well as production costs and both of these processes require optimization procedures [1, 2]. In general, optimization may be designated as making the best of things. In this case the term “best” refers to making the structure, for example, as light as possible, e.g., to minimize weight, or to make it as stiff as possible, etc. On the other hand, the stress analysis of the structure in the design process is commonly performed using the Finite Element Method [3]. It can be said that an engineering structure is usually designed, manufactured, maintained/controlled in order to guarantee that it does not contain any failures and that it can serve for the purpose for which it is intended. At room temperature and in the absence of adverse effects, a properly designed structure can support its static design load for an unlimited time [4]. Otherwise, at a sustained load of a certain level at elevated temperatures inelastic strains may occur in the material that increase with time. This phenomenon is known as creep [5]. Structure lifetime predictions and its safety during service life are key questions regarding its quality and reliability. Above implies that material availability, suitability for service conditions as well as the cost of the material should be considered. However, in engineering practice, a lot of failures may occur. These failures may be defined as any change in the size, shape or material properties of a structure that renders it incapable of satisfactorily performing its intended function. It is necessary to know why and how some engineering component has failed. In that way the main points related to the structural failure need to be mentioned and that cause of the origin of failure as well as mode of the failure manifestation. Usually, some failure causes worthy to be mentioned are: pre-existing defects or defects that initiate from imperfections, structural loading, corrosion, misuse (structure subjected to the conditions for which it was not designed), design errors, assembly error, improper maintenance, unforeseen operating conditions, yielding, creep, buckling, etc. The main attention in this research is paid to the comparison the material properties and creep resistance of selected structural and stainless steels [6–9]. Creep is usually defined as time-dependent inelastic strain under sustained load and elevated temperatures, creep may be said to be thermally activated process. Creep process at metals can be represented by creep curve consisting of three different stages and that primary (transient) stage, secondary (steady-state) stage and tertiary (accelerating) stage. Only a few percent (1–2) of creep strains is allowable in engineering practice. Dislocation climb, vacancy diffusion and grain boundary sliding are usually numbered as mechanisms of creep [10]. In addition, some data related to the minimum yield strength, tensile strength and elongation at room temperature for hot-rolled S275JR steel can be found in Ref. [11]. A study dealing with compression tests of $50CrMo_4$ steel to characterize its behavior at strains up to 150 % at appropriate strain rates are presented in [12]. Appling high temperature tension tests ductile damage evolution and fracture of a resulfurised stainless steel AISI 303 (1.4305) were analyzed in [13]. A study dealing with corrosion behavior of pipe steels used in Carbon Capture and Storage-technique (CCS) can be found in Ref. [14].

2 Data Related to Research

Materials under consideration were structural steels (1.0044/S275JR/ASTM A529; 1.7228/50CrMo4/AISI 4150) and stainless steels (1.4305/X10CrNiS18-9/AISI 303; 1.4122/X39CrMo17-1/AISI 420RM). Material 1.0044 was delivered as hot rolled bar and its applications are in many areas of engineering. Material 1.7228 can be used in statically and dynamically stressed larger cross-sections of structural components (aircraft and automotive industry, engines and machines). Material 1.4305 was delivered as a cold drown bar. It is primarily used in applications when corrosion or oxidation poses a problem. In addition material 1.4122 can be used in manufacturing of pump shafts, boat shafts for use in fresh water, then in polymer processing, compressor parts, etc. Test equipment in these investigations included: 400 kN material testing machine, the macroextensometer, a furnace (900 °C), high temperature extensometer and a Charpy impact machine. Specimens were machined from appropriate 18 mm steel rods. Material testing was performed in accordance with standards: uniaxial tests at room temperature were conducted in accordance with the standard ASTM: E8 M-11, while those at elevated temperatures in accordance with the ASTM: E21-09 standard. Creep tests were performed in accordance with ASTM: E139-11 standard, and Charpy impact tests were performed in accordance with ASTM: E23-07ae1 standard. All of the mentioned standards can be found in Ref. [15].

3 Research Results

3.1 Mechanical Properties

To compare the mechanical properties of the considered materials uniaxial tests were carried out. Using thus obtained engineering stress-strain diagrams at room and elevated temperatures (see Figs. 1, 2 and 3), it is possible to determine tensile strength, yield strength and the elastic modulus of the considered materials.

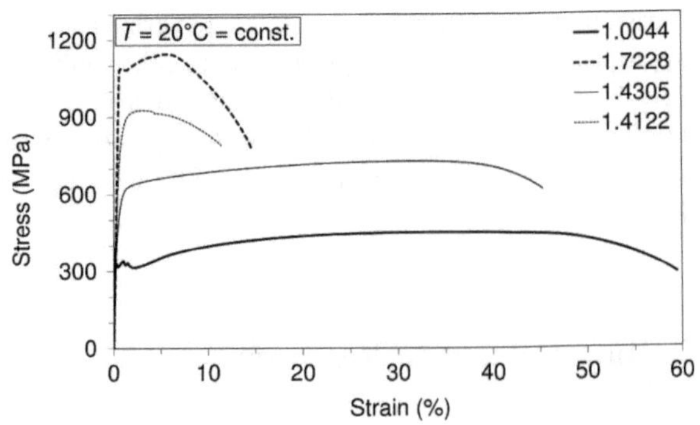

Fig. 1 Engineering stress-strain diagrams at room temperature for steels: 1.0044, 1.7228, 1.4305, 1.4122

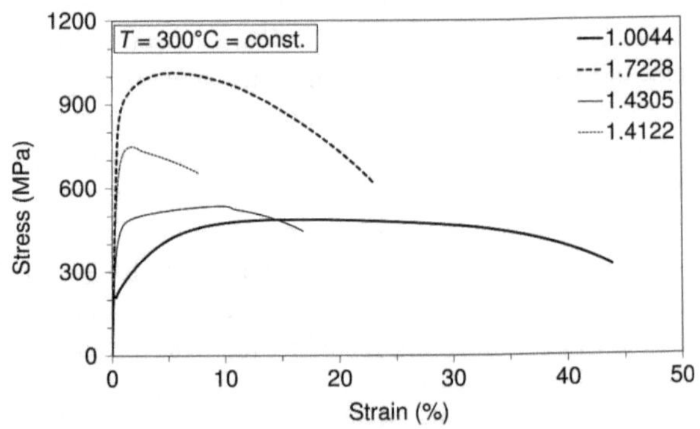

Fig. 2 Engineering stress-strain diagrams at temperature of 300 °C for steels: 1.0044, 1.7228, 1.4305, 1.4122

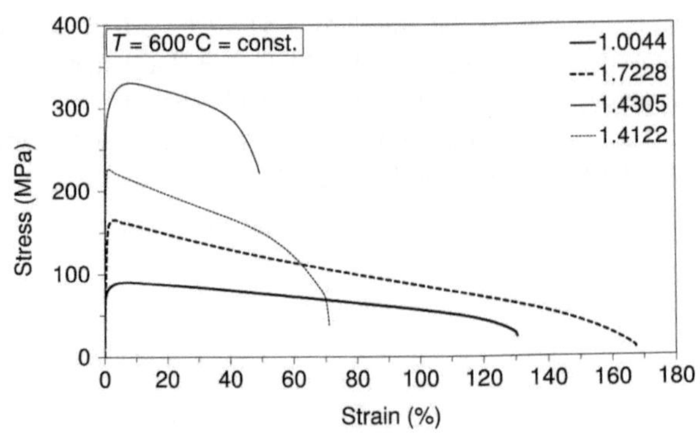

Fig. 3 Engineering stress-strain diagrams at temperature of 600 °C for steels: 1.0044, 1.7228, 1.4305, 1.4122

Based on the experimental results it is visible that all the materials at room temperature have a quite high ultimate tensile strength. The following data related to ultimate tensile strength, yield strength and modulus of elasticity at room temperature are as follows: (1.0044/452 MPa/323 MPa/211 GPa; 1.7228/1147 MPa/ 1090 MPa/204 GPa; 1.4305/728 MPa/467 MPa/187 GPa; 1.4122/927 MPa/ 746 MPa/208 GPa). Also it is visible that the lowest value of tensile strength at all test temperatures is the one of steel at 1.0044, while at the room temperature and at temperature of 300 °C steel 1.7228 has the highest ultimate tensile strength. At all of considered materials, ultimate tensile strength and yield strength decrease with temperature increase.

3.2 Short-Time Creep Tests

Several short time creep tests were carried out at selected stress levels and selected temperatures. At selected creep process, stress level is chosen to correspond to approximately the same percent of yield strength of the material under consideration that it has at the temperature of creep process. Creep tests were conducted at temperatures of 400, 500 and 600 °C. Creep curves are presented in Figs. 4, 5 and 6. As it can be seen on the basis of experiments, at a temperature of 400 °C, material 1.7228 tends to greater deformations. It should be noted, that this material was subjected to a higher level of stress at this temperature. As for creep at temperatures of 500 °C, it is evident that the materials 1.0044 and 1.4305 indicate a higher creep resistance. Regarding the creep process at 600 °C, it can be said that none of tested materials may be treated as creep resistant.

3.3 Assessment of Material Fracture Toughness Based on Experimental Impact Energy

As it is known, the yield strength of the material is a measure in structural design against plastic deformation while fracture toughness may serve as a measure against

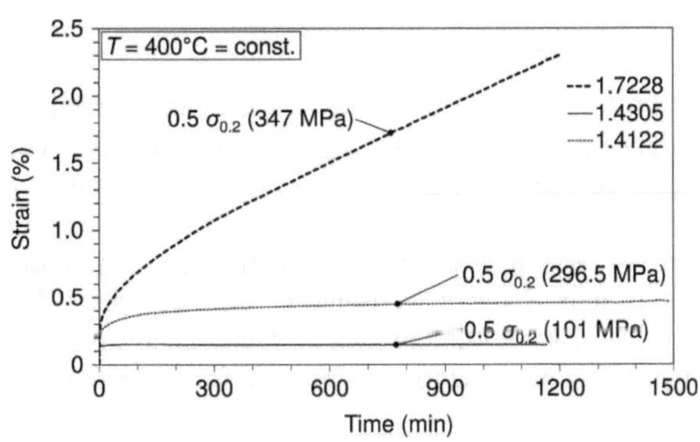

Fig. 4 Short-time creep process at temperature of 400 °C

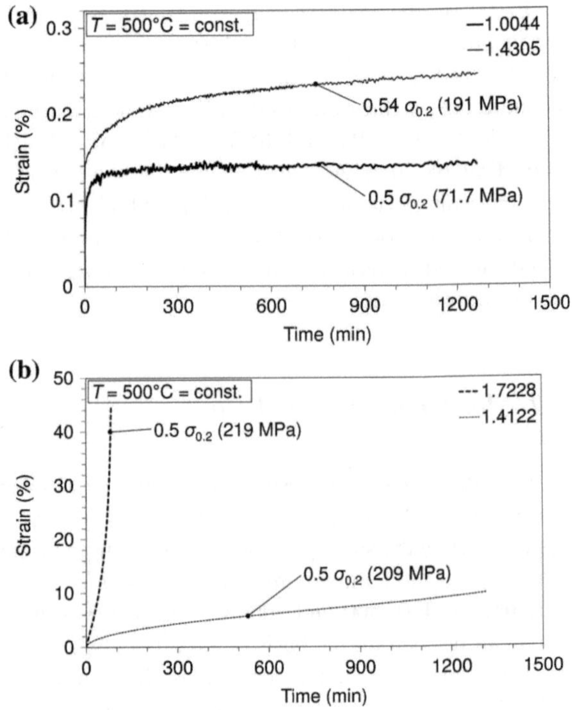

Fig. 5 Short-time creep process at temperature of 500 °C

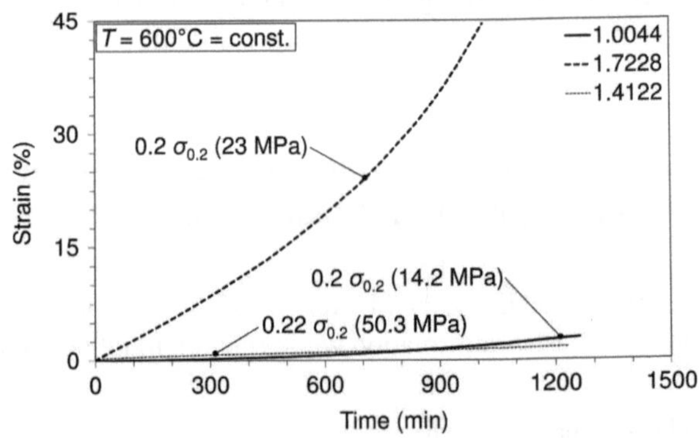

Fig. 6 Short-time creep process at temperature of 600 °C

fracture. However, the critical value of stress impact factor (SIF) is known as plane strain fracture toughness (K_{Ic}). This case implies that the fracture toughness of the material does not change with increasing thickness of the specimen. In general, the fracture toughness of the material is usually tested in laboratory conditions. Also, it can be said, that during this examination some problems may arise. On the other hand, it is also not advisable to use the results directly in engineering practice. To avoid some difficulties in experimental investigations, and for simplicity, the Charpy impact method can be used for impact energy determination. Based on the

Table 1 Charpy impact energy and fracture toughness calculation

Material: steel	Specimen 2 V-notch; temperature 20 °C	
	CVN (J)	K_{Ic} (Eq. 1) (MPa$\sqrt{m}$)
1.0044	210	245.9
1.7228	69	122
1.4305	46	94.5
1.4122	16	48.6

impact energy, an assessment of fracture toughness can be made. Using, for example, the Roberts-Newton formula that is valid regardless of temperature, fracture toughness can be calculated as follows [16]:

$$K_{Ic} = 8.47(CVN)^{0.63}. \quad (1)$$

In addition, in Table 1 some data is presented related to Charpy impact energy.

4 Conclusion

The research results presented in this paper can be useful for designers of structures that can be made of considered materials. On the basis of presented engineering stress-strain diagrams it is visible that all of the considered materials have enough high mechanical properties at room temperature but these properties decrease with an increase in temperature. Also, it is visible that material 1.7228 has the highest mechanical properties at room temperature. Regarding creep resistance, it may be said that at a temperature of 400 °C, materials 1.4305 and 1.4122 can be treated as creep resistant, while at temperature of 500 °C materials 1.0044 and 1.4305 may be treated in the same way. At the temperature of 600 °C practically none of considered materials is creep resistant. The highest value of Charpy impact energy at room temperature was measured for the case of material to be 1.0044.

Acknowledgment Research presented in this paper has been financially supported by Croatian Science Foundation under the project 6876 and by the University of Rijeka under the project 13.09.1.1.01.

References

1. Papalambros PY, Wilde DJ (2000) Principles of optimal design, 2nd edn. Cambridge University Press, Cambridge
2. Rao SS (2009) Engineering optimization, 4th edn. Wiley, New Jersey
3. Borst R, Crisfield MA, Remmers JJC, Verhoosel CL (2012) Nonlinear finite element analysis of solids and structures, 2nd edn. Wiley, Chichester

4. Boresi AP, Schmidt RJ (2003) Advanced mechanics of materials, 6th edn. Wiley, New York
5. Collins A (1993) Failure of materials in mechanical design, 2nd edn. Wiley, New York
6. Brnic J, Turkalj G, Niu J et al (2013) Analysis of experimental data on the behavior of steel S275JR—reliability of modern design. Mater Design 47:497–504
7. Brnic J, Canadija M, Turkalj G et al (2010) 50CrMo4 Steel-determination of mechanical properties at lowered and elevated temperatures, creep behavior and fracture toughness calculation. J Eng Mater-T ASME 132(2):021004–1–021004-6
8. Brnic J, Turkalj G, Canadija M et al (2012) Responses of austenitic stainless steel american iron and steel institute (AISI) 303 (1.4305) subjected to different environmental conditions. J Test Eval 40(2):319–328
9. Brnic J, Turkalj G, Lanc D et al (2013) Testing and analysis of X39CrMo17-1 steel properties. Constr Build Mater 44:293–301
10. Raghavan V (2004) Materials science and engineering. Prentice-Hall of India, New Delhi
11. Simões da Silva L, Rebelo C, Nethercot D et al (2009) Statistical evaluation of the lateral–torsional buckling resistance of steel I-beams, part 2: variability of steel properties. J Constr Steel Res 65(4):832–849
12. Diot S, Gavrus A, Guines D et al (2003) Identification of a steel compression behaviour: from quasi static approach to dynamic one. Mec Ind 4–5:519–524
13. Tinet H, Klöcker H, Le Coze J (2004) Damage analysis during hot deformation of a resulfurised stainless steel. Acta Mater 52(13):3825–3842
14. Pfennig A, Kranzmann A (2011) Reliability of pipe steels with different amounts of C and Cr during onshore carbon dioxide injection. Int J Greenh Gas Con 54:757–769
15. ASTM International (2012) Annual book of ASTM standards—metal test methods and analytical procedures vol. 03.01. ASTM International, Baltimore
16. Roberts R, Newton C (1981) Interpretive report on small scale test correlation with K_{Ic}. Dana Weld Res Council Bull 265:1–16

Investigation of the Influence of Improvement on the Effect of Strain hardening of 34CrMo4 in the Production of Seamless Steel Pressure Vessels from Pipes

V. Marušić, I. Lacković and L. Marušić

Abstract This paper analyzes the process of making a seamless pressure vessel Ø229 mm on a spinning machine. The pressure vessels are exposed to high operating pressure when in use (up to 200 bar). Therefore, the material of the pressure vessels must meet specific requirements to prevent cases of disaster and damage. It is required to pass a burst test on one out of the 200 pressure vessels. If the burst test meets the requirements and demands, a pressure vessel is randomly selected from the same batch, and is used to make the test samples for testing the mechanical properties. Only if these results meet the standard prescribed limits, the pressure vessels can be delivered to the user. The purpose these tests, by controlling the mechanical properties of samples made from the finished pressure vessels before and after heat treatment by improving 34CrMo4 steel, and comparing them with the required standard values, is to assess the technological parameters and possible approaches in future research that could contribute to the reliability of pressure vessels in use.

Before heat treatment and improvement, the analysis of test results shows that the tendency to plastic deformation is relatively small, but there is the tendency to local deformations. Local deformations indicate a possible breaking of uncontrolled pieces of pressure vessels ("fragments") in terms of burst. In the samples taken after improvements of the pressure vessel, the yield point is 46 % higher than before heat

V. Marušić
Faculty of Mechanical Engineering in Slavonski Brod, J.J. Strossmayer University of Osijek, Trg Ivane Brlić Mažuranić 2, 35000 Slavonski Brod, Croatia
e-mail: vmarusic@sfsb.hr

I. Lacković · L. Marušić (✉)
College of Slavonski Brod, Dr. Mile Budaka 1, 35000 Slavonski Brod, Croatia
e-mail: lukamarusic11@gmail.com

I. Lacković
e-mail: ivica.lackovic@vusb.hr

A. Öchsner and H. Altenbach (eds.), *Mechanical and Materials Engineering of Modern Structure and Component Design*, Advanced Structured Materials 70,
DOI 10.1007/978-3-319-19443-1_16

treatment and reaches a value of approximately 950 MPa. Tensile strength reaches a value of about 1050 MPa, which is about 17 % higher than before heat treatment. The analysis of the curve recorded during tensile testing showed that the effect of deformation strengthening is less pronounced and that it has almost linear character. Plasticity is not lower, but deformation at destruction has increased from 12.8 to 16.7 %. For the deformed neck of the improved pressure vessel, strength characteristics are almost unchanged, compared to the samples of the improved body of pressure vessel. Minimal deformations at destruction increased from 12.8 to 15.1 %, and reinforcements in this case has almost linear form.

Based on this analysis, it was concluded that the results of recording the tensile curve can be used as the indicators of the presence of anomalies that could contribute to unsatisfactory results on the burst tests. It helped as the basis for a recommendation of possible direction for the research: the selection of heat treatment parameters/properties at static tensile testing.

1 Introduction

During the design of the bottle of seamless steel pipes, bottom and throat are plastically deformed at an elevated temperature. After forming, it is necessary to use appropriate heat treatment (quenching and tempering) in order to achieve equalization of the mechanical properties of the bottom and undeformed part of the pipe, while satisfying the set boundaries of strength and ductility. Furthermore, the minimum calculated wall thickness is calculated depending on the strength of steel. Bottles are immediately thermally processed after forming quenching and then releasing, testing the mechanical properties of the samples prepared from the bottle is necessary to check if they meet the standard requirements prescribed. The burst test is performed after heat treatment. The process is performed (if necessary in the presence of inspector) in a way that the formed pressure pipe is filled with water and pressurized to the pressure which must be 50 % higher than the working pressure. In this paper there are analyzed specifics of making bottles Ø229 mm. The target of tensile tests was to record the curve of the plastic material flow, strength and ductility, and mutual comparison of the calculated values of tensile strength and the yield point it is necessary to evaluate the intensity of deformation hardening, and the ratio of total elongation and elongation at maximum load estimate propensity brittle fracture.

2 Features of Building Bottles

The schematic representation in Fig. 1 shows the technological process of making bottles from pipes on CNC machines [1]. The process consists of phases in which, after having a control input seamless pipe (1), have to be performed: closing (2) and forming the bottom (3) which needs to be checked (4), forming the throat (5), heat

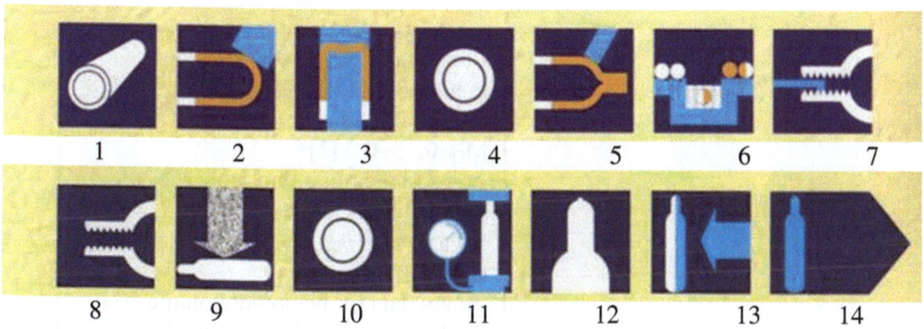

Fig. 1 Scheme of technological process of making bottles [1]

treatment of the bottle (6), tapping in throat of the bottle (7) and checking (8), blast cleaning (9) and again checking (10), testing under pressure (11) and labeling bottle (12) and paint protection (13). After the bottle is ready for delivery.

The heating is doing so that one end of pipe is inserted in a gas chamber furnace. The required length of the heating pipe is taken empirically and is about 150 mm for closing the bottom of the tube and ca 190 mm for the design of the bottle neck. The tube was maintained at a temperature between 1223 and 1273 °C for 3 min. The heated pipe from the furnace is transported to a spinning machine, and shaping is done according to a predetermined program. After completion, the bottle is removed from the chuck and visual control is performed. The shaped bottle is introduced into the flow furnace for hardening. After remaining at a temperature of $\vartheta_a \approx 870\,^{\circ}\mathrm{C}/5'$, the bottle is hardened in the emulsion, and then immediately transferred to the flow of a electro chamber furnace for releasing. The process of inserting the bottle and it being ready takes about ≈55 min (Fig. 2).

Fig. 2 Stages of making bottles **a** forming on the Spinning machine; **b** bottle at the exit of boiler-before hardening

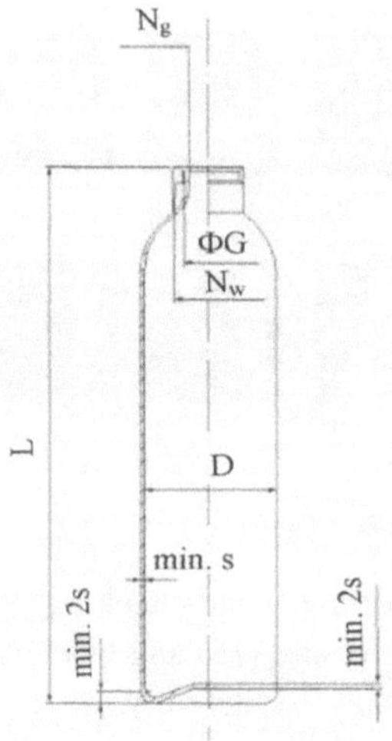

- material of the bottle 34CrMo4 : C= 0,30÷0,37; Si=0,15÷0,35; Mn=0,5÷0,8; P i S max. 0,02; Cr=0,9÷1,2 i Mo=0,15÷0,3 [3]
- prescribed minimum properties of finished bottles R_e=840 MPa; R_m=990 MPa; A_5=14 %; K=25 J/cm^2 (ISO V),
- main data of the bottle: volume 50 l; working pressure 200 bar; testing pressure 300 bar; minimum wall thickness 5,4 mm; length 1510 mm (+15/-10); weight of empty bottle 56,5 kg (+5,85/-4,52).

Fig. 3 Characteristics of seamless steel bottles Ø229 mm [2]

In Fig. 3 the manufacturer's drawing of a seamless steel bottle is shown [2]. Requirements for the pipe material, 34CrMo4, and the main technical details of the bottle are shown in Fig. 3. Furthermore it shows the chemical composition of the steel and the mechanical properties of finished bottle.

3 Burst Test

Burst test is made after heat treatment, on every 200 bottles one is examinated [1]. The procedure is performed (in the presence of inspector) so that the bottle is filled with water and pressurized to the pressure which has to be 50 % higher than working one. Bottles for high liquefied gases (oxygen, nitrogen, argon, mixtures of argon and carbon dioxide, hydrogen) must stand the test pressure of 300 bar due to the fact that for these bottles intended operating pressure is 200 bar. Bottle is pumped till the bursting. The body of the bottle must burst in the form of "fishbone" and doesn't burst in small tear parts "shrapnel". The characteristic appearance of the bottle after the test burst is shown in Fig. 4.

4 Experiment

From the viewpoint of formability it is inevitable that the manufacturing processes impacts the properties of materials [3, 4]. Bottles can impact the addiction to the differences between the properties of the body and the neck of the bottom of the bottle. The throat and the bottom of the bottle are deforming on warm from the standard pipe and standard body of the bottle stays undeformed. For the purpose of

Fig. 4 The characteristic appearance of the bottles after the burst test

testing there is selected one raw bottle-before heat treatment and one finished bottle after heat treatment, Fig. 5.

The experiment plan requires: (1) from “raw” bottle before heat treatment; (2) from body of finished bottle, after heat treatment; (3) from neck of finished bottle after heat treatment.

On samples cutted from the body of the bottle, raw and heat treated, was carried out dimensional control of wall thickness. The measurement was performed moderate caliper at 15 sites spaced at ≈10 mm. Characteristically, that the measured values, out on samples of raw bottles so and on samples from finished bottles, range between 6.6 and 6.9 mm.

Figure 6a shows of test samples, according to the requirements of the standard DIN 50125 [5], in accordance with the nominal thickness of the pipe and bottle. On Fig. 6b there is showed sample after in the jaw of machine and connected with extensometer. The tentacles are rewired with the computer, and activating the appropriate program on the screen there are printed test results: force, the relative change in length, and the line curves for plastic flow. In all tested samples the start

Fig. 5 Bottles out of which were made test samples **a** pipes; **b** bottles

(a) (b)

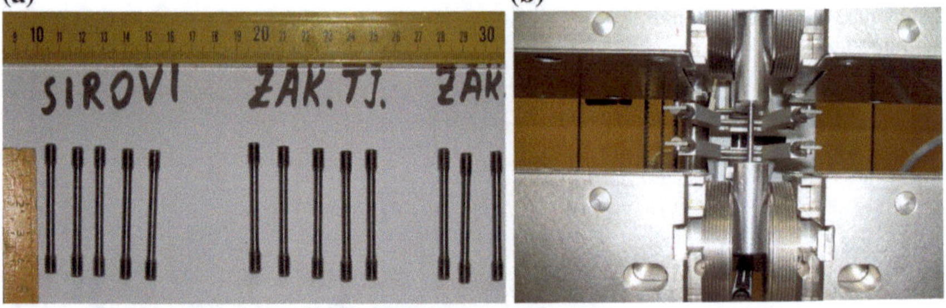

Fig. 6 Tipical shows of the samples (**a**); the sample after admission on testing machine jaws and tentacles connected with extensometer (**b**)

length L_0 was 20 mm, and the surface of initial average S_0 was 12.57 mm^2. Given the dimensions of the test samples for their capture and tightness in the jaw of testing machine there are made appropriate adapters.

There are recorded values of forces on the yield of flowing and maximal force and with them the yield of strength $R_{p0.2}$ and tensile strength R_m and the modulus of elasticity E, elongation A, and extension at maximum force A_g are calculated. Results are showed in Table 1.

5 Analysis and Conclusions

This paper points out the specifics of individual operations with emphasis on those that may directly affect the results of the control properties of the bottle before delivery.

Results of dimensional control shows that the wall thickness of each sample is bigger than the minimum required value of 5.4 mm resulting from calculation.

The tearing test with the static tensile test on samples prepared from the bottle before heat treatment (raw samples) shows that the tendency towards plastic deformation is relatively low, with a pronounced tendency towards local deformation. The study found that the delivered state (of raw bottle) tested steel (34CrMo4) has a yield strength about 650 MPa (middle value) and a tensile strength of about 900 MPa. Modulus of a elasticity is about 1.75×10^5 MPa. The relative elongation at the break is between 12 and 14.37 %. The elongation at maximum force is between 4.91 and 5.55 %.

For samples of improved body the bottle flow limit is 46.1 % than the bottles before heat treatment and reaches a value of approximately 950 MPa. Tensile strength reaches a value of approximately 1050 MPa, which is an increase of approximately 17 % compared to the situation before improving. In accordance with this change is the effect of deformation hardening which is less pronounced and has almost linear character.

Table 1 Results of tearing of test samples from pipe and bottle

Bottle type	Sample number	E	$R_{p0,2}$	R_m	A_g	A	
		GPa	MPa	MPa	%	%	
Body of raw bottle	1	200.4	640.99	911.24	5.55	13.91	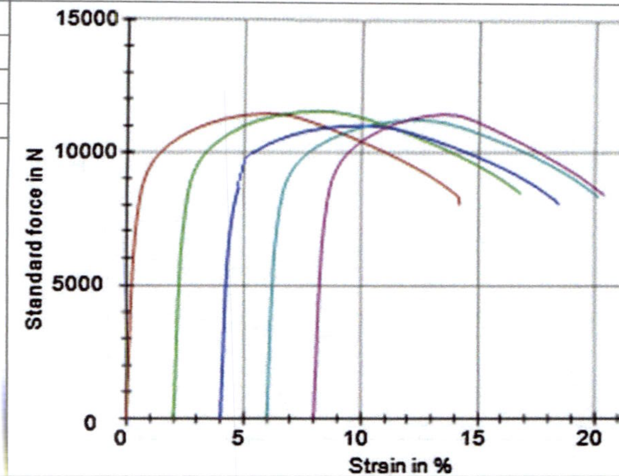
	2	151.7	685.51	919.52	5.50	14.37	
	3	176.3	626.09	875.69	5.21	14.08	
	4	177.6	630.32	892.59	5.89	13.78	
	5	174.2	650.55	910.21	4.91	12.00	

(continued)

Table 1 (continued)

Body of improved bottle	1	200.7	956.80	1049.39	5.67	16.73	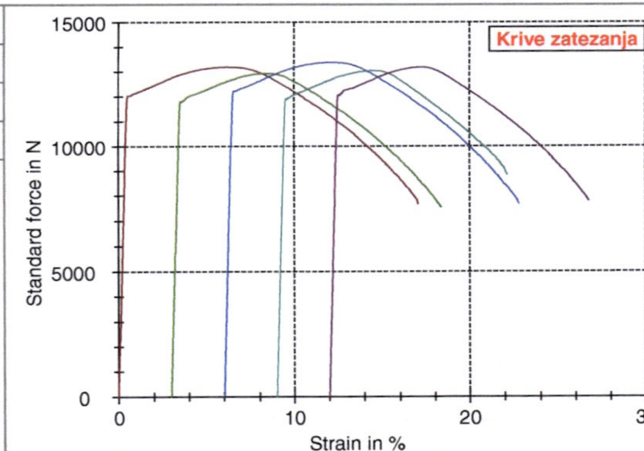
	2	195.6	940.97	1028.35	5.03	15.03	
	3	204.3	972.71	1065.05	5.65	16.48	
	4	200.1	949.00	1037.21	4.83	12.77	
	5	203.0	962.82	1048.89	4.78	14.44	
Neck of improved bottle	1	200.9	943.51	1029.34	4.83	15.14	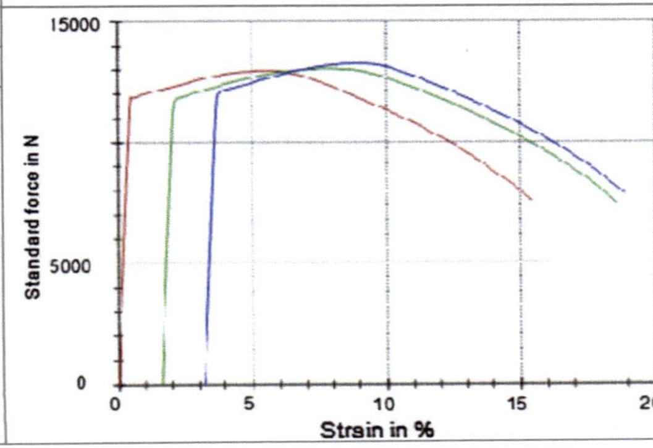
	2	195.5	942.67	1034.90	6.01	16.74	
	3	204.6	961.07	1052.77	5.38	15.43	

The bottle neck strength characteristics are almost unchanged compared to the samples of improved body of the bottle. It is noted that the minimum amount of deformation in the destruction increased from 12.8 to 15.1 %. Hardening in this case has a nearly linear form. It can be concluded that samples made from finished bottles after heat treatment don't have the tendency to brittle fracture.

Based on the results of the mechanical properties, there is calculated effect of deformation strength and they can serve as previous indication ev. present danger of brittle fracture at test bursting bottles. This has created the conditions for the continuation of research in order to determine the influence of parameters to improve the effect of deformation strength and in the other cases of forming on one or two ends of the pipe.

References

1. The Company has certificate ISO 9001, certificate ISO 11439, and type approvals for finishing of cylinders within production program. Design, finishing and testing www.csc-sb.hr/quality_control.html. 2 May 2013
2. BS EN 10083-3:2006 Steels for quenching and tempering. Technical delivery conditions for alloy steels
3. Djordjević M, Aleksandrović S, Vujinović T, Milentije Stefanović M, Lazić V, Nikolić R (2012) Computer controlled experimental device for investigations of tribological influences in sheet metal forming. Mater Eng (Materialove inžinierstvo) 19:88–94 (SK ISSN 1335-0803)
4. Grizelj B (2002) Obrada metala deformiranjem. Sveučilište J. J. Strossmayera, Strojarski fakultet Slavonski Brod
5. DIN 50125: Testing of metallic materials—tensile test pieces description: Deutsches Institut Fur Normung E. V. (German National Standard). 7 Jan 2009

Computational Modeling of Structural Problems Using Atomic Basis Functions

Vedrana Kozulić and Blaž Gotovac

Abstract This paper presents the application of the $Fup_n(x)$ basis functions in numerical modeling of different engineering problems. $Fup_n(x)$ basis functions belong to a class of atomic functions which are infinitely-differentiable functions with compact support. The collocation method has been applied in the development of numerical models. A system of algebraic equations is formed in which the differential equation of the problem is satisfied in collocation points of a closed domain while boundary conditions are satisfied exactly at the domain boundary. Using this way, the required accuracy of approximate solution is obtained simply by an increase in the number of basis functions. So, this concept represents a fully mesh free method. The properties of the atomic basis functions enable a hierarchic expansion of an approximate solution base either in the entire domain or in its segments. Presented numerical models are illustrated by examples of the torsion of prismatic bars, elasto-plastic analyses of beam bending and thin plate bending problems. The results of the analyses are compared with the existing exact and relevant numerical solutions. It can be concluded that the possibility of hierarchically expanding the number of basis functions in the domain significantly accelerates the convergence of a numerical procedure in a simple way. Values of the main solution function, e.g. displacements, and all the values derived from the main solution of the problem such as stresses, bending moments and transversal forces, are calculated in the same points and with the same degree of accuracy since numerical integration is avoided.

Keywords Numerical modeling · Mesh free method · Atomic basis functions · Collocation

V. Kozulić (✉) · B. Gotovac
Faculty of Civil Engineering, Architecture and Geodesy, University of Split, Matice hrvatske 15, 21000 Split, Croatia
e-mail: vedrana.kozulic@gradst.hr

B. Gotovac
e-mail: blaz.gotovac@gradst.hr

A. Öchsner and H. Altenbach (eds.), *Mechanical and Materials Engineering of Modern Structure and Component Design*, Advanced Structured Materials 70,
DOI 10.1007/978-3-319-19443-1_17

1 Introduction

The most widely used method for numerical analysis of structural problems is the finite element method [1]. Solving of non-linear engineering problems, in distinction from linear analyses, requires more complex numerical tool and therefore, larger number of numerical operations. For example, in elasto-plastic analyses it is interesting to detect plasticized zones and monitor their propagation parallel with the increase of load. In numerical procedures based on the weak formulation, plastic failure always is recorded before it really happens. This is the consequence of the fact that the yielding criterion [2] is not tested in the same points in which displacements are calculated.

Instead of traditional discretization into finite elements, in this paper the entire domain is analyzed at once. The approximate solution of the problem is assumed in the form of a linear combination of basis functions. Functions, which are implemented in the numerical analyses in this paper, are the Fup basis functions. They belong to a class of finite, infinitely differentiable functions named after their authors, Rvachev's basis functions or atomic functions [3–5]. The existing knowledge on atomic functions is systemized in Gotovac and Kozulić [6], basis functions are transformed into numerically applicable form, and the first steps for their use in practice are realized. The Fup basis functions belong to a space containing algebraic polynomials. These functions have good approximation properties as well as very important properties of universality [4, 6], which enables a hierarchic expansion of an approximate solution base on the domain. Because of the infinite derivability of basis functions, derivatives of all orders, required by differential equations and boundary conditions, can be used directly in the numerical procedure. Therefore, the collocation method has been applied in the development of numerical models. This concept represents a fully mesh free method [7–9] and further in the text will be referred to as FCM (Fup Collocation Method).

We created numerical models by applying an incremental-iterative procedure for monitoring the propagation of plastic zones in the cross-section. The proposed mesh free method is illustrated in examples of the elasto-plastic analyses of bending of straight beams and torsion of prismatic bars. Also, we applied FCM in solving thin plate bending problems. Numerical results obtained by the FCM are compared to exact solutions and numerical solutions obtained by the FEM.

2 Atomic Basis Functions

Atomic basis functions are infinitely-differentiable functions [3, 6] which possess the characteristics of the practical application of splines (compact support) and, at the same time, the property of universality which is a characteristic of algebraic and trigonometric polynomials. Atomic functions $y(\cdot)$ are defined as solutions of differential functional equations of the following type:

$$Ly(x) = \lambda \sum_{k=1}^{M} C_k y(ax - b_k) \tag{1}$$

where L is a linear differential operator with constant coefficients, λ is a nonzero scalar, C_k are coefficients of the linear combination, $a > 1$ is a parameter that defines the length of the compact support, and b_k are coefficients that determine displacements of the basis functions. The choice of the differential operator L determines the type of finite function. Rvachev and Rvachev [3], in their pioneering work, called these basis functions "atomic" because they span the vector spaces of all three fundamental functions in mathematics: algebraic, exponential and trigonometric polynomials.

In numerical modeling, we applied the atomic basis functions of algebraic type. The simplest function, which is the most-studied of the atomic basis functions, is the *up*(*x*) function. The function *up*(*x*) is a smooth function with compact support over $[-1, 1]$, which is obtained as a solution of a differential functional equation

$$up'(x) = 2up(2x+1) - 2up(2x-1) \tag{2}$$

with the normalized condition $\int_{-\infty}^{\infty} up(x)dx = \int_{-1}^{1} up(x)dx = 1$. The function *up* (*x*) can be expressed as an inverse Fourier transform:

$$up(x) = \frac{1}{2\pi} \int_{-\infty}^{\infty} e^{itx} \prod_{j=1}^{\infty} \left(\frac{\sin(t2^{-j})}{t2^{-j}} \right) dt. \tag{3}$$

Since Eq. (3) represents an exact but mathematically-intractable expression, Rvachev and Rvachev [4], Gotovac and Kozulic [6] provided a numerically more-adequate expression for calculating the function *up*(*x*):

$$up(x) = 1 - \sum_{k=1}^{\infty} (-1)^{1+p_1+\cdots+p_k} p_k \sum_{j=0}^{k} C_{jk} (x - 0, p_1 \ldots p_k)^j \tag{4}$$

where coefficients C_{jk} are rational numbers determined according to the following expression:

$$C_{jk} = \frac{1}{j!} 2^{j(j+1)/2} up(-1 + 2^{-(k-j)}); \quad \mathrm{j} = 0, 1, \ldots, \mathrm{k}; \quad \mathrm{k} = 1, 2, \ldots, \infty. \tag{5}$$

Calculation of the $up(-1 + 2^{-r})$; $r \in [0, \infty]$ in binary-rational points in Eq. (5), as well as all details regarding the calculation of the function *up*(*x*) values, are provided in Gotovac and Kozulić [6]. The argument $(x - 0, p_1 \ldots p_k)$ in Eq. (4) is the difference between the real value of coordinate x and its binary form in k bits, where $p_1 \ldots p_k$ are digits, 0 or 1, of the binary representation of the x coordinate.

Therefore, the accuracy of the x coordinate computation and thus the accuracy of the $up(x)$ function at an arbitrary point, depends on machine accuracy.

From Eq. (2), it can be seen that the derivatives of the $up(x)$ function can be calculated simply from the values of the function itself. The general expression for the derivative of the mth degree is

$$up^{(m)}(x) = 2^{C_{m+1}^2} \sum_{k=1}^{2^m} \delta_k up(2^m x + 2^m + 1 - 2k), \quad m \in N \tag{6}$$

where $C_{m+1}^2 = m(m+1)/2$ is the binomial coefficient and δ_k are the coefficients with value ± 1, according to the recursive formulas $\delta_{2k-1} = \delta_k, \delta_{2k} = -\delta_k$, $k \in N, \delta_1 = 1$. It can be observed that the derivatives consist of the $up(x)$ function compressed to an interval of 2^{-m+1} length, with ordinates extended by the $2^{C_{m+1}^2}$ factor.

For numerical purposes, $Fup_n(x)$ basis functions are more suitable and efficient than $up(x)$ basis functions where n is the Fup order. Index n denotes the highest degree of the polynomial that can be expressed exactly as a linear combination of $n + 2$ $Fup_n(x)$ basis functions, uniformly displaced by a characteristic interval 2^{-n}.

For $n = 0$, $Fup_0(x) = up(x)$, since $Fup_n(x)$ and its derivatives can be calculated using a linear combination of displaced $up(x)$ functions:

$$Fup_n(x) = \sum_{k=0}^{\infty} C_k(n) up\left(x - 1 - \frac{k}{2^n} + \frac{n+2}{2^{n+1}}\right) \tag{7}$$

where $C_0(n) = 2^{C_{n+1}^2} = 2^{n(n+1)/2}$. In turn, $C_k(n) = C_0(n) \cdot C'_k(n)$, where a recursive formula is used for calculating auxiliary coefficients $C'_k(n)$:

$$C'_0(n) = 1, \text{ when k} = 0; \text{ i.e., when k} > 0$$
$$C'_k(n) = (-1)^k C_{n+1}^k - \sum_{j=1}^{\min\{k;2^{n+1}-1\}} C'_{k-j}(n) \cdot \delta_{j+1}. \tag{8}$$

The $Fup_n(x)$ function support is determined as $[-(n+2)2^{-n-1}; (n+2)2^{-n-1}]$.

The basis function for numerical analyses of two-dimensional problems is obtained from the Cartesian product of two one-dimensional Fup functions defined for each direction:

$$Fup_n(x, y) = Fup_n(x) \cdot Fup_n(y). \tag{9}$$

Calculations of all required derivatives of the function $Fup_n(x, y)$ can be written in an analogue form.

3 Fup Collocation Method

It is known that functionality of the collocation method depends on the selection of basis functions and collocation points. Prenter [10] proved the stability of numerical procedures with the spline functions when collocation is performed in so-called natural knots. He developed proofs for existence and uniformity of the solution and error estimate. Since *Fup* functions can be regarded as splines of an infinite degree, it can be shown [11] that for them it is also optimal to perform collocation in natural knots of basis functions, i.e. vertices of basis functions situated in a closed domain such as e.g. for the base in x-direction formed by functions $Fup_2(x)$ shown in Fig. 1.

This selection of collocation points provides the simplest numerical procedure, the banded collocation matrix is obtained, which is diagonally dominant and thus well conditioned. This selection also implies uniformly distributed nodes set in each coordinate direction.

When practical problems are being solved with *Fup* basis functions, it is needed to map a virtual domain in which basis functions are defined into a real domain in which the problem is defined. An approximate solution base is formed on the unit virtual domain defined by the curvilinear coordinate ξ.

The compact support of the $Fup_n(x)$ basis function consists of $(n + 2)$ characteristic intervals of equal lengths $\Delta x_n = 2^{-n}$. Therefore, the unit domain can be divided into 2^n intervals of mutually equal lengths Δx_n without rest. When the unit domain is divided to N_ξ intervals of mutually equal lengths $\Delta\xi$, the calculation of $Fup_n(\xi)$ functions values and their derivatives at a general characteristic interval $\Delta\xi$ should be done in the following form:

$$Fup_n^{(m)}(\xi) = \left(\frac{1}{2^n \cdot \Delta\xi}\right)^m \cdot Fup_n^{(m)}\left(\frac{1}{2^n \Delta\xi} \cdot \xi\right) \tag{10}$$

where m is the order of the derivative.

An approximate solution is sought in the form of a linear combination of displaced basis functions given in Eq. (10):

$$\tilde{u}(\xi) = \sum_{i=-i_1}^{i_N} C_i \cdot \left\{ Fup_n\left(\frac{1}{2^n \cdot \Delta\xi}\xi - i \cdot \Delta x_n\right)\right\} \tag{11}$$

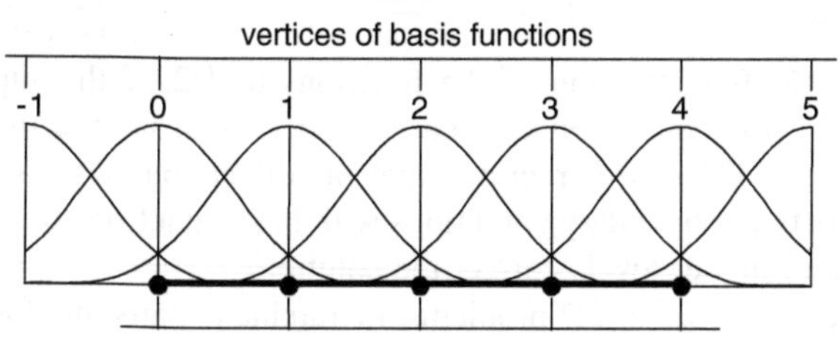

Fig. 1 Collocation points of the base formed by $Fup_2(x)$ functions

namely in collocation points:

$$u(\xi_{\mathbf{l}}) = \sum_{i=-i_1}^{i_N} C_i \cdot \left\{ Fup_n\left(\frac{\mathbf{l} - i}{2^n}\right) \right\} \tag{12}$$

where:

i	counter of basis functions
$i_1 = \left[\frac{n+1}{2}\right]$	the first basis function, the maximum integer of the fraction within square brackets
$i_N = N_\xi + \left[\frac{n}{2}\right]$	last basis function
$\mathbf{l}$	index of the collocation point, $\mathbf{l} = 0, 1, \ldots, N_\xi$
$\xi_{\mathbf{l}} = \mathbf{l} \cdot \frac{1}{N_\xi}$	coordinate ξ of the $\mathbf{l}$th collocation point.

3.1 Multilevel Approach

Atomic functions form a vector space which has the property of universality [4, 6]. It means that the vector space of nth dimension is contained within the vector space of $(n + 1)$ dimension. This very property enables a hierarchic adding of basis functions in the domain to the initial base of an approximate solution. The hierarchic expansion of an approximate solution base is realized by an algorithm in which new functions, which are all images of the same mother basis function, are added to the base of an initial solution, but displaced and compressed or stretched in comparison with the initial base.

When the distribution of basis functions mutually displaced by $\Delta\xi$ is selected, then the selected base is at the “zero level” of approximation. At the first level, functions are added, displaced by $\Delta\xi/2$ in reference to the functions of zero level, and compressed to $1/2$ of the length of basis functions support at zero level. At the second level, added basis functions are displaced by $\Delta\xi/4$ in reference to “zero level” with the support length which is $1/4$ of the length of basis function support at zero level. At higher levels of approximation, the base is built by analogy. Figure 2 shows hierarchic expansion of the vector space of basis functions $Fup_4(x)$ at the first and second levels. Compression of the functions to 1/2 of the support from the preceding level is the consequence of basic properties of atomic functions [6].

Numerical tests [11] for different densities of collocation points showed that it is sufficient to satisfy the boundary conditions with basis functions of zero level while basis functions of higher levels correct the solution.

For numerical analyses of 2D problems, hierarchic expansion of an approximate solution base is performed simultaneously for each coordinate direction.

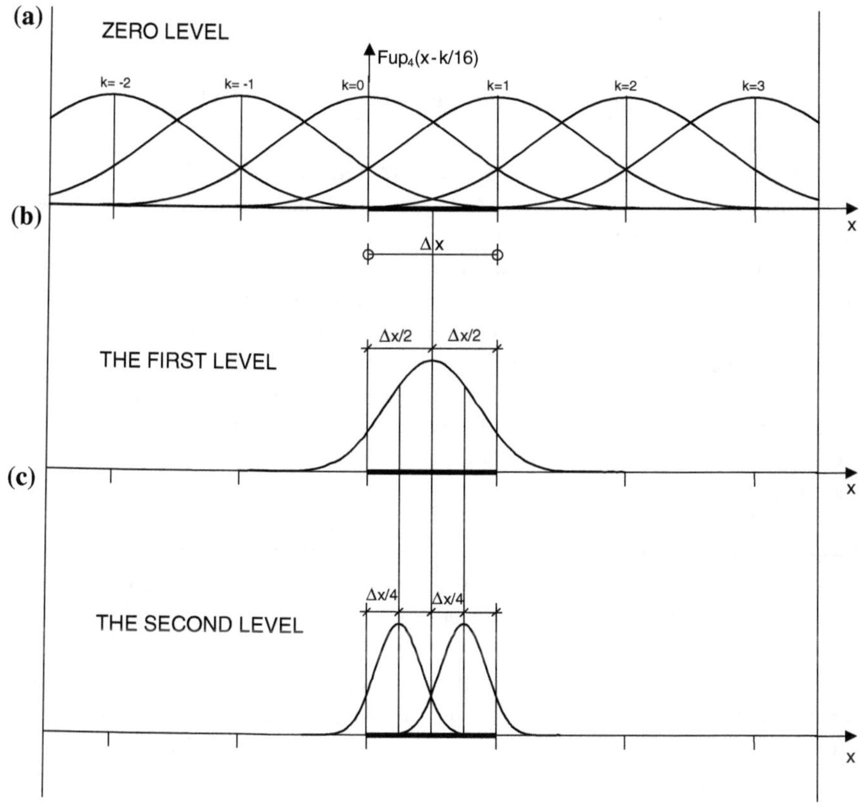

Fig. 2 Hierarchic expansion of a vector space by atomic functions $Fup_4(x)$

This multilevel approach can be applied in the entire given domain or only in a part of the domain e.g. at concentrated load locations, for singularities such as concave breaks in the edge where stress concentration occur, or in plasticization zones in elasto-plastic analyses.

4 Elasto-Plastic Analyses of Beam Bending

Bending of a beam loaded with distributed forces $q(x)$ is described by differential equation of the fourth order:

$$\frac{d^2}{dx^2}\left(EI(x)\frac{d^2w(x)}{dx^2}\right) = q(x) \tag{13}$$

or, written in numerically more acceptable form, by the system:

$$\frac{d^2M(x)}{dx^2} = -q(x); \quad EI(x)\frac{d^2w(x)}{dx^2} = -M(x) \tag{14}$$

and respective boundary conditions. In previous equations E is the Young's modulus, I is the moment of inertial of the cross-section, w is deflection of the beam whereas M is the bending moment.

With the assumption of an ideal elasto-plastic material, plastic deformations in the cross-section occur when the greatest stress exceeds the uniaxial yield stress σ_Y. The limit bending strength of the critical cross-section is reached when its elastic core disappears and the so-called plastic hinge is formed [12].

A numerical model for monitoring of yielding of the beam by an incremental-iterative procedure until failure is developed by applying FCM with a hierarchic expansion of an approximate solution base. Atomic functions $Fup_4(x)$ are used in the model.

Plastic yielding never occurs in the entire domain, but only in the part of the beam out of which the material behaves as linear-elastic. In the first incremental step of non-linear calculation, the initial density of the basis functions is selected, and then, with the increase in load, basis functions are added only between collocation points in which plastic deformations have been registered in the previous load increment. Therefore, parallel with the increase in the plasticization level of the beam, also increases the level of accuracy of an approximate solution. The matrix of the system from the previous level is contained within the matrix of the system at the next level, which makes the numerical procedure efficient.

4.1 Simply Supported Beam Subjected to a Concentrated Force in the Middle

The bending of a beam with a rectangular cross-section loaded with a concentrated force F is considered, according to Fig. 3. By an increase in the load intensity, plastic deformations are increasing in the central cross-section until plastic hinge is formed. According to the analytical solution [12], the limit value of force F, when

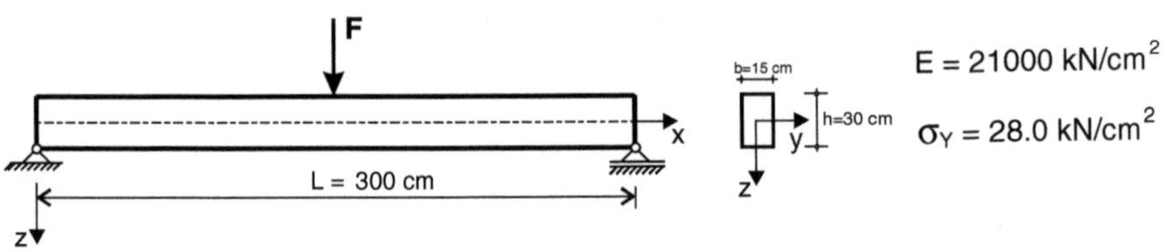

Fig. 3 Simply supported beam subjected to a concentrated force in the middle

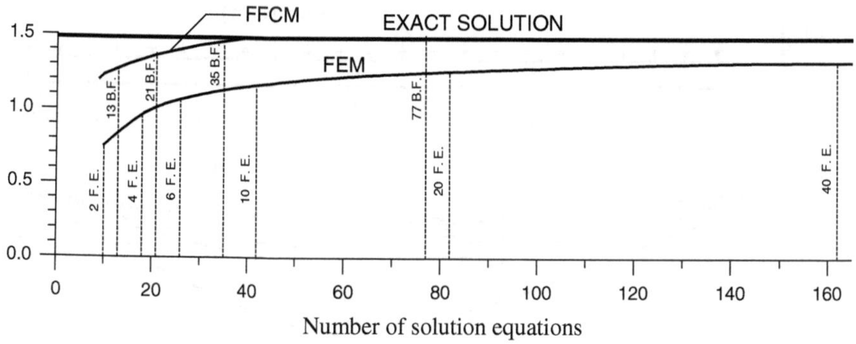

Fig. 4 Convergence of numerical solutions for simply supported beam

plastic failure occurs, for the given values E and σ_Y, is $F_{pl} = 1260$ kN. The maximum beam deflection which corresponds to the limit force F_{pl} is $w_{pl} = 1.4815$ cm.

In order to evaluate the results obtained by FCM, an example is analyzed also by the FEM [13] with three-node isoparameteric finite elements and Lagrange polynomials of the 2nd degree as basis functions. Figure 4 shows the diagrams of convergence of numerical solutions obtained by these two methods. The diagrams show obtained values of the maximum deflection w_{pl} at the moment of plastic failure depending on the number of basis functions $Fup_4(x)$ in FCM i.e. the number of finite elements in FEM.

In Fig. 4, it can be observed that at failure analytical value of the deflection at the beam's midspan [12] can never be achieved by FEM. Regardless of the number of finite elements, deflection at failure is always smaller than an exact value, while the numerical solution obtained by FCM quickly converges towards an exact solution. The reason is that in FEM deflections are calculated in nodes while in numerical integration stresses are determined in Gaussian integration points. In FCM, the criterion of plasticization is tested in collocation points i.e. in the same points in which displacements are calculated. It ensures that incremental-iterative procedure provides a good description of a real elasto-plastic behaviour of the beam.

4.2 Supported Cantilever Under Uniformly Distributed Load

A beam clamped at one end, supported at the other end and loaded with a uniformly distributed load q, as shown in Fig. 5a, is analyzed. The geometric characteristics of the beam and the material properties are the same as in previous example (see Fig. 3).

The bearing capacity of this statically undetermined structure will be completely exhausted only after the formation of two plastic hinges: on the left support and at the midspan. According to Ref. [12] the limit load, at which the mechanism is formed, is $q = 1.4571 \cdot q_0$, where q_0 is the limit load for a simply supported beam.

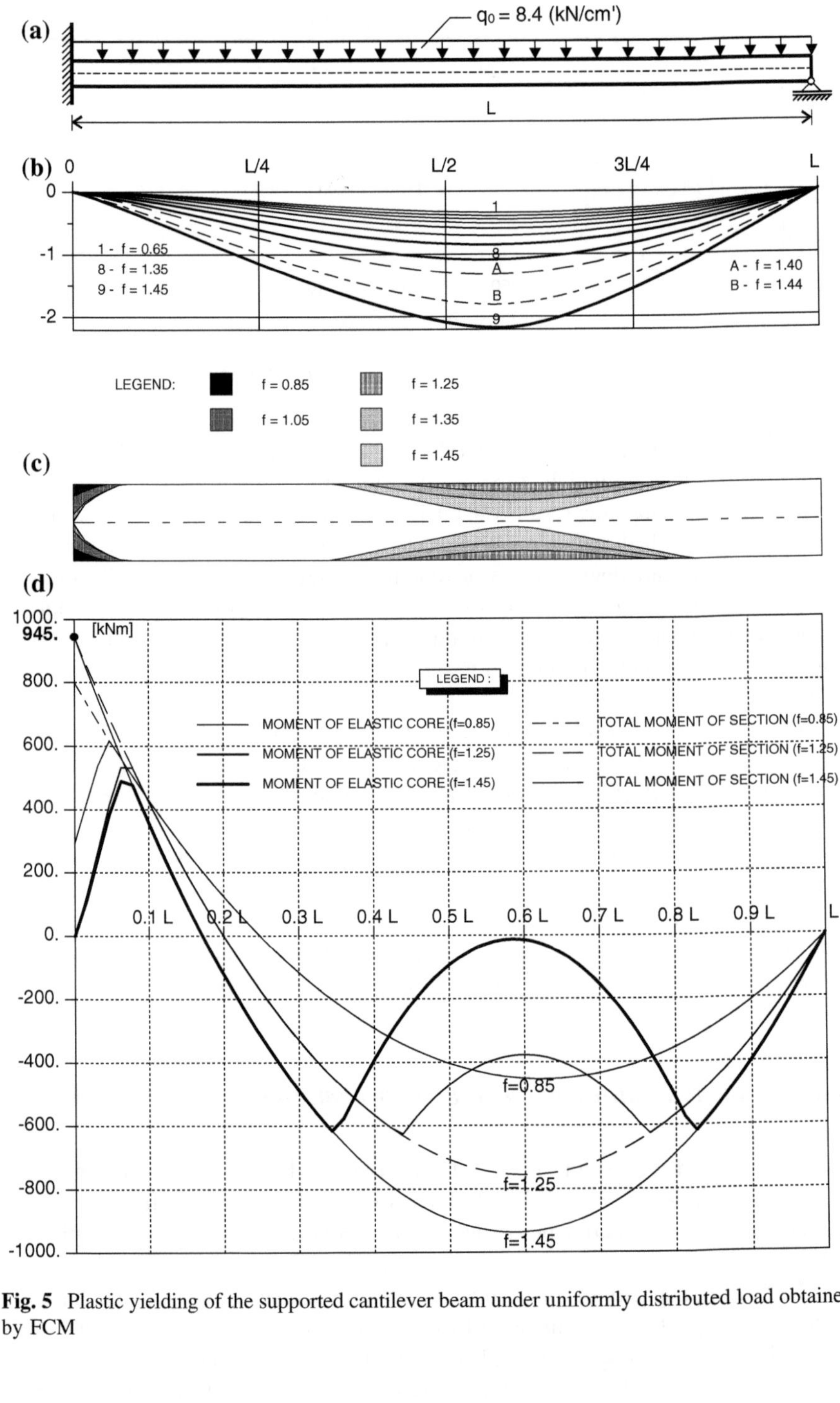

Fig. 5 Plastic yielding of the supported cantilever beam under uniformly distributed load obtained by FCM

Figure 5b gives deflection lines of the beam for different load factors *f*: from the moment when the entire beam is in a linear domain until the moment when plastic failure occurs. Figure 5c illustrates the propagation of plastic zones from the moment of first occurrence of plastic deformations at the clamped end until failure i.e. forming of the plastic hinge also at the midspan, when the height of the elastic core almost equals zero. Figure 5d gives the bending moments diagrams depending on different load intensity. The total bending moment is separated into elastic core moment and the moment that plasticized part of the cross-section can take over.

The numerical solution by FCM presented in Fig. 5 completely coincides with the analytical solution to the factor $f = 1.45$ which corresponds to 99.5 % of the value of the limit load factor $f_{pl} = 1.4571$. It shall be noted that by FEM it is not possible to monitor the beam behaviour until complete loss of the bearing capacity, but only to the load $q = 1.22 \cdot q_0$, i.e. load factor that is 83.7 % of the limit value f_{pl}, when the numerical procedure is interrupted and the appearance of the mechanism is recorded.

5 Elasto-Plastic Analyses of Prismatic Bars Torsion

The elastic torsion of a bar is a classical problem in the theory of elasticity [14, 15]. This problem may be formulated in terms of the Dirichlet boundary value problem of the Poisson equation:

$$\frac{\partial^2 \Phi(x,y)}{\partial x^2} + \frac{\partial^2 \Phi(x,y)}{\partial y^2} = -2G\vartheta; \quad \Phi|_{\Gamma} = 0. \tag{15}$$

where $\Phi(x,y)$ is the stress function, G is the shear modulus, while ϑ is the angle of twist per unit length of a bar. Shear stress components are determined according to the following expressions:

$$\tau_{xz} = \partial\Phi/\partial y; \quad \tau_{yz} = -\partial\Phi/\partial x. \tag{16}$$

Torsion rigidity of the cross-section for $\vartheta = 1$ is the double volume under the surface of the stress function Φ:

$$C_t = 2\iint \Phi dxdy. \tag{17}$$

In case of a cross-section with multiple boundary, unknown constant values of the stress function at inner boundaries Γ_i are determined based on the theorem on circulation of shear stresses. The following condition must be satisfied at each boundary of the opening:

$$-\int_{\Gamma_i} \frac{\partial\Phi}{\partial n_i} d\Gamma = 2G\vartheta A_i, \quad \mathrm{i} = 1, 2, \ldots, \mathrm{n} \tag{18}$$

where A_i is the area of each opening, n_i is the normal to the inner boundary Γ_i, while n is the number of openings.

The material starts to deform plastically when the resulting shear stress in a point reaches a critical value τ_Y. Then, the Poisson equation is satisfied in the elastic part of the domain while the yielding criterion [2]:

$$\left(\frac{\partial \Phi}{\partial x}\right)^2 + \left(\frac{\partial \Phi}{\partial y}\right)^2 = \tau_Y^2 \tag{19}$$

is satisfied in its plastic part. The greatest value of the torsion moment occurs when the entire cross-section is plasticized. It is the limit torsion moment M_{pl}. Elasto-plastic analyses includes determination of the angle of twist ϑ at which plasticization begins as well as monitoring of the expansion of plastic zones until limit moment M_{pl} is reached.

By applying Fup Collocation Method, approximate solution base is formed on the unit virtual domain defined in the system (ξ, η) according to a scheme shown in Fig. 6.

For rectangular cross-section of $a \times b$ dimensions, the differential equation of the problem and boundary condition from Eq. (15) can be written in the system (ξ, η) as:

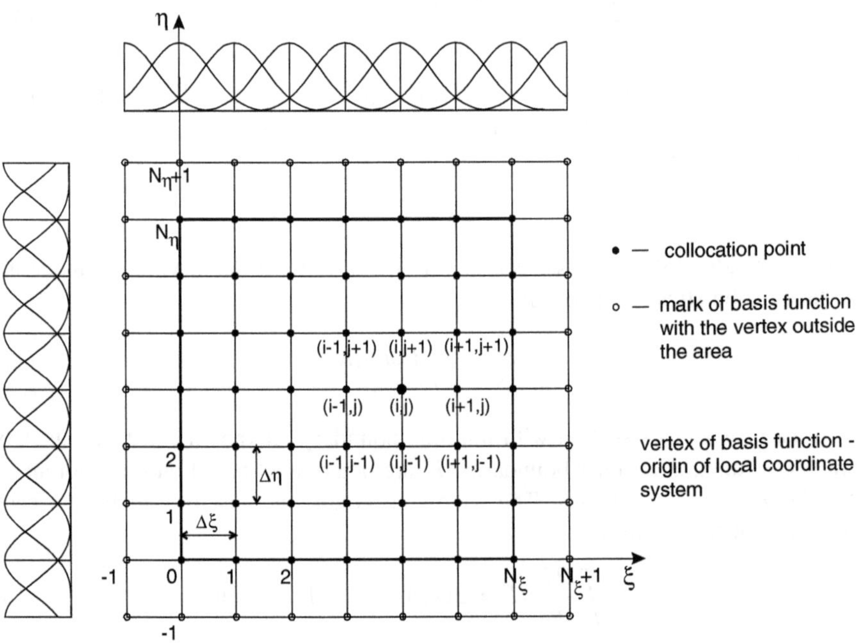

Fig. 6 Distribution of basis functions on the unit virtual 2D domain

$$\frac{1}{a^2}\frac{\partial^2\Phi(\xi,\eta)}{\partial\xi^2}+\frac{1}{b^2}\frac{\partial^2\Phi(\xi,\eta)}{\partial\eta^2}=-2G\vartheta;\quad 0\le\xi\le 1,\ 0\le\eta\le 1, \tag{20}$$

$$\Phi(\xi,\eta)=0 \quad \text{for} \quad \xi=0,\xi=1,\eta=0,\eta=1. \tag{21}$$

Collocation is performed in $(N_\xi+1)\times(N_\eta+1)$ equidistant points, while basis functions with vertices outside the domain, see Fig. 6, are retained so the basis functions set can be complete. Governing Eq. (20) is satisfied in all collocation points of the domain except in corners:

$$\sum_{i=-1}^{N_\xi+1}\sum_{j=-1}^{N_\eta+1}C_{ij}\left(\frac{1}{a^2}\frac{\partial^2F_{ij}(\xi,\eta)}{\partial\xi^2}+\frac{1}{b^2}\frac{\partial^2F_{ij}(\xi,\eta)}{\partial\eta^2}\right)=-2G\vartheta, \tag{22}$$

boundary condition (21) is satisfied in all collocation points of the domain sides:

$$\sum_{i=-1}^{N_\xi+1}\sum_{j=-1}^{N_\eta+1}C_{ij}\cdot F_{ij}(\xi,\eta)=0, \tag{23}$$

while three more conditional equations are satisfied in corners:

$$\begin{aligned}&\sum_{i=-1}^{N_\xi+1}\sum_{j=-1}^{N_\eta+1}\frac{1}{a^2}C_{ij}\frac{\partial^2F_{ij}(\xi,\eta)}{\partial\xi^2}=0\\&\sum_{i=-1}^{N_\xi+1}\sum_{j=-1}^{N_\eta+1}\frac{1}{b^2}C_{ij}\frac{\partial^2F_{ij}(\xi,\eta)}{\partial\eta^2}=0\quad.\\&\sum_{i=-1}^{N_\xi+1}\sum_{j=-1}^{N_\eta+1}\frac{1}{a^2b^2}C_{ij}\frac{\partial^4F_{ij}(\xi,\eta)}{\partial\xi^2\partial\eta^2}=0\end{aligned} \tag{24}$$

Boundary conditions are therefore exactly satisfied on the domain boundary and not only discretely in collocation points.

In the equation system (22)–(24), N_ξ and N_η denote numbers of partitions of a unit domain in directions ξ and η respectively; i and j are counters of the basis functions in ξ i.e. η directions, while $F_{ij}(\xi,\eta)$ is the basis function $Fup_2(\xi,\eta)$ with the vertex at the point (i, j). Depending on the number of partitions, function $Fup_2(\xi,\eta)$ support is condensed to $(4\Delta\xi\times 4\Delta\eta)$; $\Delta\xi=1/N_\xi$, $\Delta\eta=1/N_\eta$. Partial derivatives values of basis functions in Eqs. (22)–(24) are determined according to the following expression:

$$\frac{\partial^{(m+n)} F_{ij}(\xi,\eta)}{\partial\xi^m \partial\eta^n} = \left(\frac{1}{4\Delta\xi}\right)^m \cdot \left(\frac{1}{4\Delta\eta}\right)^n \cdot Fup_2^{(m+n)}\left(\frac{1}{4\Delta\xi}\xi - \frac{i}{4}, \frac{1}{4\Delta\eta}\eta - \frac{j}{4}\right). \quad (25)$$

Since the function $Fup_2(\xi, \eta)$ is a finite function with the support consisting of 4×4 characteristic intervals, the solution function value at collocation point (i, j) can be approximated by linear combination in the following form:

$$\Phi(\xi_i, \eta_j) = \sum_{k=i-1}^{i+1} \sum_{l=j-1}^{j+1} C_{kl} \cdot F_{kl}(\xi_i, \eta_j) \quad (26)$$

Values of all other basis functions at the point (i, j) are equal to zero. Therefore, a support domain of the point (i, j) is nine. In such a way, banded matrix of the system is obtained.

5.1 Torsion of a Prismatic Bar with a Square Cross-Section

Let us consider a bar with square cross-section length of sides $2a = 10\,\text{cm}$, $G = 1.0\,\text{kN/cm}^2$, $\tau_Y = 14.0\,\text{kN/cm}^2$. Assuming that the bar is not deformed until the yielding limit is exceeded, limit torsion moment can be determined according to expression $(8\tau_Y a^3)/3$ and, for given values, equals M_{pl} = 4666.667 kNcm. In reality, an indefinitely large angle of twist is required to obtain full plasticization of the cross-section.

A numerical model for elasto-plastic analysis of prismatic bars torsion is developed based on the FCM. Gradual plastic yielding of the cross section with the increase in the angle of twist is monitored by an incremental-iterative procedure and illustrated in Fig. 7. Plastic zones first occur at the domain boundary, and then spread towards the inside. We started the calculation with an initial density of 11 collocation points in each coordinate direction. When collocation points are detected in which the plasticity criterion is satisfied, the number of basis functions is

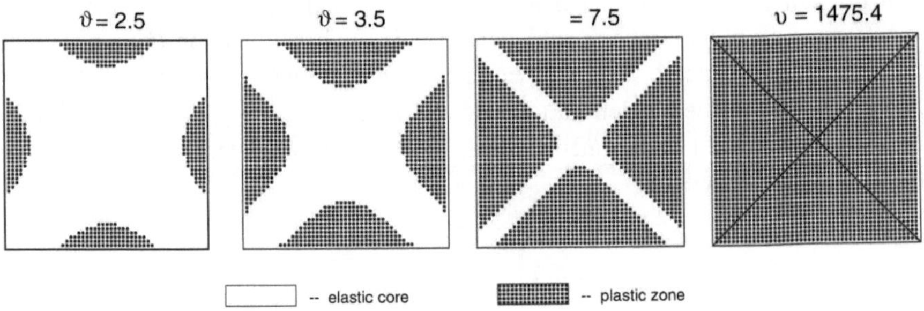

Fig. 7 Plastic yielding of a bar with a square cross-section under the torsion moment

increased only in plastic part of the cross-section while in elastic core initial density at zero level is retained. Thus, the movement of the plastic zone boundary is successfully simulated until elastic core completely disappears. In that moment, the number of basis functions was 2809.

5.2 Analyses of Curvilinear Domains by FCM

The Fup Collocation Method can be applied successfully to curvilinear domains, too.

The parametric form is extremely adequate for description of surfaces and, using the Coons formulation [16], can be written in the following form:

$$\begin{aligned} P(\xi,\eta) = {} & [(1-\xi) \quad \xi]\begin{bmatrix} Q(0,\eta) \\ Q(1,\eta) \end{bmatrix} + [Q(\xi,0) \quad Q(\xi,1)]\begin{bmatrix} 1-\eta \\ \eta \end{bmatrix} \\ & - [(1-\xi) \quad \xi]\begin{bmatrix} Q(0,0) & Q(0,1) \\ Q(1,0) & Q(1,1) \end{bmatrix}\begin{bmatrix} 1-\eta \\ \eta \end{bmatrix} \end{aligned} \tag{27}$$

where $Q(0, 0)$, $Q(0, 1)$, $Q(1, 0)$ and $Q(1, 1)$ are position vectors at the four corners while $Q(\xi, 0)$, $Q(\xi, 1)$, $Q(0, \eta)$ and $Q(1, \eta)$ are four boundary curves, see Fig. 8. Changing the parameters ξ and η in equal steps on the interval [0, 1] using Eq. (27), equidistant collocation points within the given domain are obtained.

Thus, for curvilinear domains, the partial differential equation of the torsion problem, Eq. (15), has the following collocation form:

$$\begin{aligned} \sum_{i=-1}^{N_\xi+1}\sum_{j=-1}^{N_\eta+1} C_{ij}\cdot \Bigg[& FXX\frac{\partial^2 F_{ij}(\xi,\eta)}{\partial\xi^2} + FXE\frac{\partial^2 F_{ij}(\xi,\eta)}{\partial\xi\partial\eta} + FEE\frac{\partial^2 F_{ij}(\xi,\eta)}{\partial\eta^2} \\ & + FX\frac{\partial F_{ij}(\xi,\eta)}{\partial\xi} + FE\frac{\partial F_{ij}(\xi,\eta)}{\partial\eta}\Bigg] = -2G\vartheta \end{aligned} \tag{28}$$

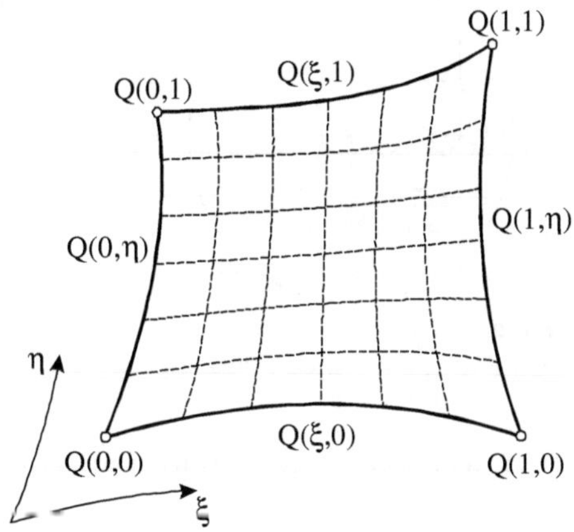

Fig. 8 A Coons surface patch

where:

$$\begin{aligned}
FXX &= \left(\frac{\partial\xi}{\partial x}\right)^2 + \left(\frac{\partial\xi}{\partial y}\right)^2; \quad FEE = \left(\frac{\partial\eta}{\partial x}\right)^2 + \left(\frac{\partial\eta}{\partial y}\right)^2; \\
FXE &= 2\cdot\left(\frac{\partial\xi}{\partial x}\frac{\partial\eta}{\partial x} + \frac{\partial\xi}{\partial y}\frac{\partial\eta}{\partial y}\right) \\
FX &= \frac{\partial\xi}{\partial x}\frac{\partial\left(\frac{\partial\xi}{\partial x}\right)}{\partial\xi} + \frac{\partial\eta}{\partial x}\frac{\partial\left(\frac{\partial\xi}{\partial x}\right)}{\partial\eta} + \frac{\partial\xi}{\partial y}\frac{\partial\left(\frac{\partial\xi}{\partial y}\right)}{\partial\xi} + \frac{\partial\eta}{\partial y}\frac{\partial\left(\frac{\partial\xi}{\partial y}\right)}{\partial\eta} \\
FE &= \frac{\partial\xi}{\partial x}\frac{\partial\left(\frac{\partial\eta}{\partial x}\right)}{\partial\xi} + \frac{\partial\eta}{\partial x}\frac{\partial\left(\frac{\partial\eta}{\partial x}\right)}{\partial\eta} + \frac{\partial\xi}{\partial y}\frac{\partial\left(\frac{\partial\eta}{\partial y}\right)}{\partial\xi} + \frac{\partial\eta}{\partial y}\frac{\partial\left(\frac{\partial\eta}{\partial y}\right)}{\partial\eta}.
\end{aligned} \tag{29}$$

Partial derivatives of elements of the inverse mapping matrix in expressions of Eq. (29) are determined by derivations of parametric equations of a surface (Eq. 27), while partial derivatives of the basis functions are determined according to Eq. (25).

Example Plastic yielding of a bar with a cross-section in the form of an eccentric ring.

Elastic and elasto-plastic analyses of a bar with a cross-section in the form of an eccentric ring, shown in Fig. 9a, were made by FCM. An analytic solution exists for this shape of a cross-section [15].

A real domain of a cross-section is mapped into the virtual unit domain using Eq. (27) where sides (1) and (2), see Fig. 9b, are described using the parametric equations of a circle; sides (3) and (4) overlap in a real domain.

Convergence of torsion rigidity value C_t and stress function value Φ on the inner boundary Γ_1 with an increase in the number of collocation points is given in

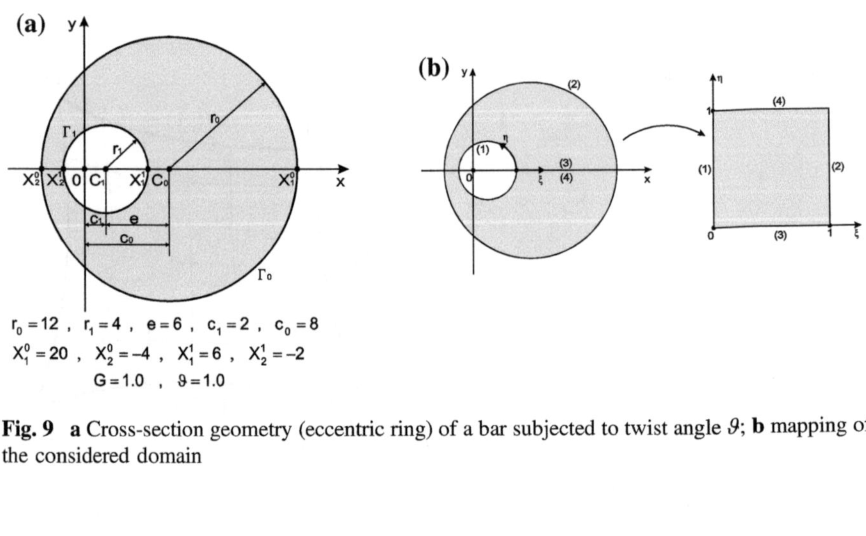

Fig. 9 **a** Cross-section geometry (eccentric ring) of a bar subjected to twist angle ϑ; **b** mapping of the considered domain

Table 1 Numerical results of elastic analyses depending on N_ξ and N_η

Number of coll. points: $(N_\xi + 1) \times (N_\eta + 1)$	$\Phi\|_{\Gamma_1}$	$\frac{\Phi - \Phi_{exact}}{\Phi_{exact}}$	C_t	$\frac{C_t - C_{texact}}{C_{texact}}$
$N_\xi = 10$, $N_\eta = 20$	41.387	5.32 %	28,345.72	2.57 %
$N_\xi = 20$, $N_\eta = 40$	40.279	2.50 %	27,976.24	1.24 %
$N_\xi = 50$, $N_\eta = 100$	39.649	0.90 %	27,768.75	0.48 %
$N_\xi = 100$, $N_\eta = 200$	39.445	0.38 %	27,701.30	0.24 %
Exact solution [15]	39.297	–	27,634.63	–

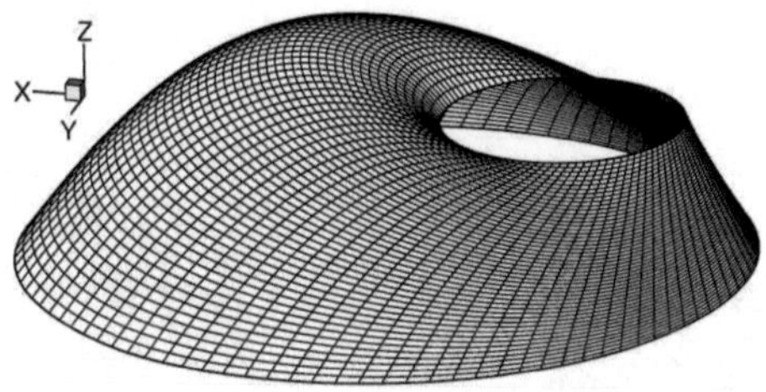

Fig. 10 Stress function surface $\Phi(x, y)$ for elastic state of a bar

Table 1. Figure 10 shows the stress function surface obtained by the FCM for $\vartheta = 1.0$.

For the purpose of elasto-plastic analysis, twist angle ϑ increases to the full plastic yielding. The theoretical value of the limit torsion moment M_{pl} for the yield stress value $\tau_Y = 14.0$ is 37708.746. Figure 11 shows surface shapes and isolines of the stress function Φ obtained by FCM for different load increments, from fully elastic to fully plastic state.

6 Analyses of Thin Plate Bending

Bending of thin homogenous isotropic plates is described by the following differential equation:

$$\nabla^4 w = \frac{p(x, y)}{D} \tag{30}$$

and respective boundary conditions, where D is flexural rigidity, p is normal pressure while w is deflection of the plate.

Since plate behaviour is described by a partial differential equation of the fourth order, in numerical modeling using the FCM it is most adequate to select $Fup_4(\xi, \eta)$ as basis functions. This basis function has 6 characteristic intervals per each coordinate direction, so two series of vertex points are outside the domain which are

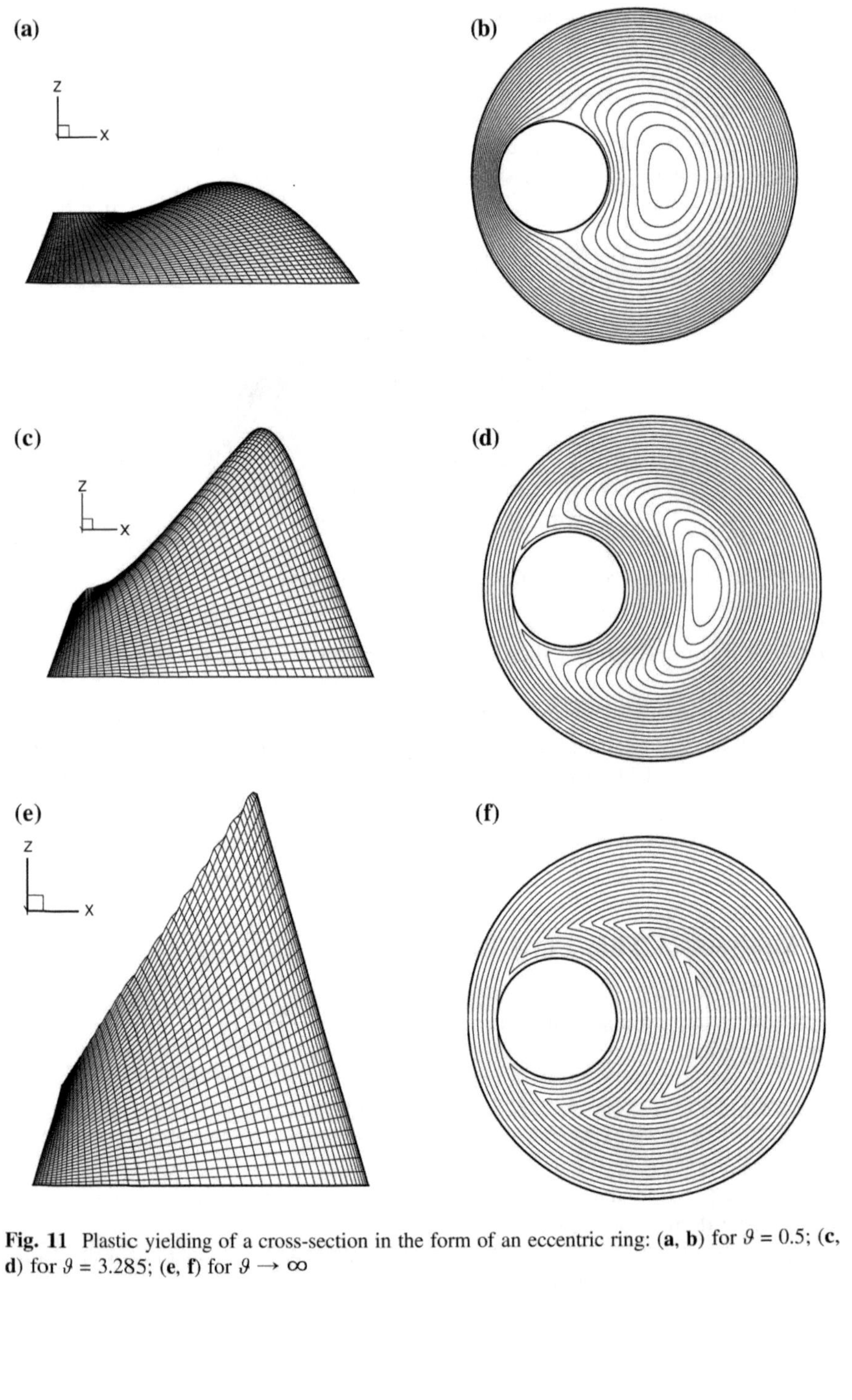

Fig. 11 Plastic yielding of a cross-section in the form of an eccentric ring: (**a**, **b**) for $\vartheta = 0.5$; (**c**, **d**) for $\vartheta = 3.285$; (**e**, **f**) for $\vartheta \to \infty$

required for the complete *Fup* approximation. For all external basis functions, corresponding collocation points are located at the boundary.

Therefore, an approximate solution of the problem is sought using $(N_\xi + 1) \times (N_\eta + 1)$ collocation points i.e. $(N_\xi + 5) \times (N_\eta + 5)$ basis functions. When the FCM is applied to a rectangular plate of $(a \times b)$ size, collocation form of the differential Eq. (30) is:

$$\sum_{i=-2}^{N_\xi+2} \sum_{j=-2}^{N_\eta+2} C_{ij} \left[\frac{1}{a^4} \frac{\partial^4 F_{ij}(\xi,\eta)}{\partial \xi^4} + \frac{2}{a^2 \cdot b^2} \frac{\partial^4 F_{ij}(\xi,\eta)}{\partial \xi^2 \partial \eta^2} + \frac{1}{b^4} \frac{\partial^4 F_{ij}(\xi,\eta)}{\partial \eta^4} \right] = \frac{p}{D} \tag{31}$$

In Eq. (31), $F_{ij}(\xi, \eta)$ is the basis function $Fup_4(\xi, \eta)$ with the vertex in point (*i, j*), while partial derivatives of basis functions are determined according to the following expression:

$$\frac{\partial^{(m+n)} F_{ij}(\xi,\eta)}{\partial \xi^m \partial \eta^n} = \left(\frac{1}{16\Delta\xi}\right)^m \cdot \left(\frac{1}{16\Delta\eta}\right)^n \cdot Fup_4^{(m+n)} \left(\frac{1}{16\Delta\xi}\xi - \frac{i}{16}, \frac{1}{16\Delta\eta}\eta - \frac{j}{16} \right) \tag{32}$$

Numerical solving of the given problem by the FCM is reduced to searching of C_{ij} unknown coefficients of linear combination by solving the system of Eq. (31) complete with respective boundary conditions. For a plate simply supported on all four edges, the boundary conditions are:

$$\begin{gathered} w(\xi,0) = w(\xi,1) = w(0,\eta) = w(1,\eta) = 0 \\ \frac{\partial^2 w}{\partial \xi^2}(0,\eta) = \frac{\partial^2 w}{\partial \xi^2}(1,\eta) = \frac{\partial^2 w}{\partial \eta^2}(\xi,0) = \frac{\partial^2 w}{\partial \eta^2}(\xi,1) = 0 \end{gathered} \tag{33}$$

For a plate clamped on all four edges, boundary conditions are:

$$\begin{gathered} w(\xi,0) = w(\xi,1) = w(0,\eta) = w(1,\eta) = 0 \\ \frac{\partial w}{\partial \xi}(0,\eta) = \frac{\partial w}{\partial \xi}(1,\eta) = \frac{\partial w}{\partial \eta}(\xi,0) = \frac{\partial w}{\partial \eta}(\xi,1) = 0 \end{gathered} \tag{34}$$

Conditional equations at the corner of the plate are obtained by Cartesian product of operators in boundary conditions on plate edges intersecting in that corner.

The numerical model has been tested on the example of homogeneous isotropic thin square plate with side length *a* and different boundary and loading conditions. Numerical results obtained by FCM are compared with analytical solutions [17] in Table 2. The presented method shows excellent corresponding with the exact results. As it can be seen in Table 2, deviations of numerical values from the analytic ones is of the same order also for the deflection in the centre of plate, which represents a main solution, as well as for bending moments which are derived values.

Table 2 Deflections and bending moments of square plate with different boundary and loading conditions, $\nu = 0.3$

Number of collocation points (N + 1) × (N + 1)	Simply supported plate				Clamped plate					
	Uniform load p		Central concentrated load P		Uniform load p			Central concentrated load P		
	Center deflection W_{max}	Moment at plate center M_x	Center deflection W_{max}	Moment at plate center M_x	Center deflection W_{max}	Moment at plate center M_x	Moment at side center m_x	Center deflection W_{max}	Moment at plate center M_x	Moment at side center m_x
N = 4	0.004284	0.05037	0.01127	0.19287	0.001312	0.02438	−0.04757	0.004833	0.1359	−0.1038
N = 8	0.004115	0.04848	0.01145	0.25896	0.001285	0.02335	−0.05012	0.005379	0.2049	−0.1181
N = 16	0.004075	0.04803	0.01155	0.32954	0.001271	0.02303	−0.05099	0.005542	0.2758	−0.1234
N = 32	0.004066	0.04792	0.01158	0.40098	0.001267	0.02294	−0.05124	0.005591	0.3473	−0.1251
N = 40	0.004064	0.04791	0.01160	0.42410	0.001266	0.02293	−0.05128	0.005598	0.3703	−0.1254
Analytic	0.004062	0.04790	0.01160	∞	0.001260	0.02310	−0.05130	0.005600	∞	−0.1257
Multiplier	pa^4/D	pa^2	Pa^2/D	P	pa^4/D	pa^2	pa^2	Pa^2/D	P	P

Figure 12 shows bending moments M_{xx} and transverse forces Q_x on clamped uniformly loaded square plate. It can be observed that boundary curves of bending moment and transverse force diagrams are extremely smooth which is the consequence of high smoothness of applied basis functions, which also means that boundary conditions are exactly satisfied.

Figure 13 shows distributions of the bending moment and transverse force on clamped square plate loaded by concentrated force in the middle. It can be concluded that the numerical solution obtained by FCM accurately describes real behaviour of the plate even in the case of a concentrated load.

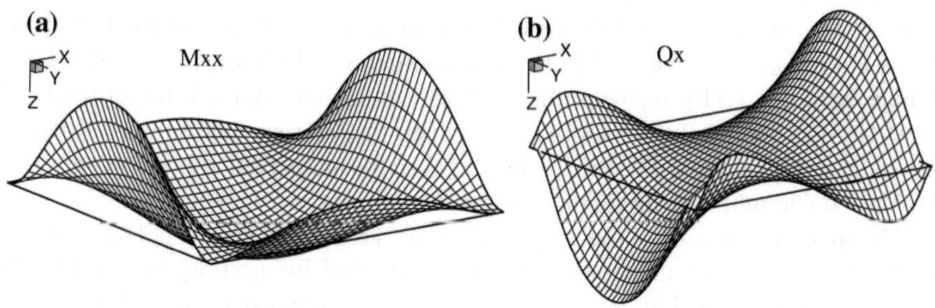

Fig. 12 Clamped plate under uniformly distributed load: **a** bending moments M_{xx}; **b** transverse forces Q_x

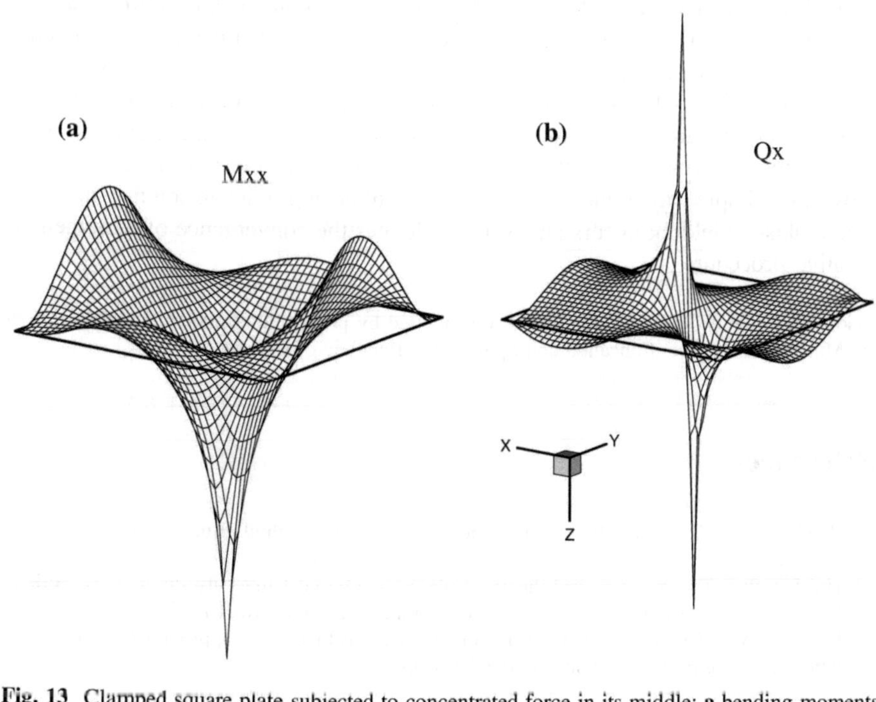

Fig. 13 Clamped square plate subjected to concentrated force in its middle: **a** bending moments M_{xx}; **b** transverse forces Q_x

7 Conclusions

This paper presents the numerical modeling of engineering problems by the mesh free collocation method based on using atomic basis functions. These functions are not commonly used in numerical analyses. So, the primary purpose of this work is to popularize application of atomic functions in numerical solving of different engineering problems. Numerical models developed in this work are illustrated by examples of the torsion of prismatic bars, elasto-plastic analyses of beam bending and thin plate bending problems. The numerical examples show that the new method efficiently simulates the real non-linear behaviour of the structure by comparing with the exact solutions. More accurate results are attained with the FCM in comparison with the Finite Element Method which always records plastic failure before it really happens. In the Fup Collocation Method, the criterion of plasticity is tested in the same points for which the values of the solution function are calculated i.e. in collocation points. Thus, the numerical procedure with the FCM is stable until plastic failure occurs.

The main advantages of the proposed mesh free method can be pointed out: (1) an arbitrarily accurate numerical solution is obtained by arbitrary increase in the number of basis functions over the domain; (2) predefined mesh and numerical integration are avoided: there are no stress discontinuity problems existing on interfaces between the finite elements; (3) values of the main solution function and all values derived from the main solution are calculated in the same points with the same level of accuracy; (4) all fields derived from the main solution can be expressed by continuous functions on the entire domain; (5) using the parametric formulation for the description of a given domain geometry, the method is adapted to analyses of curvilinear domains; (6) the method is convenient for multilevel approach i.e. hierarchic increase in number of basis functions on the domain or its parts without intrusion into the rest of the domain; (7) the multilevel approach provides a simple way to increase the accuracy of an approximate solution in places where plastic yielding occurs and also accelerates the convergence of incremental-iterative procedure.

Acknowledgments This research has been supported by project number 083-0831541-1534 of the Ministry of Science, Education and Sport of the Republic of Croatia.

References

1. Zienkiewicz OC, Taylor RL (2002) The finite element method. Butterworth-Heinemann, Oxford
2. Hill R (1985) The mathematical theory of plasticity. Oxford University Press, New York
3. Rvachev VL, Rvachev VA (1971) On a finite function. DAN URSR A6:705–707
4. Rvachev VL, Rvachev VA (1979) Non-classical methods for approximate solution of boundary-value problems. Naukova dumka, Kiev

5. Kravchenko VF, Rvachev VA, Rvachev VL (1995) Mathematical methods for signal processing on the basis of atomic functions. Radiotech Electron 40(9):1385–1406
6. Gotovac B, Kozulić V (1999) On a selection of basis functions in numerical analyses of engineering problems. Int J Eng Model 12:25–41
7. Atluri SN (2005) Methods of computer modeling in engineering and the sciences, vol I. Tech Science Press, Irvine
8. Griebel M, Schweitzer MA (eds) (2003) Meshfree methods for partial differential equations. Springer, Berlin
9. Liu GR, Gu YT (2005) An introduction to Meshfree methods and their programming. Springer, Dordrecht
10. Prenter PM (1989) Splines and variational methods. Wiley, New York
11. Kozulić V, Gotovac B (2000) Numerical analyses of 2D problems using Fupn(x, y) basis functions. Int J Eng Model 13(1–2):7–18
12. Nadai A (1963) Theory of flow and fracture of solids, vol 2. McGraw-Hill, Pittsburgh
13. Owen DRJ, Hinton E (1980) Finite elements in plasticity: theory and practice. Pineridge Press, Swansea
14. Timoshenko SP, Goodier JM (1961) Theory of elasticity. McGraw-Hill, New York
15. Lurie AI (1970) Theory of elasticity. Nauka, Moskva
16. Yamaguchi F (1988) Curves and surfaces in computer aided geometric design. Springer, Berlin
17. Timoshenko SP, Woinowsky-Krieger S (1959) Theory of plates and shells, 2nd edn. McGraw-Hill, New York

Simulation of Plastic Deformation Behaviors of Bulk Metallic Glasses with Micro- and Nano-sized Pores

Hong-Ying Zhang and Guang-Ping Zheng

Abstract Based on the phase-field model for deformations in bulk metallic glasses (BMG), shear banding in BMG with micro- and nano-sized pores is simulated and the thermo-plastic deformation behaviors are investigated. In the simulations, we use the free-volume concentration w_0 at the pore surface as a measure of the roughness of the pore. We obtain the critical w_0 when shear bands are initiated from the pore surface under different loading conditions. The effect of local heating due to shear banding on the critical w_0 is also quantitatively determined. By considering the heat conduction around pores, shear banding around vacuum pores or pores filled with helium gas are found to be quite different. It is shown that the nano-sized pores act as sinks or sources for shear bands when the pore surfaces are tailored. The simulations indicate that engineering BMG with nano-sized pores is effective in improving their ductility.

Keywords Metallic glasses · Shear bands · Phase-field modeling · Nano-sized pores

1 Introduction

Bulk metallic glasses (BMG), which are topologically disordered solids without long-range translational order, as seen in crystalline materials, have high fracture toughness, high strength-to-density ratio, good thermal conductivity and corrosion resistance. However, shear deformation always confined in the narrow localized shear zones in BMG makes these outstanding materials not practical in structural

H.-Y. Zhang · G.-P. Zheng (✉)
Department of Mechanical Engineering, Hong Kong Polytechnic University, Hung Hom, Kowloon, Hong Kong, China
e-mail: mmzheng@polyu.edu.hk

A. Öchsner and H. Altenbach (eds.), *Mechanical and Materials Engineering of Modern Structure and Component Design*, Advanced Structured Materials 70,
DOI 10.1007/978-3-319-19443-1_18

application. Thus, a lot of efforts have been made to develop new kinds of BMG exhibiting ductility [1–6]. It has been of great interest that a certain volume fraction of pores are effective in enhancing the plasticity of BMG without much degradation of its mechanical strength, which is demonstrated first in porous BMG or BMG foams with micro-sized pores [7, 8]. Recently it is found that artificially patterned pores on BMG surfaces or in BMG heterostructures could hinder the catastrophic spreading of shear bands during plastic deformations [9–12], especially those pores with sizes below several hundred nanometers [11, 12]. The mechanical properties of the BMG containing micro-sized or nano-sized pores are much related to the concentration of the pores. In particular, the enhanced ductility of BMG comes from the pores acting as powerful stress concentrators as well as multi-axis stress conditions which result in the multiplication of shear bands.

Shear localization, as the only mechanism for the catastrophic failure of glassy alloys, has attracted significant attention. For examples there are various investigations on this issue using conventional continuum mechanics [13, 14], free-volume-exhaustion mechanism [15] and experimental methods [16]. Although those experiments and simulations have captured the shear banding behaviors during the deformation of BMG, few have obtained the details of the strain/stress fields around the shear bands, and the understanding of the deformation mechanisms of these glassy alloys is still incomplete. In particular, though the study on the shear banding and its relations with pores is especially important and valuable in the development of BMG with superior mechanical properties, the understanding of the deformation mechanisms of BMG containing micro- or nano-sized pores is still limited.

In this work, the effects of pores on shear banding in BMG are investigated based on a phenomenological phase-field model, which has been successfully employed for simulation of shear banding, shear-band propagation and branching in conventional BMG [17]. On one hand, how the shear bands interact with the pores and the effects of heat transfer around the pores on shear banding will be elucidated from the simulations. On the other hand, the results can be compared with experimental data available on porous BMG and BMG with patterned pores or arrays of pores. This paper is arranged as follows. In Sect. 2, the phase-field model used for the simulation is described. In Sect. 3, the effect of pores on plastic and thermoplastic deformations in BMG is systematically investigated in the model system with a micro-sized pore. In Sect. 4, shear banding in BMG with an array of nano-sized pores is elucidated.

2 Simulation Methods

According to the atomic-scale deformation mechanism of BMG [17], the free-energy functional of glassy alloys under external loading can be expressed in terms of a local free-energy density $f(\varepsilon_{ij}, w)$ that is a function of the defect field $w(r)$ at the local position $\boldsymbol{r}$ only, the energy for deviations from spatial uniformity of defect distribution, and the kinetic energy associated with the local displacement ***u*(*r*)**, as follows:

$$F = \int \left\{ f(w, \varepsilon_{ij}) + \frac{\rho}{2}[\dot{\vec{u}}]^2 + \frac{\kappa}{2}|\nabla w|^2 \right\} dV, \tag{1}$$

where w is the normalized local density of deformation defects and ρ is the mass density of the glassy alloy; $\boldsymbol{\varepsilon_{ij}}$ is the local strain tensor; κ is the 'interfacial' energy between the regions with different densities of deformation defects. And when a crack appears, it becomes the surface energy. The local free-energy density of a metallic glass can be described as [17],

$$f(w, \varepsilon_{ij}) = e\left[\varepsilon_{ij}\right] + \frac{a_0}{2}w^2 + \frac{b_0}{3}w^3 + \frac{c_0}{4}w^4 + \left(\frac{a_1}{2}w^2 + \frac{b_1}{3}w^3\right)\left(e\left[\varepsilon_{ij}\right] - e_0\right). \tag{2}$$

The coefficients a_0, b_0, c_0 depend on the energy landscape of deformation defects, independent of the strain. There should exist an energy barrier between the un-activated (w = 0) and activated (w = 1) states of deformation defects in the un-deformed BMG. When w increases to a critical value w_s = 0.8 under plastic deformation, deformation defects are activated. Deformation defects can also be activated by thermal effects [18]. The deformation defects activated by temperature can be described as shown in Eq. (2) by defining the relation between the coefficient a_0 and the temperature T as $a_0 = a' - b'T/T_g$, where T_g is the glass transition temperature and a', b' are constants. a_1 and b_1 are constants which are described below.

Phase-field modeling of shear banding in BMG is based on the free-energy functional F of a glass alloy under external loading as discussed in detail in Ref. [17]. According to the Ginzburg-Landau formulism, the equations of motions for the displacement field ***u*** and the scaled field of deformation defects w are described as follows,

$$\tau_w \frac{\partial w}{\partial t} = -\frac{\delta F}{\delta w} = \kappa \nabla^2 w - (a_0 w + b_0 w^2 + c_0 w^3) - w(a_1 + b_1 w)(e\left[\varepsilon_{ij}\right] - e_0) \tag{3a}$$

and,

$$\rho \frac{\partial^2 \vec{u}}{\partial t^2} = -\frac{\delta F}{\delta \varepsilon_{ij}} = \mu \nabla \cdot \left\{ \left[1 + w^2\left(\frac{a_1}{2} + \frac{b_1}{3}w\right)\nabla \vec{u} \right] \right\}, \tag{3b}$$

where τ_w is the characteristic time of activation of these defects under local shear. μ is the shear modulus; The strain energy of metallic glass $e[\boldsymbol{\varepsilon_{ij}}]$ at the elastic limit is defined as e_0, which is an important physical parameter of BMG.

Moreover, the local heating due to the localized plastic straining can be described by the heat conduction equation,

$$k\nabla^2 T + \beta\frac{\partial Q}{\partial t} = C_p\rho\frac{\partial T}{\partial t}, \tag{4}$$

where Q is calculated as $Q = -T(\partial F/\partial T)$ representing the heat related to the generation or annihilation of deformation defects. Therefore Q is non-zero only near the shear bands in the model system under tensile deformation. Based on energy conservation, Q could be much closer to the mechanical work if the speed of the shear band is slow. The parameter β is the coefficient representing the percentage of conversion of mechanical work into heat, or the Taylor-Quinney coefficient. $\beta = 1$ means the adiabatic heating, while $\beta = 0$ represents isothermal deformation process. k is the thermal conductivity and C_p is the heat capacity.

The shear banding and its propagation in BMG can be described by solving Eq. (3a, 3b) assuming adiabatic conditions in BMG. Adding Eqs. (4) to (3a, 3b) can help us to investigate the deformation behaviors when the local heating effect of shear banding is considered. The material properties are listed as follows: $T_g = 625\,\text{K}$; the Young's modulus $E = 95\,\text{GPa}$; Poisson's ration $\nu = 0.35$; the mass density $\rho_0 = 6050\,\text{kg/m}^3$. The elastic strain limit under uniaxial tension is $\varepsilon_0 = 0.02$. The characteristic time of deformation defect activation is $\tau_w = 0.25\,\text{ns}$. Another important parameter, the deformation defect activation energy is generally determined by the strain-rate-sensitivity of plastic flow at a different temperature below T_g. In this work, ΔG is chosen as 4.6 eV at $T = 300\,\text{K}$ for Zr-based BMG, equivalent to the energy barrier for the deformation defects to be activated. The coefficients in Eq. (2) can be described as $a_0 = 4(2 - T/T_g)\Delta G$, $b_0 = -24\Delta G$, $c_0 = 16\Delta G$. The aim of choosing $a_1 = -6$ and $b_1 = 6$ is to stabilize the activated deformation defect state at $w = 1$ under plastic deformation.

The boundary condition of Eq. (3a, 3b) at the pore surface is $w = w_0$. Assume w represents the deformation defect density at the pore surface, w_0 between 0 and 1 could be a measure of the roughness of the pore surface. By solving Eqs. (3a, 3b)–(4), shear banding in BMG consisting of pores could be investigated. We consider a Zr-based BMG plate with dimensions of 10 × 10 × 1 μm^3. Plastic deformations of the model systems with two types of pores are investigated. First, the effect of pores on shear banding in BMG is systematically investigated in a model system with micro-sized pores, as shown in Fig. 1a. The pore diameter is r_0 = 1 μm and the initial crack length is l_0 = 0.2 μm. The results can be compared with experimental data available mostly in porous BMG containing pores with several microns. Second, BMG containing an array of nano-sized pores are employed to investigate how the patterning of the pores could improve ductility, as shown in Fig. 1b.

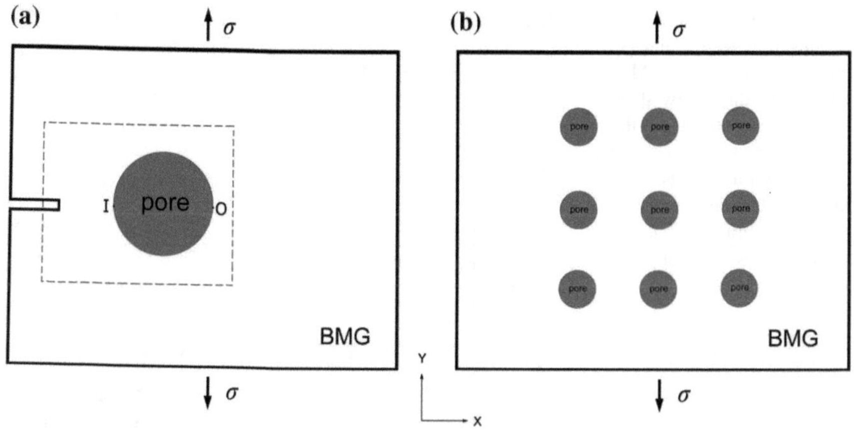

Fig. 1 **a** Schematic of the model BMG containing a micro-pore. **b** Schematic of the model system containing an array of pores. The length of the initial crack is l_0 = 0.2 μm. Tensile stress is perpendicular to the initial crack

3 Simulation of the Effects of Pores on Shear Banding in BMG

3.1 Shear Banding Without Considering Local Heating

We first investigate the interaction between shear band initiated from the initial crack and the pore in BMG without considering local heating. Thus the deformation of BMG is simulated using Eq. (3a, 3b). As shown in Fig. 2a, shear bands generated from an initial crack are discontinuous due to the presence of the pore. From the simulation, it is found that there are two ways that the discontinuous shear band could extend in BMG, as shown in Fig. 2a, b which are denoted as mode-A and mode-B shear banding, respectively. When the pore surface is smooth ($w_0 \rightarrow 0$), shear band is impeded by the pore which has been observed in experiment [8], resulting in mode-A shear banding. When the pore surface is rough ($w_0 \rightarrow 1$), shear band can be initiated from the side of the pore opposite to where the shear band touches the pore, resulting in mode-B shear banding; and the BMG are still brittle. As indicated in Fig. 2a, BMG containing pores could have enhanced fracture toughness if they are deformed via mode-A shear banding.

Stress analyses on mode-A and mode-B shear banding are shown in Fig. 3. In mode-B shear banding, shear stress on the pore surface is heterogeneous. Shear stress tends to build up at the side of the pore surface opposite to where the incident shear band touches the pore; and shear stresses at other places on the pore surface are relaxed as shown in Fig. 3d. The shear band could be initiated at the place marked as 'O' where the stresses concentrate on the pore surface. In mode-A shear banding, shear stresses are built up on several places on the pore surface and are relatively homogeneous as shown in Fig. 3b, and no obvious stress concentration

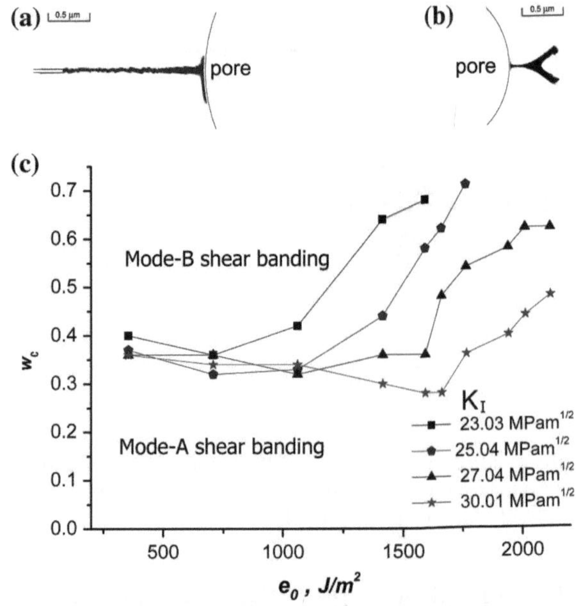

Fig. 2 **a** Mode-A shear banding: the shear band is impeded by the pore. **b** Mode-B shear banding: new shear bands are initiated from the pore surface. **c** The critical w_0 (represented as w_c) as a function of strain energy at the elastic limit e_0 under different stress intensity factors K_I

can be observed. The stress distribution after the interaction between the incident shear band and the pore can be further revealed by the change of pore geometry under the deformation caused by the stress field of the incident shear band. As shown in Fig. 3c, in mode-B shear banding, the shear deformation of the pore surface is mainly at the side opposite to where the shear band touches the pore, which is marked as 'O'. Thus shear band could be generated at 'O'. On the contrary, in mode-A shear banding, shear deformation of the pore surface occurs at its side close to the shear band, which is marked as 'I'. The shear band is thus impeded by the pore surface near the side marked as 'I', as shown in Fig. 3a.

Three factors, the applied tensile stress, the strain energy at the elastic limit of BMG e_0, and the roughness of the pores w_0, actually control these modes of shear banding. Under a fixed applied stress, mode-A shear banding changes into mode-B shear banding when the roughness of the pores w_0 changes from 0 to 1, and there exists a critical w_0 represented as w_c for such a transition of shear banding mode. In Fig. 2c, the w_c as a function of the strain energy at the elastic limit e_0 is shown for a given applied stress σ. According to the curves shown in Fig. 2c for various stress intensity factors $K_I = \sigma\sqrt{\pi \cdot l_0}$, it is seen that w_c is not significantly affected by e_0 and the applied stress when the strain energy at the elastic limit e_0 is smaller than 1062 J/m^2, and that the critical roughness of the pore surface w_c is about 0.3–0.4. That means BMG with e_0 smaller than 1062 J/m^2 could be tougher if the pore surface is made to have a roughness $w_0 < 0.3$, providing an important guideline for the design of BMG containing pores with enhanced fracture toughness.

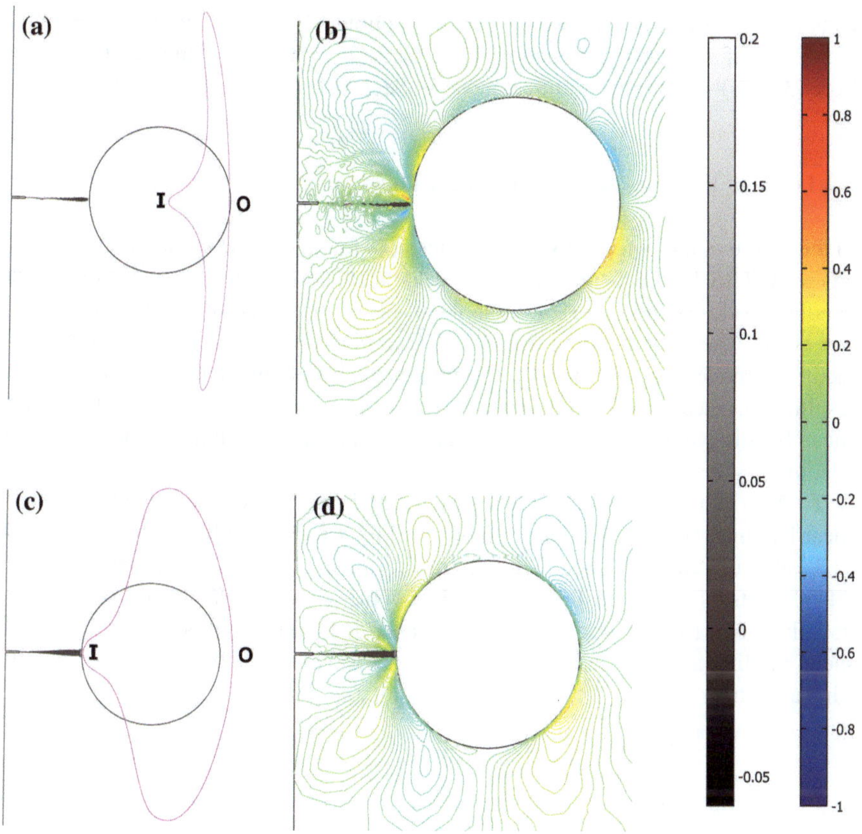

Fig. 3 **a** Pore surface before (colored in *black*) and after (colored in *pink*) mode-A shear banding. **b** Contour plots of shear stresses in mode-A shear banding. **c** Pore surface before (colored in *black*) and after (colored in *pink*) mode-B shear banding. **d** Contour plots of shear stresses in mode-B shear banding. The *color bar* is for the contour plots of shear stresses. The *gray bar* corresponds to the value of $1 - w$

3.2 *The Effects of Heat Conduction Around Pores on Shear Banding*

Although in what stage it is relevant to shear banding remains unclear, local heating is an important characteristic of shear banding. In BMG containing pores, local heating could play a dominate role in shear banding since shear bands will be significantly affected by the heat conduction conditions around the pores. For example, shear bands among vacuum pores could be generated by adiabatic deformation, while shear bands around pores filled with gases are formed in a relatively homogeneous thermal environment. To investigate the shear banding by considering local heating, we solved the coupled Eqs. (3a, 3b)–(4). The initial condition of the heat conduction Eq. (4) is T = 300 K. Other conditions of Eqs. (3a, 3b)–(4) are the same as those in Sect. 3.1.

3.2.1 Shear Banding Around Vacuum Pores

In the case that the pores are vacuum holes, it is found that the shear banding modes are related to β, the percentage of mechanical work to heat conversion. Shear banding modes B and C are shown in the insets of Fig. 4. The decreases of β from 1 to 0 change the shear banding from mode C to mode B under a fixed stress intensity factor. In mode-C shear banding, the pore acts as sink and source of shear bands, resulting in shear band multiplication. Hence porous BMG deformed in mode-C shear banding could have better ductility compared with that in mode-B shear banding.

To determine the conditions that BMG containing pores is deformed in mode-C shear banding, we plot diamond symbols in Fig. 4 indicating the critical β (denoted as β_c) at which mode-B shear banding transforms into mode-C shear banding under different stress intensity factors K_I. Hence the curve of $K_I \sim \beta_c$ denotes the boundary of brittle-to-ductile transition, and distinguishes mode-B and mode-C shear banding as discussed above. From Fig. 4 it is remarkable to observe that the adiabatic shear banding ($\beta \to 1$) results in ductile deformation in porous BMG.

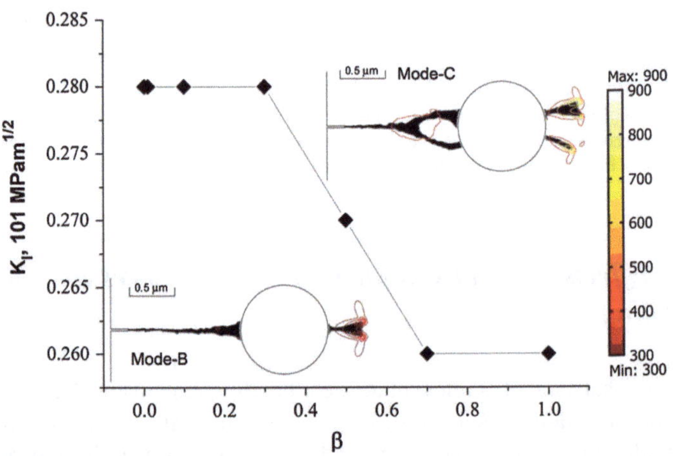

Fig. 4 Brittle to ductile transition controlled by the coefficient β of mechanical work to heat conversion in porous BMG with vacuum pores. The critical β (denoted as β_c) at which mode-B shear banding transforms into mode-C shear banding under a stress intensity factor K_I is plotted as *diamond symbol*. The *color bar* is for the contour plots of temperatures. The insets show the brittle porous BMG in mode-B shear banding and the ductile porous BMG in mode-C shear banding

3.2.2 Shear Banding Around Pores Filled with Helium

In fact those pores in BMG are usually made by blowing gas bubbles such as helium bubbles into the melts of alloys [19]. Thus the pores could contain gas which provides conduction paths for the heat generated by shear banding around the pores. We consider porous BMG consisting of pores with a diameter of 1 μm. The pores are filled with helium at 100 kPa which have the physical properties as follows, thermal conductivity 0.152 W/m K, heat capacity 5193 J/kg K, and density 0.178 kg/m^3. The shear banding around pores filled with helium is shown in Fig. 5a–d, assuming $\beta = 0, 0.001, 0.1$, and 1, respectively. The stress intensity factor is K_I = 32 MPa$^{1/2}$ and surface roughness of the pore is w_0 = 0.48. While shear banding in BMG with vacuum pores under the same pore surface roughness condition and stress intensity factor K_I are shown in Fig. 5e–h. Comparing the features of shear banding and the temperature distribution around shear bands which are

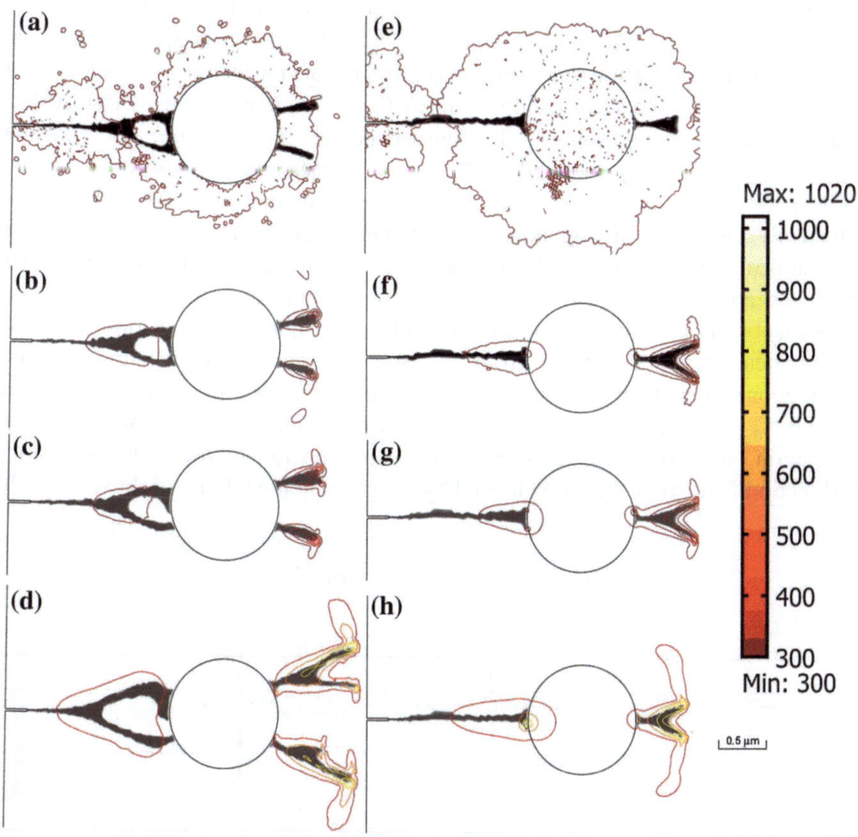

Fig. 5 Shear banding in BMG with vacuum pores, **a–d**: $\beta = 0, 0.001, 0.1$, and 1, respectively; and shear banding in BMG with pores filled with helium, **e–h**: $\beta = 0, 0.001, 0.1$, and 1, respectively. The *color bar* is for the contour plots of temperatures

close to the vacuum pores and pores filled with helium, we can find that the helium filled in the pores significantly affects the shear banding behaviors in BMG containing pores, while the temperature distribution around shear bands is not much affected. Obviously, BMG containing pores filled with helium deforms in mode-B shear banding and its ductility could be less than that of BMG containing vacuum pores.

4 Plastic Deformation Behaviors of BMG with an Array of Nano-sized Pores

When there are a lot of pores especially those patterned inside BMG, the interaction between the pores and a shear band could be more complicated. As the first attempt in elucidating the plastic deformation of BMG with patterned pores, we consider BMG with a 5-by-5 array of nano-sized pores, as shown in Fig. 1b. The diameter of the pores is 200 nm. A stress σ with $\sigma\sqrt{\pi \cdot l_0} = 40$ MPa$^{1/2}$ is applied along the Y-direction. Local heating due to shear banding is not considered in the simulation.

Figure 6 shows the shear bandings in BMG with patterned pores whose surface roughness or w_0 varies from 0 to 0.3. Shear bands are initiated from a pore located at the center of the array (denoted as '**C**') whose surface roughness or w_0 is assumed to be 1. As shown in Fig. 6a, when the pores are smooth (w_0 = 0.05) they could effectively block the propagation of shear bands inside the array of pores. The mechanism of such mode-A shear banding has been well discussed in Sect. 3.1. However, an abnormal shear-band multiplication occurs around the site '**C**' where the shear bands are initiated, as shown in Fig. 6a. Under tensile loading, conventional shear bands generated from the site '**C**' should propagate along the X-direction perpendicular to the loading direction since the shear stresses close to the site '**C**' are greatly intensified along the X-direction. As we can observe in the simulation, the conventional shear bands are generated from site '**C**' along X-direction first. After they are blocked by the pores at the left and right sides of the

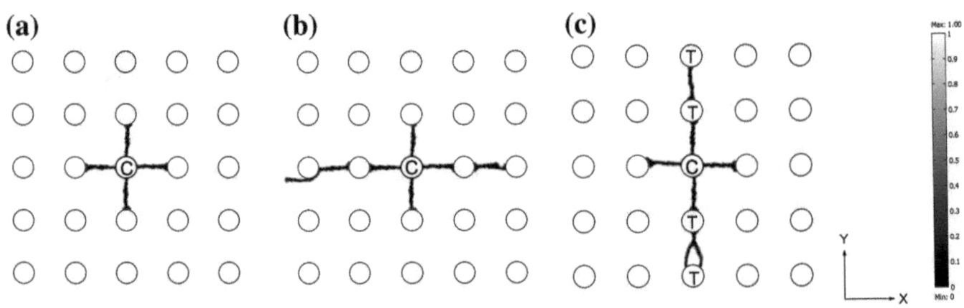

Fig. 6 Shear banding in BMG with an array of nano-sized pores. **a** w_0 = 0.05; **b** w_0 = 0.15; **c** w_0 = 0.3 for the pores marked as '*T*'. The rest pores are with w_0 = 0.05. The nano-sized pores are separated by 1.5 μm. The *gray bar* corresponds to the value of 1 − *w*

site ‘C’, the shear stresses close to the site ‘C’ are intensified along the Y-direction, resulting in the generation of abnormal shear bands along the loading direction.

If the roughness of the pores increases, mode-B shear banding could occur. Figure 6b shows a mixed mode-A and mode-B shear banding in BMG with pores whose surface roughness are homogeneous (w_0 = 0.15). The abnormal shear banding is still in mode A and blocked by the pores. The simulation results suggest that the plastic deformation of BMG with patterned pores can be controlled by tailoring the roughness of the pore surfaces. Figure 6c demonstrates such an idea that the shear band propagation can be controlled by the pores with heterogeneous surface roughness. If the surface roughness of the pores is w_0 = 0.05 except those marked as ‘T’, whose surface roughness is w_0 = 0.3, the abnormal shear banding tends to occur along the pores with large surface roughness. Hence the simulations indicate that engineering BMG with nano-sized pores is effective in improving their ductility.

5 Conclusions

Shear banding in BMG containing micro- and nano-sized pores are investigated by phase field modeling. We find that shear banding in BMG is affected by the surface roughness of the pores, the gas filled in the pores and the elastic limit and Young’s modulus of the BMG. The modeling provides quantitative measures on these parameters that determine different modes of shear banding. It is found that the pores act as sinks and sources for shear bands when the BMG containing vacuum pores is deformed in a way similar with adiabatic shear banding, resulting in improved ductility of the BMG. If the pores are filled with gas, BMG may be brittle. The mechanisms of shear banding are found to be consistent with those in BMG consisting of an array of nano-sized pores. The simulations indicate that engineering BMG with nano-sized pores is effective in improving their ductility.

Acknowledgments The authors are grateful for the supports provided by the Science and Technology Innovation Commission of Shenzhen and the Research Funds of Hong Kong Polytechnic University (Project No. A-PL98).

References

1. Schroers J, Veazey C, Johnson WL (2003) Amorphous metallic foam. Appl Phys Lett 82: 370–372
2. Wada T, Inoue A (2003) Fabrication, thermal stability and mechanical properties of porous bulk glassy Pd-Cu-Ni-P alloys. Mater Trans 44:2228–2231
3. Brothers AH, Dunand DC (2005) Ductile bulk metallic glass foams. Scripta Mater 52:335–339
4. Brothers AH, Dunand DC (2005) Processing and structure of open-celled amorphous metal foams. Adv Mater 17:484–486

5. Wada T, Inoue A (2004) Formation of porous Pd-based bulk glassy alloys by a high hydrogen pressure melting-water quenching method and their mechanical properties. Mater Trans 45:2761–2765
6. Wada T, Inoue A, Greer AL (2005) Enhancement of room-temperature plasticity in a bulk metallic glass by finely dispersed porosity. Appl Phys Lett 86:251907
7. Wada T, Inoue A, Greer AL (2007) Mechanical properties of porous bulk glassy alloy prepared in high-pressure hydrogen atmosphere. Mater Sci Eng A 449–451:958–961
8. Inoue A, Wada T, Louzguine-Luzgin DV (2007) Improved mechanical properties of bulk glassy alloys containing spherical pores. Mater Sci Eng A 471:144–150
9. Sarac B, Schroers J (2013) Designing tensile ductility in metallic glasses. Nat Commun 4:2158
10. Kumar G, Tang HX, Schroers J (2009) Nature 457:868
11. Lee MH, Sordelet DJ (2006) Nanoporous metallic glass with high surface area. Scripta Mater 55:947–950
12. Jayaraj J, Park BJ, Kim DH, Kim WT, Fleury E (2006) Nanometer-sized porous Ti-based metallic glass. Scripta Mater 55:1063–1066
13. Stief PS, Spaepen F, Hutchinson JW (1982) Strain localization in amorphous metals. Acta Metall 30:447–455
14. Vaks VG (1991) Possible mechanism for formation of localized shear bands in amorphous alloys. Phys Lett A 159:174–178
15. Yang B, Morrison ML, Liaw PK, Buchanan RA, Wang G, Liu CT, Denda M (2005) Dynamic evolution of nanoscale shear bands in a bulk-metallic glass. Appl Phys Lett 86:141904-1–141904-3
16. Wright WJ, Saha R, Nix WD (2001) Bulk metallic glasses III. Deformation mechanisms of the Zr40Ti14Ni10Cu12Be24 bulk metallic glass. Mater Trans 42:642–649
17. Zheng GP, Li M (2009) Mesoscopic theory of shear banding and crack propagation in metallic glasses. Phys Rev B 80:104201-1–104201-6
18. Pampillo CA (1972) Localized shear deformation in glassy metal. Scripta Metall 6:915–918
19. Wada T, Wang X, Kimura H, Inoue A (2009) Supercooled liquid foaming of a Zr–Al–Cu–Ag bulk metallic glass containing pressurized helium pores. Mater Lett 63:858–860

The Influence of Process Parameters on the Temperature Profile of Friction Stir Welded Aluminium Alloy 6063-T6 Pipe Butt Joint

Azman Ismail, Mokhtar Awang and Shaiful Hisham Samsudin

Abstract The temperature profile of friction stir welded aluminum alloy 6063-T6 pipe joints will be investigated in this paper. A pipe with an outside diameter of 89 mm and a wall thickness of 5 mm will be used as test pipe piece for this experiment on closed butt joint configuration by utilizing a Bridgeport 2216 CNC milling machine and orbital clamping unit which is specially-designed to cater for this task and function. Several samples will be prepared with varying process parameters such as rotational speed and travel speed. A very simple tool was used with a flat shoulder and a cylindrical pin. An infra-red thermometer will be employed to assess the temperature profile of the friction stir welded pipe butt joints during the experiment. The correct selection of process parameters will lead to a better joining condition of the welded joint. Several good samples were produced by this experiment setting.

Keywords Temperature profile · AA6063-T6 pipe · Butt joint · Friction stir welding · Bridgeport 2216 CNC milling machine

A. Ismail (✉)
Universiti Kuala Lumpur Malaysian Institute of Marine Engineering Technology, Jalan Pantai Remis, 32200 Lumut, Perak, Malaysia
e-mail: azman@unikl.edu.my

M. Awang · S.H. Samsudin
Department of Mechanical Engineering, Universiti Teknologi PETRONAS, Bandar Seri Iskandar, 31750 Tronoh, Perak, Malaysia
e-mail: mokhtar_awang@petronas.com.my

S.H. Samsudin
e-mail: shaiful_samsudin@petronas.com.my

A. Öchsner and H. Altenbach (eds.), *Mechanical and Materials Engineering of Modern Structure and Component Design*, Advanced Structured Materials 70,
DOI 10.1007/978-3-319-19443-1_19

1 Introduction

This solid state joining process was invented by Wayne Thomas from The Welding Institute, United Kingdom in 1991. This process is called as friction stir welding (FSW). FSW utilizes heat from friction to soften the adjoining section and then these sections are stirred together soundly as shown in Fig. 1 [1].

This welding process requires no filler metal and shielding gas, producing no arc and fumes. This FSW was initially developed to cater to the problem found in arc welding for aluminium such as distortion, shrinkage, and porosity. The implementation of FSW prevents such problems from occurring. This welding technique has been used for many applications due to lightweight construction, cost saving and environmental protection [2].

Pipe joining for FSW introduced complex challenges due to its tubular shape. Not many studies have been done for pipes instead of flat panels. Therefore, in order to run the experiment successfully, a proper fixture is vital. Several successful methods were introduced by previous studies [1, 3–6]. This will become the basic reference of the new built fixture called the orbital clamping unit (OCU). Hence, it is necessary to enable the available Bridgeport 2216 unit, a CNC milling machine, to run as a FSW unit for pipe joining.

It is important to understand the temperature profile in the pipe piece as it determines the success of the joint to be made, residual stress imposed, grain size and the strength of the welds [7]. The welding parameters for a successful FSW process will be discussed and the temperature profile at the tool will be measured in this present study.

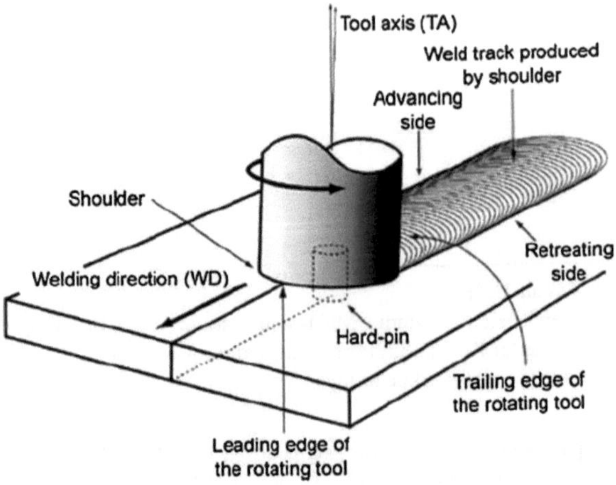

Fig. 1 FSW process

2 Experimental Setup

An experimental setup is shown in Fig. 2. The pipe with an outside diameter of 89 and 5 mm wall thickness was used in this present study. The tool was made of surface-hardened high carbon steel H13 with 20 mm diameter of shoulder length, a

Fig. 2 **a** FSW setup—front view. **b** FSW—side view

Table 1 FSW welding parameters

FSW sample	Welding parameters	
	Rotation speed (rpm)	Travel speed (mm/s)
FSW#1	900	1.2
FSW#2	1200	1.2
FSW#3	1500	1.2
FSW#4	1500	1.8
FSW#5	1500	2.4

Table 2 Chemical composition of AA6063-T6 pipe

Element	Percent present
Manganese (Mn)	0.0–0.10
Iron (Fe)	0.0–0.35
Magnesium (Mg)	0.45–0.90
Silicon (Si)	0.20–0.60
Zinc (Zn)	0.0–0.10
Titanium (Ti)	0.0–0.10
Chromium (Cr)	0.0–0.10
Copper (Cu)	0.0–0.10
Aluminium (AI)	Balance

Table 3 Mechanical properties of AA6063-T6 pipe

Property	Value
Proof stress	170 Min MPa
Tensile strength	215 Min MPa
Elongation A50 mm	6 Min %
Hardness Brinell	75 HB
Elongation A	8 Min %

pin diameter of 5 and 3.8 mm of pin length. The tool was positioned with a 6 mm forward offset from the centerline [1, 8, 9].

The welding parameters used were shown in Table 1. The plunge depth and dwell time used were 4 mm and 30 s respectively. Chemical composition and mechanical properties are shown in Tables 2 and 3 respectively [10]. The infrared (IR) thermometer was used to measure the temperature profile at the rotating tool for further analysis. All samples were inspected based on the AWS D17.3 acceptance level [11].

3 Results and Discussion

An IR thermometer was used to measure the temperature profile of the full weld cycle. The IR thermometer was shot on the rotating tool shank. The outer surface of the aluminium pipe was too reflective therefore the tool shank was used as point of

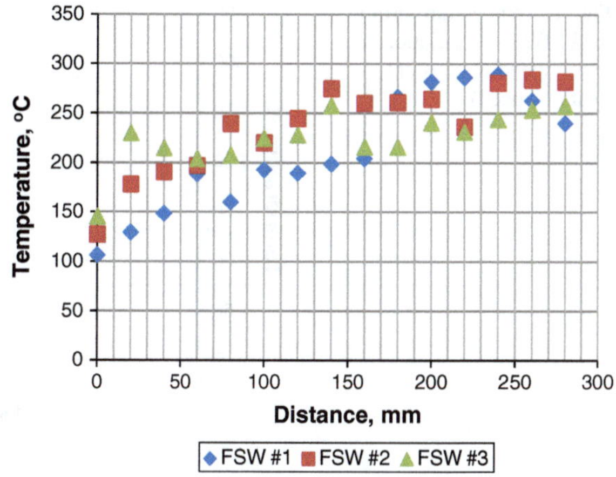

Fig. 3 Temperature profile for FSW#1–3 (variation in rotation speed at constant travel speed)

measurement of temperature [1]. Figures 3 and 4 show the temperature profile measured for certain welding parameters, with the increment of rotation speed at constant travel speed and with the increment of travel speed at constant rotation speed, respectively.

Based on the Fig. 3, the increment of rotation speed increases as the temperature generated from this friction process along the weld joint increases. Higher rotation speed generally generates higher temperatures. However, the recorded temperature profile varies due to pipe eccentricity, therefore causing a variation in contacts between the tool's shoulder and the outer pipe surface thus giving different temperature readings during the experiment. It was also dependent on the tool's offset setting from the axis of rotation. The measured temperature varies between 106.4 and 289.1 °C. A constant temperature was not detected during the experiment. As noted, the temperature increases in variation throughout the weld cycle. This quite similar temperature pattern was observed by a previous study [1, 5, 7].

Based on the Fig. 4, the temperature is decreasing with the increment of travel speed and off course these readings differ due to the same reason as before; i.e. the pipe eccentricity which affects the friction contact between tool's shoulder and the outer pipe surface. The measured temperature is between 130.9 and 285.9 °C. The increment of travel speed causes less time spent at a certain temperature level hence causing the reduction in generated temperatures.

For both experiment settings, the variation in temperature did affect the weld surface roughness quality as shown in Fig. 5. The issues of secondary heating can be seen on both settings as shown in Figs. 2 and 3 respectively as the tool starts and stops at the same point in order to complete the weld cycle, which previously underwent heat treatment.

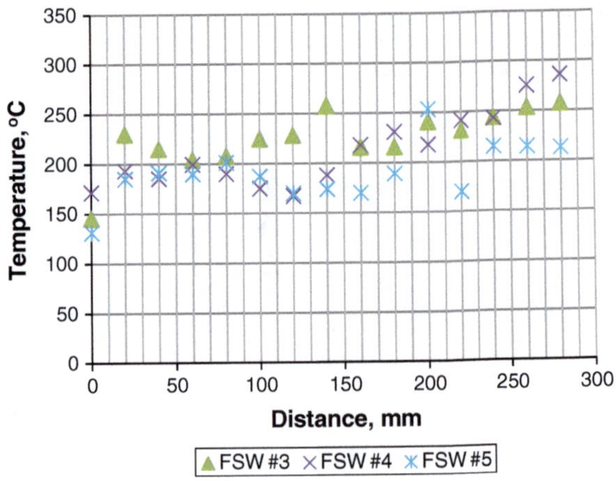

Fig. 4 Temperature profile for FSW#3–5 (variation in travel speed at constant rotation speed)

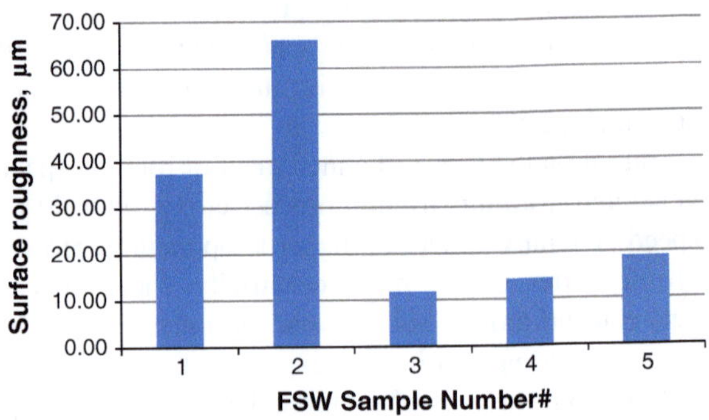

Fig. 5 Surface roughness of FSW samples

4 Conclusion

Based on the present study, the following conclusions can be made;

1. With the increment of rotation speed at constant travel speed, the temperature will increase, which was measured to be between 106.4 and 289.1 °C.
2. With the increment of travel speed at constant rotation speed, the temperature will decrease, which was measured to be between 130.9 and 285.9 °C.
3. The plowing effect can be achieved by offsetting the tool from its axis of rotation.
4. The variation in temperature did affect the weld surface quality (roughness).
5. The variation in temperature measurements for both welding parameters are due to pipe eccentricity which caused contact fluctuation in heat generation.
6. Secondary heating occurred in friction stir welded pipe joining.
7. Tool-surface contact fluctuated within an acceptable range during the weld cycle.

Acknowledgments The authors thank the Universiti Kuala Lumpur for providing the conference grant, 452-520435(004) and the Department of Mechanical Engineering, Universiti Teknologi PETRONAS for providing the required facilities and assistances.

References

1. Lammlein DH, Gibson BT, DeLapp DR et al (2010) Friction stir welding of small diameter pipe: an experimental and numerical proof of concept for automation and manufacturing. Proc Inst Mech Eng Part B:1–16
2. Kumar A, Fairchild DP, Macia ML et al (2011) Evaluation of economic incentives and weld properties for welding steel pipelines using friction stir welding. In: Proceedings of the international offshore polar engineering conference, pp 460–467
3. Packer SM, Matsunaga M (2004) Friction stir welding equipment and method for joining X65 pipe. In: Proceedings of the international offshore polar engineering conference, pp 55–60
4. Defalco J, Steel R (2009) Friction stir process now welds steel pipe. Weld J Am Weld Soc 88 (5):44–48
5. Gercekcioglu E, Eren T, Yildizh K et al (2005) The friction behavior on the external surface of the friction stir welding of AA6063-T6 tubes. The 5th international conference on tribology, pp 225–228
6. Doos Qasim M, Wahab Bashar Abdul (2012) Experimental study of fricton stir welding of 6061-T6 aluminium pipe. IJMERR 1(3):143–156
7. Hwang Y-M, Kang Z-W, Chiou Y-C, Hsu H-H (2008) Eperimental study on temperature distributions within the workpiece during friction stir welding of aluminum alloys. Int J Mach Tool Manu 48(7–8):778–787
8. Ismail A, Awang M, Fawad H, Ahmad K (2013) Friction stir welding on aluminum alloy 6063 Pipe. In: Proceedings of the 7th Asia Pacific IIW international congress, pp 78–81
9. Ismail A, Awang M (2014) Surface hardness of friction stir welded AA6063 pipe. MATEC web of conferences, vol. 13, 04025, pp 1–5
10. Aalco Metals Ltd, Aluminium alloy 6063-T6, www.aalco.co.uk
11. AWS D17.3 (2010) Specification for friction stir welding of aluminium alloys for aerospace application. American National Standard Institute

Influence of Cement Type and Mineral Additions, Silica Fume and Metakaolin, on the Properties of Fresh and Hardened Self-compacting Concrete

Sandra Juradin and Dražan Vlajić

Abstract Proportioning and mixing self-compacting concrete is a challenging task because the concrete mixture has to be stable and has to have the ability to fill formwork and to bypass obstacles under the influence of its own weight. Besides that, the final product has to be quality hardened concrete. It is known that even a little alteration of any component can significantly change characteristics of fresh and hardened concrete. In this work, the influences of the type of cement and additions, namely silica fumes and metakaolin on the workability and compressive strength of self-compacting concrete, are experimentally examined. For this purpose, several mixtures of self-compacting concrete were prepared and tested. The properties of fresh mixture were determined by the slump flow method, visual assessment of stability, T50 time, V-funnel method, L-box method and J-ring method. Also, in the hardened state, compressive strengths after 7 and after 28 days were determined. Results obtained in this work were compared with the results of other authors.

Keywords Cement type · Silica fume · Metakaolin · Workability · Compressive strength of concrete

1 Introduction

The basic property of SCC, which has to be achieved, is the ability to fill the formwork. That can be achieved only if the concrete can "flow" under its own weight and fill all types of formwork and if the concrete has good viscosity and resistance to segregation. Therefore, mixture proportioning of SCC is a challenging

S. Juradin (✉) · D. Vlajić
Faculty of Civil Engineering, Architecture and Geodesy, University of Split,
Matice Hrvatske 15, 21000 Split, Croatia
e-mail: Sandra.Juradin@gradst.hr
URL: http://www.gradst.hr

A. Öchsner and H. Altenbach (eds.), *Mechanical and Materials Engineering of Modern Structure and Component Design*, Advanced Structured Materials 70,
DOI 10.1007/978-3-319-19443-1_20

task. Even a small "error" in mixture proportioning of SCC leads to problems in self compacting, because this type of concrete is very sensitive even to the slightest deviation in terms of quality and components dosage.

The main requirement in the process of mixture proportioning has to be the amount of coarse aggregate, paste amount, water-cement ratio and the amount of additions. One of the first methods of mixture proportioning of SCC was developed by Okamura and Ozawa at the University of Tokyo [1, 2]. The basic principles of proportioning are:

- Volume of the coarse aggregate in concrete has to be 50 % of all solids in concrete
- Volume of the fine aggregate has to be around 40 % of the mortar volume
- Water-cement ratio by volume has to be 0.9–1.0; depending on the properties of the cement
- By changing the water-cement ratio and the amount of superplasticizer the effect of self-compacting is achieved.

Petersson and Billberg [3] and Billberg [4] have developed the CBI method according to which concrete is compounded of solids (aggregate) and liquid and of paste. Paste fills the voids in the aggregate skeleton and makes the layer which coats the aggregate particles. To determine the minimum volume of the paste, there are requirements which need to be fulfilled. They are defined as design criteria, void content and blocking criteria.

Mixture design of SCC very often includes testing of mortar properties which are representative enough. Because of that, quality tests can be minimized on concretes.

Erdem et al. [5] compared the rheological properties of SCC with the representative equivalent mortar in which coarse aggregate was replaced with sand. The method of equivalent mortar was used to determine, more easily, the properties of SCC including density of binders and properties of fresh materials. The authors found the correlation between mortar and SCC flow using the factor of correlation r:

$$r = \frac{V_{SCC}}{A_{SCC} \cdot (D_{max})_{SCC}} = \frac{V_{MSCC}}{A_{SMCC} \cdot (D_{max})_{MSCC}} \quad (1)$$

whereas

V_{SCC}	the volume of SCC in slump cone
V_{MSCC}	the volume of equivalent mortar in mini slump cone
A_{SCC}	the area of SCC spread
A_{MSCC}	the area of equivalent mortar spread
$(D_{max})_{SCC}$	the maximum aggregate size in SCC mixture
$(D_{max})_{MSCC}$	the maximum aggregate size in equivalent mortar mixture.

The correlation factor for their tested concretes is roughly 1, but it is important to underline that mixtures were made with additive HRWR (High-Range-Water-

Reducer) and liquid viscosity modifier (VMA). "A small incremental increase in the HRWR can result in a considerable decrease in yield stress and an increase in SCC slump flow. This can reduce the number of trial batches for concrete mixture design needed to ensure high deformability and stability, thus resulting in time savings and reduction in manpower and materials" [5].

Materials used in preparation of SCC are mostly the same as those used in preparation of regular concrete, but in different ratios. Besides cement, aggregate and water, chemical admixtures (superplasticizer, viscosity modifier) and mineral additions like filler, flying ash, pigments, silica fume, metakaolin, etc. are added to the mixture. Silica fume and metakaolin are additions which actively participate in the process of hydration of cement, causing the pozzolanic reaction. The pozzolanic reaction is the chemical reaction between calcium hydroxide and pozzolan. The products of reaction are the products of hydrotation-like C-S-H gel. Metakaolin is a relatively new addition to concrete. Its properties and the pozzolanic reaction can be compared to those of silica fume, but metakaolin is less expensive. The size of an average particle of metakaolin is 2 μm which is much smaller than a cement particle, yet not as fine as particles of silica fume (0.2 μm). Previous studies show that results of mixtures that contain metakaolin are similar to those which contain silica fume [6].

Wild et al. [7] have proved that calcium hydroxide, expressed as a percentage of total Portland cement, in mortars with metakaolin and the equivalent paste shows the minimum after 14 h. This has been explained as a peak in the pozzolanic reaction which shows that more calcium hydroxide is reduced from the paste in the reaction with metakaolin, than new is made by cement hydration. After a year, there is still a significant amount of calcium hydroxide, even in pastes where 15 % of cement is replaced with metakaolin. So it is suggested that the level of replacement needs to be over 15 % to use up completely all calcium hydroxide. Said-Mansour et al. [8] have come to similar conclusions about 3 main factors which affect the behaviour of metakaolin in concrete: filler effect, hydration acceleration and the pozzolanic reaction. Ding and Zongjin [9] have shown "that metakaolin offers much better workability than silica fume for given mixture proportions".

Madandoust and Yasin Mousavi [10] have shown that the addition of metakaolin increases the strength of SCC, especially the early strength at 3–14 days of age. That is in accordance with some earlier studies which showed that the biggest contribution to early strength of these concretes came from the pozzolanic reaction of metakaolin. Also, higher compressive strength is achieved with a lower water-cement ratio. It can be concluded that concrete with addition of metakaolin has a similar hydration progress during time as concrete with addition of silica fume. After workability testing of fresh concrete, the authors have concluded that the spreading of fresh concrete is reduced with a larger usage of metakaolin (20 %). They presume that this can be explained with the fact that the particles of metakaolin have a considerably higher specific area than the particles of Portland cement [11]. These kinds of results are consistent with the known fact that the addition of metakaolin increases the need of superplasticizer. Results of the L-box testing were worse in samples with addition of metakaolin, but they were still in a satisfying range.

Based on the testing results, the authors have concluded that the optimal amount of metakaolin is 10 % of the cement mass in order to satisfy specific requirements of fresh SCC.

Hassan et al. [12] examined the influence of metakaolin and silica fume on SCC with different percentages of addition compared to one control mix without any additions. Metakaolin was added in amounts of 3, 5, 8, 11, 15, 20 and 25 % of cement mass. Silica fume was added in amounts of 3, 5, 8 and 11 % of cement mass. Based on the time T50 result and V-funnel test they concluded that metakaolin increased viscosity of fresh SCC and that the addition of silica fume has no effect. Also, they showed that the increasing amount of metakaolin caused a rise in time T50 and it was in allowed boundaries according to EFNARC [13]. Metakaolin increased the ability to flow through and around obstacles.

When the amount of metakaolin increased from 0 to 25 %, the results of the L-box test raised from 0.63 to 0.89. Also, it was noticed that the addition of metakaolin increased the need for the superplasticizer dosage. However, when compared with silica fume, the addition of metakaolin requires less superplasticizer.

The compressive strength of SCC containing MK increased as MK content increased from 0 to 25 % (as a partial replacement of cement). On the other hand, the optimum percentage of SF in terms of compressive strength was 8 %, and it was similar to that of 8 % MK (both increased the strength of the control mixture by about 14 %). However, raising the amount of MK from 8 to 25 % only enhanced the compressive strength by 7 % (with respect to 8MK).

Other studies [14–16] also showed that SF and MK increased compressive strength of concrete, reduced shrinking, increased chloride resistance and resistance to freezing. Besides that, by replacing certain amounts of cement with SF or MK, the price of concrete can be reduced.

The purpose of this paper is to analyse the influence of those additions on self-compacting concrete. The control mixture was determined by the CBI method and by experiments on equivalent mortar. Eleven mixtures were prepared into which altered types of the cement (cement type I and type III) and percentage of partial replacement of cement by SF and MK were added.

2 Experimental Investigation

2.1 Introduction, Used Materials and Mixtures

The goal of the experimental investigation is to determine the influence of the type of the cement and the amount of SF and MK on the properties of fresh self-compacting concrete and on its compressive strength after 7 and 28 days. For this purpose, 9 different mixtures of SCC were prepared.

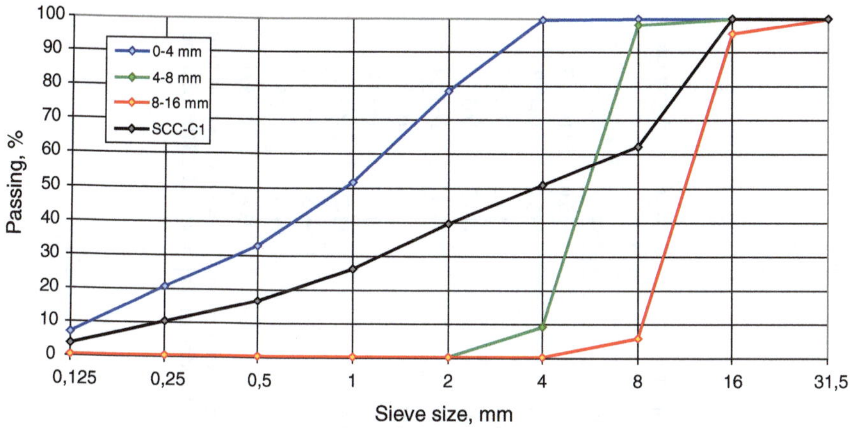

Fig. 1 The granulometric curve of the aggregates and of the reference SCC mixture—C1

Materials that were used:

- cement, types I and III (CEM I 42,5R and CEM III/A 42,5 N LH),
- the aggregate was crushed limestone, with the composition and grain size distribution of which is shown in Fig. 1. In mixtures, three fractions, 0–4, 4–8, 8–16 mm were used,
- silica fume with a specific surface area according to Blaine greater than 15,000 cm^2/g and the specific weight was 2.3 g/cm^3,
- metakaolin, density 2.6 g/cm^3 and specific surface area according to Blaine was around 24,000 cm^2/g,
- filler, which was obtained by recycling old concrete, had s specific surface area according to Blaine 7891 cm^2/g and density 2.45 g/cm^3,
- polycarboxylate superplasticizer.

The composition of each mixture is given in Table 1.

Control mixtures C1 and C3 did not contain any mineral additions. The only difference between them was in the type of the cement that was used. The number after the letter C, in the mixture label, stands for the type of cement. Letters "S" and "M" stand for mineral addition (S-silica fume, M-metakaolin) and the number at the end stands for the percentage of partial replacement of cement by silica fume or metakaolin. All mixtures had the same w/c ratio—0.42. According to the standards HRN EN 206-1, the calculation of the water–cement ratio when silica fume is used as an addition, was done considering the k-concept:

$$w/c = \frac{\text{water}}{\text{cement} + 2 \cdot \text{silica fume}} \quad (2)$$

So, in the mixtures which contain silica fume or metakaolin, the amount of water was determined according to the water-cement ratio which was calculated in this

Table 1 Mixture designs and labels for tested mixtures (mass for 1 m^3 concrete)

Label of mixtures	Cement		Silica fume	Metakaolin	Water	w/c	Superplasticizer	Filler	Aggregate
	CEM I	CEM III							
	kg		kg	kg	kg		kg	kg	kg
C1	400.0	–	–		168.0	0.42	4.0	136.0	1640.3
C3	–	400.0	–		168.0	0.42	4.0	136.0	1640.3
C1S5	381.0	–	19.0		176.0	0.42	4.0	114.2	1637.5
C3S5	–	381.0	19.0		176.0	0.42	4.0	114.2	1637.5
C1S10	363.6	–	36.4		183.5	0.42	4.0	93.8	1634.9
C3S10	–	363.6	36.4		183.5	0.42	4.0	93.8	1634.9
C1M5	381.0	–	–	19.0	176.0	0.42	4.0	116.3	1637.8
C3M5	–	381.0	–	19.0	176.0	0.42	4.0	116.3	1637.8
C1M10	363.6	–	–	36.4	183.3	0.42	4.0	98.3	1635.4
C1M10*	363.6	–	–	36.4	152.7	0.42	4.0	98.3	1755.4
C3S10*	–	363.6	36.4		165.8	0.38	4.0	93.8	1725.0

* All mixtures had the same w/c ratio 0,42 and their w/c = water/(cement +2* addtition) except mixtures marked with asterisk. The water-cement ratio of C1M10* also was 0.42 but it was determined as w/c = water/cement (we didn't taken into account the proportion of addition). The mixture C3S10* with the w/c ratio 0.38 had the lowest value of w/c and in that was differs from C3S10.

way. The concept (2) was also used in mixtures with the addition of metakaolin in order to achieve better workability of the concrete, keeping in mind that used metakaolin particles were finer then those of silica fume. Exceptions were the mixtures C1M10*, where the water-cement ratio was 0.42 but it was determined as in the case when we used only cement without additions, and the mixture C3S10*, where the water-cement ratio was 0.38.

Fresh self-compacting concrete was tested on the flow ability, passing ability, and segregation resistance with the following methods: slump-flow test, T50 time, V-funnel, L-box and J-ring test. The compressive strength of hardened concrete was determined after 7 and 28 days. The results were analysed and compared to the previous results from the listed literature.

2.2 *Designing of the Control Mixture C1*

The composition of the control mixture C1 was determined according to the CBI method and equivalent mortar method [4, 5]. As previously mentioned, the control mixture C1 was without mineral additions, and grain size distribution of this concrete, shown in Fig. 1, represent the distribution of all concretes. Initial content of this mixture, when 1 % of additive was used, is shown in Table 2.

The volume of paste (cement, water, sand, additive, water and air) in concrete was determined from the first design of the mixture. The amount per volume of the sand (<0.25 mm) was determined from the grain size distribution, see Fig. 1 and Table 2.

The mass ratio of gravel (>4 mm) and the total aggregate (for mixture C1 the mass ratio is 0.5) was calculated using Table 2. With that value, and according to void content criteria and blocking criteria, the minimum required paste volume was determined (see Fig. 2). The difference between the recommended and assumed volume of paste was compensated by adding the filler. The result was a final mixture whose composition is given in Table 1.

Because the method of equivalent mortar was used, the total area of aggregate (fractions 0–4, 4–8 and 8–16 mm) in the concrete was replaced with the fraction 0–4 mm. The tests on mortar showed the best dosage of additive for given values of spreading. For those tests, the amount of additive was 0.8 and 1 % of cement mass. The mixtures had to fulfil two main requirements: stability and spreading size

Table 2 Initial composition of mixture C1

C1	Cement	Water	Superplasticizer	Aggregate			Air	Total
				0–4 mm	4–8 mm	8–16 mm		
Mass (kg)	400	168	4.0	895.85	179.17	716.68	–	2362.9
Volume (dm^3)	132.9	168	4.0	333	66.6	266.4	30	1000

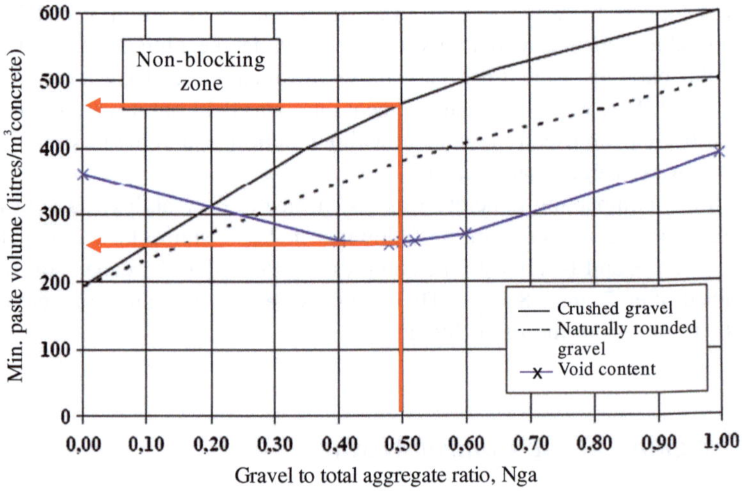

Fig. 2 Required minimum paste volume (using diagram of void content criteria and blocking criteria) [4]

Table 3 Results of measurement on equivalent mortar of mixture C1

Equivalent mortar for mixture C1		
Amount of admixtures (%)	0.8	1
D_1 (mm)	260	300
D_2 (mm)	270	310
SF (mm)	265	305

270–305 mm. The results of the spreading size of mortar had to provide results of spreading size of concrete in the range from 600 to 730 mm [17]. Measured values are in Table 3 and in Fig. 2.

According to the results of spreading and stability of mixture, the selected amount of additive was 1 % of the binder mass. That is, at the same time, the maximum recommended dosage for this additive (Fig. 3).

For all other mixtures the adopted dosage of additive was 1 %, but the filler mass was determined for each mixture in accordance to void content criteria and blocking criteria in order to acquire the minimum necessary volume of the paste (diagram, see Fig. 2). Final mixtures are shown in Table 1.

3 Experimental Results and Discussion

Concrete workability All mixtures were made in volumes of 25 dm^3, in the laboratory, in the mixer with the capacity of 50 dm^3. Mixing was conducted as follows. First, cement, filler, aggregate and mineral addition were mixed together.

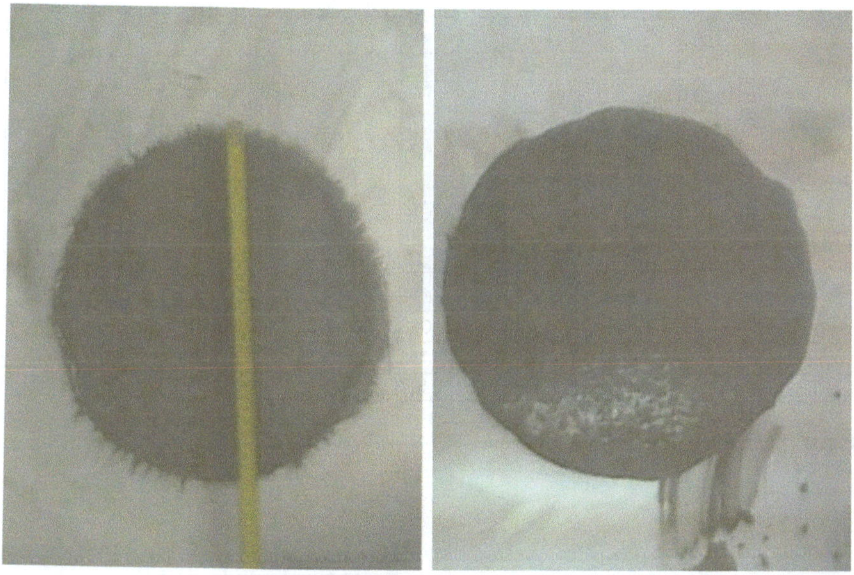

Fig. 3 Tests on equivalent mortar for 0.8 and 1 % of admixture

Later, while mixing, water and superplasticizer were added. After that, mixing was continued for a few minutes.

The workability was examined on fresh concrete using the slump-flow method, visual check of stability, T50 time, and with V-funnel, L-box and J-ring. The tests were carried out in accordance with the European standards HRN EN 12350 8-12. Based on test results, concrete mixtures can achieve different labels of consistency according to the classes of European association EFNARC, see Table 4.

In Figs. 4, 5, 6, 7, 8 and 9 the achieved results and effects occurred during the tests are shown.

Table 4 Slump-flow, viscosity and passing ability classes with respect to EFNARC [13]

Slump-flow (mm)	SF1	550–650
	SF2	660–750
	SF3	760–850
T50 (s)	VS1	≤2
	VS2	>2
V-funnel (s)	VF1	≤9
	VF2	9–25
L-box	PL1	≥0.80 (with 2 rebar)
	PL2	≥0.80 (with 3 rebar)
J-ring	PJ1	≤10 (with 12 rebar)
	PJ2	≤10 (with 16 rebar)

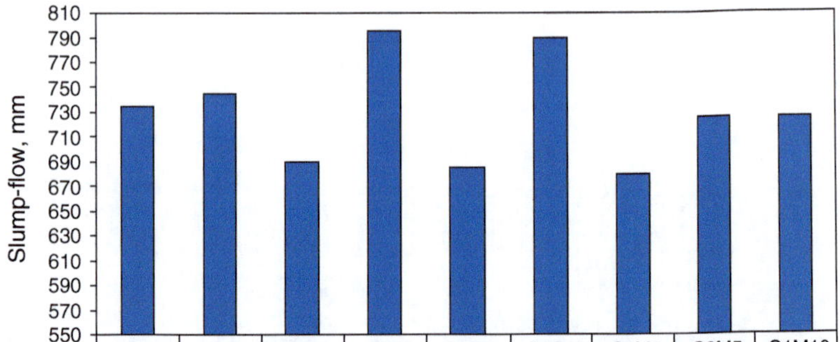

	C1	C3	C1S5	C3S5	C1S10	C3S10	C1M5	C3M5	C1M10
slump-flow, mm	735	745	690	795	685	790	680	725	725
with respect to EFNARC	SF2	SF2	SF2	SF3	SF2	SF3	SF2	SF2	SF2

Fig. 4 Slump-flow results for all mixtures

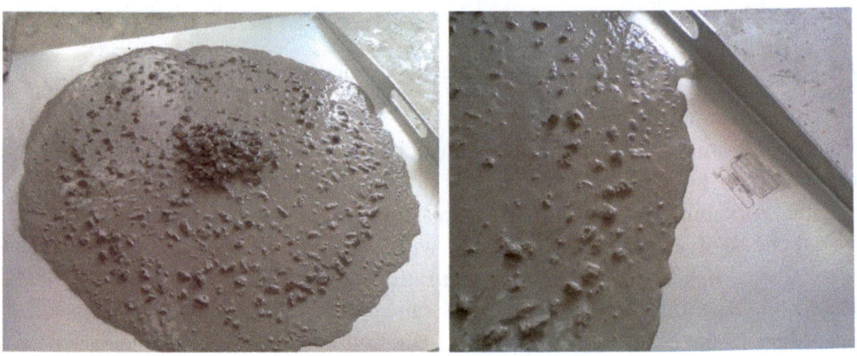

Fig. 5 Slump-flow test: the mixtures C3S10 (*left*) and C3S5 (*right*)

Fig. 6 Slump-flow test: the mixtures C1S5 (*left*) and C1S10 (*right*)

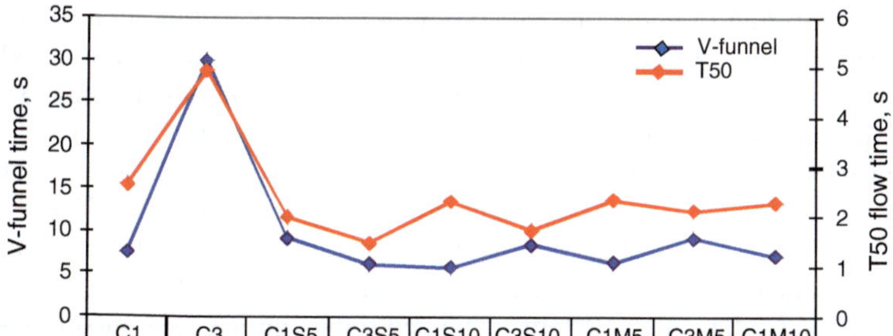

	C1	C3	C1S5	C3S5	C1S10	C3S10	C1M5	C3M5	C1M10
V-funnel time, s	7.49	30	9.19	6.22	5.86	8.43	6.4	9.19	7.18
with respect to EFNARC	VF1	-	VF2	VF1	VF1	VF1	VF1	VF2	VF1
T50 flow time, s	2.66	4.94	2	1.5	2.32	1.75	2.37	2.12	2.28
with respect to EFNARC	VS2	VS2	VS1	VS1	VS2	VS1	VS2	VS2	VS2

Fig. 7 Flow ability: V-funnel method and T50 time

Fig. 8 Aggregate blocking on bars when the mixture C3 was tested (*left*) and water segregation without aggregate blocking when mixture C3S5 was tested (*right*)

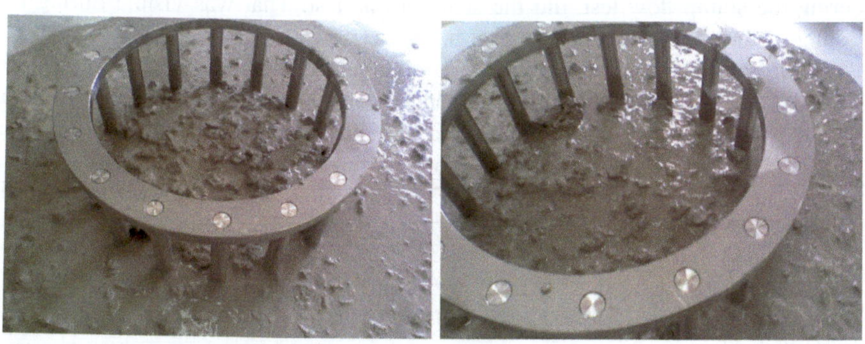

Fig. 9 J-ring test: the mixtures C3S5 (*left*) and C1M10 (*right*)

According to Fig. 4, the slump flow test and the classification of mixtures, only the mixtures C3S5 and C3S10 achieved class SF4. Other mixtures had results for class SF2. The same amount and type of addition had different effects on different types of cement. Silica fume lowered the workability of cement CEM I, and increased spreading of cement CEM III. With visual observation, segregation and water separation on edges were determined in the mixtures C3S5 and C3S10 (see Fig. 5). The most stable were the mixtures C1S5 and C1S10 (see Fig. 6).

So, although replacing a part of the cement CEM I with silica fume resulted with lower spreading size, mixtures remained stable, unlike those mixtures where CEM III was used and had the opposite effect. Replacing 5 % of cement mass with metakaolin lowered the spreading of concrete but all mixtures remained stable. In the mixture C1M10 where the level of replacement was 10 %, water separation and segregation occurred.

Since we had the results of spreading size on equivalent mortar for mixture C1, it was possible to determine the correlation coefficient according to expression (1), see Table 5.

Based not only on results from Table 5, but also on some other tests and achieved results from literature [5], it is clear that the type of additive and adding VMA significantly contribute to the correlation coefficient. In all tests which were carried out without adding a viscosity modifier, the correlation coefficient was less than 1, and it was in a range from 0.23 to 0.66.

Results of the T50 time and V-funnel test are compatible, see Fig. 7.

As it is shown in Fig. 7, the maximum deviation occurred in mixture C3. During the V-funnel test of this mixture, the aggregate blocked the funnel exit and that was the reason why the measured time was 30 s. Metakaolin and silica fume lowered the flow time. The exception was the mixture C1S5. The unstable mixtures C3S5 and C3S10 had the lowest time T50. That shows the test has to be considered in relation to the results of the slump flow test.

Similar to the V-funnel test, blocking the effect in the mixture C3 happened in the L-box test also, see Fig. 6 (on the left). The mixtures which segregated water during the slump flow test, did the same in this test. That was visible during the testing of the mixture C3S5 (see Fig. 6, on the right).

According to Table 6 results, only 4 mixtures satisfied the L-box test—C3S5, C3S10, C3M5 and C1M10. The L-box and J-ring test results are given in Table 6.

The results in Table 6 show that silica fume had a larger influence on mixtures with cement type III and metakaolin on mixtures with cement type I. Neither mixture satisfied the norms for EFNARC classification when they were tested with the J-ring. Mixtures C3S5, C1M5 and C1M10 had the best results (see Fig. 9).

Table 5 The correlation coefficient according to expression (1) for mixture C1

Mixture	r_{SCC}	r_{MSCC}	$r_{MSCC}-r_{SCC}$
C1	0.81	1.47	0.66

Table 6 Passing ability and classification according to EFNARC [13]

Mixture	PL		PJ	
	Measured	Class	Measured	Class
C1	0.78	–	30.5	–
C3	0.69	–	33	–
C1S5	0.72	–	29.25	–
C3S5	0.95	PL2	16.75	–
C1S10	0.65	–	28.5	–
C3S10	0.86	PL2	25	–
C1M5	0.77	–	16	–
C3M5	0.93	PL2	23.75	–
C1M10	0.95	PL2	16.75	–

Analysis of mixture constituents of self compacting concrete In Table 7 the constituents of SCC mixtures and evaluation according to EFNARC [13] are shown.

Since [18] points out that these proportions are in no way restrictive and many SCC mixes will fall outside this range for one or more constituents, it can be concluded that tested concretes meet the suggested criteria. However, fulfilling the required criteria does not guarantee the workability of self-compacting concrete.

Compressive strength was determined after 7 and 28 days, and the obtained results are shown in Fig. 10.

Table 7 The constituents of SCC mixtures and evaluation according to EFNARC [13]

Constituent	C1	C3	C1S5	C3S5	C1S10	C3S10	C1M5	C3M5	C1M10	Typical range
Vol of coarse aggregate/m^3 (%)	30.5	30.5	30.4	30.4	30.4	30.4	30.4	30.4	30.4	27–36
Mass sand versus total aggregate (%)	50.0	50.0	50.0	50.0	50.0	50.0	50.0	50.0	50.0	48–55
Paste (%) (vol)	38.7	38.7	38.8	38.8	38.9	38.9	38.8	38.8	38.9	30–38
Powder (kg/m^3) (mass)	596.7	596.7	574.8	574.8	554.3	554.3	576.9	576.9	558.8	380–600
Water (kg/m^3)	168.0	168.0	176.0	176.0	183.5	183.5	176.0	176.0	183.3	150–210
Water/powder ratio by vol	0.80	0.80	0.86	0.86	0.93	0.93	0.86	0.86	0.90	0.85–1.1
Water/powder ratio by mass	0.28	0.28	0.31	0.31	0.33	0.33	0.31	0.31	0.33	0.28–0.37

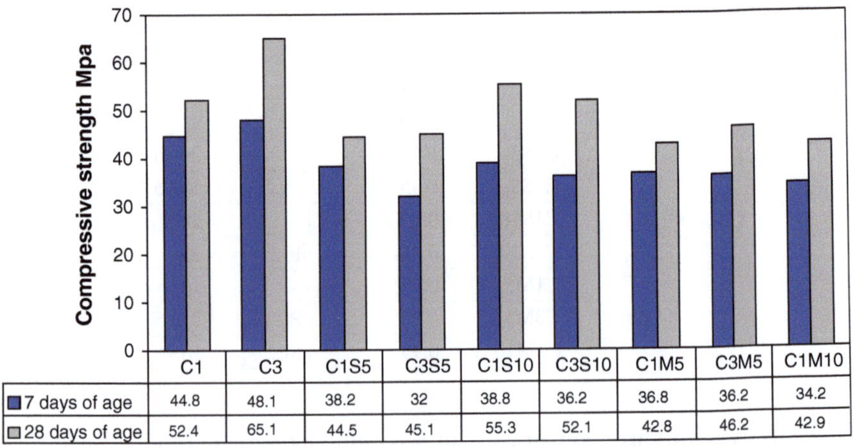

Fig. 10 Compressive strength after 7 and 28 days

Earlier studies show [7–12, 15, 16] that adding silica fume and metakaolin increases 28-day compressive strength of both classic and self-compacting concrete. Also, according to [11] metakaolin has an effect on early strength. In this work, a slight increase in 28-day compressive strength, compared to the control mixture C1, was noticed only when 10 % of mass of cement type I was replaced with silica fume. All other mixtures had 17 % lower compressive strength. Nearly all mixtures with cement type III had lower compressive strength, compared to mixture C3, by around 30 %. Only the mixture C3S10 had lower strength, by 15 %.

Results of additional mixtures Two additional mixtures, whose composition is shown at the bottom of Table 1, were also examined in their fresh and hardened state. The test of fresh mixture with lowest water-cement ratio (0.38)—C3S10* shows that this mixture cannot be fully considered self-compacting concrete. Test samples of concrete-cubes were still made without vibrating. The compressive strength after 7 and 28 days was measured and the results can be seen in Fig. 11.

The second mixture C1M10* did not show the self-compacting properties, but classic concrete ones. It did not have the ability of self-compacting and the test samples were made with vibrating.

The compressive strength results after 7 and 28 days are shown in Fig. 11 together with the results of mixtures with the same composition but which were made with a larger amount of water, according to the k-concept. Although, it can't be considered as SCC, it can be seen that the mixture C3S10* achieved 30 % higher 28 day compressive strength than the mixture C3S10. Mixture C1M10* achieved 64 % higher 28-day compressive strength than the C1M10 mixture. Both results are in accordance with the studies presented in [11]. Thus, it can be concluded that the amount of water influenced greatly the compressive strength of concrete. Also, it

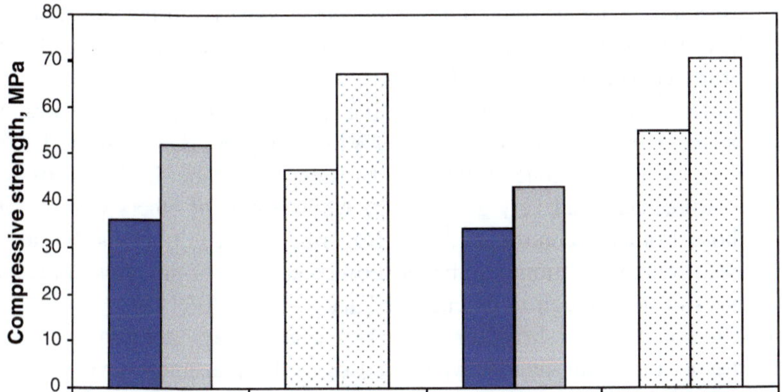

	C3S10	C3S10*	C1M10	C1M10*
7 days of age	36.2	46.9	34.2	54.9
28 days of age	52.1	67.2	42.9	70.4

Fig. 11 Compressive strength of additional mixtures

can be assumed that by reducing the amount of water and with use of a superplasticizer and viscosity modifier, compressive strength would be even higher than those in the reference mixtures C1 and C3, and prepared concrete would be classified as self-compacting.

4 Conclusions

The goal of this work is to determine the influence of type of the cement, silica fume and metakaolin on the properties of fresh and hardened self-compacting concrete. For this purpose, nine different mixtures were prepared that differ in type of used cement, and in level of replacement of cement by mineral addition. Water-binder ratio, mass of binder materials (cement and mineral addition), ratio of coarse and fine aggregate and mass of superplasticizer stayed the same in all mixtures.

- Measured values of spreading size showed that mixtures with silica fume or metakaolin achieved lower results than those achieved by control mixtures without any mineral additions. Similar tendencies were demonstrated in earlier works where it was proven that an increase in level of replacement of cement with silica fume or metakaolin increased the need for amount of superlasticizer in order to achieve the same values of spreading size. Having in mind, that in this work, the maximum amount of superlasticizer was used, it is necessary either to change the type of superplasticizer or/and to add a viscosity modifier.
- Addition of silica fume and metakaolin generally enhances the ability of concrete to fill the formwork because they increase the speed of flow, which was

especially evident for mixtures with cement type III. If the speed of flow is too low, we should be cautious, because that can mean an occurrence of segregation of the coarse aggregate and water.

- Earlier works show that the passing ability of obstacles increases when the percentage of silica fume or metakaolin increases. Based on the L-box and J-ring test results, general conclusions cannot be made. In the L-box test, some mixtures achieved very good results while results of others were in range with those from the control mixtures. Nevertheless, it is obvious that all mixtures with mineral additions had better results than control mixtures in the J-ring test, but still not enough to be classified according to EFNARC.
- The compressive strength results show that 28-days strength is higher only in the mixture with cement type I and cement replacement levels of 10 % with silica fume. Early strength is also higher in control mixtures. Cause for this can be in larger quantity of water as a result of different calculation of water-cement ratio, which was shown on two additional mixtures. They had very good compressive strength but weak workability. This again confirms the need for an increase in dosage of superplasticizer and addition of viscosity modifier. Besides that, because cement CEM III is represented as a cement with considerable strength growth after 28 days of age, the test should be repeated after, at least, 90 days of age.

References

1. Okamura H, Ozawa K (1995) Mix-design for self-compacting concrete. Concr Libr of JSCE 25:107
2. Okamura H, Ouchi M (2003) Self-compacting concrete. J Adv Concr Technol. doi:10.3151/jact.1.5
3. Petersson Ö, Billberg P (1999) In: Proceedings of first international RILEM symposium on self-compacting concrete, Stockholm, Cachan, France, p 333
4. Billberg P (1999) Self-compacting concrete for civil engineering structures—the Swedish experience. Stockholm
5. Erdem TK, Khayat KH, Yahia A (2009) Correlating rheology of self-consolidating concrete to corresponding concrete-equivalent mortar correlating rheology of self-consolidating concrete to corresponding concrete-equivalent mortar. ACI Mater J. doi:10.14359/56462
6. Mitrović A et al (2005) Metakaolin - nova generacija dopunskih cementnih materijala. Građevinski materijali i konstrukcije. http://www.dimk.rs/
7. Wild S, Khatib JM, Jones A (1996) Relative strength, pozzolanic activity and cement hydration in superplasticised metakaolin concrete. Cem Concr Res. doi:10.1016/0008-8846(96)00148-2
8. Said-Mansour M et al (2011) Influence of calcined kaolin on mortar properties. Constr Build Mater. doi:10.1016/j.conbuildmat.2010.11.017
9. Ding JT, Zongjin L (2002) Effects of metakaolin and silica fume on properties of concrete. ACI Mater J. doi:10.14359/12222
10. Madandoust R, Yasin Mousavi S (2012) Fresh and hardened properties of self-compacting concrete containing metakaolin. Constr Build Mater. doi:10.1016/j.conbuildmat.2012.04.109

11. Kadri EH et al (2011) Influence of metakaolin and silica fume on the heat of hydration and compressive strength development of mortar. Appl Clay Sci. doi:10.1016/j.clay.2011.06.008
12. Hassan AA et al (2012) Effect of metakaolin and silica fume on the durability of self-consolidating concrete. Cem Concr Comp. doi:10.1016/j.cemconcomp.2012.02
13. BIBM, CEMBUREAU, ERMCO, EFCA, EFNARC (2005) The European guidelines for self-compacted concrete. http://www.efnarc.org/pdf/SCCGuidelinesMay2005.pdf
14. Zongin L, Bin M, Chui SNC (1999) Systematic study of properties of extrudates with incorporated metakaolin or silica fume. ACI Mater J. doi:10.14359/660
15. Turk K, Turgut P, Karatas M, Benli A (2010) In: 9th international congress on advances in civil engineering, Trabzon, Turkey
16. Siddique R, Klaus J (2009) Influence of metakaolin on the properties of mortar and concrete: a review. Appl Clay Sci. doi:10.1016/j.clay.2008.11.007
17. Domone P (2006) Mortar tests for self-consolidating concrete. Concr Int. http://www.concreteinternational.com/pages/featured_article.asp?ID=15266
18. De Schutter G (2005) Guidelines for testing fresh self-compacting concrete. http://www.researchgate.net/publictopics.PublicPostFileLoader.html?id=549a98cdd685cc92758b4675&key=93965f79-353b-4d7e-ab50-8d4962569a9d

On the m-Term Best Approximation of Signals, Greedy Algorithm

Martin G. Grigoryan

Abstract Described the class of signals such that the error between signal and m-term best approximant with respect to orthonormal basis has the following behavior : o(log m)$^{-1/4}$.

Signal processing applications of orthogonal bases most often rely on their ability to efficiently approximate certain types of signals with just a few vectors.

Linear approximations project the signal on m vectors selected a priori. The approximation can be made more precisely by choosing the m orthogonal vectors depending on the signal properties.

Non-linear algorithms outperform linear projections by approximating each signal with vectors selected adaptively within a basis. Let $\{\varphi_n(x)\}$ be an orthogonal basis in L^2, and let $\{f_m(x)\}$ be the projection of f over the m vectors whose indices are in A_m

$$f_m(x) = \sum_{k\epsilon A_m} \langle f, \varphi_k \rangle \varphi_k(x), \quad \text{where} \quad \langle f, \varphi_k \rangle \geq \int_0^1 f(t)\varphi_k(t)dt = c_k(f).$$

The approximation error is the sum of the remaining coefficients

$$r_m(f) = \|f - f_m\|_2 = \left(\sum_{k \in A_m} |\langle f, \varphi_k \rangle|^2 \right)^{\frac{1}{2}}$$

To minimize this error, the indices in A_m must correspond to the m vectors having the largest inner product amplitude $|\langle f, \varphi_k \rangle|$. They are the vectors that best correlate $f(x)$. So they can be interpreted as the "main" features of $f(x)$. The resulting $r_m(f)$ is necessarily smaller than the error of the linear approximation, which selected the

M.G. Grigoryan (✉)
Yerevan State University, Yerevan, Republic of Armenia
e-mail: gmarting@ysu.am

A. Öchsner and H. Altenbach (eds.), *Mechanical and Materials Engineering of Modern Structure and Component Design*, Advanced Structured Materials 70,
DOI 10.1007/978-3-319-19443-1_21

m approximation vectors independently of $f(x)$. Let us sort $\{|c_k(f)|\}_{k\geq 1}$ in decreasing order $|c_{n_k}| \geq |c_{n_{k+1}}|$. The best non-linear approximation is

$$f_m^{best}(x) = \sum_{k=1}^{m} c_{n_k}(f)\varphi_{n_k}(x).$$

Let $\Psi = \{\psi_n\}_{n=1}^{\infty}$ be a normalized basis in Banach space X. Then for each element $f \in X$ there exists a unique series by system $\{\psi_n\}_{n=1}^{\infty}$ converging to f in the norm of space X:

$$f = \sum_{n=0}^{\infty} c_n(f)\psi_n,$$

Let an element $f \in X$ be given. We call a permutation $\sigma = \{\sigma(n)\}_{n=1}^{\infty}$ of nonnegative integers decreasing and write $\sigma \in D(f, \psi)$, if

$$|c_{\sigma(n)}(f)| \geq |c_{\sigma(n+1)}(f)|, \quad n = 1, 2, \ldots.$$

In the case of strict inequalities here $D(f, \Psi)$ consists of only one permutation.

We define the mth greedy approximant of f with regard to the basis Ψ corresponding to a permutation $\sigma \in D(f, \Psi)$ by formula.

Consider

$$G_m(f) := G_m(f, \Psi, \sigma) := \sum_{n=1}^{m} c_{\sigma(n)}(f)\psi_{\sigma(n)}.$$

This nonlinear method of approximation was considered in [1] and is known as greedy algorithm.

We say that the greedy approximant of element f by system Ψ converges, if for some $\sigma \in D(f, \Psi)$ we have

$$\lim_{m\to\infty} \|G_m(f, \Psi, \sigma) - f\|_X = 0.$$

In this direction a number of interesting results have been obtained in [1–9].

We define the best m-term approximation with regard to $\Psi = \{\psi_n\}_{n=1}^{\infty}$ as follows

$$R_m(f) := R_m(f, \Psi)_X := \inf \left\| \sum_{n=1}^{m} a_n\psi_n - f \right\|_X$$

where inf is taken over coefficients a_n and sets of indices Λ with cardinality $|\Lambda| = m$.

It is clear that in the case $X = L^2$ and Ψ is an orthonormal basis we have

$$\|G_m(f, \Psi, \sigma) - f\| = R_m(f, \Psi) = \sum_{n=m+1}^{\infty} \left|c_{\sigma(n)}(f)\right|^2.$$

For >0, we denote $G(\ln^q)$ the space of signals $f(x) \in L^2\,[0, 1]$ whose Fourier coefficients have a finite $l^2(\ln^q)$ norm:

$$\|f\|_{G(\ln^q)} = \left(\sum_{k=1}^{\infty} c_k^2(f) \ln^q k\right)^{\frac{1}{2}}$$

i.e.

$$G(\ln^q) = \left\{f(x) \in L^2[0, 1]; \quad \text{with } \sum_{k=1}^{\infty} c_k^2(f) \ln^2 k < \infty\right\}.$$

We set

$$G \searrow (\ln^q) = \left\{f(x) \in L^2[0, 1]; \quad \text{with } \sum_{k=1}^{\infty} c_{n_k}^2(f) \ln^2 k < \infty\right\},$$

where $\{n_k\}$ is the permutation of natural numbers such that $|c_{n_k}| \geq |c_{n_{k+1}}|, \forall k \geq 1$ and

$$\|f\|_{G\searrow(\ln^q)} = \left(\sum_{k=1}^{\infty} c_{n_k}^2(f) \ln^q k\right)^{\frac{1}{2}}.$$

In [10] it is proved the Theorem: If $\sum_{k=1}^{\infty} |c_k(f)|^p < \infty, p < 2$ then hold Jachson inequality:

$$R_k(f) \leq \frac{\|f\|_{B_p}}{\frac{2}{p} - 1} \cdot \frac{1}{k^{\frac{2}{p}} - 1}, \quad \text{where} \quad \|f\|_{B_p} = \left(\sum_{k=1}^{\infty} |c_k(f)|^p\right)^{\frac{1}{p}}$$

and $R_k(f) = o\left(\frac{1}{k^{\frac{2}{p}-1}}\right)$.

Conversely, if $R_k(f) = O\left(\frac{1}{k^{\frac{2}{p}-1}}\right)$ then $\sum_{k=1}^{\infty} |C_k(f)|^q < \infty$, for all $p < q$.

In this paper we prove

Theorem 1 *If a signal* $f(x) \in G(\ln^q), q > 1$ *then*

$$R_k(f) \leq \frac{\|f\|_{G\searrow}}{q-1} \cdot \frac{1}{(\ln k - \ln 2)^{q-1}}$$

and $R_k(f) = o\left(\frac{1}{(\ln k)^{q-1}}\right)$.

We need the following elementary result:

Lemma 1 *Let m be an arbitrary natural number. Given any finite sequence* $\{x_k\}_{k=1}^n$ *of non negative integers and a monotonically increasing finite sequence* $\{y_k\}_{k=1}^n$. *Then*

$$\sum_{k=1}^{m} x_{n_k} y_k \leq \sum_{k=1}^{m} x_k y_k, \tag{1}$$

where $\{n_k\}_{k=1}^m$ *is that permutation of the positive integers which* $x_{n1} \geq x_{n2} \geq \cdots \geq x_{n_m}$.

Proof Let $m = 2$ and let $x_m \geq x_1$ and $y_1 < y_2$. We have

$$0 \leq (x_2 - x_1)(y_2 - y_1) = x_2 y_2 + x_1 y_1 - (x_2 y_1 + x_1 y_2),$$

hence

$$\sum_{k=1}^{2} x_{n_k} y_{n_k} = x_{n_1} y_1 + x_{n_2} y_2 = x_2 y_1 + x_1 y_2 \leq \sum_{k=1}^{2} x_k y_k.$$

It is not hard to see that, using the mathematical induction methods we can obtain inequality (1) for each natural m. □

Lemma 2 *Given any sequences* $\{x_k\}_{k=1}^m$ *and* $\{y_k\}_{k=1}^m$, *with*

$$x_k \geq 0, \quad \text{and} \quad 0 < y_1 < y_2 < \cdots < y_k < \cdots \tag{2}$$

then

$$\sum_{k=1}^{\infty} x_{n_k} y_k \leq \sum_{k=1}^{\infty} x_k y_k,$$

where $\{n_k\}_{k=1}^m$ *be that permutation of the natural numbers* $1, 2, \ldots$, *which* $x_{n_1} \geq x_{n_2} \geq \cdots, (\{x_{n_k}\} \searrow)$.

Proof We may assume that

$$\sum_{k=1}^{\infty} x_k y_k < \infty.$$

Let $\{n_k\}_{k=1}^{m}$ be that permutation of natural numbers 1, 2, … which

$$x_{n_1} \geq x_{n_2} \geq \cdots x_{n_k} \geq \cdots \tag{3}$$

For any natural number s we set

$$N_s = \max\{n_k; 1 \leq k \leq s\}.$$

Using (2), (3) and Lemma 1, with $m = N_s$, for $\{x_k\}_{k=1}^{N_s}$ and $\{y_k\}_{k=1}^{N_s}$, we get

$$\sum_{k=1}^{N_s} x_{n_k} y_k \leq \sum_{k=1}^{N_s} x_k y_k \leq \sum_{k=1}^{\infty} x_k y_k.$$

Since $x_k \geq 0$ and $y_k \geq 0$ we obtain

$$\sum_{k=1}^{s} x_{n_k} y_k \leq \sum_{k=1}^{\infty} x_k y_k, \quad \text{for all } s \geq 1,$$

which completes the proof of lemma 2. □

From lemma 2 we obtain the following:

Theorem 2 $G(\ln^q) \subset G \searrow (\ln^q)$ *for all* $q > 0$, *and* $\|f\|_{G\searrow} \leq \|f\|_G$.

Proof Using Lemma 2 with $x_k = c_k^2(f)$ and $y_k = \ln^q k$, $q > 0 \ \forall\, k \geq 1$ we have if $f(x) \in G(\ln^q)$ then $f(x) \in G \searrow (\ln^q)$ and $\|f\|_{G\searrow} \leq \|f\|_G$ (see definitions of $G(\ln^q), G \searrow (\ln^q)$). It is not hard to see that there exists a signal $f_0(x) \in G \searrow (\ln^q)$ but $f_0(x) \notin G(\ln^q), q > 0$. □

Theorem 3 *If a signal* $f(x) \in G \searrow (\ln^q), q > 1$ *then*

$$R_k(f) \leq \frac{\|f\|_{G\searrow}}{q-1} \cdot \frac{1}{(\ln k - \ln 2)^{q-1}}$$

and

$$R_k(f) = o\left(\frac{1}{\ln^{q-1} k}\right)$$

Conversely, if $R_k(f) = O\left(\frac{1}{\ln^q k}\right)$ *then* $f(x) \in G(\ln^q)$ *for any* $p < q - 1$.

Proof Let $\forall\, f(x) \in G \searrow (\ln^q), q > 0$. From the definition of $G \searrow (\ln^q)$ we get $\|f\|_{G\searrow} = \sum_{k=1}^{\infty} c_{n_k}^2(f) \ln^q k < \infty$, where $\{n_k\}$ is a permutation of the natural numbers $1, 2, \ldots$, which

$$|c_{n_k}| \geq |c_{n_{k+1}}| \geq \cdots \tag{4}$$

We put

$$\lambda_m(f) = \sum_{k=m}^{\infty} c_{n_k}^2(f) \ln^q k.$$

From this and (4) we have

$$k c_{n_{2k}}^2(f) \ln^q k \leq \sum_{s=k}^{2k-1} c_{n_s}^2(f) \ \ln^q s < \lambda_k(f).$$

Hence

$$c_{n_{2k+1}}^2(f) \leq c_{n_{2k}}^2(f) \leq \frac{\lambda_k(f)}{k \ln^q k}.$$

From an approximation's error we obtain

$$R_k(f) = \sum_{s=k}^{\infty} c_{n_s}^2(f) \leq \lambda_{[\frac{k}{2}]}(f) \sum_{s=[\frac{k}{2}]} \frac{1}{s \ln^q s}$$

$$\leq \lambda_{[\frac{k}{2}]}(f) \int_{[\frac{k}{2}]}^{\infty} \frac{dx}{x \ln^q x} \leq \lambda_{[\frac{k}{2}]}(f) \cdot \frac{1}{(q-1)(\ln [\frac{k}{2}])^{q-1}}.$$

Since $\lim_{k\to\infty} \lambda_k(f) = 0$ and $\lambda_k(f) \leq \|f\|_{G\searrow(\ln^q)}$. We get

$$R_k(f) = o\left(\frac{1}{\ln^{q-1} k}\right)$$

$$R_k(f) \leq \frac{\|f\|_{G\searrow}}{q-1} \cdot \frac{1}{(\ln k - \ln 2)^{q-1}}, \quad \forall k > 2.$$

Conversely suppose that there exists a $C > 0$ such that

$$R_k(f) \leq C \frac{1}{\ln^q k}, \quad q > 0.$$

Since

$$R_k(f) = \sum_{i=k+1}^{\infty} |c_{n_i}(f)|^2 \geq \sum_{i=k+1}^{2k} |c_{n_i}(f)|^2 \leq k|c_{n_{2k}}(f)|^2,$$

then

$$|c_{n_{2k+1}}(f)|^2 \leq |c_{n_{2k}}(f)|^2 < C\frac{1}{k \ln k}.$$

Hence, if $p < q - 1 (q - p > 1)$

$$\sum_{k=1}^{\infty} c_{n_k}^2(f)(\ln k)^p \leq 2 \cdot \sum_{k=1}^{\infty} \frac{1}{k(\ln k)^{q-p}} < \infty,$$

which completes the proof of Theorem 3. □

Acknowledgments This work was supported by the State Committee Science MES RA, in frame of the research project No. SCS 13-1A313.

References

1. Konyagin SV, Temlyakov VN (1999) A remark on Greedy approximation in Banach spaces. East J Approx 5:1–15
2. Temlyakov VN (1998) Greedy algorithm and m-term Trigonometric approximation. Constr Approx 14:569–587
3. Wojtaszczyk P (2000) Greedy algorithm for general biorthogonal systems. J Approx Theory 107:293–314
4. Grigoryan MG (2010) Uniform convergence of the Greedy algorithm with respect to the Walsh system. Studia Math 198(2):197–206
5. Grigoryan MG (1991) On convergence of Fourier series in complete orthonormal systems in the L^1 metric and almost everywhere. Mat USSR-Sb 70:445–466
6. Episkoposian SA (2007) On the divergence of Greedy algorithms with respect to Walsh subsystems. Nonlinear Anal Theory Methods Appl 66(8):1782–1787
7. Grigoryan MG, Gogyan SL (2006) On nonlinear approximation with respect to the Haar system and modifications of functions. Ann Math 32:49–80
8. Grigoryan MG, Zink RE (2006) Greedy approximation with respect to certain subsystems of the Walsh orthonormal system. Proc Am Math Soc 134:3495–3505
9. Episkoposian SA, Grigoryan MG (2012) Convergence of Greedy algorithm by generalized Walsh system. J Math Anal Appl 389:1374–1379
10. DeVore R, Lorentz G (1993) Constructive approximation. V. 303 of comprehensive studies in math. Springer, Berlin

On the Best Approximation [illegible]

[illegible]

Then [illegible]

Hence [illegible]

[illegible]

References

[illegible]

[illegible] Theory [illegible]

[illegible]

[illegible]

[9] [illegible] approximations [illegible]

Effect of Simultaneous Plasma Nitriding and Aging Treatment on the Microstructure and Hardness of Maraging 300 Steel

Adriano Gonçalves dos Reis, Danieli Aparecida Pereira Reis, Antônio Jorge Abdalla, Jorge Otubo, Susana Zepka, Antônio Augusto Couto and Vladimir Henrique Baggio Scheid

Abstract Simultaneous nitriding and aging heat treatment of maraging 300 steel was carried out inside a DC-pulsed plasma nitriding reactor. A single heat treatment cycle was done, as the plasma nitriding and age hardening processes occur at the same ranges of temperatures and times. Samples of maraging 300 steel, in the solution annealed and solution annealed and aged conditions, were tested. Plasma nitriding and aging, carried out at 480 °C for 3 h, increased the surface hardness up to 1140 HV, producing case depths of 50 μm since ε-Fe_3N and γ′-Fe_4N nitrides were formed in the hardened surface layer. It is observed that the microstructure of the core material remains unaltered as the typical martensite plate-like microstructure of maraging steels. The core hardness of solution annealed samples increased from 331 to 597 HV after the plasma nitriding treatment proving the

An erratum to this chapter is available at DOI 10.1007/978-3-319-19443-1_37

A.G. dos Reis (✉) · D.A.P. Reis · J. Otubo · S. Zepka
Instituto Tecnológico de Aeronáutica, 12228-900 São José dos Campos, Brazil
e-mail: areis@ita.br

D.A.P. Reis
e-mail: danieli.reis@unifesp.br

D.A.P. Reis
Universidade Federal de São Paulo, 12231-280 São José dos Campos, Brazil

A.J. Abdalla · V.H.B. Scheid
Instituto de Estudos Avançados, 12228-001 São José dos Campos, Brazil
e-mail: abdalla@ieav.cta.br

A.A. Couto
Instituto de Pesquisas Energéticas e Nucleares and Mackenzie, 05508-900 São Paulo, Brazil
e-mail: acouto@ipen.br

A.G. dos Reis
Rua Matias Peres, 364, Bairro Floradas de São José, São José dos Campos SP 12230-082, Brazil

A. Öchsner and H. Altenbach (eds.), *Mechanical and Materials Engineering of Modern Structure and Component Design*, Advanced Structured Materials 70,
DOI 10.1007/978-3-319-19443-1_22

possibility of nitriding and aging at the same treatment cycle. The pre-aged samples did not show any overaging or martensite reversion to austenite after the simultaneous plasma nitriding and aging treatments, that could be showed by the core hardness of 620 HV and can be related to the time of total aging exposure of 6 h, including pre-aging and plasma nitriding.

Keywords Maraging steel · Plasma nitriding · Case depth · Microstructure · Hardness

1 Introduction

Maraging refers to the aging of martensite, a hard microstructure commonly found in steels. Martensite is easily obtained in these steels owing to the high nickel content. The only transformation that occurs at ordinary cooling rates is martensite formation. The martensite without carbon is quite soft, but heavily dislocated. These steels are based on the Fe–Ni binary alloy with additions of various alloying elements such as cobalt, molybdenum, titanium and aluminum. Hardening and strengthening of these steels are subsequently produced by heat treating (aging), caused by precipitation of extremely fine, coherent intermetallic compounds such as Ni_3X (X = Ti, Mo) [1–3]. Maraging 18 %Ni grades are denoted by numbers such as 200, 250, 300 or 350, the number specifying the level of the yield strength in ksi that can be obtained in the steel with appropriate heat treatments [2]. Its high strength to weight ratio, good weldability, and easy machinability in the solution annealed condition and dimensional stability during aging make this material an ideal choice for critical applications in aerospace industries, such as rocket motor casing [1–3]. It has been reported that depending upon the aging duration, the steel undergoes systematic characteristic microstructural changes. The early aging period is characterized by recovery of martensitic structure and hardening due to precipitation of hexagonal the intermetallic precipitates, that takes place rapidly due to fast diffusion of titanium atoms [4–6]. The intermediate aging period is characterized by reversion of austenite accompanied by precipitation of hexagonal Fe_2Mo intermetallic phase. These two processes, occurring at the intermediate aging period, affect hardening in the opposite manner; thus, overall hardening levels off after reaching a maximum [4]. A decrease in hardening, observed during longer aging durations, is attributed essentially to the formation of reverted austenite and precipitate coarsening. The amount of reverted austenite has been reported to increase with an increase in aging temperature and time [1, 2]. Maraging 300 steels are normally subjected to solution annealing at 820 °C for 1 h followed by aging at 480 °C for 3 h. This heat treatment results in the best combination of mechanical properties, i.e., ultrahigh strength coupled with good fracture toughness due to precipitation of intermetallic phases in low-carbon soft martensitic matrix. Additional time exposure at this aging temperature does not improve the hardness substantially, and after

10–15 h of exposure, it has observed a decrease in the hardness due to the reasons described before [3].

There are some specific demands that requires high strength and good wear resistance, such as slat track, high speed gear and torsion shaft for aeronautical components [7]. Nitriding is a surface treatment process involving the introduction of nitrogen into the surface of steel, which produces a modified layer with excellent properties such as high hardness, good wear and corrosion resistance [8, 9]. However, conventional nitriding of this steel is usually carried out at high temperature for a long time, which is beyond the aging temperature and time of 18Ni maraging steel and would result in the overaging and reversion of martensite to austenite of the core. Maraging steel possesses high strength, and good wear resistance can be achieved by plasma nitriding, that can be carried out at the same temperature or lower than the aging temperature, avoiding the overaging or reversion of martensite to austenite of the core [10].

In the present investigation, a combined plasma nitriding and aging at 480 °C for 3 h treatment of a maraging 300 steel was studied. Specially focusing on the possibility of increasing the hardness of the core by aging, at the same time that hard surface layers were created during the thermo-chemical treatment, were kinetic and microstructure control can easily be done, improving the resistance and decreasing the cost of production of the material. Microstructural changes were evaluated by of optical microscope (OM), X-ray diffraction (XRD) and energy dispersive spectroscopy (EDS). Hardness and case depth were evaluated and explained with the help of microstructural observations.

2 Experimental Procedure

The samples were solution annealed at 820 °C for 1 h. The chemical composition of the maraging 300 steel solution annealed is shown in Table 1. Some of the samples were subjected to an additional age hardening treatment at 480 °C for 3 h. The samples in both conditions were nitrided in the furnace under vacuum. Plasma was obtained by passing the gas mixture of H_2 and N_2 gases in the ratio of 3:1 under vacuum. Plasma nitriding was carried out at 480 °C for 3 h for both conditions.

Microstructural examination was carried out on Carl Zeiss H-PL optical microscope. Cross sections from nitrided and un-nitrided samples were ground, polished (final polishing step: 1 μm diamond suspension) and etched using $FeCl_3$ 10 % (5 g $FeCl_3$ in 45 mL of water) at room temperature for about 2 s.

Phases present in the un-nitrided and formed in the nitrided samples were characterized by XRD on a Panalytical X′ Pert Powder diffractometer using Cu–K_α

Table 1 Chemical composition (wt%) of the maraging 300 steel studied

Element	Ni	Co	Mo	Ti	Al	C	S	P	Si	Mn	Fe
wt%	19.00	9.37	4.94	0.63	0.08	0.008	0.002	0.004	0.06	0.01	Balance

radiation, $\lambda = 0.1542$ nm, in conventional θ/2θ Bragg-Brentano symmetric geometry. The diffraction angle range ($30° < 2\theta < 90°$) was scanned in steps of 0.0170° with a counting time of 15.24 s per step. For the identification of the phases, based on the positions of the diffraction peaks, data of the HighScore database were used.

Energy dispersive spectroscopy (EDS) analyses, to measure the nitrogen profile in the surface after nitriding were performed by means of an x-act SDD spectrometer incorporated with scanning electron microscope (SEM) model VEGA 3/ TESCAN.

Hardness from the surface, profile and core was measured using Vickers hardness tester (FutureTech FM-700) with a load of 100 gF and a dwell time of 9 s. Hardness values were measured at five places for each sample. The core hardness before and after plasma nitriding treatment was measured for previously solution annealed and solution annealed and aged samples in order to compare the behavior of the core when exposed to the thermal cycle during the plasma nitriding treatment.

3 Results and Discussions

Optical micrographs of un-nitrided and plasma nitrided samples are shown in Fig. 1. The nitriding behavior is the same for samples with starting microstructure on the solution annealed and aged state. These micrographs show a uniform and

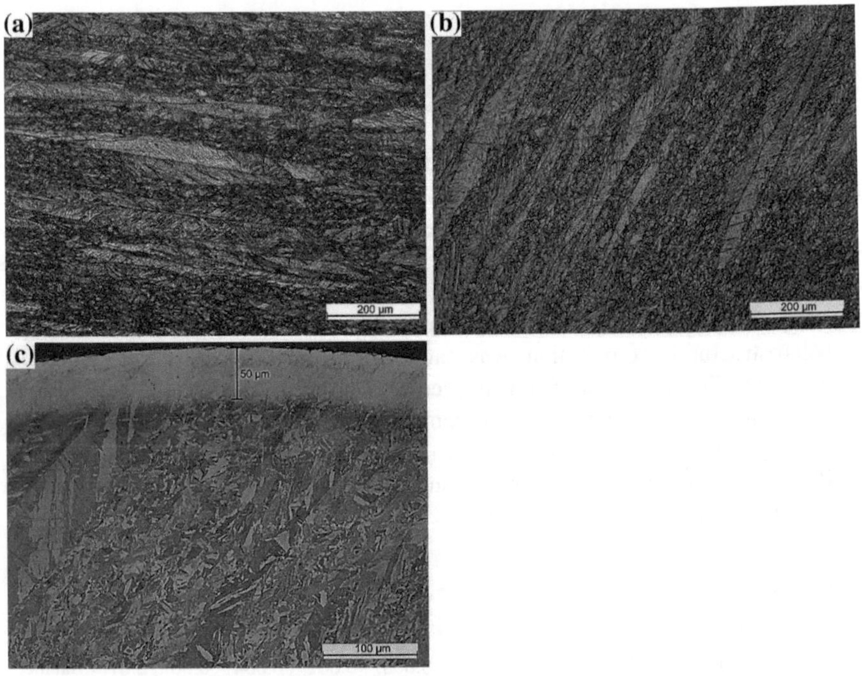

Fig. 1 Optical micrographs of maraging 300 **a** solution annealed, **b** solution annealed and aged, **c** solution annealed and plasma nitrided and **d** solution annealed, aged and plasma nitrided

continues case hardened nitrided layer and the typical martensite plate-like microstructure of maraging steels can be seen in the un-nitrided core [1–3]. The thickness of 50 μm was found in the nitrided layer for all nitrided samples. In the light optical micrographs (Fig. 1), no difference in the microstructure of the un-nitrided core of the age-hardened sample can be observed as compared to that of the only solution annealed sample. No hints of austenite can be seen in the micrographs, which is compatible with the XRD phase analysis discussed below (Fig. 2).

XRD patterns obtained from the surface of un-nitrided and plasma nitrided samples are shown in Fig. 2. The un-nitrided samples exhibit diffraction peaks only due to the martensitic phase α′-Fe. The presence of intermetallic precipitates in the aged sample is not revealed by separate reflections in the aged's X-ray diffraction patterns, likely due to the coherent nature of these precipitates [10, 11]. After plasma nitriding, α′-Fe peaks disappeared or are overlapped, giving place to peaks indexed as ε-Fe_3N and γ′-Fe_4N nitrides, corresponding to the compound layer.

Figure 3 shows concentration depth profile of nitrogen measured by energy dispersive spectroscopy (EDS) on the samples plasma nitrided. The composition of the outer layer reached to around 7 wt% nitrogen for both conditions. These compositions agree very well with ε and γ′ nitrides, respectively [11]. The data demonstrates a gradual decrease of the nitrogen concentration when moving from the compound layer toward the substrate, and about 50 μm from the top surface, the

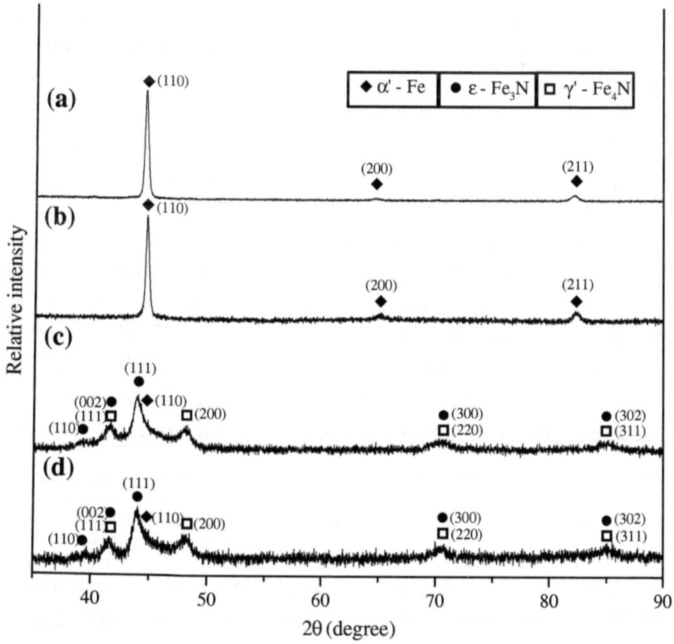

Fig. 2 X-ray diffraction pattern of **a** solution annealed, **b** solution annealed and aged, **c** solution annealed and plasma nitrided and **d** solution annealed, aged and plasma nitrided

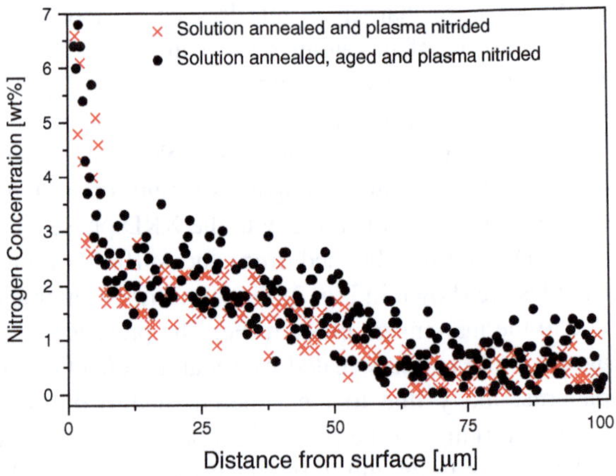

Fig. 3 EDS depth profile of nitrogen analyzed from solution annealed and plasma nitrided and solution annealed, aged and plasma nitrided

nitrogen concentration starts decrease to zero, confirming the information of nitride layer thickness observed by optical micrograph.

The hardening effect was evaluated by surface hardness and hardening profiles. Considering the starting core hardness for solution annealed condition as 331 HV and for aged condition as 604 HV, Table 2 shows that the surface hardness increases for both conditions and the presence of the compound layer is responsible for an additional hardening effect, achieving a hardness up to 1140 HV. Figure 4 shows the hardness profile after the plasma nitriding for both conditions. The continuous decrease of hardness from surface to the core of the sample suggests the presence of a diffusion zone in which precipitates of nitrides of iron and other metals were formed at the grain boundaries as well as within the grains. These precipitates distort the lattice and pin crystal dislocations and thereby increase the hardness of the surface layer of the ion-nitrided samples [10].

The possibility to use the plasma technology for a simultaneous aging and nitriding treatments is dependent of the core hardness response. Core hardness attained after the plasma nitriding treatments are shown in Table 3. Considering the starting core hardness, the treatment was effective to increase the core hardness for the solution annealed sample from 331 HV up to 597 HV, proving that the simultaneous aging and nitriding treatments are possible to be done. On the other hand, the aged sample retain the core hardness after the treatment avoiding the overaging process and reversion of martensite to austenite for the tested cycle,

Table 2 Surface hardness values of un-nitrided and plasma nitrided samples

	Solution annealed	Solution annealed and aged	Solution annealed and plasma nitrided	Solution annealed, aged and plasma nitrided
Surface hardness (HV)	331 ± 5	604 ± 18	1010 ± 6	1140 ± 4

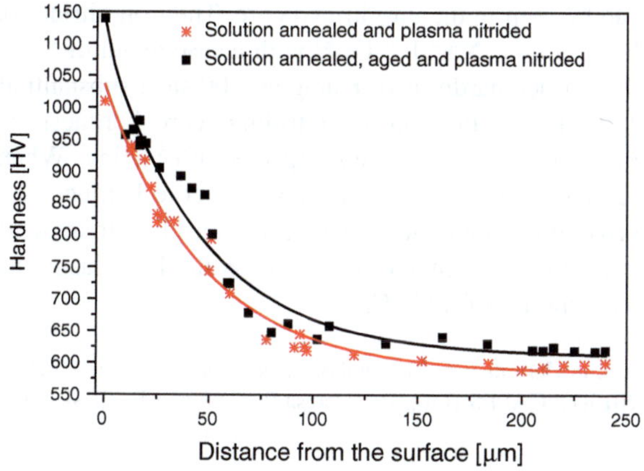

Fig. 4 Hardness profiles for samples plasma nitrided

Table 3 Core hardness values of un-nitrided and plasma nitrided samples

	Solution annealed	Solution annealed and aged	Solution annealed and plasma nitrided	Solution annealed, aged and plasma nitrided
Core hardness (HV)	331 ± 5	604 ± 18	597 ± 2	620 ± 3

showing that the use of additional heating during the nitriding process affect slightly the hardness, increasing the core hardness up to 620 HV. This fact is related to the precipitation growth. The increase in the strength and hardness of maraging steels is a function of the precipitate fraction and size. The growth of the precipitate and the increase in the precipitation fraction is a function of time and temperature, and at 480 °C the aging of maraging 300 is rapid and intense [1]. In the solution annealed, aged and plasma nitrided sample, after 6 h of aging (3 h of aging plus 3 h of plasma nitriding), the precipitate remains distributed in the matrix as an extremely dense dispersion, and its average diameter can be slightly grown from the previous aging treatment. In the case of 2000 MPa grade cobalt-free maraging steel, after aging for 6 h, the Ni_3Ti average diameter has grown to about 4–5 nm when compared to 3–4 nm after aging for 3 h [1]. Therefore, the core hardness increasing around 4 % only does not justify the additional costs related to 3 h of heating.

4 Conclusions

In the present investigation, maraging steel (300 Grade) solution annealed and aged was plasma nitrided at 480 °C for 3 h. An uniform and continues case hardened nitrided layer and the typical martensite plate-like microstructure of maraging steels can be seen in the un-nitrided core. The iron nitride formed in the hardened surface layer is ε-Fe_3N and γ′-Fe_4N with a case depth of 50 μm and these nitrides improve the surface hardness of maraging 300 steel substantially achieving a hardness up to 1140 HV. After plasma nitriding, core hardness of solution annealed samples increased up to the levels expected after aging. When solution annealed and aged samples are nitrided no decrease in core hardness can be observed. These results prove that simultaneous aging and nitriding treatments may be done using a single cycle thermochemical treatment on the DC-plasma nitriding reactor using the low temperature of 480 °C.

Acknowledgments The authors acknowledge the research agencies CNPq (Proc.no 141274/2013-1), CAPES (Proj. Pró-Defesa 014/08) and FAPESP for financial support.

References

1. Sha W, Guo Z (2009) Maraging steels: modelling of microstructure, properties and applications. Woodhead Publishing Limited, London
2. Slunder CJ, Hall AM (1968) The metallurgy, behavior, and application of the 18-percent nickel maraging steels—a survey. Battelle Memorial Institute, NASA SP-5051
3. Schmidt M, Rohrback K (1991) Heat treating of maraging steels. In: Handbook Committee (ed) ASM handbook v. 4: Heat treating, 10th edn. ASM International, Materials Park
4. Vasudevan VK, Kim SJ, Wayman CM (1990) Precipitation reactions and strengthening behavior in 18 Wt Pct nickel maraging steels. Metall Mater Trans A 21A:2655–2668
5. Sha W, Smith GDW (1994) Improved thermochemical calculations for maraging steels and comparison with atom probe measurements of phase composition. Metall Mater Trans A 25A:2538–2539
6. Floreen S, Decker RF (1979) Maraging steels. In: Decker RF (ed) Source book on maraging steels. ASM International, Metals Park
7. Blanter ME, Kovaleva LA (1969) Mechanical properties and nature of strengthening of maraging steels. Met Sci Heat Treat 10:30–33
8. Wen DC (2009) Microstructure and corrosion resistance of the layers formed on the surface of precipitation hardenable plastic mold steel by plasma-nitriding. Appl Surf Sci 256:797–804
9. Oliveira SD, Tschiptschin AP, Pinedo CE (2007) Simultaneous plasma nitriding and ageing treatments of precipitation hardenable plastic mould steel. Mater Des 28:1714–1718
10. Shetty K, Kumar S, Raghothama Rao P (2009) Effect of ion nitriding on the microstructure and properties of maraging steel (250 grade). Surf Coat Technol 203:1530–1536
11. Hossini SR, Ashrafizadeh F (2011) Compositional depth profile investigation of plasma nitriding by multiple analyses techniques. Vacuum 85:920–926

State Analysis and Development Perspectives of the Algeria's Railway Network

Hakim Siguerdjidjene

Abstract The current rail transport in Algeria does not meet the needs of the national economy because some of the lines have been designed and built according to a set of old standards, which results in them being characterized by a relatively low productivity and lack of work efficiency. This is why the reconstruction and strengthening of the Algerian railway network is becoming necessary if not mandatory. The further development of the railway linking and improving the quality of its operation will result in improved performance and speed. This is why the development of the Algerian rail network is a very important and specific task addressing this issue must ensure a high competitiveness. The development of an approach in order to increase the railway lines efficiency in the Algerian context and the process of reconstructing them needs a thorough analysis the state of a number of railway lines with low profitability, in order to organize their reconstruction, open the possibility of having easier access and efficient facilities, thus reaching a rational scheme.

Keywords Development · Rational scheme · Competitiveness · Productivity · Speed of trains · Railway network

1 Introduction

Currently, in most areas of Algeria's railway network, the maximum speed of freight is no more than 60 km/h while the passenger trains do not reach 120 km/h. Improving the competitiveness of railways in passenger and freight traffic is the main part related to the modernization of infrastructure, the introduction of more powerful locomotives and increasing train speeds.

H. Siguerdjidjene (✉)
Laboratory of Materials and Sustainability of Bouira, Faculty of Science, M'hamed Bougara University, Boumerdes, Algeria
e-mail: Siguer1@hotmail.com; siguer.h@gmail.com

A. Öchsner and H. Altenbach (eds.), *Mechanical and Materials Engineering of Modern Structure and Component Design*, Advanced Structured Materials 70,
DOI 10.1007/978-3-319-19443-1_23

However, to increase the capacity of railways through modernization and reconstruction of the way requires a thorough examination and study of the state of existing lines to create a classification system for railway stations and lines in Algeria.

In this paper, we propose a new approach in order to solve most of techno-economic problems related to the management process of increasing the railway power. This approach is based on the classifying method used in the network at the design stage with relation to new—and reconstruction of existing—lines, and track facilities. That operates on two parameters: (1) the speed of passenger and freight trains and (2) freight traffic density before and after the modernization and reconstruction of the all the categories. That firmly depends on choosing effective measures to solve the problem of power increase of the railway network.

In addition, the classification of the line at the present stage was examined in the Pevzner et al. [1], added to some suggestions for improving the system of line classification developed in VNIIZhT [2]. Currently and according to the latest railway guidelines, [3] presented a line classification system on the Status of the traveling facilities keeping system in 6 group. In turn, the International Union of Road recommends classifying working conditions for the two indicators characterizing the force action. The lines open to international traffic, must have a category at least equal to (B1). In other countries, like France, the classification criteria adopted by the efficient volume of traffic depends on the level of (T) axle load and speed. In Canada, the classification is done based on the annual volume of traffic and speed. On the roads of the Netherlands, there are three classes. The first class includes a line with permissible speed of over 130 km/h, the second goes from 100 to 130 km/h, and the third addresses lines with less than 100 km/h. Thus, different countries have their classification done based on individual and specific parameters. In the same way Akkerman G. L and Kravtchinko O. A [4] examined the assessment of the possibility to use the conception biclothoïde in the projects of the railroad sections in curve for the grand lines to high speed in the program of universal complex mechanism. In the other side Mitrakov A. C and Charokhov C. G [5] have developed a parameters determination methodology of the forced declivity system of temporary transportation wagons tracing in sections ways curve.

2 Specificity of the Increased Power Problem

The initial description of the existing railway lines, the final reconstruction state and the intermediate reinforcement steps are reflected upon in the following section.

As a general rule, rangelands are not identical in times, what practically exclude the possibility of using the train crossing without stopping. Thus, a total reconstruction of all permanent mechanisms is needed. The short length of arrival-departure track and low power of locomotives determine the low norm mass of merchandise convoy.

All this explains the yield of many railways in the country which differs depending on the status and parameters of the line which is in the range $5 \leq \Gamma \leq 15$ million tons/year.

For the ordering problem, it is appropriate to introduce the classification of existing railway lines based on technical characteristics and performance of the railway lines into three groups:

1. Lines with distance between railway tracks equals to 1055 mm and has a length of 1000 km on which freight traffic does not exceed 5 million tons/year and the speed of movement of freight trains does not exceed 50 km/h;
2. In the mixed-line configuration, the gap between tracks is between 1055 and 1435 mm, in normal cases. Over the length of 388 km the transit of goods does not exceed 7 million tons/year and the speed of movement of trains passenger does not exceed 80 km/h;
3. The normal-line railway has a gap between tracks of 1435 mm, with a total length of 3854 km, while it is 3045 km including the single line with the traffic of goods not exceeding 15 million tons/year and the rate of traffic based on type of line not reaching 100 km/h.

According to these characteristics we can designate alternative directions of reconstruction and avoid building inefficient railways. The choice of such essential directions is generally determined by two main causes:

- The parameters of the existing permanent installations;
- The final aim of the reconstruction and modernization.

For the purpose of the railway line storage requirements after the reconstruction, it is useful to classify them according to the most important characteristic:

- The goods transit of 10th year Γ (million tons/year).
- The moving speed of freight and passenger trains.

In correspondence with this division of the railway reconstruction we divide the project standards into four categories.

It is clear that the aim of the measures chosen tackle the decrease in capital investment and goods transportation with minimal expenses and a reduction in time of transport to the destinations. To solve such a problem, it is necessary to find the most appropriate measures to achieve the essential aims mentioned above and avoid unnecessary spending.

3 Analysis of the Power Increase Problem

The analysis showed that for the Algerian context, it was only rational to consider a small number of increases in transit capacity and transition measures:

- Extension of the useful length of the *arrival-departure* track $L_{ar\text{-}dep}$, meters;
- Increase the power of the locomotive F_{pl}, Newton;
- Removal of the speed limit on the course that limits the ability of crossing the line to a single track, km/h;
- Increase the throughput η_{tr} and the transport performance capacity (carrying capacity) Γ, millions of tons per year;

For the correction of the axe position separated by solving the question on longer routes stops, it is necessary to consider:

- The financial resources for longer routes as well as their servicing;
- The investment in the locomotive fleet, the savings in operating expenses for locomotive brigades.

For the Algerian conditions, the prospects lengths of arrival-departure tracks should cover all possible range values indicated above.

At the present time, the speed factor is of great importance in the movement of goods and passengers to lower transportation expenses and increase throughput capacity. The maximum permissible speed of freight trains is set depending on:

- How sectoral normative content structures and devices;
- For each site, depending on the technical equipment and the infrastructure and rolling stock.

The maximum speed of passenger and freight trains is determined by the following formula:

$$\mathrm{V_{max}} = \sqrt{(\mathrm{h} + \Delta\mathrm{h}) \cdot \mathrm{R}/11.8} = A\sqrt{R}, \qquad (1)$$

The minimum curve radii should be determined by the formula:

$$\mathrm{R_{min}} = \frac{11.8\mathrm{v^2}_{\mathrm{max}}}{\mathrm{h} + \mathrm{a_{nc}} \cdot 153} \qquad (2)$$

where

$\Delta\mathrm{h} = \frac{an \cdot S}{g}$	Drawback cant, mm;
h	The smallest value of elevation, mm;
R	Radius of curvature, m;
A	The parameter accepts equal to 4.7.

The unliquidated lateral acceleration per person ($\mathrm{a_{nc}}$):

$$\mathrm{a_{nc}} \frac{V^2}{R} - \frac{g \cdot h}{S}, \qquad (3)$$

Table 1 Minimum allowable curve radii

h, mm	The values R_{min} (m) at the unliquidated lateral acceleration value a_{nc} (m/s^2) and speeds V_{max} (km/h)		
	$a_{nc} = 0.7$ $V_{max} = 120$	$a_{nc} = 0.7$ $V_{max} = 160$	$a_{nc} = 0.6$ $V_{max} = 200$
100	820	1460	2460
110	780	1390	2340
120	750	1330	2230
130	715	1270	2130
140	680	1220	2040
150	660	1170	1950

where

V Train speeds, m/s;
S The track width, m;
G Acceleration of gravity, m/s^2.

For creating comfortable conditions for passengers unliquidated acceleration Accepted a_{nc} at speeds up to 160 km/h must equal to 0.7 m/s^2, and from the speed at 161 to 200 km/h, the a_{nc} should be equal to 0.6 m/s^2 and the cant practice *hp* equal to 150 mm.

The results of the calculations are listed in Table 1. On the railways combined movement of freight and passenger trains, for passenger train speeds at 120–160 km/h and the unliquidated lateral acceleration value $a_{nc} = 0.7$ m/s^2 the minimum allowable curve radii (R_{min}) in the way is located at the maximum cant practice *hp* equal to 150 mm and should not be less 1170 m, for the unliquidated lateral acceleration value $a_{nc} = 0.6$ m/s^2 and speed of passenger trains $V_{max} = 200$ km/h, the minimum value of R_{min} is located at the same maximum cant practice -*hp*- equals to 150 mm and not less than 1950 m.

The railway power is determined by its Throughput and carrying capacity. Throughput is the number of η_{tr} pairs of trains that the road could miss in a day. Carrying capacity is determined by the power of road freight transport. This number of tons of cargo Γ that the road can carry per unit time in each direction.

The maximum possible throughput of single-track lines in parallel not package displacement-time diagram (train schedule), pairs of trains per day is calculated:

$$\eta_{tr} = \frac{1440}{T} \tag{4a}$$

$$T = t_{\mathrm{a}} + t_{\mathrm{r}} + \tau_1 + \tau_2 + t_{\mathrm{acc.dec}} \tag{4b}$$

where

η_{pass} Track capacity, pair train per day;
T Period graph movement of trains;

$t_{acc. dec}$	The sum of acceleration and deceleration time;
t_a, t_r	Running time of the train on the stretch in the direction there and back respectively, min;
τ_1, τ_2	Stationary intervals.

The transport performance (carrying capacity) Γ in million tons/year of the railway line and area can be realized at a certain combination of the basic parameters of permanent structures, equipment and process technology transportation and can be determined using the following formulae:

$$\Gamma = \frac{365 \cdot Q_{moy} \cdot \rho}{\gamma} \eta_{tr}, \tag{5a}$$

$$\Gamma = 365 \cdot Q_{moy} \cdot \rho \frac{1440}{\gamma \cdot T}, \tag{5b}$$

$$\eta_{tr} = \frac{1440V}{60 \cdot S}, \tag{5c}$$

$$\Gamma = \frac{6052\mathrm{V} \cdot \mathrm{Q}}{\mathrm{S}}, \tag{5d}$$

where

η_{tr}	available throughput of cargo moving in a given direction, trains/days;
γ	Coefficient of irregularity equal to 1.1;
Q_{moy}, Q	The average gross and net weigh of the convoy (train);
ρ	Ratio of the gross train weight to net weight equal to 0.76;
V	Train speed in km/h;
S	Interval between trains in km.

It follows from formulae (5b) and (5d) that there is a proportional relationship between the productivity of the railway line and the train speed and the gross weight of train. On the other side there is an inverse relationship between the productivity of the railway line and the rolling time of a pair of train journeys in the limited term and the interval between trains (Fig. 1).

As it is shown in Fig. 1 when the net average weight of the convoy freight is Q = 1400 T and its speed V = 50 km / h and the interval between the trains are S = 85 km, the increase of the weight convoy freight until Q = 2000 T to the same speed with a reduction of the half distance S/2 = 42.5 km leads to an increase the convoy productivity of 5 million tons/year to 14 million tons/year, what represents an increase of 35 %. Otherwise increasing the speed of the freight train until V = 120 km leads to the increase in productivity of 12 to 34 million tons/year, that is a contribution of 35 %, which allows us to affirm the linearity of the increase in productivity (carrying capacity) Γ of the railway according to the parameters in the

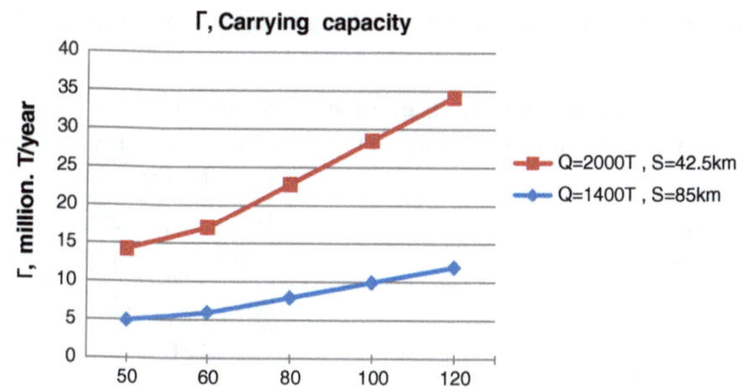

Fig. 1 The dependence of the performance (carrying capacity) Γ of the railway at the speed, the net average weight and the interval between of the convoy (train)

speed V (km / h), the average net weight of the train (train) Q (Ton) and the distance S (km) between trains.

On the other hand, the reduced freight traffic density area Γ_{red}, million of gross ton-km/km per year can be identified for the classification of lines in the control track facilities and to solve economic problems of transport is determined using the following formulae:

$$\Gamma_{frt} = \sum_{i=1}^{k} \Gamma_i l_i, \tag{6a}$$

$$\Gamma_{pas} = \sum_{i=1}^{k} n_i l_i \cdot 10^{-6} \tag{6b}$$

$$\Gamma = \alpha_{fr}\Gamma_{frt} + \beta_{pas}\Gamma_{pas} \tag{6c}$$

$$\Gamma_{red} = \Gamma \cdot \mu$$

where

Γ	The calculated freight traffic density in the area, million gross. Ton-km/km per year
k	The number of halt points on which embarkation and disembarkation of passengers;
μ	Coefficient taking into account local operating conditions of a particular stretch of track
n_i	The number of passengers departed from the i_{th} item on the year;
l_i	The length of the line, km;
$\Gamma_{frt}, \Gamma_{pas}$	Respectively freight traffic density of freight and passenger trains on a reported basis in the area, one million of gross ton-km/year;
α_{fr}, β_{pas}	coefficient of reduction for freight and passenger trains, taking into account the impact on the way the disorder freight and passenger wagon and locomotives, as well as the speed of movement.

Table 2 Categories of Algeria's railways at the design stage, new and reconstruction of existing lines

Category of railways	Transit (traffic) merchandise to the 10th year Γ (million (tons/year)	Speed of passenger trains v_{max} (km/h)	Speed of freight trains v_t (km/h)
I	Over 25	$160 < v_{max} \leq 200$	$80 < v_{max} \leq 120$
II	15–25	$120 < v_{max} \leq 160$	$60 < v_{max} \leq 80$
III	$5 < \Gamma \leq 15$	$80 < v_{max} \leq 120$	$50 < v_{max} \leq 60$
IV	$\Gamma \leq 5$	$v_{max} \leq 80$	$v_{max} \leq 50$

The Proposals for the Category of railways Algerian at the design stage of new, modernization and reconstruction of existing lines are given in Table 2.

4 Conclusion

Research devoted to increasing the railway lines power is based on the classification of the Algeria's railway lines into categories before and after modernization and reconstruction. This is the key performance indicator and the most important index. The performance (carrying capacity) Γ, million tons/year of railway line and area is found directly related to the speed of movement of passengers and freight trains. Also it is true that the long term preservation of high speeds, in turn, has a positive effect on the class of way stability, despite more frequent load variations of Γ (carrying capacity). Subsequently, if necessary, adjustments can be made to class ways by Γ (carrying capacity) if change occurs.

References

1. Pevzner VO, Petropavlovckaya IB et al (2014) Railway lines classification problem as in the day and age. Vestnik VNIIZhT, Moscow, p 5
2. Turbin IV, Gavrilenkov AV, Kantor II, et al (1989) Research and exploration railways. Moscow, Transport
3. Anisimov VA (1994) The simplified mathematical model for the analysis and planning of the power of railways in arbitrary dynamic conditions for transport projection of railways in difficult conditions. Mejvoz sbornik. Naotch. Transport, Academy. The state of the East Khabarovsk communications channels
4. Akkerman GL, Kravtchinko OA (2012) In: International scientific and practical conference. Rol track facilities in railway infrastructure, Moscow, MIIT
5. Mitrakov AC, Charokhov CG (2014) In: 10th International conference in scientific and practical Trans-Mech-Art-Chem, Moscow, MIIT

Characterization of the Superalloy Inconel 718 After Double Aging Heat Treatment

Katia Cristiane Gandolpho Candioto, Felipe Rocha Caliari, Danieli Aparecida Pereira Reis, Antônio Augusto Couto and Carlos Angelo Nunes

Abstract The application of heat treatment, by solid solution and precipitation hardening, is very important to optimize the mechanical properties of superalloys. The main phases present in Inconel 718 are: gamma prime γ' face ordered $Ni_3(Al, Ti)$; gamma double prime γ'' bct ordered Ni_3Nb; eta η hexagonal ordered Ni_3Ti; delta δ orthorhombic Ni_3Nb intermetallic compounds and other topologically closed-packed structures such as μ and Laves phases. δ, μ and Laves phases have low ductility, which causes losses in mechanical and corrosion properties (Fu et al. in Mater Sci Eng A 499:215–220, 2009 [1]). The heat treatment applied to Inconel 718, precipitation hardening, has two steps: solid solution and aging treatment. In first step the secondary (hardening) phases are dissolved along the matrix, as well as carbides. It is important to note that after 650 °C (Durand-Charre in The microstructure of superalloys. CRC Press, Boca Raton, 1997 [2]) and with long exposure times, γ'' transforms in the stable phase δ, which results in a loss of mechanical resistance. In this study we aim to characterize the microstructure and phases in superalloy Inconel 718 during the steps of heat treatment with double aging. The double aging treatment

K.C.G. Candioto (✉) · C.A. Nunes
Escola de Engenharia de Lorena-USP, Lorena, São Paulo, Brazil
e-mail: katcg_candioto@usp.br

C.A. Nunes
e-mail: cnunes@demar.eel.usp.br

F.R. Caliari · D.A.P. Reis
Universidade Federal do Estado de São Paulo, São José dos Campos, São Paulo, Brazil
e-mail: felipercaliari@yahoo.com.br

D.A.P. Reis
e-mail: danieli.reis@unifesp.br

A.A. Couto
Instituto Pesquisas Energéticas e Nucleares, São Paulo, São Paulo, Brazil
e-mail: acouto@ipen.br

A.A. Couto
MACKENZIE—Universidade Presbiteriana do Mackenzie, São Paulo, São Paulo, Brazil

A. Öchsner and H. Altenbach (eds.), *Mechanical and Materials Engineering of Modern Structure and Component Design*, Advanced Structured Materials 70,
DOI 10.1007/978-3-319-19443-1_24

performed followed the steps of solid solution to 1095 °C/1 h and double aging at 955 °C/1 h to 720 °C/8 h + 620 °C/8 h. The characterizations were performed through the techniques of XRD and SEM/EDS. It was possible to obtain the microstructural and phases characterizations before and after heat treatment in all steps.

Keywords Inconel 718 · Heat treatment · Double aged · Solid solution · Laves phases

1 Introduction

Inconel 718 is a precipitation-hardening superalloy developed by International Nickel Co in the 50s [3]. This alloy is a Nb-modified Fe–Cr–Ni-base superalloy and has been widely used in gas turbines and related applications due to its good mechanical properties and structural stability at elevated temperatures (~650 °C). Superalloys are generally applied in heat treatment equipment, aeronautics gas turbines, nuclear power plants, medical components, chemical and petrochemical industries [4]. It is used under high homologous temperatures ($T_h > 0.5$), showing high stress-rupture and good oxidation resistance, good creep and low cycle fatigue behavior. In 1989 [5], the Inconel 718 alloy represented 45 % of all wrought nickel-iron base superalloys produced in the world.

Inconel 718 has crystallographic lattices face-centered cubic (fcc), body-centered cubic (bcc), hexagonal close-packed (hcp) and body-centered tetragonal (bct), among others. These niquel-iron superalloys are made of austenitic fcc matrix γ (gamma phase), as well as secondary phases: gamma prime γ' face ordered $Ni_3(Al, Ti)$; gamma double prime γ'' bct ordered Ni_3Nb; eta η hexagonal ordered Ni_3Ti; delta δ orthorhombic Ni_3Nb intermetallic compounds and other topologically closed-packed structures such as μ and Laves phases. δ, μ and Laves phases have low ductility, causing losses in mechanical and corrosion properties and providing grain size control [4]. In addition, these phases appear in alloys containing high levels of bcc transition metals (Ta, Nb, Cr) [5].

Superalloys have their microstructure characteristics improved by using heat treatment techniques. Solution treatment, usually the first step in heat treatment of precipitation hardening alloys, aims to: recrystallize, homogenize and dissolve phases in fcc matrix structure, dissolving carbides in grain boundaries and the grain-growth results in high creep-rupture resistance [5]. The purpose of aging treatments is the increasing of the strength through the precipitation of additional quantities of secondary phases, from the supersaturated matrix, developed by solution treating. For Inconel 718 more than one phase can precipitate, so double aging is applied in order to enhance the formation of both γ' and γ'' phases. Secondary phases γ' and γ'' play a main role in the strengthening mechanism of Inconel 718, mainly γ'', a coherent disk-shaped precipitate. Wang et al. [6] highlighted the structural stability dependence with secondary phases production on aging heat treatments, as well the competition of γ'/γ'' and delta phase δ formation. By comparing three kinds of double aging techniques, the optimized technique of double aging under 720 °C/8 h + 620 °C/8 h was identified due to the fact that it can get higher contents of the γ' and γ'' phases, the lowest content of the δ phases and optimum structure in Inconel 718 alloy using this technique. In this context, the purpose of this preliminary study is to characterize the microstructure and phases in superalloy Inconel 718 during the steps of heat treatment with double aging.

2 Experimental Procedure

2.1 Heat Treatments

The material used for the present study was multi-alloy bars. The superalloy was obtained by VIM/VAR process. A solid solution of 1095 °C for 1 h treatment (air cooling) was applied initially and followed by double aging treatment at 955 °C/1 h (air cooling) to 720 °C/8 h (furnace cooling) + 620 °C/8 h (air cooling). The heat treatments were conducted at Escola de Engenharia de Lorena (EEL-USP), using a Lindberg/Blue M—Tube Furnace 100 V/50 A/5 kW. The heat treatment is described in Fig. 1.

2.2 Characterization

The microstructures of the superalloys Inconel 718 in all steps of the heat treatment were analyzed via scanning electron microscopy/back-scattered electrons mode

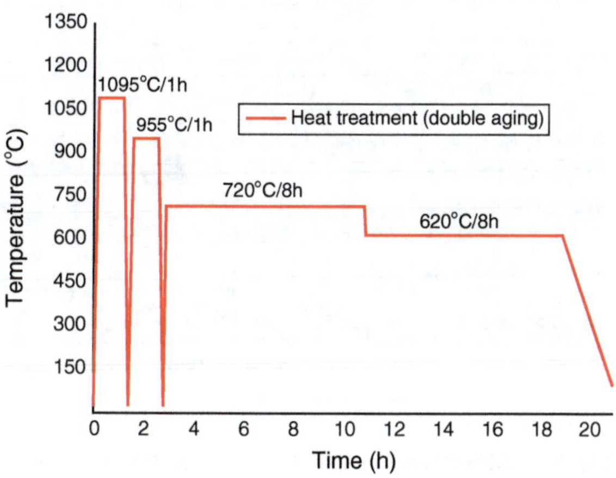

Fig. 1 Heat treatment of double aging

Table 1 Inconel 718 composition

Analyse %wt														
Ni	Cr	Ti	C	Al	Nb	Mo	Mn	Si	P	Cu	Co	B	Ta	Fe
52.83	18.39	0.95	0.03	0.48	5.05	3.01	0.03	0.03	0.004	0.03	0.23	0.001	0.01	bal

(SEM/BSE) using an acceleration voltage of 25 kV in a LEO 1450VP SEM model equipped with Oxford Instruments energy-dispersive X-ray spectrometry system at Escola de Engenharia de Lorena (EEL-USP).

Conventional X-ray diffraction (XRD) was used to characterize the materials produced after each processing step. The XRD experiments were performed in a Panalytical Empyrean model at room temperature under Cu–Kα and Ni filter at Escola de Engenharia de Lorena (EEL-USP).

3 Results and Discussion

The composition of the superalloy used in this study is summarized on Table 1.

The heat treatment improved an increase of the grain size. Figures 2, 3, 4 5 show the micrographs of the Inconel in all steps of heat treatment. The presence of

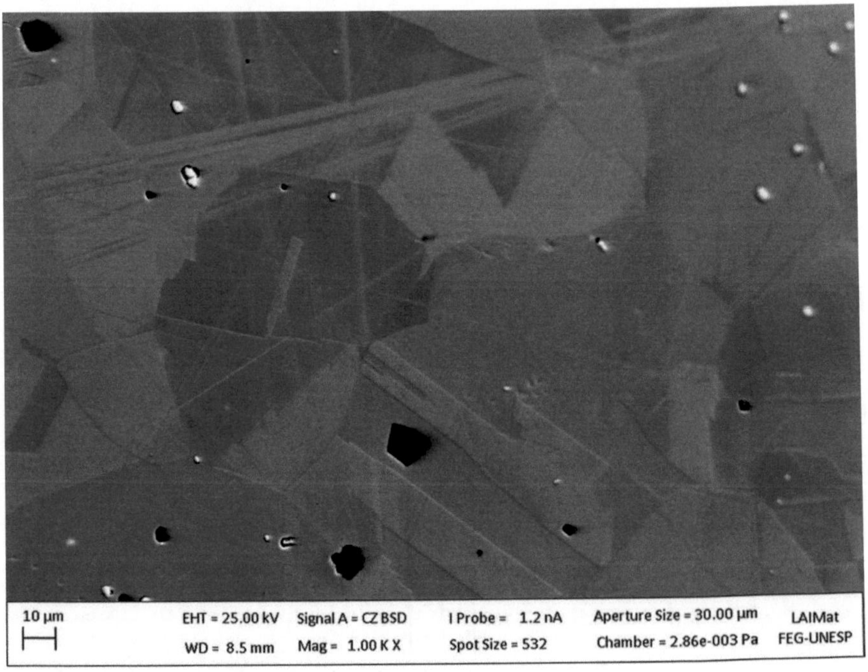

Fig. 2 SEM/BSE micrograph of the Inconel 718 as received

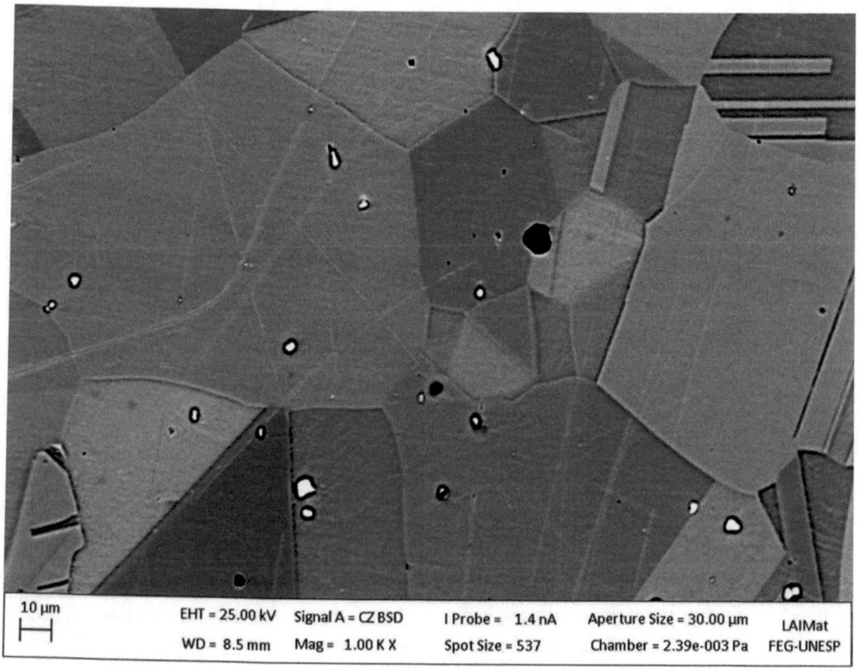

Fig. 3 SEM/BSE micrograph of the Inconel 718 after solid solution

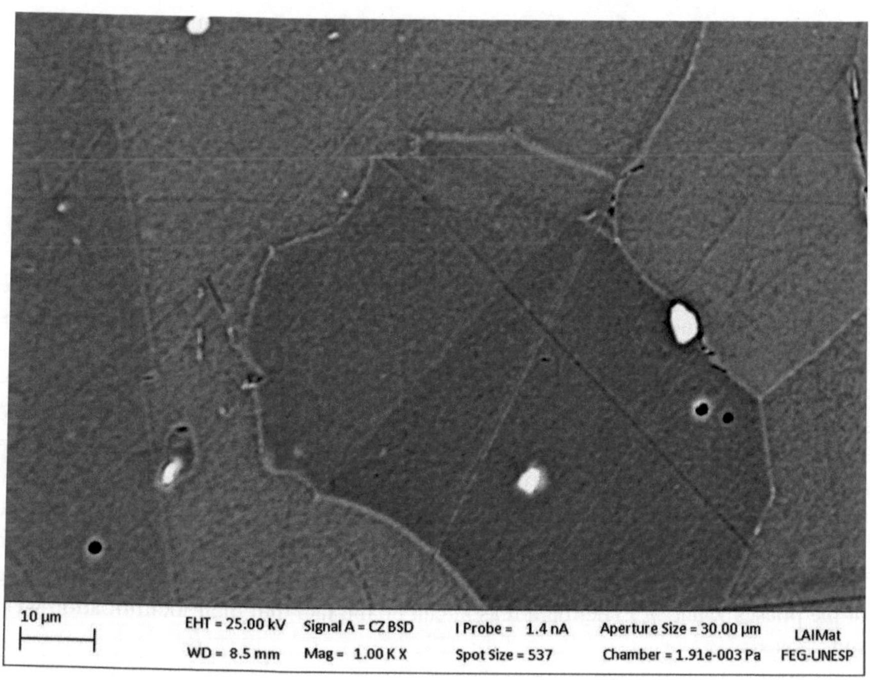

Fig. 4 SEM/BSE micrograph of the Inconel 718 before aging

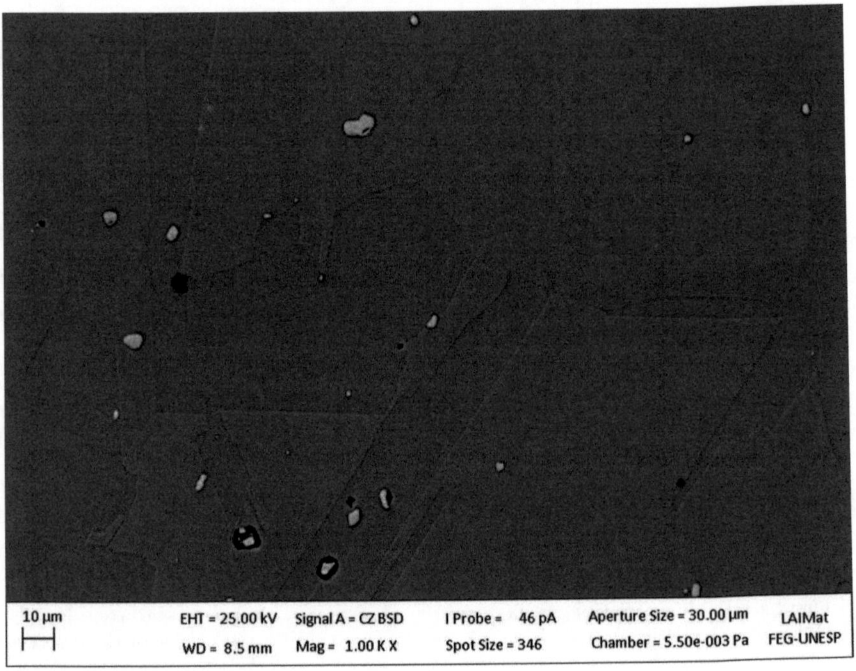

Fig. 5 SEM/BSE micrograph of the Inconel 718 after double aging

Table 2 EDS of the Inconel 718 precipitates in all steps of heat treatment

Inconel 718	Al (%at)	Ti (%at)	Cr (%at)	Fe (%at)	Ni (%at)	Nb (%at)
As received	0.71	5.58	20.49	18.38	47.54	7.3
Solid solution	–	7.41	11.47	9.42	22.03	49.67
Before aging	–	–	1.07	0.95	2.07	82.88
After double aging	0.95	5.45	15.94	13.65	33.80	30.21

precipitates in all stages can be observed. The results of the EDS obtained of the precipitates indicates that the heat treatment improved the formation of precipitates with base composition of Ni/Nb (Table 2).

Figure 6 show the XRD obtained of the Inconel 718 in all steps of heat treatment. The results of XRD show the presence of the gamma phase (γ-CFC) in the Inconel 718 as received. In the Inconel 718 solid solution the presence of gamma phase (γ-CFC), NbC (MC-CCC) and Ni_3Nb (δ-orthorhombic) is observed. Before the double aging, it can be observed that the same phases are present in the superalloy. Similarly it can be observed that after double aging there is only the presence of gamma phase (γ-CFC) again. It is not possible to separate the presence of the phases γ' and γ''. The main reason can be the fact that their identification were superposed.

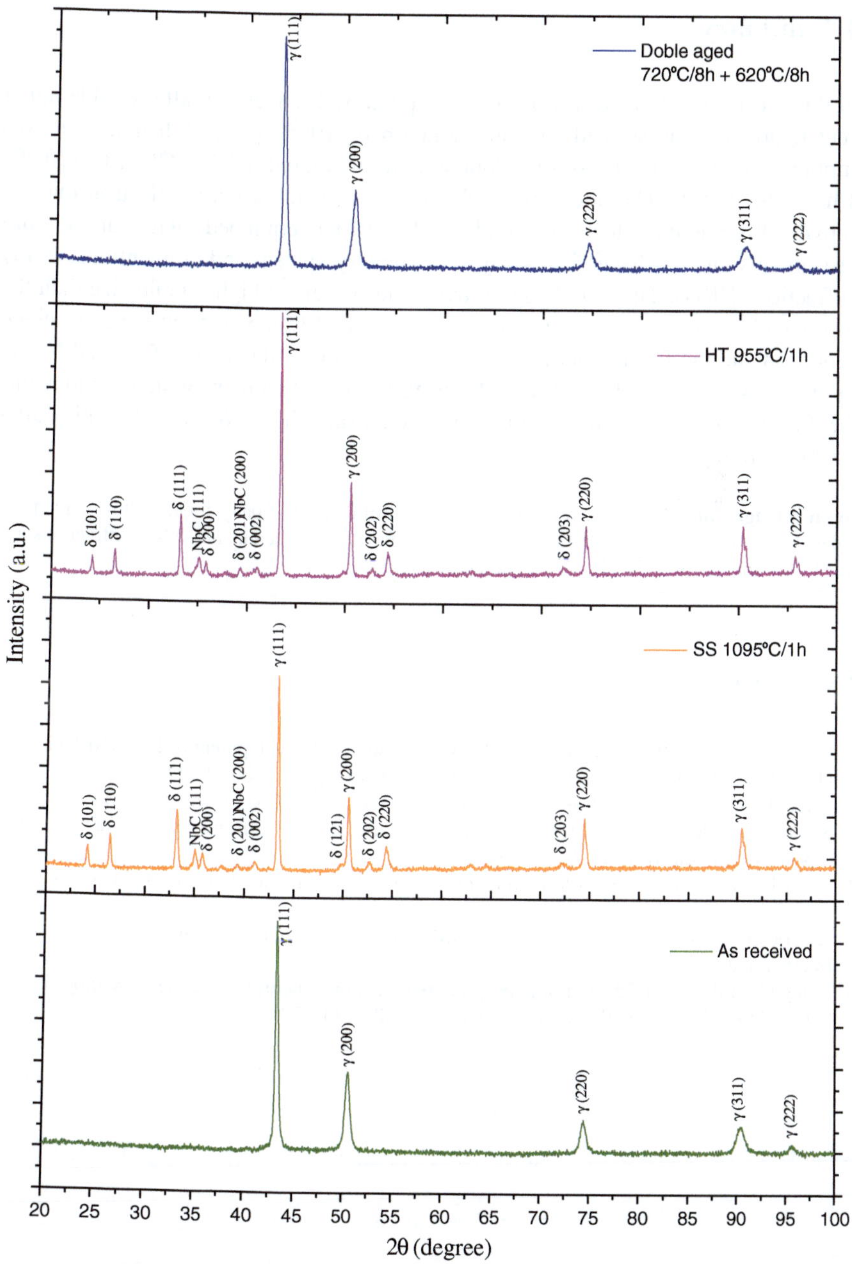

Fig. 6 XRD of the Inconel 718 in all steps of heat treatment

4 Summary

In this work the characterization of the superalloy Inconel 718 after double aging heat treatment was studied. A solid solution of 1095 °C for 1 h treatment was applied initially and followed by double aging treatment at 955 °C/1 h to 720 °C/8 h + 620 °C/8 h. The alloys were characterized using scanning electron microscopy/back-scattered electrons mode (SEM/BSE) equipped with an Oxford Instruments energy-dispersive X-ray spectrometry system and conventional X-ray diffraction (XRD). The double aging treatment presented higher grain size than the Inconel 718 as received. The gamma phase (γ-CFC) was present in the as received condition and the formation of gamma phase (γ-CFC), NbC (MC-CCC) and Ni_3Nb (δ-orthorhombic) could be observed during the heat treatment which demonstrate the formation of the gamma phase (γ-CFC) from Ni_3Nb (δ-orthorhombic) after double aging.

Acknowledgments The authors acknowledge the Multialloy company for the alloys and specimens, Escola de Engenharia de Lorena (EEL/USP) for the heat treatments, XRD diffractions and SEM/EDS images and CAPES for the financial support.

References

1. Fu SH, Dong JX, Zhang MC et al (2009) Alloy design and development of INCONEL718 type alloy. Mater Sci Eng A 499(1–2):215–220. doi:10.1016/j.msea.2007.11.115
2. Durand-Charre M (1997) The microstructure of Superalloys. CRC Press, Boca Raton
3. Silva ALVC, Mei PR (2010) Aços e ligas especiais. In: Aços ferramenta, inoxidáveis e ligas especiais, 3rd edn. Blücher, São Paulo
4. Donachie M, Donachie S Jr (2002) Superalloys: a technical guide, 2nd edn. ASM International, USA
5. American Society for metals (1991) ASM handbook volume 4: heat treatment. ASM International, USA
6. Wang C, Li R (2004) Effect of double aging treatment on structure in Inconel 718 alloy. J Mater Sci 39:2593–2595. doi:10.1023/B:JMSC.0000020036.96777.9c

Development of an Innovative 3D Simulator for Structured Polymeric Fibrous Materials and Liquid Droplets

Contribution to the Experimental Characterization and Optimization of Deposition Time and Spreading Area Using an Innovative 3D Optic System and a Fibrous Porous Materials Simulator

Joana M.R. Curto, António O. Mendes, Eduardo L.T. Conceição, António T.G. Portugal, Paulo T. Fiadeiro, Ana M.M. Ramos, Rogério M.S. Simões and Manuel J. Santos Silva

Abstract An innovative 3D approach is used to develop and optimize structured polymeric fibrous materials. Using our own materials simulator we are able to design a new material with optimized properties such as porosity or thickness. In this work, for the first time we developed and tested a methodology to design a new fibrous material for which the penetration of a liquid drop can be predicted and optimized. The deposition of a liquid drop into a porous solid material is an

J.M.R. Curto (✉) · A.M.M. Ramos · R.M.S. Simões · M.J. Santos Silva
Fibre Materials and Environmental Technologies, University of Beira Interior, Covilhã, Portugal
e-mail: jmrc@ubi.pt

A.M.M. Ramos
e-mail: ammr@ubi.pt

R.M.S. Simões
e-mail: rmss@ubi.pt

M.J. Santos Silva
e-mail: mjssilva@ubi.pt

A.O. Mendes · P.T. Fiadeiro
Remote Sensing Unit, University of Beira Interior, Covilhã, Portugal
e-mail: aomendes@ubi.pt

P.T. Fiadeiro
e-mail: fiadeiro@ubi.pt

A. Öchsner and H. Altenbach (eds.), *Mechanical and Materials Engineering of Modern Structure and Component Design*, Advanced Structured Materials 70,
DOI 10.1007/978-3-319-19443-1_25

important step in many materials applications like printing or filtration. The ability to control and predict the liquid kinetic deposition and the spreading area, are also determinant steps when developing new advanced materials, like structured porous materials, and new applications, like the transport of therapeutic molecules for medicinal purposes. We have used a methodology that includes an innovative 3D model for fibrous materials, an experimental plan to obtain 3D structures with different fibers, and an optic 3D prototype for collecting 3D data of the liquid drop deposition along time. We were able to quantify the time of deposition and the spreading area over time. Using the simulation model for fibrous materials we were able to simulate and produce in the laboratory an optimized structure for the ideal spreading area of the liquid droplet.

Keywords Liquid drop · Droplet deposition time · Droplet spreading area · 3D fibrous structured materials · Simulation and optimization of structured materials

1 Introduction

Polymeric fibrous materials are formed using fibers as building blocks. By choosing fiber dimensions and fiber mechanical properties, like fiber flexibility, new materials can be designed to have optimized properties, like the desired porosity for an application. The optimization method is able to determine the combination of fibers among the ones that are available, with certain dimensions and flexibility in order to obtain the desired porosity (see Fig. 1a–d). This porosity and distribution of pores can be the one that optimizes the deposition time or the spreading area of the liquid drop into the fibrous structure. To develop novel applications using cellulose fibrous materials it is important to optimize the time of deposition, for the liquid drop, and the spreading area [1–3]. With the available fibers, changing the fiber flexibility and the mixture of fibers used to obtain the fibrous structure, a new 3D optimized material is obtained. For materials made from cellulose natural fibers, like the ones we have chosen for this study, the structured materials are formed using a fiber deposition step, water filtration and pressing processes, according to

E.L.T. Conceição · A.T.G. Portugal
Chemical Process Engineering and Forest Products Research Centre,
Chemical Engineering Department, University of Coimbra, Coimbra, Portugal
e-mail: etc@eq.uc.pt

A.T.G. Portugal
e-mail: atp@eq.uc.pt

J.M.R. Curto
Department of Chemistry, University of Beira Interior, Av. Marquês d'Ávila e Bolâma n.º 54,
6200-001 Covilhã, Portugal

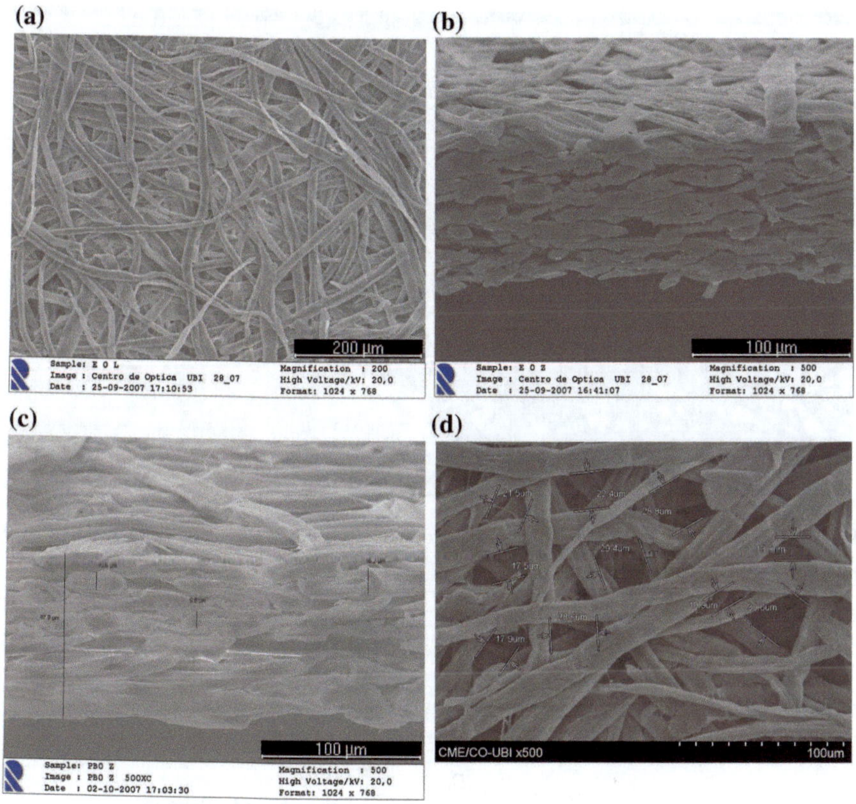

Fig. 1 Scanning electron microscope image of the 3D fibrous structures for different fiber dimensions and fiber flexibilities

ISO standard procedure for laboratory paper sheets. The structure formation processing steps have been modeled and implemented in Matlab resulting in a validated fibrous materials simulator (see Fig. 2) [4–10]. The fibrous materials simulator is a multiscale simulator that includes a fiber model with fiber structure, dimensions and flexibility and has been used to simulate and optimize fibrous materials for different scales and polymeric materials [5, 6]. To obtain the structure that has a certain time of liquid deposition into the structure or to optimize the spreading area, the desired structure and porosity can be obtained using a combination of different fibers, with different dimensions and mechanical properties. In this work we have used an experimental plan to produce laboratory 3D fibrous structures with different fiber dimensions, flexibilities and mixture of fibers. For collecting 3D information about the liquid droplet deposition we have used an optic 3D prototype developed at the University of Beira Interior (see Fig. 3) [11–13]. In Comparison to other methods, this one has the advantage of collecting 3D data of the liquid drop and accurate deposition images along time (see Fig. 4). We were able to quantify the time of deposition and spreading area along time.

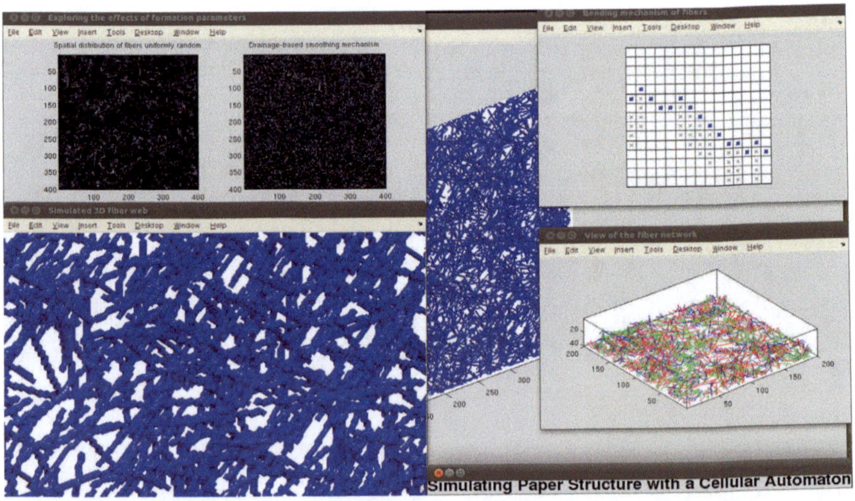

Fig. 2 Image of the fibrous materials simulator with Fiber deposition and 3D structure with different fibers

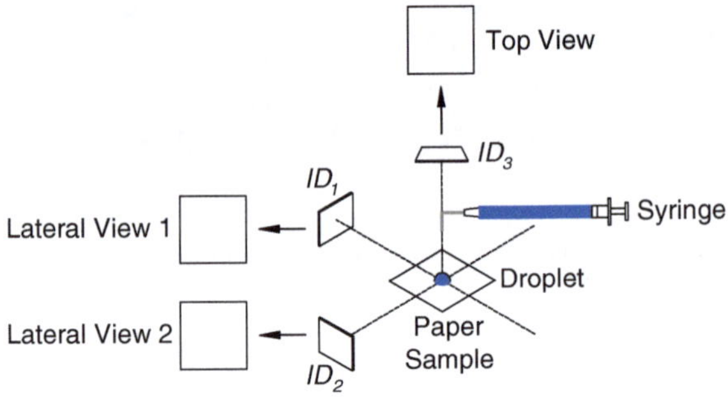

Fig. 3 Scheme of the optical system used for analysis over time of the interaction of liquid droplets on different kinds of paper samples

Using our simulation model for fibrous materials we were able to simulate the structure produced in the laboratory and to have information about porosity and pore distribution in the surface. The developed fibrous material model has the advantage of being a 3D model that includes a detailed fiber structure model, with fiber collapse and structure of the fiber void, not considered in other simulation models [14–16]. The 3D simulation structures are obtained by a formation mechanism similar to the one occurring in the laboratory, fiber by fiber, and the obtained 3D structure can be used to calculate useful information about the structure, like its porosity, pore distribution, superficial topography, thickness, etc. The methodology developed for the optimization study done in this work starts with the experimental characterization of the fibrous 3D structures and the 3D droplet deposition on these

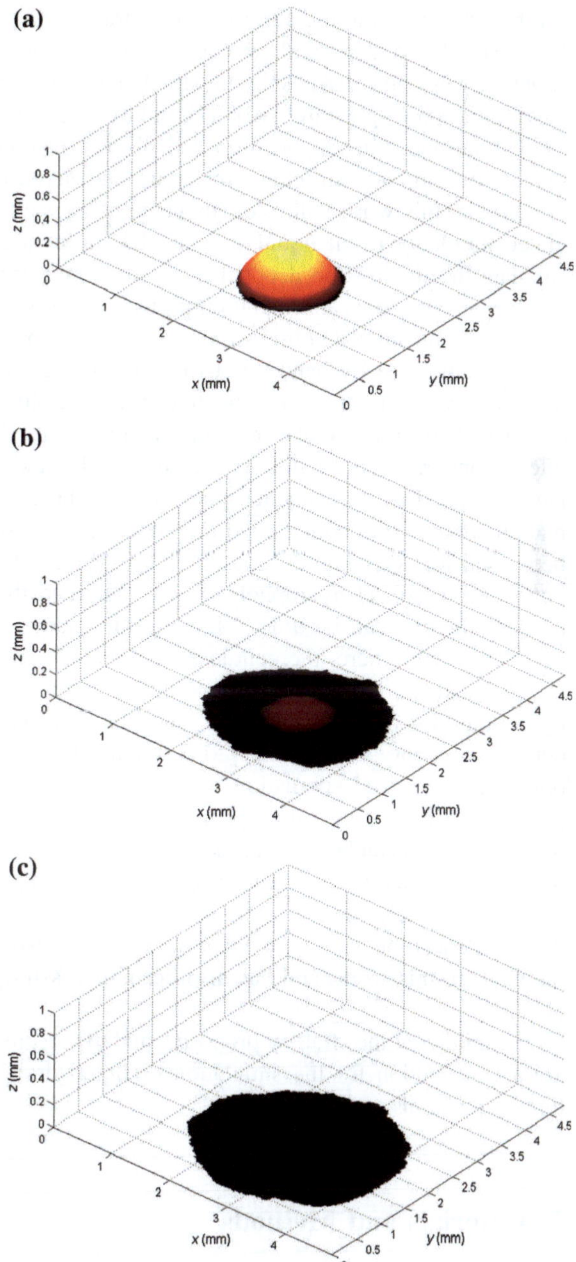

Fig. 4 View of the 3D model created for one of the analyzed paper samples, namely, the sample A-3-14 (mix fibers—20 % long fiber with medium fiber flexibility). **a** For the initial time (t ≈ 0.00 s), **b** for an intermediate time (t = 0.50 s), and **c** for the final time (t = 1.00 s)

structures. The results of the 3D structures are complemented with simulated structures and more information about porosity and its values on the top surface layer. Using a simulation study, an optimized structure for a certain spreading area can be proposed, and the corresponding structure is produced in the laboratory. A liquid droplet is deposited into this structure, and the results indicate that the structure was optimized, using the simulation model and the experimental data, to have the desired spreading area. The ability to optimize the spreading area, deposition time and other liquid drop and fibrous material interactions can be useful tools to develop new structured materials, and advanced applications, so we intend to continue this work with. For future work we have planned to present the integrated simulation of the liquid drop and 3D fibrous structure. Our ultimate goal is to simulate the liquid drop deposition on structured fibrous materials, including the key fibrous structures and liquid drop physical properties. For future work our goal is to be able to simulate other scales, like for example at nanoscale level, which we have already done for nano fibrous material using our simulator [5, 6] and can be integrated with liquid drop deposition using the strategy presented here. Another aspect that we have to take into consideration is the hydrophobic and hydrophilic interactions between the fibrous structures and liquid droplets. For future work we have designed an experimental plan that will be dedicated to collect more information about hydrophobic liquid drop deposition. The optic equipment has already been used for other liquids, and the image acquisition and drop deposition can be optimized to collect hydrophobic liquid data, and the optimization of the fibrous porous structure can be done following the strategy that we present here. In the results presented here, the cellulose fibrous materials analysed have OH^- groups [17–19] and the liquid drop is a coloured water solution, so the interactions are hydrophilic for both, liquid droplet and fibrous material. The study of liquid drop deposition, theory and practice [20–35], indicates that one of the key parameters to obtain an integrated simulation model is to have accurate information about the porous structure. The results presented here indicate that our simulation fibrous materials model can be a useful tool, and the work we have already done proves that it can be extended for different scales and materials [5, 6]. The other crucial aspect, is to have an accurate and programmable experimental optic device to collect 3D droplet deposition data (see Fig. 3), which is also presented and proved to be the case, in this work, and has also been successfully used for other applications [11–13].

We think that the strategy proposed, and the results to be presented in this work, are a contribution for the simulation of liquid drop deposition on 3D structured fibrous materials.

2 Materials and Methods

2.1 Materials

Cellulose fibers of *Eucalyptus globulus* and *Pinus pinaster.*

2.2 Methods

The network structures are formed with fibers from *Eucalyptus globulus* bleached kraft pulp that are fibers coming from a Hardwood tree, after a delignification, fiber separation and bleaching process. It is considered a "short" fiber and is the most important wood fiber produced in Portugal and Brazil.

The softwood fibers, or "long" fibers are from a *Pinus pinaster* kraft softwood pulp. This pulp is called reinforcement pulp because it is added to the structure to increase the structure's strength, and also the "runnability" of the process, to avoid the rupture of the wet web during the process of making the fibrous structured material.

To evaluate the impact of different fibers on structural properties, isotropic laboratory fibrous structures were prepared and tested, according to structural properties and ISO standards. Formation and pressing of fibrous structures was done according to ISO 5269:2005. Characterization of fibrous structures was done using ISO 536 for the basis weight and ISO 534:2011 for thickness, with an Adamel Lhomargy MI 20 micrometer. An experimental design plan was executed in order to quantify the influence of fiber dimensions and fiber flexibility, for the two pulp fibers, with different dimensions and beating degrees, to obtain different fiber flexibilities. The fibers' dimensions were determined automatically by image analysis of a diluted suspension (20 mg/l) in a flow chamber in Morfi (Equipment developed and commercialized by TECHPAP, Grenoble, France). The pulps were beaten in a PFI mill at 1000, 3000 and 6000 revolutions under a refining intensity of 3.33 N/mm. PFI is a ISO standard laboratory beating equipment, consisting of a rotor and a stator, who provides a pulp mechanical treatment called beating, that changes fiber flexibility. Fiber flexibility was determined according to the Steadman and Luner method [36], using CyberFlex from CyberMetrics. The experimental method includes the formation of a very thin and oriented fiber network on the wire of a small head-box simulating the paper formation, followed by transfer under controlled pressure conditions to a glass slide with metal wires. The laboratory made structures were characterized regarding their structural properties: thickness, basic weight, density and porosity. Images of the fibrous structures were obtained through scanning electron microscopy, using the SEM Hitachi S-3400 N and a gold sputter to make the fibrous structures conductive.

The experimental system [11–13] that was used in the current work to study the interaction over time of liquid droplets on different kinds of fibrous structures is schematically presented in Fig. 3. Basically, the system works by ejecting microliter droplets (approximately 0.33 μl) through a syringe filled with the liquid in study, toward the surface of a given fibrous structure sample, which is placed on a proper holder of the system for its convenient fixation. During the event, three image detectors, identified in the scheme of Fig. 3 as ID1, ID2 and ID3, provide image registration of the interaction of the liquid droplet with the paper sample in study. The three image detectors are placed in the system perpendicularly to each other, two of them (ID1 and ID2) being horizontally aligned in the system, allowing to obtain lateral views of the event, and the last image detector (ID3) is aligned in the system vertically, allowing to obtain a top view of the event (see Fig. 3). Typically, in the experiments carried out with this methodology, 215 images are registered per second by each one of the three image detectors during a maximum period of 10 s, using a field of view of 4.8 × 4.8 mm^2.

The images are then stored in a computer for their corresponding processing in order to determine a set of specific parameters for evaluation of the liquid-paper interaction. In the current work, our focus was placed in the determination of the spreading area, the area of the exposed droplet, and the difference between the two. For the construction of the graphical representations of these three parameters, a time step of 0.25 s was considered, which showed to be adequate for the six different studied cases that will be presented in the next section of this work. In addition, one other parameter was also determined for the six studied cases, namely the total time of droplet deposition, which corresponds to the time range between the instant of impact of the liquid drop in the surface of the paper sample in study and the instant of its complete deposition. This last instant corresponds to the time on which the droplet deposited at the surface of the paper is no longer seen by the lateral views of the system, since its height reaches the value zero at this point. To determine this last parameter, a much lower time step was considered, using the totality of the acquired images for its proper discrimination.

Concerning the samples to be analyzed with the optical system, six different paper samples were selected with the following identifications and characteristics: A-14 (Maximum Fiber Flexibility), A-2-14 (Minimum Fiber Flexibility), A-3-14 (Mix Fibers—20 % Long Fiber with Medium Fiber Flexibility), A-4-14 (Mix Fibers—10 % Long Fiber with Maximum Fiber Flexibility), A-5-14 (Mix Fibers—10 % Long Fiber with Medium Fiber Flexibility), and A-6-14 (Medium Fiber Flexibility). Regarding the liquid used in the conducted experiments, it was defined to be deionized water dyed with an innocuous blue colorant. This last element was used simply to achieve a better discrimination of the droplets in the images, for their proper identification relatively to the background of the images.

3 Results and Discussion

3.1 Pulp Fiber Characterization

The *Eucalyptus globulus* and *Pinus pinaster* fibers used have different dimensions and coarseness (see Table 1).

3.2 Three Dimensional Fibrous Structure Characterization

The 3D fibrous structures were characterized in terms of porosity. The experimental and simulation results are presented in Table 2.

Table 1 Fibre dimensions and coarseness for *Eucalyptus globulus* and *Pinus pinaster*

	Eucalyptus globulus	*Pinus pinaster*
Fibers (million/g)	33.898	5.663
Length (weighted in length) (mm)	0.8	2.0
Width (μm)	18	34
Coarseness (mg/m)	0.067	0.22

Table 2 Characterization of 3D fibrous structures

Fibre flexibility	Porosity of *Pinus pinaster* experimental fibrous structures	Porosity of *Pinus pinaster* simulated fibrous structures
Minimum fiber flexibility 1000 PFI rv A-2-14	0.60	0.60
Intermediate fiber flexibility 3000 PFI rv A-6-14	0.56	0.56
Maximum fiber flexibility 6000 PFI rv A-14	0.54	0.54
Natural fiber flexibility without beating reference sample	0.67	0.66

SEM images were obtained for different 3D Fibrous structures, with different fiber dimensions, for *Pinus pinaster* and *Eucalyptus globulus* fibers, for different fiber flexibilities. The samples presented are examples from top views, as in Fig. 1a, d and thickness cuts, Fig. 1b, c from laboratory made 3D structures (60 g/m^2).

3.3 Results for the Deposition of a Liquid Drop into the Fibrous Structure

3.3.1 Influence of Fiber Flexibility

Fiber flexibility has been identified as a key parameter to change the 3D fibrous structure and the apparent porosity [9, 15, 16]. In order to obtain different fiber flexibilities, a controlled laboratory mechanical treatment called beating has been done. The response of the liquid drop deposition into the 3D fibrous structure made from fibers having different fiber flexibilities indicate that the spreading area is smaller (see Table 3; Figs. 4 and 5) and that the total time of deposition is higher for the maximum flexible fiber (see Table 4), corresponding to a less porous structure. The 3D structures made in the laboratory were simulated using the 3D fibrous simulator and the same porosity was obtained (see Table 2).

Table 3 Values of initial, intermediate and final times, and corresponding values of spreading area and area of exposed droplet obtained for the samples A-2-14 (minimum fiber flexibility), A-6-14 (medium fiber flexibility), and A-14 (maximum fiber flexibility)

Paper	Time	Time (s)	Spreading area (mm^2)	Area of exposed droplet (mm^2)
A-2-14 minimum fiber flexibility	Initial	t ≈ 0.00	1.503	1.503
	Intermediate	t = 0.75	5.257	1.490
	Final	t = 1.25	6.817	0.000
A-6-14 medium fiber flexibility	Initial	t ≈ 0.00	1.760	1.760
	Intermediate	t = 2.25	4.881	2.021
	Final	t = 4.50	7.078	0.000
A-14 maximum fiber flexibility	Initial	t ≈ 0.00	1.759	1.759
	Intermediate	t = 5.00	4.663	2.221
	Final	t = 10.0	6.513	2.229

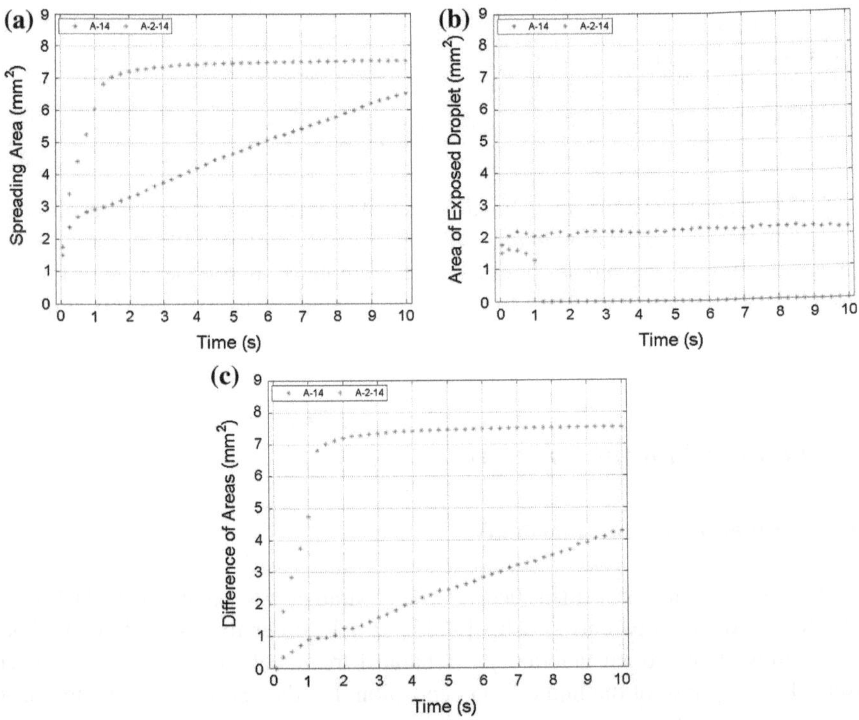

Fig. 5 Graphical representation over time obtained for the samples A-14 (maximum fiber flexibility), and A-2-14 (minimum fiber flexibility). **a** For the spreading area, **b** for the area of exposed droplet, and **c** for the difference of areas

Table 4 Summary of values of total time of droplet deposition for each of the six analyzed samples A-2-14 (minimum fiber flexibility), A-6-14 (medium fiber flexibility), A-14 (maximum fiber flexibility), A-5-14 (mix fibers—10 % long fiber with medium fiber flexibility), A-4-14 (mix fibers—10 % long fiber with maximum fiber flexibility), and A-3-14 (mix fibers—20 % long fiber with medium fiber flexibility)

Paper	A-2-14 minimum fiber flexibility	A-6-14 medium fiber flexibility	A-14 maximum fiber flexibility	A-5-14 (mix fibers) 10 % long fiber with medium fiber flexibility	A-4-14 (mix fibers) 10 % long fiber with maximum fiber flexibility	A-3-14 (mix fibers) 20 % long fiber with medium fiber flexibility
Total time of deposition (s)	t = 1.130	t = 4.437	t > 10	t = 0.772	t = 0.847	t = 0.916

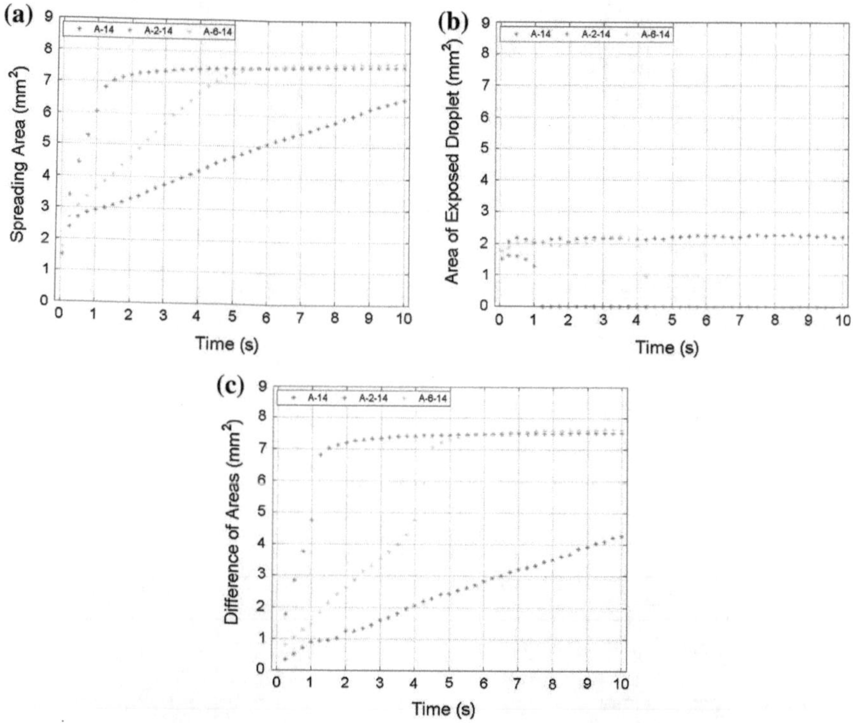

Fig. 6 Graphical representation over time obtained for the samples A-14 (maximum fiber flexibility), A-2-14 (minimum fiber flexibility), and A-6-14 (medium fiber flexibility). **a** For the spreading area, **b** for the area of exposed droplet, and **c** for the difference of areas

To evaluate the model predictive capacity, the simulator was used to propose an optimized 3D structure that will result in an intermediate spreading area between 1.5 and 1.8 mm^2, for the initial time. The optimized 3D structure has been produced in the laboratory, with the defined fiber flexibility and the liquid drop experiment has been done using the same conditions. Figure 6 presents the results for the liquid drop deposition spreading area and it is verified that the model is able to determine the right structure to obtain the desired spreading area. In this case, for the initial time, the intermediate value of 1.7 mm^2 is obtained (see Figs. 6, 7, 8 and 9; Table 3).

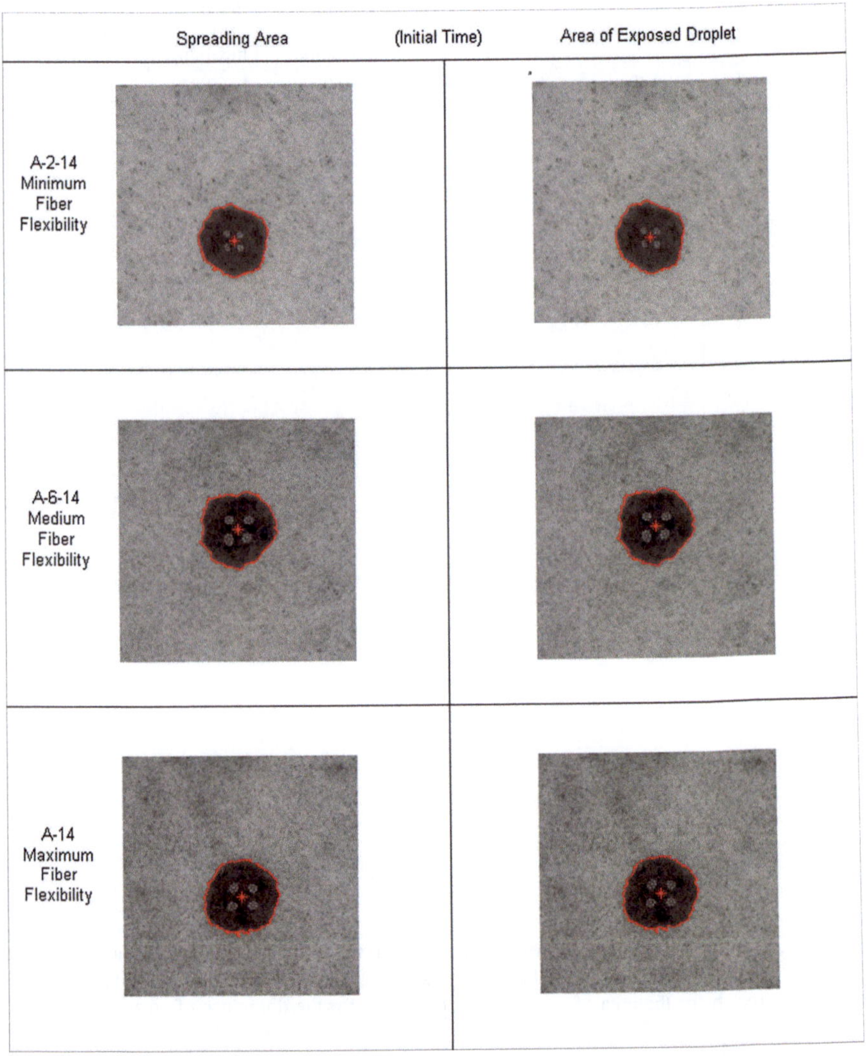

Fig. 7 Images of the top view, and also of the centroid and contour line (represented in *red color*) of the spreading area, and the area of exposed droplet for the samples A-2-14 (minimum fiber flexibility), A-6-14 (medium fiber flexibility), and A-14 (maximum fiber flexibility) at the initial time

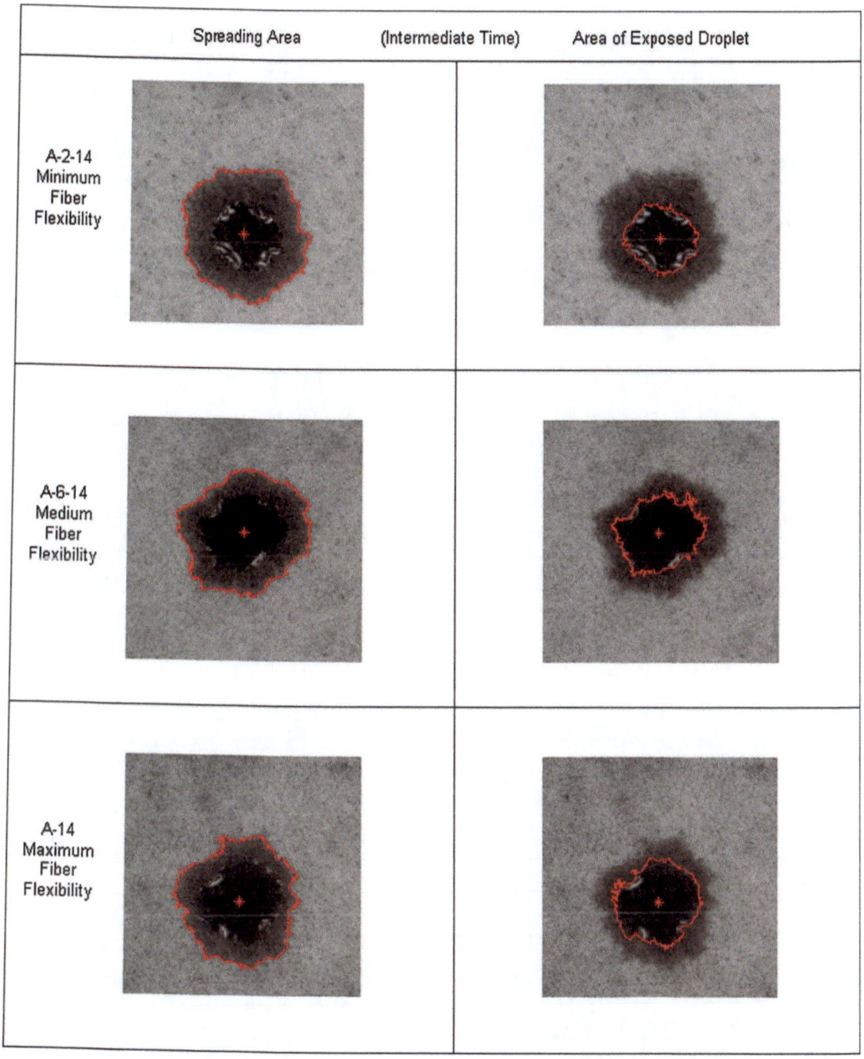

Fig. 8 Images of the top view, and also of the centroid and contour line (represented in *red color*) of the spreading area, and the area of exposed droplet for the samples A-2-14 (minimum fiber flexibility), A-6-14 (medium fiber flexibility), and A-14 (maximum fiber flexibility) at the intermediate time

To study the fiber flexibility influence, real and simulated fibrous structures of *Eucalyptus globulus* with a basis weight of 60 g/m^2 were produced. To obtain different fiber flexibilities the fibers have been beaten at three beating levels (1000, 3000 and 6000 PFI revolutions), and laboratory made structures have been made according to standard ISO standards.

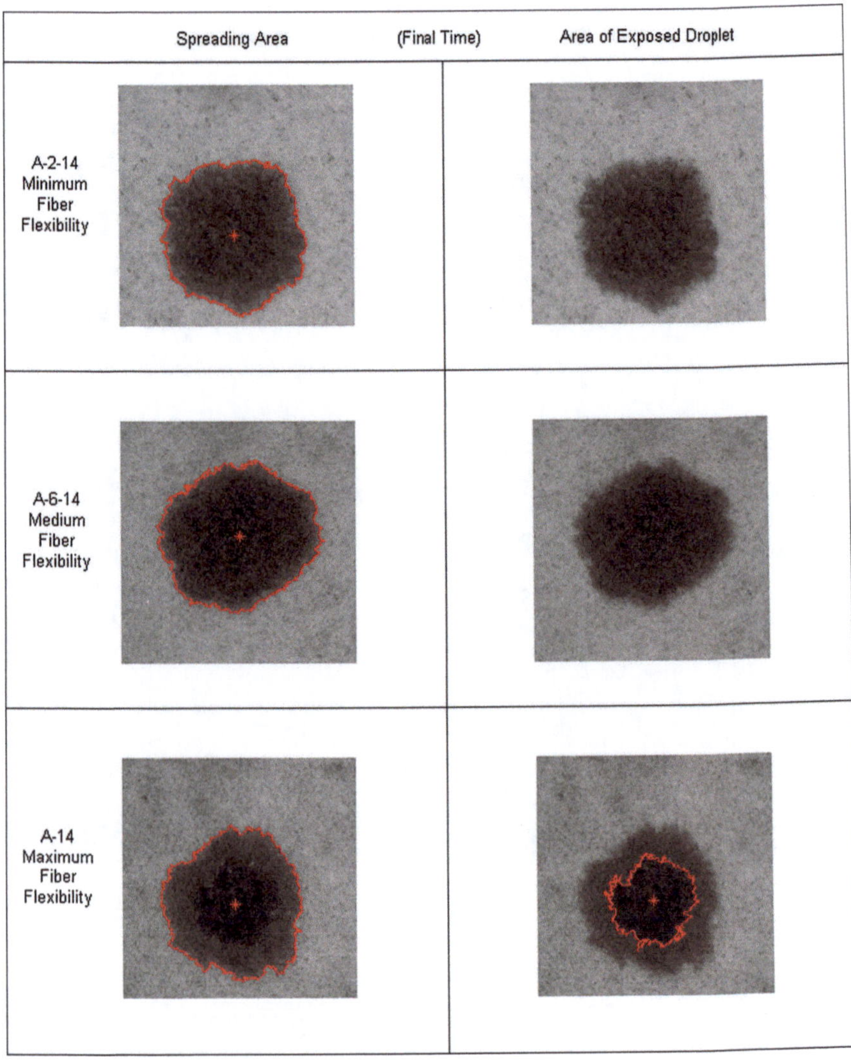

Fig. 9 Images of the top view, and also of the centroid and contour line (represented in *red color* when applicable) of the spreading area, and the area of exposed droplet for the samples A-2-14 (minimum fiber flexibility), A-6-14 (medium fiber flexibility), and A-14 (maximum fiber flexibility) at the final time

The analysis of the results for the liquid drop spreading area and time of deposition indicate that it is possible to design a 3D structure to obtain an intermediate spreading area or deposition time using the simulation model to determine the fiber flexibility that will correspond to the desired 3D structure.

3.3.2 Influence of the Mixture of Fibers

To study the influence of the mixture of different fibers on the liquid drop deposition three samples were selected.

The computer simulated structures obtained for a basis weight of 60 g/m^2 have the same porosity as the laboratory made structures and the same behaviour regarding the liquid drop deposition. To illustrate the results obtained for the deposition of a liquid drop into these structures, three examples of figures were selected (Figs. 10, 11, 12 and 13). The time of deposition and spreading area were identical for all the studied mixtures (see Table 5).

For the 3D computational simulation the porosity values were very similar and we can conclude that the structures obtained with the mixture of different fibers are very similar, and have the same results for spreading area and deposition time (see Figs. 10, 11, 12 and 13).

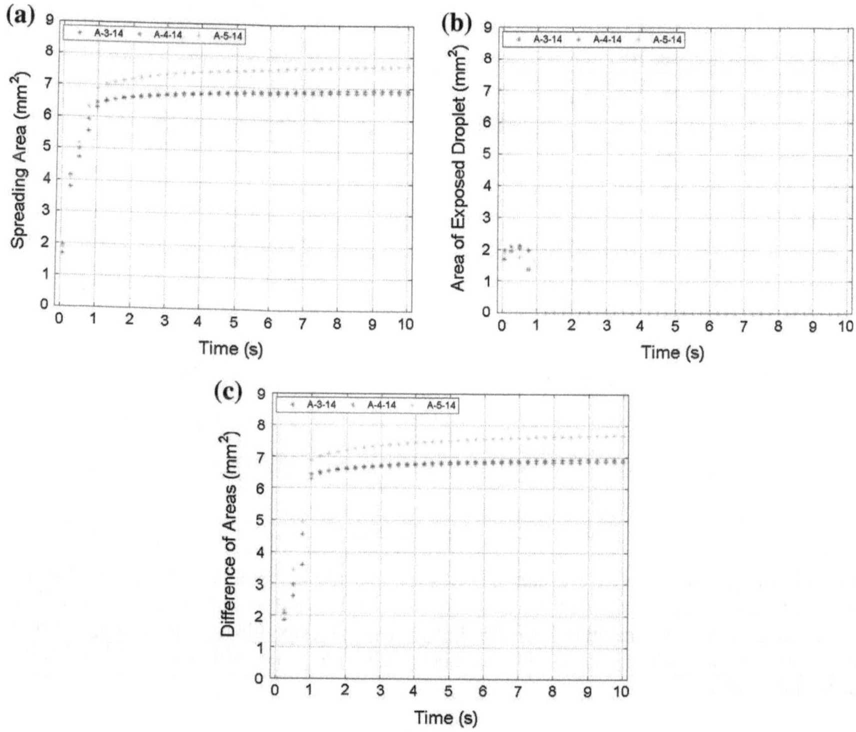

Fig. 10 Graphical representation over time obtained for the samples A-3-14 (mix fibers—20 % long fiber with medium fiber flexibility), A-4-14 (mix fibers—10 % long fiber with maximum fiber flexibility), and A-5-14 (mix fibers—10 % long fiber with medium fiber flexibility). **a** For the spreading area, **b** for the area of exposed droplet, and **c** for the difference of areas

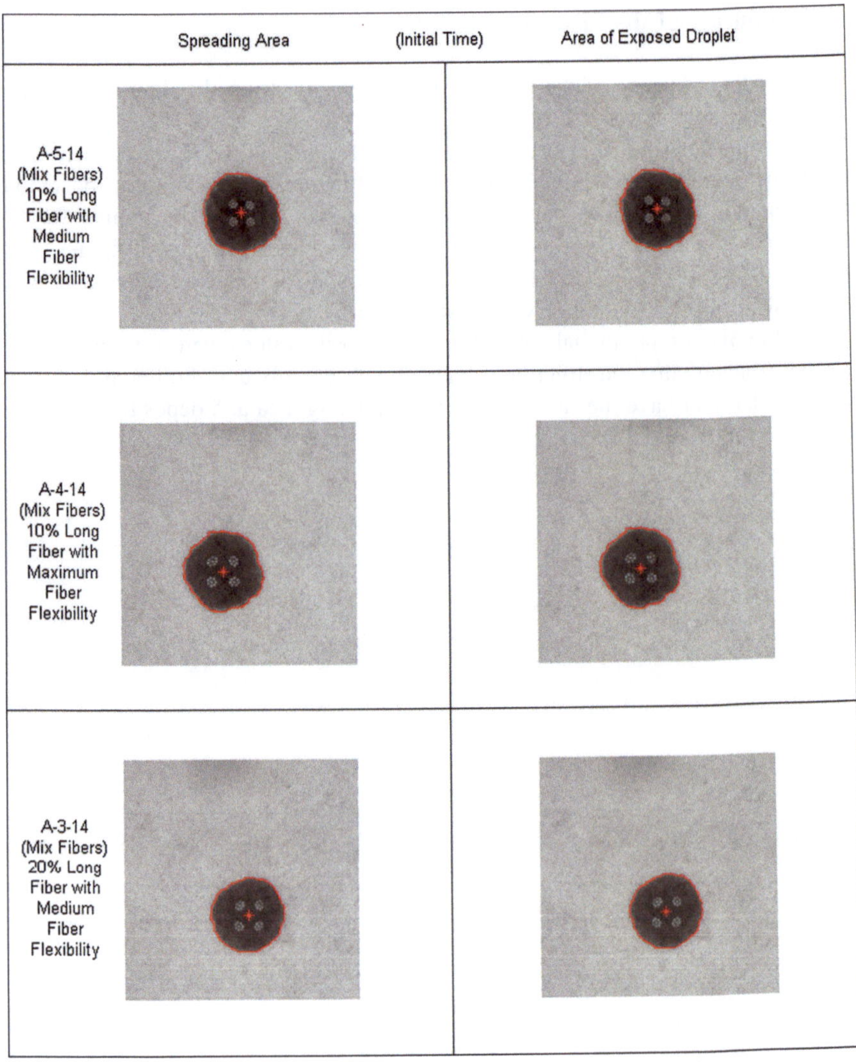

Fig. 11 Images of the top view, and also of the centroid and contour line (represented in *red color*) of the spreading area, and the area of exposed droplet for the samples A-5-14 (mix fibers—10 % long fiber with medium fiber flexibility), A-4-14 (mix fibers—10 % long fiber with maximum fiber flexibility), and A-3-14 (mix fibers—20 % long fiber with medium fiber flexibility) at the initial time

It can be concluded that to change the parameters for liquid drop deposition, spreading area and time of deposition, the key factor is to change fiber flexibility. Changing fiber dimensions by using fiber mixtures has no effect to obtain the desired spreading area or deposition time.

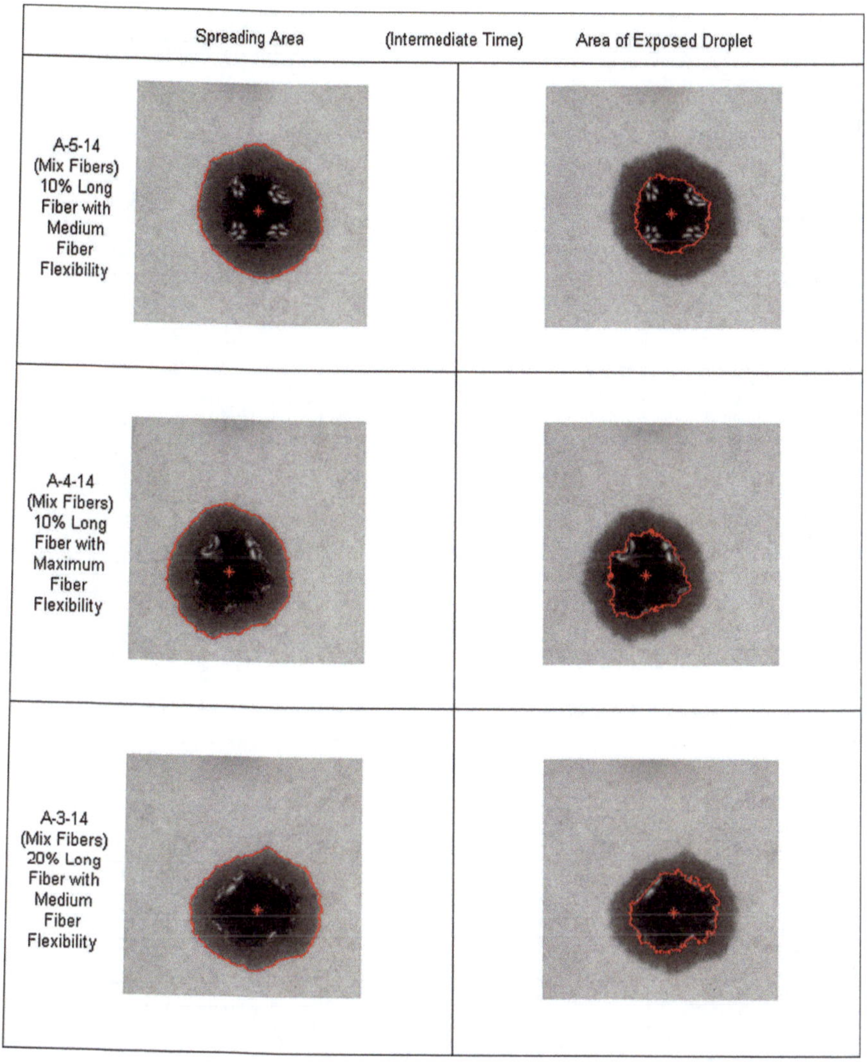

Fig. 12 Images of the top view, and also of the centroid and contour line (represented in *red color*) of the spreading area, and the area of exposed droplet for the samples A-5-14 (mix fibers—10 % long fiber with medium fiber flexibility), A-4-14 (mix fibers—10 % long fiber with maximum fiber flexibility), and A-3-14 (mix fibers—20 % long fiber with medium fiber flexibility) at the intermediate time

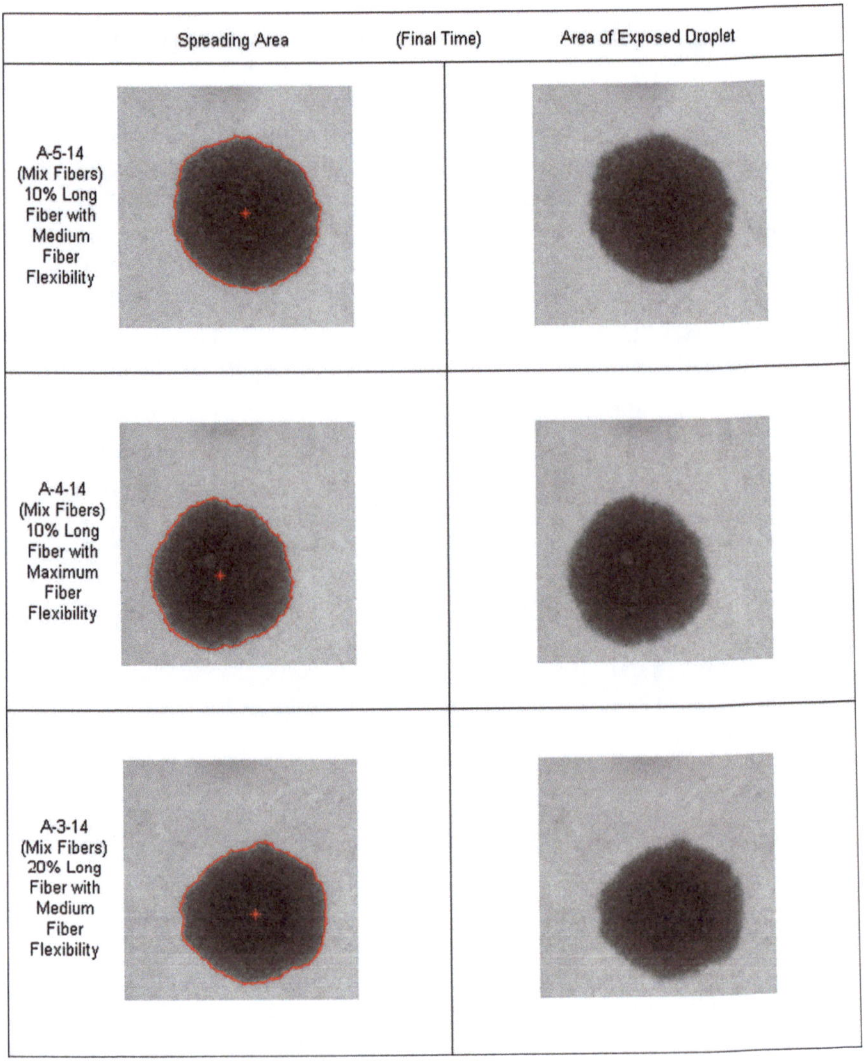

Fig. 13 Images of the top view, and also of the centroid and contour line (represented in *red color*) of the spreading area, and the area of exposed droplet for the samples A-5-14 (mix fibers—10 % long fiber with medium fiber flexibility), A-4-14 (mix fibers—10 % long fiber with maximum fiber flexibility), and A-3-14 (mix fibers—20 % long fiber with medium fiber flexibility) at the final time

Table 5 Values of initial, intermediate and final times, and corresponding values of spreading area and area of exposed droplet obtained for the samples A-5-14 (mix fibers—10 % long fiber with medium fiber flexibility), A-4-14 (mix fibers—10 % long fiber with maximum fiber flexibility), and A-3-14 (mix fibers—20 % long fiber with medium fiber flexibility)

Paper	Time	Time (s)	Spreading area (mm^2)	Area of exposed droplet (mm^2)
A-5-14 (mix fibers) 10 % long fiber with medium fiber flexibility	Initial	t ≈ 0.00	1.875	1.875
	Intermediate	t = 0.50	5.164	1.742
	Final	t = 1.00	6.895	0.000
A-4-14 (mix fibers) 10 % long fiber with maximum fiber flexibility	Initial	t ≈ 0.00	1.961	1.961
	Intermediate	t = 0.50	4.986	2.017
	Final	t = 1.00	6.454	0.000
A-3-14 (mix fibers) 20 % long fiber with medium fiber flexibility	Initial	t ≈ 0.00	1.681	1.681
	Intermediate	t = 0.50	4.720	2.105
	Final	t = 1.00	6.306	0.000

3.4 Conclusions

The purpose of this article was to study the liquid drop deposition on 3D fibrous structures. A combination of a three dimensional materials model with experimental data for a 3D liquid drop is presented. The proposed simulation model includes key fiber dimensions and properties and the liquid drop deposition time and spreading area. The fibrous structured material was built by the sequential deposition of individual fibers. The model is detailed up to the point where it includes fiber morphology and behaviour in the z-direction. Both fiber wall thickness and fiber lumen can be changed independently in order to allow the implementation of different degrees of fiber dimensions and fiber collapse, which are important fiber structure details, not present in other fibrous materials models, that are decisive to have representative simulations of the fibrous porous structure. The simulated results represent a good estimative of the real laboratory made fibrous structured materials and can be used to optimize the fibrous material structure.

The integration of the liquid drop deposition data into the fiber model presented proved to be a good tool for studying the influence of fiber properties such as fiber flexibility and fiber dimensions on the time of deposition and spreading area of a liquid drop into a structured fibrous material. The methodology proposed, and the results that have been presented in this work, constitute a contribution for the optimization of liquid drop deposition on 3D structured fibrous materials, and an important step for the development of an integrated 3D simulator for liquid drop deposition on 3D structured fibrous materials.

References

1. Yano H, Sasaki Md, Shams I et al (2014) Wood pulp-based optically transparent film: a paradigm from nanofibers to nanostructured fibers. J Adv Optical Mater 1–4
2. Eichhorn SJ et al (2009) Review: current international research into cellulose nanofibers and nanocomposites. J Mater Sci 45:1–33
3. Miao C, Hamad WY (2013) Cellulose reinforced polymer composites and nanocomposites: a critical review. Cellulose 20:2221–2262
4. Curto JMR, Conceição ELT, Portugal ATG et al (2011) Three dimensional modeling of fibrous materials and experimental validation. Materialwiss Werkstofftech 42(5):370–374
5. Conceição ELT, Curto JMR, Simões RMS et al (2010) Coding a simulation model of the 3D structure of paper. In: Proceedings of the 2th international symposium on computational modeling of objects represented in images, Comp Image 2010, USA. Springer-Berlin, Lecture Notes in Computer Science 60(20):299–310
6. Curto JMR (2012) 3D computational simulation and experimental characterization of polymeric stochastic network materials: case studies in reinforced *Eucalyptus* office paper and nanofibrous materials. Ph.D. thesis. University of Beira Interior, Portugal
7. Curto JMR, Hekmati, AH, Drean JY et al (2011) Three dimensional polyamide-6 nanowebs modelling and simulation. In: Proceedings of the 11th world textile conference, Autex2011, vol 2, Mulhouse, France, June, pp 639–643
8. Curto JMR, Conceição ELT, Portugal ATG et al (2009) The fiber coarseness and collapsibility influence on a three dimension paper model. In: Proceedings of 63rd Appita annual conference and exhibition, Melbourne 19–22 April, Australia, pp 303–310
9. Curto JMR, Conceição ELT, Portugal ATG et al (2010) The Influence of eucalyptus's and reinforcement's fibers flexibility on paper properties: experimental and 3D paper model evaluation. In: Proceedings of the 5th ICEP—International Colloquium on Eucalyptus Pulp held in May 9–12, Porto Seguro, Brazil
10. Curto JMR, Conceição ELT, Portugal ATG et al (2013) Characterization and optimization of paper using 3D computational simulation and synchrotron X-ray micotomogaphy. In: Proceedings of the XXII Tecnicelpa international forest, pulp and paper conference, 2–4 Oct, Tomar, Portugal
11. Mendes AO, Fiadeiro PT, Ramos AMM et al (2013) Development of an optical system for analysis of the ink-paper interaction. Mach Vis Appl 24:1733–1750
12. Sousa SCL, Mendes AO, Fiadeiro PT et al (2014) Dynamic interactions of pigment-based inks on chemically modified papers and their influence on inkjet print quality. Ind Eng Chem Res 53(12):4660–4668
13. Fiadeiro PT, Mendes AO, Ramos et al (2013) Study of the ink-paper interaction by image analysis: surface and bulk inspection. In: Proceedings of SPIE 8th Iberoamerican optics meeting and 11th Latin American meeting on optics, lasers and applications, 8785, 8785BV-1/8785BV-8, doi:10.1117/12.2024991
14. Provatas N, Haataja M, Seppal E et al (1997) Structural properties of disordered fiber networks. Phys A 329:304–313
15. Niskanen K, Nilsen N, Hellen E et al (1997) KCL-PAKKA simulation of 3D structure of paper. In: Proceedings of 11th FRS Cambridge, pp 1273–1293
16. Alava M, Niskanen K (2006) The physics of paper. Rep Prog Phys 69(3):669–723
17. Fengel D, Wegner G (1989) Wood chemistry, ultrastucture, reactions. Walter de Gruyter, Berlin
18. Gullihsen J, Paulapuro H (1999) Papermaking chemistry, papermaking science and technology. Published by Finnish Paper Engineers Association and TAPPI. ISBN:952-5216-04-7
19. Fardim P, Moreno T, Holmbon B (2005) Anionic groups on cellulosic fiber surfaces investigated by XPS, FTIR-ATR and different sorption methods. J Colloid Interf Sci 290:383–391

20. Apel-Paz M, Marmur A (1999) Spreading of liquids on rough surfaces. Colloids Surf A Physicochem Eng Aspects 146(1–3):273–279
21. Zhdanov SA, Starov VM, Sobolev VD et al (2003) Spreading of aqueous SDS solutions over nitrocellulose membranes. J Colloid Interface Sci 264(2):481–489
22. Starov VM, Zhdanov SA, Kosvintsev SR et al (2003) Spreading of liquid drops over porous substrates. Adv Colloid Interface Sci 104(1–3):123–158
23. Kannangara D, Zhang H, Shen W (2006) Liquid-paper interactions during liquid drop impact and recoil on paper surfaces. Colloids Surf A Physicochem Eng Aspects 280(1–3):203–215
24. Clarke A, Blake TD, Carruthers K et al (2002) Spreading and imbibition of liquid droplets on porous surfaces. Langmuir 18(8):2980–2984
25. Girard F, Attané P, Morin V (2006) A new analytical model for impact and spreading of one drop: application to inkjet printing. Tappi J 5(12):24–32
26. Wågberg L, Westerlind C (2000) Spreading of droplets of different liquids on specially structured papers. Nord Pulp Paper Res J 15(5):598–606
27. Clarke A, Blake TD, Carruthers K et al (2002) Spreading and imbibition of liquid droplets on porous surfaces. Langmuir 18:2980–2984
28. Modaressi H, Garnier G (2002) Mechanism of wetting and absorption of water droplets on sized paper: effects of chemical and physical heterogeneity. Langmuir 18:642–649
29. Starov VM, Zhdanov SA, Kosvintsev SR et al (2003) Spreading of liquid drops over porous substrates. Adv Colloid Interface Sci 104:123–158
30. Desie G, Deroover G, De Voeght F (2004) Printing of dye and pigment-based aqueous inks onto porous substrates. J Imaging Sci Technol 48(5):389–397
31. Berg J, Daniel RC (2006) Spreading on and penetration into thin, permeable print media: Application to ink-jet printing. Adv Colloid Interface Sci 126:439–469
32. Girard F, Attané P, Morin V (2006) A new analytical model for impact and spreading of one drop: application to inkjet printing. Tappi J 5(12):24–32
33. Kannangara D, Zhang H, Shen W (2006) Liquid-paper interactions during liquid drop impact and recoil on paper surfaces. Colloids Surf 280:203–215
34. Alleborn N, Raszillier H (2007) Dynamics of films and droplets spreading on porous substrates. Tappi J 6(3):16–23
35. Hilpert M, Ben-David A (2009) Infiltration of liquid droplets into porous media: effects of dynamic contact angle and contact angle hysteresis. Int J Multiph Flow 35:205–218
36. Steadman RK, Luner P (1992) An improved test to measure the wet fiber flexibility of pulp fibers, Chap. V. Empire State Paper Research Institute Report 79

The Effect of Vacancy Defects on the Evaluation of the Mechanical Properties of Single-Wall Carbon Nanotubes: Numerical Simulation Study

Nataliya A. Sakharova, Jorge M. Antunes, André F.G. Pereira, Marta C. Oliveira and José V. Fernandes

Abstract A three-dimensional finite element model is used in order to evaluate the tensile and bending rigidities and, subsequently, Young's modulus of non-chiral and chiral single-walled carbon nanotubes containing vacancy defects. It is shown that the Young's modulus of single-walled carbon nanotubes with vacancies is sensitive to the presence of vacancy defects in nanotube: it decreases with increasing of the density of vacancy defects.

Keywords Carbon nanotubes · Numerical simulation · Young's modulus · Vacancy defects

---

N.A. Sakharova (✉) · J.M. Antunes · A.F.G. Pereira · M.C. Oliveira · J.V. Fernandes
CEMUC, Department of Mechanical Engineering, University of Coimbra, Rua Luís Reis Santos, Pinhal de Marrocos, 3030-788 Coimbra, Portugal
e-mail: nataliya.sakharova@dem.uc.pt

J.M. Antunes
e-mail: jorge.antunes@dem.uc.pt

A.F.G. Pereira
e-mail: andre.pereira@dem.uc.pt

M.C. Oliveira
e-mail: marta.oliveira@dem.uc.pt

J.V. Fernandes
e-mail: valdemar.fernandes@dem.uc.pt

J.M. Antunes
Escola Superior de Tecnologia de Abrantes, Instituto Politécnico de Tomar, Rua 17 de Agosto de 1808, 2200 Abrantes, Portugal

A. Öchsner and H. Altenbach (eds.), *Mechanical and Materials Engineering of Modern Structure and Component Design*, Advanced Structured Materials 70,
DOI 10.1007/978-3-319-19443-1_26

1 Introduction

Carbon nanotubes (CNTs) have attracted great research interest, because of their extraordinary mechanical, optical, thermal properties [1]. Although various experimental studies have been carried out to evaluate the mechanical properties of nanotubes, there is inconsistency in the experimental results reported in the literature, owing to complexity of the characterization of nanomaterials at the atomic scale. Another reason of such result's scattering can be related with the presence of nanotube defects. Due to manufacturing constraints, it is almost impossible to produce CNTs with a perfect structure. It should be noted that the quality of the CNTs used in the experimental studies has a significant influence on the results. The numerical simulation of the mechanical properties of single-walled carbon nanotubes (SWCNTs) with defects is an important task, providing data that can be compared with experimental results. The structural defects of nanotubes, such as single and multiple vacancies, show suitable effects for various applications of nanotubes. For example, strength of the nanocomposites as interfacial bonding sites, storage of hydrogen and transition of nanotubes from one diameter to another in forming of the nanotube heterojunctions. In recent years, studies regarding the effect of the defects on the CNT mechanical properties (see, for example, [2–5]) have been performed. Nevertheless, it is important to better understand the influence of defects in nanotubes on the mechanical behaviour, in order to take advantages.

There are three main categories of the theoretical approaches for modelling the CNTs behaviour: the atomistic approach (ab initio, classical molecular dynamics, tight-binding molecular dynamics), the continuum approach and the nanoscale continuum approach [6]. Atomistic modelling approaches provide good predictions of the CNTs mechanical properties, but they are time-consuming and involve complex mathematical formulation. For this reason, in recent years, the atomistic methods have been progressively replaced by continuum methods, which allow effective simulation of large atomic systems. The basic assumption of the continuum mechanics-based approaches (CM) is the modelling of CNT as a continuum structure [7]. The nanoscale continuum modelling (NCM) has been considered as an adequate compromise to overcome the disadvantages of molecular dynamics (MD) simulations, i.e. enormous computational efforts, and the lack of sensitivity of CM modelling. In the NCM modelling, the carbon–carbon (C–C) bond is replaced by a continuum element (such as a truss, spring or beam) well-described in continuum mechanics (see, for example [8–10]). The beam elements are well accepted in NCM modelling since equivalent beams used to replace the C–C bond have led to accurate results [10, 11].

The present study aims to contribute towards the study of the mechanical behaviour of single-walled carbon nanotubes containing different percent of vacancy defects, using the equivalent continuum beam approach. Three dimensional (3D) finite element (FE) method was used in order to evaluate the tensile and bending rigidities, and subsequently, Young's modulus of various defect SWCNTs, as non-chiral (zigzag, θ = 0°, and armchair, θ = 30°) and one family of chiral (θ = 8.9°) ones, for a wide range of diameters.

2 Atomic Structure of SWCNTs

A simple way to describe SWCNT is as a rolled-up graphene sheet giving rise to a hollow cylinder, which surface is composed by hexagonal carbon rings (see, for example [12]). A schematic illustration of an unrolled hexagonal graphene sheet is shown on Fig. 1. The symmetry of the atomic structure of SWCNTs is characterized by the chirality, which is defined by the chiral vector $\overrightarrow{C_h}$:

$$\overrightarrow{C_h} = n\overrightarrow{a_1} + m\overrightarrow{a_2} \tag{1}$$

where $\overrightarrow{a_1}$ and $\overrightarrow{a_2}$ are the unit vectors of the hexagonal lattice and n and m are integers.

The length of the unit vector $\vec{a}$ is defined as $a = \sqrt{3}a_{C-C}$ with the equilibrium carbon–carbon (C–C) covalent bond length a_{C-C} usually taken to be 0.1421 nm. The nanotube diameter, D_n, is defined as:

$$D_n = \frac{a\sqrt{n^2 + nm + m^2}}{\pi} \tag{2}$$

The chiral angle, θ, is given by [12]:

$$\theta = \sin^{-1}\frac{\sqrt{3}m}{2\sqrt{n^2 + nm + m^2}} \tag{3}$$

Fig. 1 Schematic representation of the hexagonal lattice of the two-dimensional graphene sheet along with the definition of the chiral vector and description of SWCNTs formation

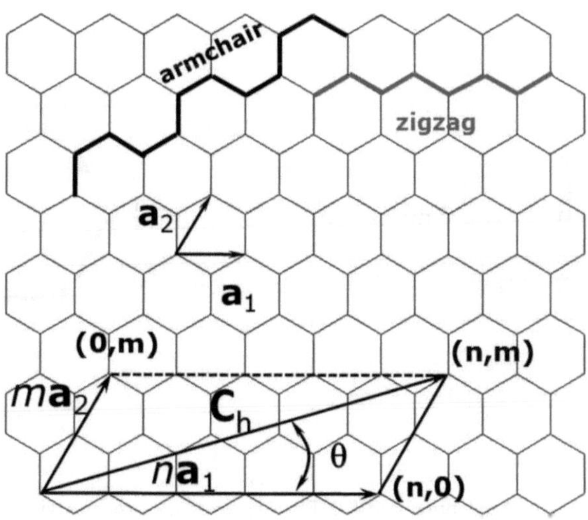

Three main symmetry groups of SWCNTs exist. When $n = m$, the structure (n, n) is called armchair configuration; when $m = 0$, the structure (n, 0) is called zigzag; when $n \neq m$, the structure (n, m) is called chiral. These three major categories of SWCNTs can also be defined based on the chiral angle, θ. For the two limiting chiral angles of 0° and 30°, the nanotubes are referred as armchair and zigzag, respectively. For θ different from 0° and 30°, the nanotubes designed as chiral.

3 Numerical Simulation and Analysis

3.1 Configurations of SWCNTs and FE Modelling

In the current work, the 3D FE model as proposed by Li and Chou [9] and developed by Tserpes and Papanikos [10] was adopted for assessing the mechanical properties of SWCNTs. The displacement of individual atoms of CNT under an external force is constrained by the C–C bonds. Since the C–C bonds are considered as connecting load-carrying elements, and the carbon atoms as joints of connecting elements, CNTs can be simulated as space-frame structures (Fig. 2). The modelling establishes the equivalences between the bond length, a_{C-C}, and the equivalent beam length, l, and between the nanotube wall thickness, t_n, and the beam element thickness. In fact, in the present study, the SWCNTs are modelled as space-frame structures and the wall thickness, t_n, should be identified in this context. Thus, assuming the cross-sectional area of the beam element to be circular, the wall thickness, t_n, corresponds to the beam element diameter, d. The meshes for SWCNT structures used in finite element analyses were constructed using the academic software CoNTub 1.0 [13], which permits building the CNT structures for

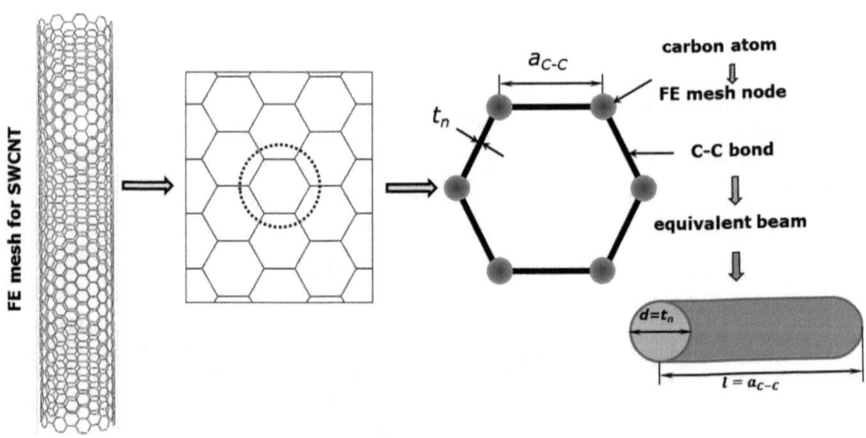

Fig. 2 Modelling of SWCNT as a space-frame structure with beam elements replacing C–C bonds

designing and investigation of new nanotube-based devices. This code is able to generate ASCII files, describing atom positions and the connectivity, which can be entered as input in available commercial or in-house FE codes, in order to perform the simulation of mechanical tests. In order to convert ASCII files, from the CoNTub 1.0 program, into the format usable by the commercial FE code ABAQUS®, the in-house application, *InterfaceNanotubes*, was developed. The geometrical characteristics of SWCNTs used for the present FE analyses are summarized in Table 1, where the number of nodes and elements of the finite element meshes of the SWCNTs is also presented, for the length of nanotube considered: 20 nm. This length corresponds to the minimum recommended value from which the mechanical properties are not influenced by the length of the SWCNT [14, 15]. The finite element meshes for three types of SWCNTs are shown in the Fig. 3. As it is known, vacancy defects result from missing carbon atoms in the CNT walls. The atoms in percentage of 0.5, 1.0, 2.0, 5.0 and 10.0 % were removed from all SWCNTs presented in the Table 2. The vacancy defects considered in the current study had the following configurations: single vacancies (one atom missing), double vacancies (two neighbouring atoms missing) and four vacancies together. By removing the carbon atoms from the nanotube, all of the bond interactions (beams) between atoms are removed. All of the vacancy defects were situated randomly in the nanotube.

As previously proposed [9], direct relationships can be established between the structural mechanics parameters, i.e. tensile and bending rigidities, E_bA_b, E_bI_b (where E_b, A_b and I_b are the Young's modulus, the cross-section area and the

Table 1 Geometrical characteristics of the SWCNTs studied

SWCNT type		(*n*, *m*)	D_n (nm)	θ°	Number of nodes[a]	Number of elements[a]
Non-chiral	Armchiar	(5, 5)	0.678	30	1620	2414
		(10, 10)	1.356		3240	4829
		(15, 15)	2.034		4860	7244
		(20, 20)	2.713		6840	9659
	Zigzag	(5, 0)	0.392	0	930	1384
		(10, 0)	0.783		1860	2769
		(15, 0)	1.175		2790	4154
		(20, 0)	1.566		3720	5539
Chiral	Family *θ* 8.9	(5, 1)	0.436	8.9	1044	1554
		(10, 2)	0.872		2088	3109
		(15, 3)	1.308		3132	4664
		(20, 4)	1.744		4176	6219
		(25, 5)	2.180		5220	7774
		(30, 6)	2.616		6264	9329

[a]The number of nodes and elements of SWCNTs finite element meshes correspond to nanotube length of 20 nm

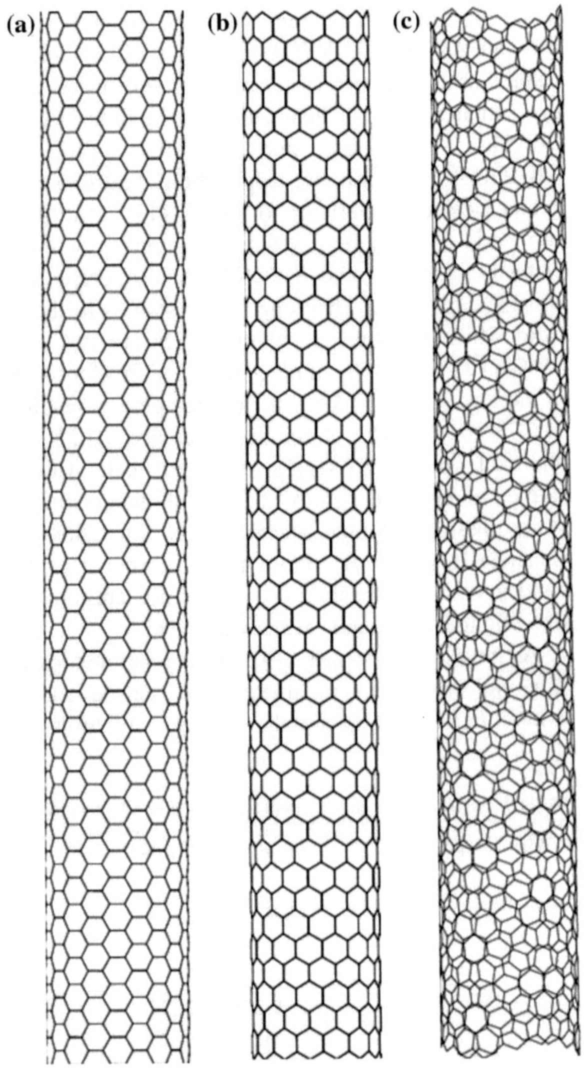

Fig. 3 FE meshes of SWCNTs: **a** (10, 10), **b** (15, 0) and **c** (15, 3)

moment of inertia of the beam, respectively), and the bond stretching and bond bending force constants, k_r, k_θ:

$$\frac{E_b A_b}{l} = k_r \tag{4}$$

$$\frac{E_b I_b}{l} = k_\theta, \tag{5}$$

Table 2 Input parameters for FE modelling of SWCNTs

Parameter	Value	Formulation
Force constant, k_r	6.52×10^{-7} N nm^{-1}	–
Force constant, k_θ	8.76×10^{-10} N nm rad^{-2}	–
C–C bond/beam length ($l = a_{C-C}$)	0.1421 nm	–
Diameter (d)	0.147 nm	$d = 4\sqrt{k_\theta/k_r}$
Cross section area, A_b	0.01688 nm^2	$A_b = \pi d^2/4$
Moment of inertia, I_b	2.269×10^{-5} nm^4	$I_b = \pi d^4/64$
Young's modulus, E_b	5488 GPa	$E_b = k_r^2 l/4\pi k_\theta$
Tensile rigidity, E_bA_b	92.65 nN	$E_bA_b = k_r l$
Bending rigidity, E_bI_b	0.1245 nN nm^2	$E_bI_b = k_\theta l$

where l is the bond length generally considered equal to 0.1421 nm. Consequently, Eqs. (4) and (5) establish the basis for application of continuum mechanics to the analysis of the mechanical behaviour of CNTs and provide the input for simulation of the CNTs as space-frame structures. The values of force constants and input data for the FE model are given in Table 2.

Numerical simulation of conventional mechanical tests of tension and bending was carried out in order to study the effect of the nanotube length and diameter of SWCNTs on their mechanical properties, focusing on the tensile and bending rigidities. The boundary and loading conditions for each case are presented in Fig. 4. The FE analysis was performed using commercial FE code ABAQUS®.

In order to simulate mechanical behaviour of a SWCNT in tension, an axial force, F_x, is applied at the one nanotube's end, leaving the other end fixed. The tensile rigidity of the nanotube, EA, is determined as:

$$EA = \frac{F_x L}{u_x} \tag{6}$$

where L is the nanotube length and u_x is an axial displacement taken from the FE analysis.

Likewise, for simulating bending, a transverse force, F_y, is applied at one of the nanotube's extremity, leaving the other fixed. The bending rigidity of the nanotube, EI, is determined as:

$$EI = \frac{F_y L^3}{3u_y} \tag{7}$$

where u_y is a transverse displacement taken from the FE analysis.

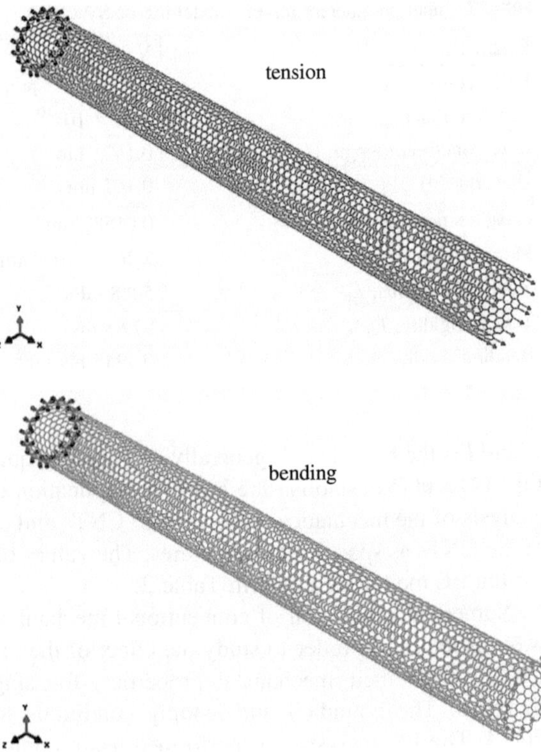

Fig. 4 Loading and boundary conditions for (10, 10) armchair SWCNT

3.2 Young's Modulus of SWCNTs

The nanotube rigidities, *EA* and *EI*, are required for the evaluation of the nanotube Young's modulus, *E*. Considering a hollow cylindrical profile for the equivalent beam, i.e. a geometry which is similar to the CNT, the cross-sectional area and its moment of inertia for the equivalent hollow cylinder can be written:

$$A = \frac{\pi}{4}\left[(D+t)^2 - (D-t)^2\right] = \pi D t \qquad (8)$$

$$I = \frac{\pi}{64}\left[(D+t)^4 - (D-t)^4\right] \qquad (9)$$

where D and t are the mean diameter and the thickness of the equivalent hollow cylinder, respectively.

Assuming $t = t_n$ (where t_n is the nanotube wall thickness), the value of D can be derived from Eqs. (8) and (9):

$$\frac{EI}{EA} = \frac{1}{8}\left(D^2 + t_n^2\right) \Rightarrow D = \sqrt{8\left(\frac{EI}{EA}\right) - t_n^2} \tag{10}$$

Thus, the Young's modulus of the equivalent beam can be calculated using the following expression taking into account the rigidities in tension and bending:

$$E = \frac{EA}{A} = \frac{EA}{\pi t_n \sqrt{8\left(\frac{EI}{EA}\right) - t_n^2}} \tag{11}$$

4 Results and Discussion

4.1 Rigidities of SWCNTs with Vacancy Defects

First of all, a study of the location along the length of the nanotube (except the ends of the nanotube) and configuration of vacancy defects on the tensile and bending rigidities of SWCNTs was carried out. It was found that the location of the vacancies does not have any effect on the rigidity results. Examples of studied configurations of double vacancies and two double vacancies side by side for armchair and zigzag nanotubes are shown in the Fig. 5. The results concerning the effect of the vacancy configuration on the tensile and bending rigidities of SWCNTs are shown in the Fig. 6 for single vacancies, double vacancies and two double vacancies side by side. Also, the vacancy defect configuration does not cause significant influence on the values of both rigidities. For example, the presence of 8 vacancies in different configurations, i.e. 8 single vacancies, 4 double ones, or two double vacancies side by side, in the (10, 10) SWCNT does not change considerably the nanotube rigidities.

Therefore, a study was carried out for SWCNTs with different percentage (0.5, 1.0, 2.0, 5.0 and 10.0 %) of vacancy defects, including single vacancies, double vacancies and two double vacancies side by side, simultaneously. Figure 7 shows the evolution of the tensile and bending rigidities of non-chiral and chiral SWCNTs with the percentage of the vacancies. Both rigidities decrease with increasing of the amount of the vacancies in SWCNT.

The values of the tensile, *EA*, and bending, *EI*, rigidities as a function of the SWCNT diameter, D_n, are plotted in Fig. 8a, b for ideal SWCNTs and for SWCNTs with different percentage of the vacancies. The values of both rigidities decrease with increasing of the percent of vacancies in the nanotube. It can be seen that the evolution of the tensile rigidity, *EA*, is unified for all ideal SWCNTs studied as well for chiral and non-chiral SWCNTs with the same content of the vacancy defects. This also applies for the evolution of the bending rigidity, *EI*. The FE results concerning the evolution of tensile and bending rigidities with nanotube diameter can be fitted by a quasi-linear trend for the case of tensile rigidity, *EA*, and close to a

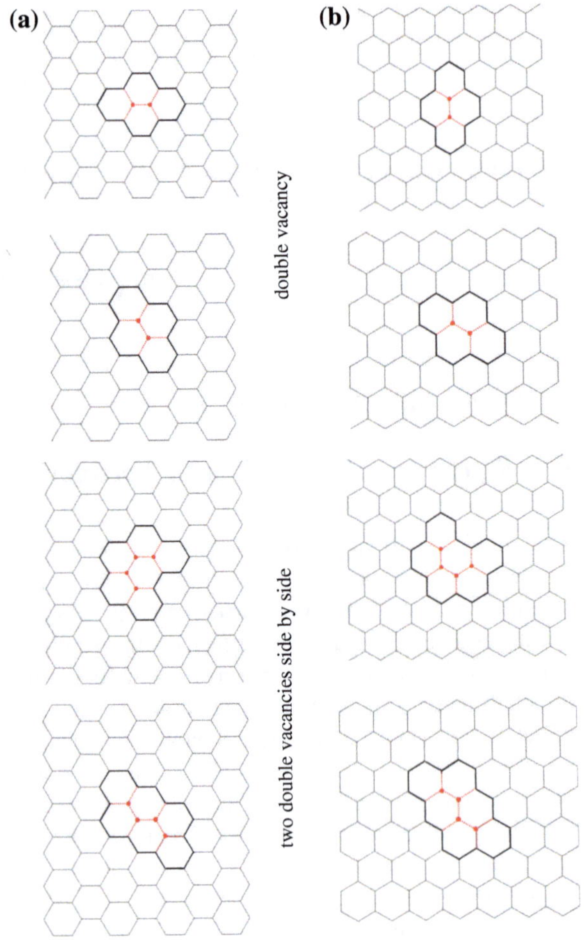

Fig. 5 Examples of configurations of the double vacancies and two double vacancies side by side used in the analysis: **a** armchair, **b** zigzag. The axis of the nanotube is horizontal

cubic power trend for the case of bending rigidity, EI, as shown in Fig. 9a, b. These dependencies can be simply described:

$$EA = \alpha(D_n - D_0) \tag{12}$$

$$EI = \beta(D_n - D_0)^3 \tag{13}$$

These equations are of the same type as the ones previously proposed [12], but replacing the SWCNT diameter, D_n, for the chiral index, n. The fitting parameters α, β and D_0 obtained in the current work for ideal and defect SWCNTs are given in Table 3.

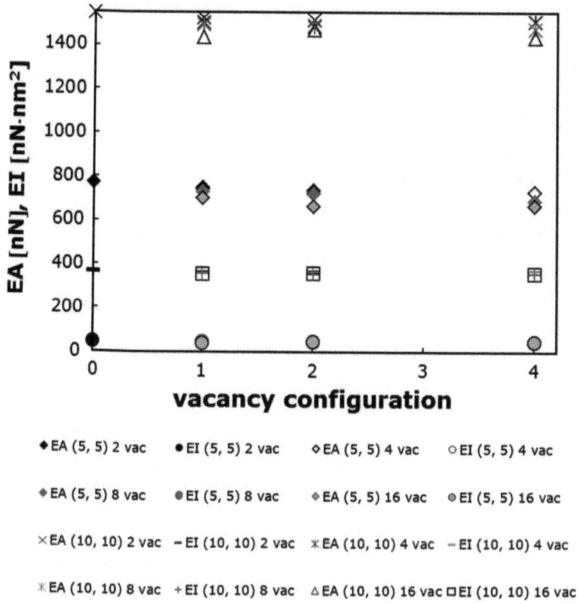

Fig. 6 Effect of the vacancy configuration on the tensile and bending rigidities of SWCNTs

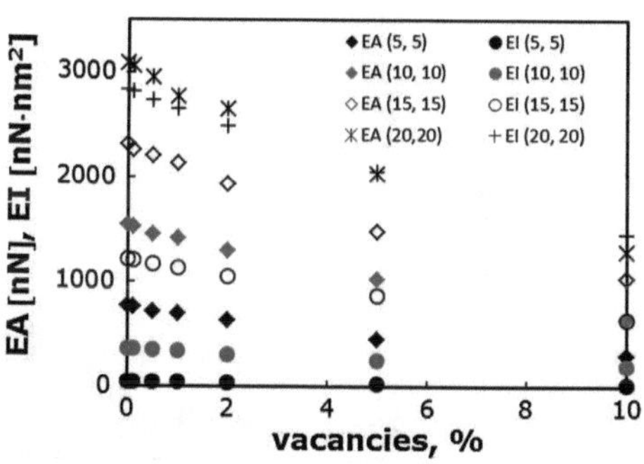

Fig. 7 Evolution of the tensile, *EA*, and bending, *EI*, rigidities with the percentage of vacancies in SWCNTs: example for armchair nanotubes

4.2 Young's Modulus of SWCNTs with Vacancy Defects

Equations (12) and (13) for the tensile and bending rigidities enable writing Eq. (11) as follows:

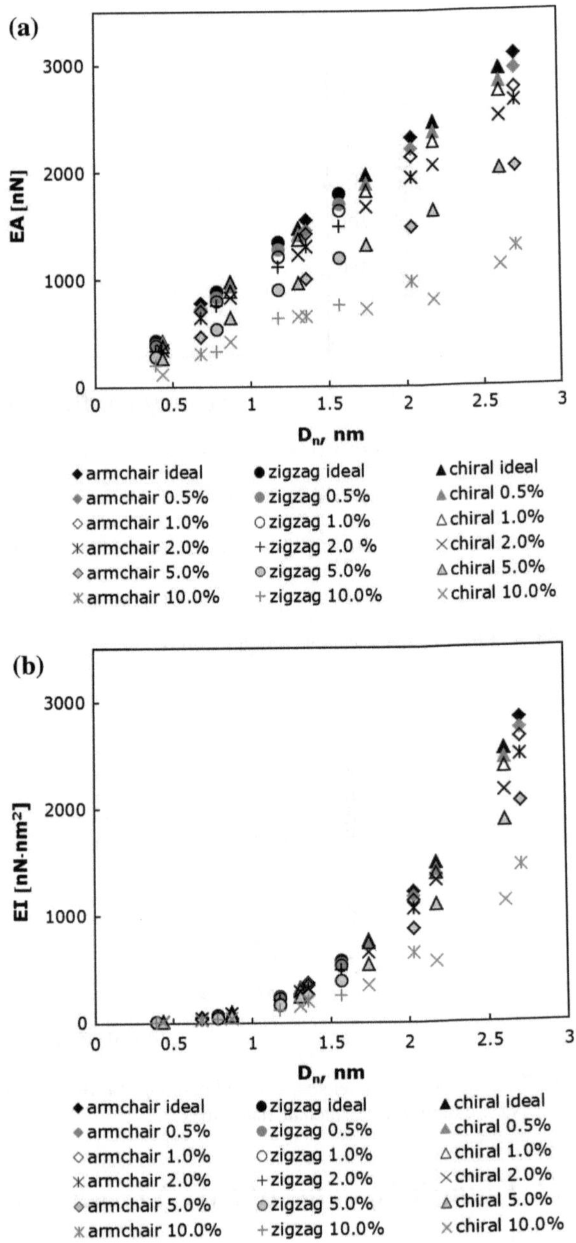

Fig. 8 Evolution of nanotube rigidities as a function of the nanotube diameter, D_n, for ideal armchair, zigzag and chiral $\theta = 8.9°$ SWCNTs and for SWCNTs containing different percentage of the vacancies: **a** the tensile rigidity, *EA*; and **b** bending rigidity, *EI*

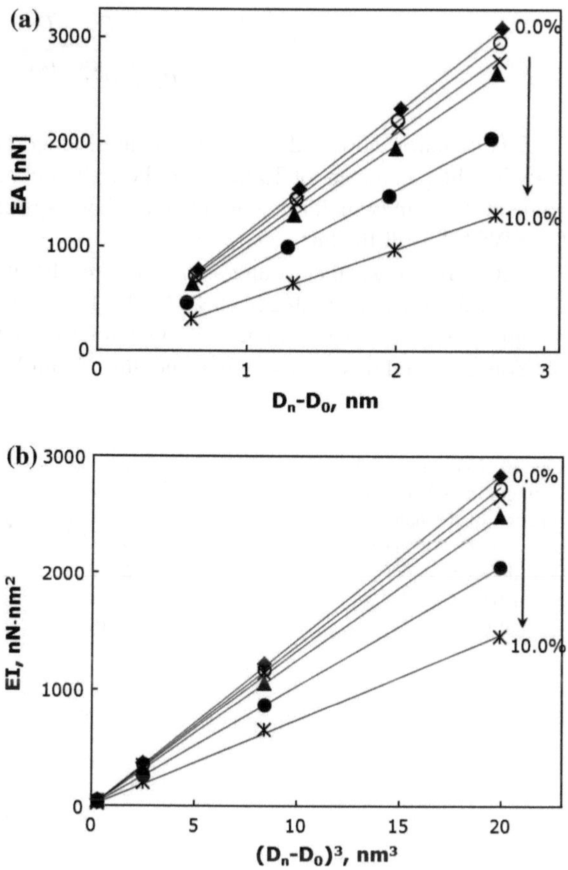

Fig. 9 Evolution of the rigidities for ideal and with defect SWCNTs: **a** tensile rigidity, *EA*, as a function of $(D_n - D_O)$; and **b** bending rigidity, *EI*, as a function of $(D_n - D_O)^3$

Table 3 Fitting parameters α, β and D_O

Vac (%)	0.0 (%)	0.1 (%)	0.5 (%)	1.0 (%)	2.0 (%)	5.0 (%)	10.0 (%)
Armchair							
α (nN/nm)	1139.8	1126	1097.7	1042.9	986.34	769.58	486.76
β (nN/nm)	142.77	142.3	141.19	136.17	130.8	113.2	77.593
D_O (nm)	0.003	0.006	0.024	0.019	0.040	0.081	0.046
Zigzag							
α (nN/nm)	1164.9	1153.4	1116.5	1070	977.93	797.85	505.47
β (nN/nm)	156.38	156.47	154.79	150.43	137.95	118.07	71.242
D_O (nm)	0.021	0.025	0.039	0.039	0.038	0.042	0.042
Chiral							
α (nN/nm)	1151.1	1142	1106.1	1076.9	980.7	793.81	418.27
β (nN/nm)	150.73	150.06	145.73	144.44	132.32	117.89	62.411
D_O (nm)	0.042	0.044	0.044	0.062	0.063	0.092	0.018

$$E = \frac{EA}{A} = \frac{\alpha(D_n - D_0)}{\pi t_n \sqrt{8\frac{\beta(D_n - D_0)^2}{\alpha} - t_n^2}} \quad (14)$$

This equation allows determining Young's modulus of any type of SWCNT, knowing the parameters of Table 3 and the wall thickness, t_n. The most widely used value of 0.34 nm (which is equal to the interlayer spacing of graphite) is adopted for the SWCNT wall thickness, t_n.

Figure 10 shows the Young's modulus results as a function of the nanotube diameter for ideal and with defect SWCNTs. For all SWCNTs studied, the Young's modulus values decrease with increasing nanotube diameter, but the rate of decrease of Young's modulus slows down and the Young's modulus tends to stabilize.

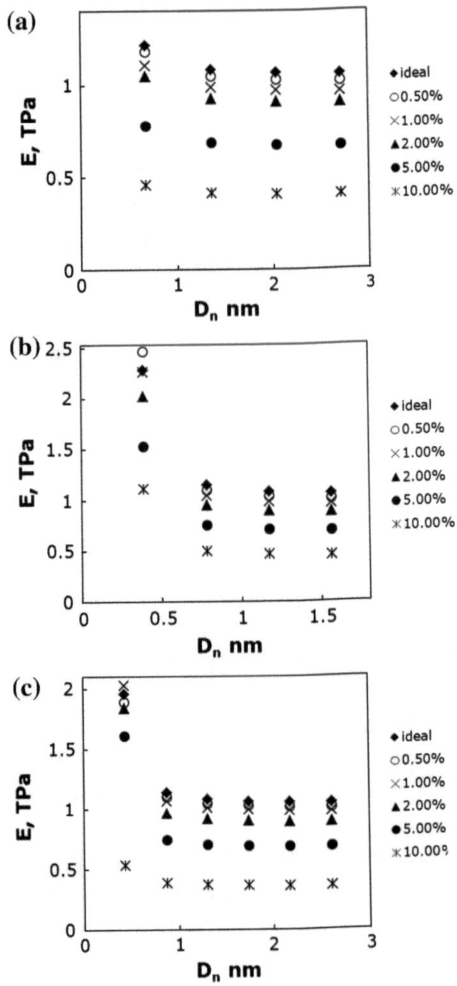

Fig. 10 Evolution of the Young's modulus as a function of the nanotube diameter, D_n for ideal and with defect SWCNTs: **a** armchair, **b** zigzag and **c** chiral θ = 8.9°

For example, in case of armchair nanotubes, the stabilized values (Fig. 10a) are about 1.080 TPa for ideal SWCNTs, 1.017, 0.965, 0.903, 0.650 and 0.405 TPa, for SWCNTs with 0.5, 1.0, 2.0, 5.0 and 10.0 % of vacancies, respectively. The Young's modulus results obtained for ideal SWCNTs are in good agreement with the results available in the literature. The stabilized value of Young's modulus for defect SWCNTs decreases with increasing of the content of the vacancies in the nanotube. In order to clarify this trend, the evolution of the Young's modulus with the percentage of the vacancy defects is shown in the Fig. 11 for all types of SWCNTs studied. Significant decreasing of the Young's modulus, of about 40 % of loss when compared with the value obtained for ideal SWCNT, is observed for the case of the

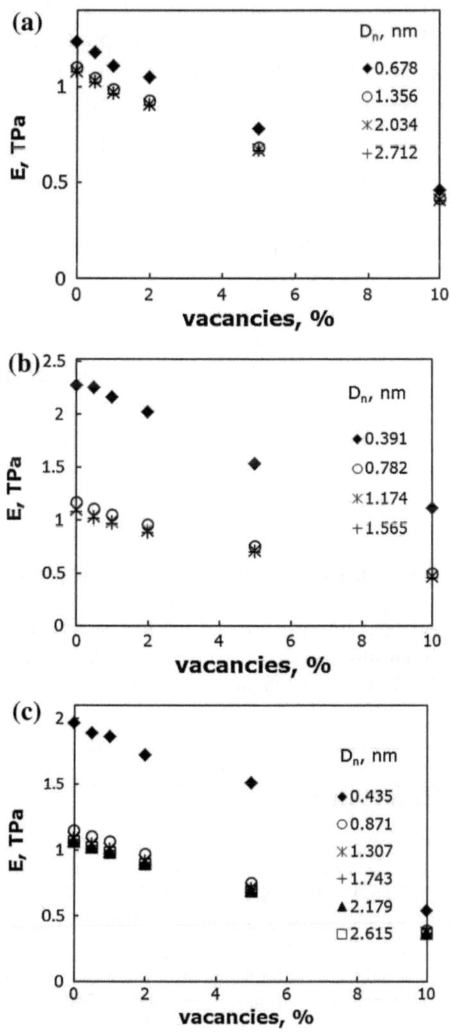

Fig. 11 The Young's modulus as a function of the percentage of the vacancy defects

nanotube containing 10.0 % of vacancies. The decrease in the value of Young' modulus due to the presence of vacancies was previously reported for few cases, such as zigzag (12, 0) and armchair (7, 7) SWCNTs containing single, double and triple vacancies [4]. The current systematic study allows improving the information concerning the influence of the presence and density of defects on the mechanical properties of SWCNTs, in order to better grasp the dispersion generally observed in experimental results.

5 Conclusions

The tensile and bending rigidities and the Young's modulus of non-chiral and chiral SWCNTs with defects were predicted using three-dimensional finite element method within the framework of nanoscale continuum modelling. The main conclusions can be drawn as follows:

- The tensile and bending rigidities of non-chiral and chiral SWCNTs do not seem to be influenced by the type of vacancy configuration (single, double or two double vacancies side by side);
- The tensile and bending rigidities of non-chiral and chiral SWCNTs are sensitive to vacancy defects and density: both rigidities decrease with the density of single vacancies in the SWCNT;
- Equations allowing direct relation between both rigidities and the nanotube diameter have been obtained for ideal and with defects SWCNTs (with different percentage of vacancy defects in SWCNTs); these relationships allow evaluating the SWCNT Young's modulus;
- The Young's modulus of SWCNTs with vacancies decreases with increasing the percentage of vacancies in nanotube.

Acknowledgments This research work is sponsored by national funds from the Portuguese Foundation for Science and Technology (FCT) via the projects PTDC/EME–TME/122472/2010 and PEst-C/EME/UI0285/2013 and by FEDER funds via "Programa Operacional Factores de Competitividade"—COMPETE, under the project CENTRO-07-0224_FEDER-002001 (MT4MOBI). All financial support is gratefully acknowledged.

References

1. Robertson J (2004) Realistic applications of CNTs. Mater Today 7:46–52
2. Hou W, Xiao S (2007) Mechanical behaviors of carbon nanotubes with randomly located vacancy defects. J Nanosci Nanotechnol 7:1–8
3. Scarpa F, Adhikari S, Wang CY (2009) Mechanical properties of non-reconstructed defective single-wall carbon nanotubes. J Phys D Appl Phys 42:142002
4. Parvaneh V, Shariati M (2011) Effect of defects and loading on prediction of Young's modulus of SWCNTs. Acta Mech 216:281–289

5. Parvaneh V, Shariati M, Torabi H (2012) Bending buckling behavior of perfect and defective single-walled carbon nanotubes via a structural mechanics model. Acta Mech 22:2369–2378
6. Rafiee R, Moghadam RM (2014) On the modeling of carbon nanotubes: a critical review. Compos Part B-Eng 56:435–449
7. Pantano A, Parks DM, Boyce MC (2004) Mechanics of deformation of single-and multi-wall carbon nanotubes. J Mech Phys Solids 52:789–821
8. Odegard GM, Gates TS, Nicholson LM et al (2002) Equivalent continuum modeling of nano-structured materials. Compos Sci Technol 62:1869–1880
9. Li C, Chou TW (2003) A structural mechanics approach for the analysis of carbon nanotubes. Int J Solids Struct 40:2487–2499
10. Tserpes KI, Papanikos P (2005) Finite element modeling of single-walled carbon nanotubes. Compos Part B-Eng 36:468–477
11. Lu X, Hu Z (2012) Mechanical property evaluation of single-walled carbon nanotubes by finite element modeling. Compos Part B-Eng 43:1902–1913
12. Dresselhaus MS, Dresselhaus G, Saito R (1995) Physics of carbon nanotubes. Carbon 33:883–891
13. Melchor S, Dobado JA (2004) CoNTub: an algorithm for connecting two arbitrary carbon nanotubes. J Chem Inf Comp Sci 44:1639–1646
14. Papanikos P, Nikolopoulos DD, Tserpes KI (2008) Equivalent beams for carbon nanotubes. Comput Mater Sci 43:345–352
15. Sakharova NA, Antunes JM, Oliveira MC et al (2014) Modelling of single-wall carbon nanotubes mechanical behaviour. In: E Oñate, J Oliver, A Huerta (eds) Proc (e-book) 11th World Congr Comput Mech (WCCM XI), Barcelona, Spain, 20–25 July 2014, pp 2568–2579. ISBN: 978-84-942844-7-2

Structure and Properties of Zn–Al–Cu Alloys with Alloying Additives

Krupińska Beata

Abstract In the present work the investigations concerning the influence of the modification with cerium on the microstructure and properties of the Zn–Al–Cu alloy was shown. The investigations were realized for Zn–Al–Cu samples with approximately 0.1 % Ce in the form of Al–Ce master alloys added. The heat treatment was performed in a resistance furnace with a chamotte-graphite crucible. For the remelting, with cerium as an alloying element, an argon protective atmosphere was used. The treated alloy was cast into the metal casting dies. To determine the influence of the modification on the structure and properties of those alloys, the following investigations were carried out: transmission and scanning microscopy, EDSX-ray analysis, and hardness analysis. The modification of the chemical composition, properly realized, leads to the improvement of mechanical properties of the produced castings. That is why, it is important to recognize the dependencies between the structure of the casts and chemical composition changes connected with the addition of modifiers. Investigations concerning both the optimal chemical composition and production method of Zn–Al–Cu alloys modified by chosen rare-earth metals as well as the improved properties in comparison with traditional alloys and methods contribute to better understanding the mechanisms influencing the improvement of mechanical properties of the new developed alloys.

Keywords Innovative cast materials · Metallography · Modification · Cast zinc alloys · Microstructure

K. Beata (✉)
Institute of Engineering Materials and Biomaterials, Silesian University of Technology, Konarskiego St. 18a, 44-100 Gliwice, Poland
e-mail: beata.krupinska@polsl.pl

A. Öchsner and H. Altenbach (eds.), *Mechanical and Materials Engineering of Modern Structure and Component Design*, Advanced Structured Materials 70,
DOI 10.1007/978-3-319-19443-1_27

1 Introduction

Cast zinc alloys are divided into three main groups: dual zinc-aluminum alloys, triple zinc-aluminum alloys with copper and dual zinc alloys with copper. Zinc alloys with aluminum contain up to 40 % Al [1, 2].

Over 20 % of the produced Zn is used for cast alloys and wrought alloys. The low melting temperature and good castability allows the casting of these alloys, production of products with a low wall thickness and low porosity, whereas the non-flammability can be helpful for applications in construction. The disadvantage thereof is, however, in particular in quick castings, the dimensional instability and mechanical characteristics due to the phase transitions in the solid state. Sheets and strips of zinc and its alloys are characterized by high anisotropy of mechanical properties and the functional properties [1–5].

Depending on the cast process conditions, there are prepared in the material in a variety of structures due to solidification of liquid metal. The conditions that have an impact on the microstructure, grain size, interdendritic spacing and thermal conditions. The relationship between the microstructure and the properties of the investigated cast in the works of [6], which stated the dependence of grain size on the plastic properties in accordance with Hall-Petch relationships. The experimental results confirmed the dependency of the mould on the heat transfer coefficient and distance of the dendrite arms, and therefore the Zn alloy properties.

Technical difficulties in the conventional cast process where there is a strengthening of the material associated with the segregation and the emergence of clusters, on the boundaries of interfacial reactions, increased porosity and a low incidence of interfacial coherence largely reduced the possibility of conventional methods and materials application for improving the preparation of functional properties. The use of inductive and mechanical blending methods influences the low segregation effect and creates very good mechanical properties, but there are difficulties involved in obtaining repeatability of the mentioned properties, the total elimination of the occurrence of blowholes and porosity, and to obtain a homogeneous structure largely contribute to a reduction in the application of conventional alloys production [7, 9].

In study [8], it was found that the addition of rare earth metals causes a reduction of the particle consolidation because it hinders the diffusion of carbon. This prevents the coalescence of particles to ensure homogeneity of the microstructure.

Modifiers can be added to the Zn alloy in form of Zn or Zn master alloys with Me or Al–Me [2], but the Al master alloy may cause formation of oxide phases in the melt which may interfere with the crystallization process of the alloys.

Properly performed chemical modification leads to improved functional properties of the produced castings. Therefore it is very important to know how the cast structure changes with the change in chemical composition by adding liquid metal modifiers.

2 Material and Investigation Methodology

For statement of the interdependence between the chemical composition and the structure of the Zn–Al–Cu zinc cast alloy modified Ce, see Table 1.

The casts were made in a resistance furnace with chamote-graphite crucibles. While performing a protective Argon atmosphere was used. The alloys were cast into metal moulds. The alloys were modified with a Al–Ce master alloy.

In order to determine the relationship between the crystallization kinetics of the alloy and the chemical composition and microstructure of cast zinc alloy Ce-modified the following investigations were performed:

- Thermo-derivative analysis using the UMSA device equipped with a computer-controlled cooling system, which allows to flexibly set the cooling rate applied to the Zn–Al–Cu alloys. The samples for thermo-derivative analysis were prepared with a diameter of Ø30 mm and a height of 35 mm. There have also been made holes for thermocouples, in the samples were the thermal node occurs for this type and arrangement of the sample geometry. The investigated samples were characterized with a mass of 160 ± 3 g for this type of geometrical dimension.
- Microstructure and chemical composition investigations using EDS microanalysis on the scanning electron microscope Zeiss Supra 25.
- Alloy structure analysis the using MEF4A optical microscope supplied by Leica together with the image analysis software as well as with an electron scanning microscope using the Zeiss Supra 25 device on high resolution mode.
- The examinations of thin film's microstructure and phase identification were made on the high resolution transmission electron microscope Titan 80-300 from FEI with the STEM scanning mode.
- Hardness of the modified Zn–Al–Cu alloys—using the Rockwell hardness tester supplied by Zwick ZHR 4150.
- Wear resistance investigations were performed using a reciprocating test "ball-on-plate" test. As the ball there was used a counterpart ball made from 100Cr6 bearing steel with a diameter of 6 mm. The surface of the samples before the test was polished with 1200 μm sandpaper. The average surface roughness of the samples before the test was 0.06 mm. Parameters of the wear resistance test are shown in Table 2.

Table 1 Chemical composition of the Zn alloys Zn–Al–Cu–Ce

Sample description	Elements as compounds of the modeled Zn casts
Cast H Zn–Al–Cu–Ce	Zn, Al (8.04 % mas. concentration), Cu (0.64 % concentration), Ce (up to 0.68 % mas. concentration)

Table 2 Parameter of the wear resistance test

Parameter	Value
Linear speed (cm/s)	4
Load (N)	10
Distances (m)	50
Length of the test stretch (cm)	4

3 Investigation Results

Modification of Zn–Al–Cu alloys with cerium causes morphological changes of the α' phase of the dendritic and globular nature of precipitates (Fig. 1) in the tweeded structure. As a result of the quantitative EDS microanalysis investigation performed results with an X-rays scattered energy EDS spectrometer, the presence of the major alloying elements Al, Cu alloy and the additive in form of Ce was confirmed (Figs. 2 and 3). The results of diffraction investigations using a high resolution transmission microscope confirm the presence of a phase Zn_2Ce (see Fig. 4).

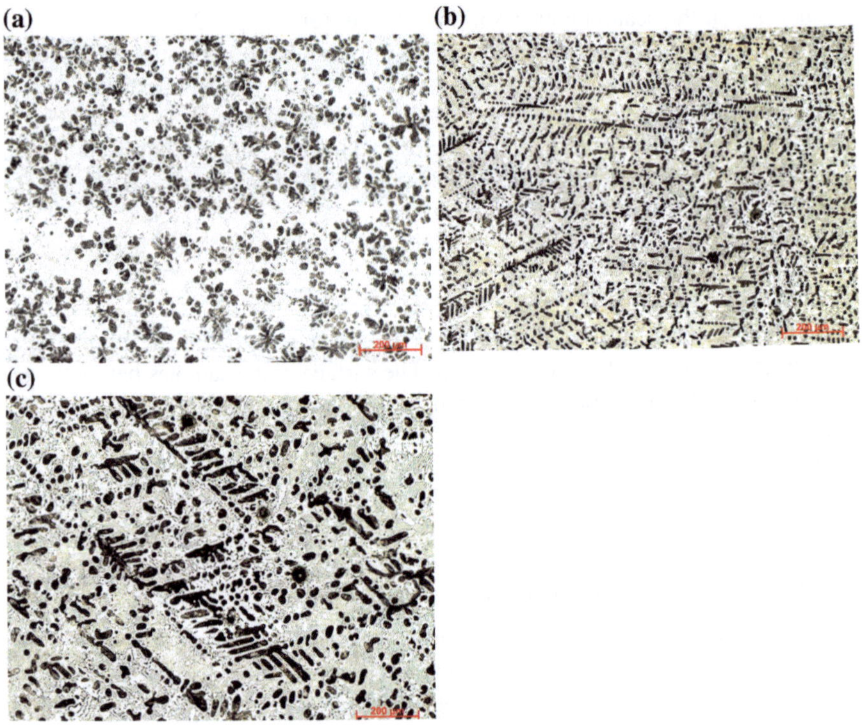

Fig. 1 Microstructure of the Zn–Al–Cu alloy (**a**, **b**) modified alloy with Al–Ce forced cooling (**c**) modified alloy with Al–Ce free cooling

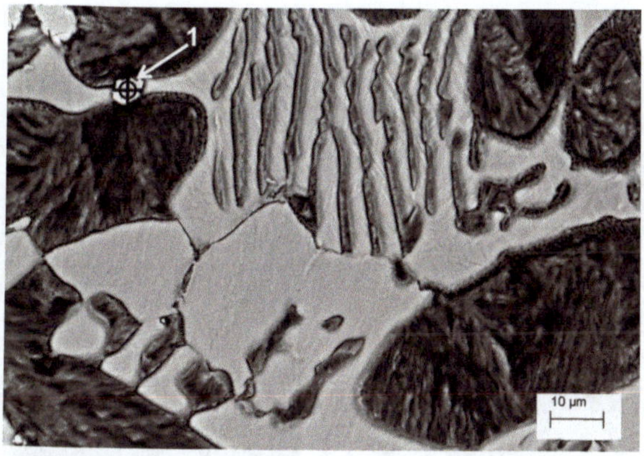

Fig. 2 Microstructure of the modified cast Zn–Al–Cu Ce

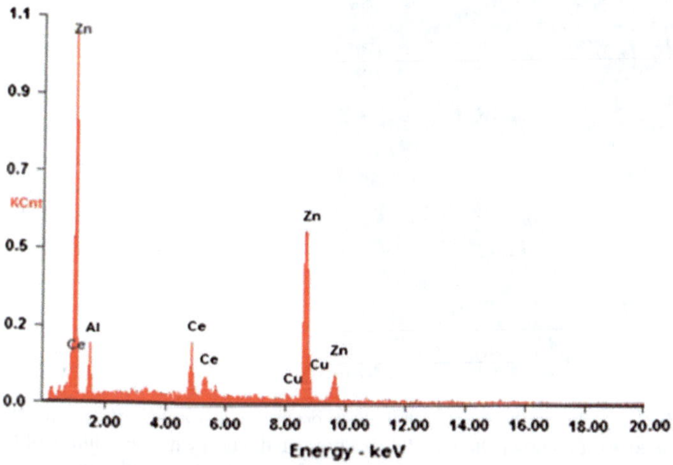

Fig. 3 The X-ray diffraction-graph of Zn–Al–Cu with Ce phases composition (point 1 on Fig. 2)

Based on the TEM investigation results presented in Figs. 4 and 5, an occurrence of the Zn_2Ce phase was confirmed in the investigated Zn alloy matrix. For the investigated zinc alloy a crystalline microstructure of the alloy after the controlled solidification process is visible too. In Fig. 5a the microstructure of the Zn alloy is shown using the bright field technique, which can help determine the grain size of the Zn, which can be determined in sizes of up to 1000 μm. A smaller crystalline structure could also have better mechanical properties. Also the Zn_3Ce phase was detected by the TEM diffraction technique (Fig. 6). Both the increase of the cooling rate and the Ce addition causes an increase of the hardness of the investigated alloys of about 11 % when compared to the non-modified alloys.

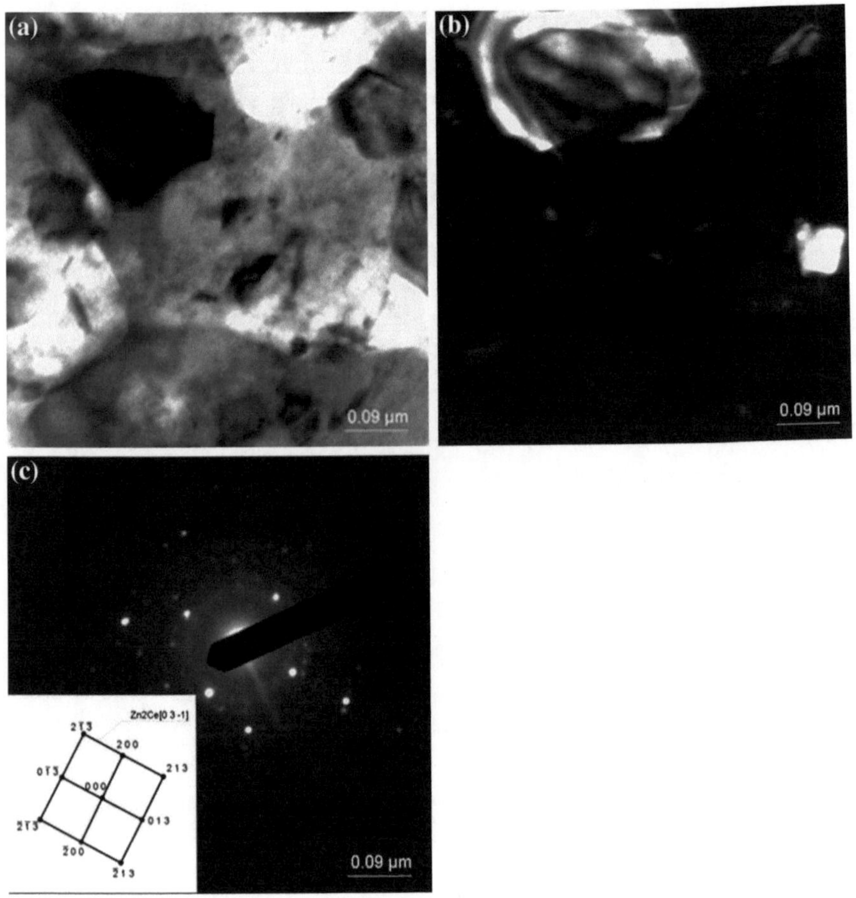

Fig. 4 **a** Zn–Ce grains in the investigated alloy, bright field, **b** Zn–Ce grains in the investigated alloy, dark field, **c** diffraction pattern of the area presented on (**a**) with the solution of the diffraction pattern confirming the Zn matrix phase as well as the Zn_2Ce phase with the zone axis [0 3-1]

Wear Resistance

As a result of the carried out wear resistance investigations using the "ball-on-plate" method there was observed an increase of about 20 % of the modified alloys when compared to the primary alloys comparing to the primary alloys Zn–Al–Cu (see Fig. 7).

4 Conclusions

Based on the investigations carried out, it was found out that with an increasing concentration of cerium during the thermal analysis performed at a constant cooling rate, the temperature will decrease. Also the nucleation of the α phase and

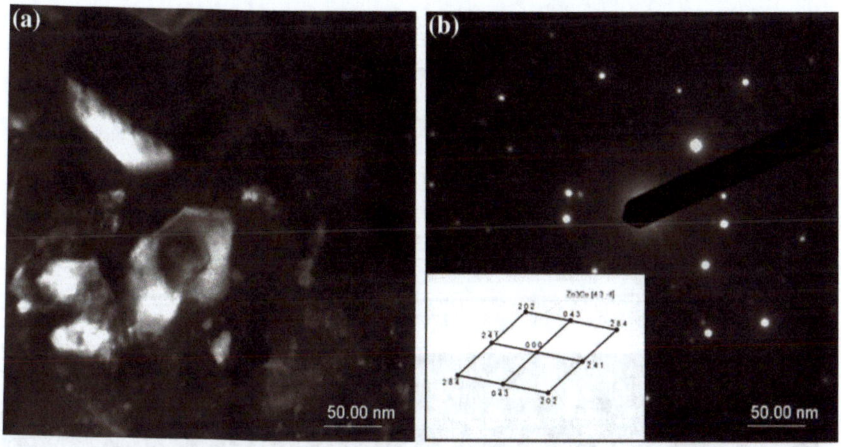

Fig. 5 **a** Zn–Ce grains in the investigated alloy, dark field, **b** diffraction pattern of the area presented on (**a**) with the solution of the diffraction pattern confirming the Zn matrix phase as well as the Zn_3Ce phase with the zone axis [4 3-4]

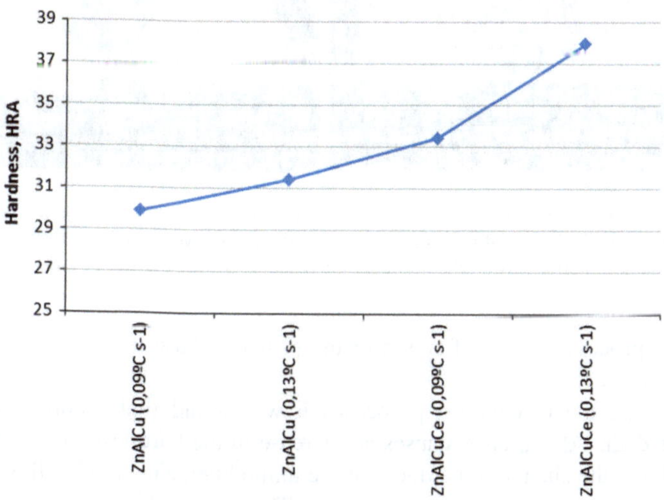

Fig. 6 Results of the hardness measurement of the base alloy Zn–Al–Cu, modified with Zn–Al–Cu–Ce

crystallization end temperature increased when compared to the alloy without any modification.

An increase in the Ce addition in the studied alloys causes both changes in the morphology of the precipitation phases and tweeded structure of the alloy. Also with an increase of the cooling rate there is a rise in the temperature of the a solid solution nucleation of the investigated alloys as a result of changes of the dynamic

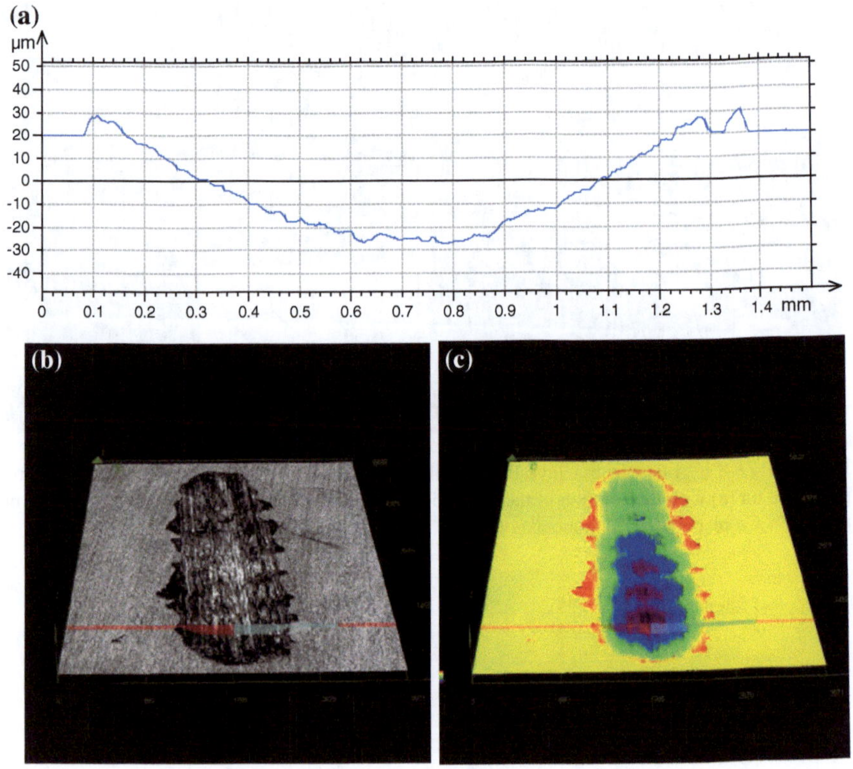

Fig. 7 Wear resistance investigations of the Zn–Al–Cu alloy with Ce using the ball on plate method, (**a**) cross-section profile with the wear path, (**b**, **c**) wear path on the surface of the Zn–Al–Cu sample performed on the confocal microscope OLIMPUS LEXT4000 OLS 4000

nucleation process; also the fraction of the primary phase ε participation includes Zn_2Ce changes.

Concerning the mechanical properties it was found that modification of the investigated Zn–Al–Cu alloy causes an increase in the hardness. In this paper it is presented that the change in hardness of the unmodified Zn–Al–Cu alloy as well as modified Zn–Al–Cu–Ce alloy is significant. The average hardness value based on the particular investigation results, is also characterized by a lower value of the standard deviation for the Zn–Al–Cu alloy modified with Ce, which means that the entire alloy has a similar harness value and that the modification is not area related.

Acknowledgments The research was funded by the Science Centre National No. 2640/B/T02/2011/40 headed by Dr. Beata Krupińska.

References

1. Skrzypek SJ, Przybyłowich K (2012) Engineering metals and their alloys. Publisher AGH, Cracow
2. Górny Z, Sobczak J (2005) Non-ferrous metals based novel materials in foundry practice. ZA-PIS, Cracow
3. Krupińska B, Labisz K, Rdzawski Z (2012) Light and electron microscope investigations of cast Zn-Al alloys. Arch Mater Sci Eng 55(1):29–36
4. Krupińska B, Labisz K, Rdzawski Z (2011) Crystallisation kinetics of the Zn-Al alloys modified with lanthanum and cerium. J Achievements Mater Manufact Eng 46(2):154–160
5. Osorio WR, Garcia A (2002) Modeling dendritic structure and mechanical properties of Zn-Al alloys as a function of solidification conditions. Mater Sci Eng A 325:103–111
6. Szajnar J (2009) The impact of selected physical factors on the crystallization process and structure of castings. Arch Foundry Eng 9–1 M
7. Chena TJ, Haoa Y, Sunb J, Lia YD (2003) Effects of Mg and RE additions on the semi-solid microstructure of a zinc alloy ZA27. Sci Technol Adv Mater 4:495–502
8. Ozgowicz W, Labisz K (2011) Analysis of the state of the fine-dispersive precipitations in the structure of high strength steel Weldox 1300 by means of electron diffraction. J Iron Steel Res Int 18(1):135–142

The Benefits of Using Tyre Rubber Aggregate in Concrete Specimens

Hadda Hadjab, Oussama Boulekfouf and Ahmed Arbia

Abstract The present experimental paper studies the mechanical and physical properties of concrete in which tyre rubber aggregate is incorporated. From this study, one can conclude that the presence of tyre rubber aggregates (RG) significantly reduces the velocity of ultrasonic waves in concrete added with RG. This composite has the potential to dampen vibration, which may be of particular use of sound insulation. Furthermore, the incorporation of RG significantly reduces the kinetics of ultrasonic pulses in concrete. This reduction is mainly attributed to the decrease in density the composite and the presence of water and air pores in the material. With such a damping capacity of waves in general and more specifically mechanical vibrations, it is worth considering the use of this composite in the applications such as noise walls of arterial roads, as the foundation of rails gear wheel or foundation for vibrating machines, etc. From this study, we conclude, that the benefit of using waste tyre rubber could be greatly appreciated to preserve environment and to discover other new concrete composites.

Keywords Concrete · Tyre rubber · Ultrasonic · Wave

1 Introduction

The ultrasound has been used for thirty years to assess the quality of in situ concrete and most importantly for the detection of defects and microcracks in the structure. This can be done by analyzing the wave velocity, which is an important factor to

H. Hadjab (✉) · O. Boulekfouf · A. Arbia
Faculty of Civil Engineering, University of Sciences and Technology
Houari Boumediene, BP 32, Bab Ezzouar 16111, Algiers, Algeria
e-mail: hhadjab@usthb.dz

O. Boulekfouf
e-mail: Oussama-tp@hotmail.fr

A. Arbia
e-mail: ahmedmoon@hotmail.fr

A. Öchsner and H. Altenbach (eds.), *Mechanical and Materials Engineering of Modern Structure and Component Design*, Advanced Structured Materials 70,
DOI 10.1007/978-3-319-19443-1_28

describe the quality of concrete, since it is reduced by the presence of microcracks. The propagation velocity of ultrasonic waves gives a fair idea about the quality of concrete. Concrete is considered very poor if its speed is low and it is excellent if it is large. This makes it possible to distinguish between different types of concrete only on the basis of the knowledge of the speed. This speed is influenced by knowledge of the material in the presence or absence of microcracks.

This research work aims to evaluate the influence of the addition of tyre rubber aggregates [1–8] on ultrasonic waves and thermal conductivity of concrete test specimens with different percentages of tyre rubber aggregates.

The use of tyre rubber aggregates (RG) can help us to solve the environmental problem which can be summarized as following:

- Store tires include an environmental risk because of the extent of pollution in case of fire. The tires are petroleum products; their combustion is very difficult to stop.
- The smoke from a burning tire contains dioxins, furans and polycyclic aromatic hydrocarbons.
- The proliferation of snakes and mosquitoes (tires stored outdoors without protection against the elements constitute a privileged proliferation of mosquito larvae and potential vectors of viruses such as Chikungunya).
- The impairment and social function of the land concerned (brake on development projects of common industrial area …).

2 Materials and Methods

2.1 Composition of Concrete

The composition or formulation of the concrete is the set of operations leading to determine the proportions of the various constituents of the desired concrete with certain characteristics (resistance, consistency …).

The determination of its proportions is approximate and is carried out according to the Dreux-Gorisse Method.

2.2 Formulation of Concrete by the Dreux-Gorisse Method

This design method allows to determine the optimal amounts of materials (E water, cement C, S sand, gravel and gravel G) needed to make a cubic meter of concrete based on several criteria such as handling, consistency, strength, size and type of aggregate and the cement and water.

Table 1 Composition for 1 m^3 of concrete

Constituents	Masses (kg)
Sand	644
Gravel 3/8	206.7
Gravel 8/15	1004.6
Cement	400
Water	221

2.3 The Addition of Adjuvant

Studies conducted by Cuong [4] have identified the difficulties of formulating a composition based cement incorporating rubber granules, and this is the case also for the work where the support was a mortar Bonnet [2] or the even most recent of which covered a high performance concrete (HPC) Garros [7]. The incorporation of rubber aggregates is particularly damaging vis-à-vis the handling and they are very sensitive to segregation since they have a tendency to rise to the top surface of the concrete volume Cuong [4]. To correct this problem Cuong used a superplasticizer (SIKA VISCOCRETE 3030).

In our case we used a superplasticizer called SIKA VISCOCRETE 12 TEMPO, an adjuvant which enhances stability, and limits the segregation of concrete, making the formulas less susceptible to variations in water and constituents.

Sika VISCOCRETE TEMPO 12 is a superplasticizer/High Water reducer, versatile concrete ready, according to NF EN 934-2. Its density is 1.06 ± 0.0 and pH range is between 4.5 and 6.5.

Implementation: the TEMPO SIKA VISCOCRETE 12 is added, either simultaneously with the mixing water, or delayed in the concrete previously dampened with a fraction of the mixing water.

The characteristic of the used concrete are summarized in Tables 1, 2, 3, 4. The compression set up is represented in Photograph 1. The obtained compression stress related to tyre rubber aggregates percent in Table 5. The influence of TRA on compression stress are represented in Fig. 1

3 Method of Measuring Ultrasonic Speeds

3.1 Principle of Ultrasonic Testing

The principle of ultrasound is that an electrical signal is converted to a strain wave (see Photograph 2). By a piezoelectric transducer the wave propagates in the specimen of concrete or mortar and is captured by the receiving transducer. The propagation time and the speed of ultrasonic waves are then deducted (Tables 6 and 7) and the change in ultrasonic velocity wave by Fig.2.

Table 2 Composition of C25RG

Constituents	Masses (kg)
Sand	644
Gravel 3/8	155.02
Gravel 8/15	1004.6
RG	51.67
Cement	400
Water	188
Adjuvant	4

Table 3 Composition of C50RG

Constituents	Masses (kg)
Sand	644
Gravel 3/8	103.35
Gravel 8/15	1004.6
RG	103.35
Cement	400
Water	188
Adjuvant	4

Table 4 Composition of C100RG

Constituents	Masses (kg)
Sand	644
Gravel 3/8	0
Gravel 8/15	1004.6
RG	206.7
Cement	400
Water	188
Adjuvant	4

Photograph 1 Compression test set up

Table 5 Compression stress values versus tyre rubber aggregates percentage

Composition of concrete	BT	B25RG	B50RG	B100RG
Compressive strength (MPa)	26.6	24.8	17.16	15.07

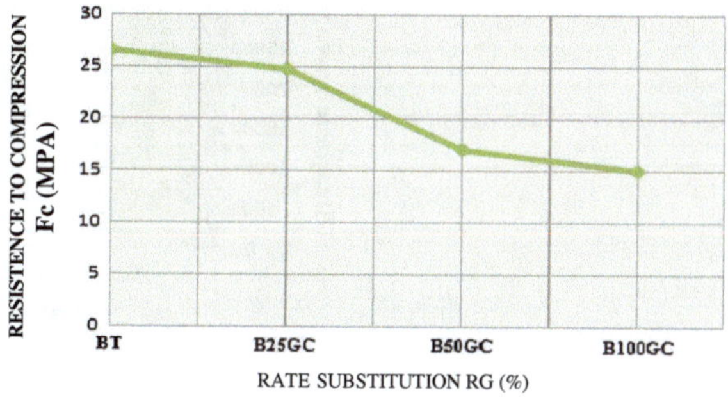

Fig. 1 Tyre rubber aggregates influence on compression stress

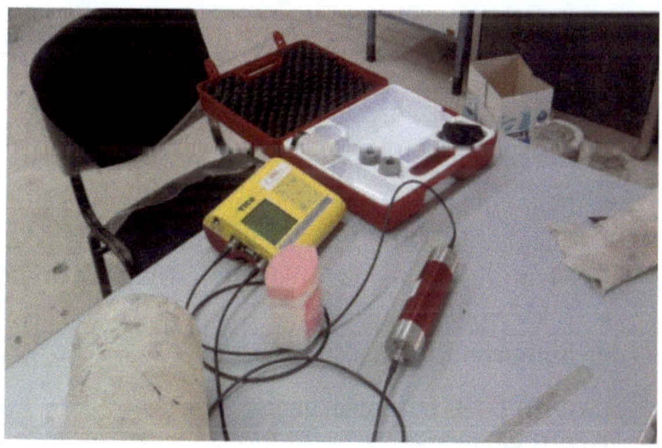

Photograph 2 Ultrasonic testing test

Table 6 Values of ultrasonic wave velocities in the rate of incorporation of RG

Composition of concrete	BT	B25RG	B50RG	B100RG
"P" primary wave velocity (m/s)	4570	4340	4130	3037

Table 7 Modulus of elasticity versus dosage in RG

Composition of concrete	BT	B25GC	B50GC	B100GC
Modulus of elasticity (GPa)	37.26	32.77	29.42	15.22

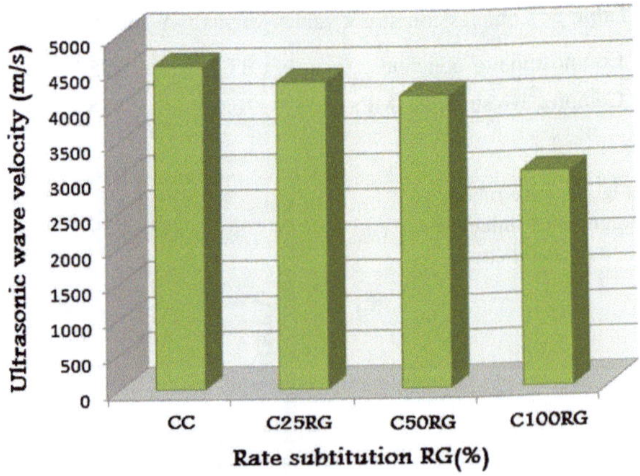

Fig. 2 Change in velocity of ultrasonic waves in the rate of incorporation of RG

3.2 Materials Used

For the measurement of ultrasonic velocities, the following equipment is used (Photograph 2):

1. Unit display: the display unit has the following characteristics:
   - Non-volatile memory with a capacity of 250 measurements.
   - Graphic LCD of 128 × 128 dots
   - Measuring range approximately from 15 to 6550 μs
   - 1 kV voltage pulse
   - Temperature range −10 to +60 °C
2. Two (2) 54 kHz transducers
3. Two BNC cables
4. Calibration bar
5. Box of 150 ml of coupling agent (or petroleum jelly).

4 Procedure

4.1 Measurement Process

1. Measure the distance between the measuring points (this is the length of the specimen).
2. Apply a layer of the coupling agent on the contact surface of the transducers and test pieces (thin layer to smooth concrete, for a thick concrete rough).
3. Pressing the START button to begin measuring.

4.2 Method of Calculation

Because of the heterogeneity of the material, the interpretation of the ultrasonic signals is not easy. Generally, tests are often based on the measurement of velocities of longitudinal ultrasonic waves. The speed of the wave through an elastic solid is given by the following expression

$$V = \sqrt{\frac{E(1-\upsilon)}{\rho(1+\upsilon)(1-2\upsilon)}} \quad (1)$$

V wave velocity measured in km/s.
E dynamic modulus of elasticity in GPa.
υ Poisson's ratio.

4.3 Results and Interpretation

The results of P wave velocities found are summarized in Table 8.

It was found that the propagation velocity of ultrasonic waves decreases by increasing the rate of incorporation of RG. The results are illustrated in Fig. 2

We also note that the modulus of elasticity decreases by increasing the percentage of RG, a trend logically expected given the results previously presented on the compressive strength. The results obtained are listed in Table 6 and illustrated in Fig. 2.

The addition of RG induced a significant decrease in the modulus of about 59 % for concrete B100RG, 21 % for concrete B50RG and about 12 % for concrete B25RG compared to the reference concrete (Fig. 2; Tables 6, 7 and 8).

Numerous studies confirm that the incorporation of RG induced a significant drop in the modulus of elasticity (Fig. 3) and Table 7.

The adverse effect of the incorporation of RG on the modulus of elasticity was introduced by Güneyisi [8] and the results illustrated show a decrease of 83 % of the modulus of elasticity when half of the volume is replaced by granular rubber aggregates (Figs. 4, 5 and 6).

In a recent study by Ganjian [6], aggregates rubber whose maximum size is 25 mm replace 5, 7.5 and 10 wt% of natural coarse aggregate. This study shows a 30 % reduction of the modulus of elasticity of concrete when passing reference to

Table 8 Thermic conductivity for different concrete composition

Composition of concrete	BT	B25GC	B50GC	B100GC
Thermal conductivity λ (W/m K)	1.63	1.44	1.18	1.08

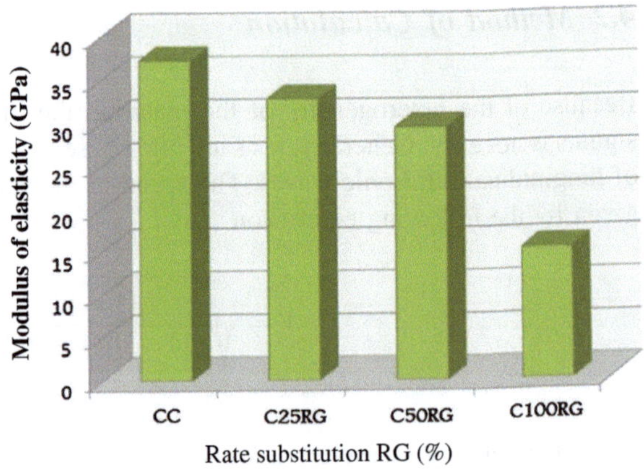

Fig. 3 Effect of dosage of RG on the modulus of elasticity

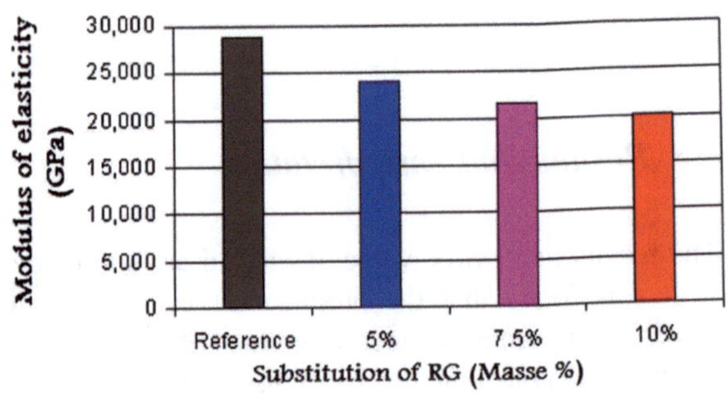

Fig. 4 Influence of RG on elasticity modulus

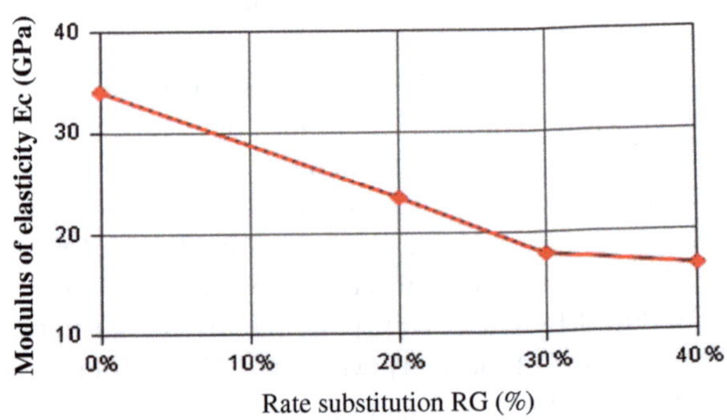

Fig. 5 Influence of dosage of RG on the elasticity modulus 28 days, Cuong

the concrete incorporating 10 % RG (see Fig. 7). The study by Cuong [4] showed that the addition of RG induced a fall in the modulus of elasticity of about 51 % substitution rate of 40 % compared to the reference concrete (see Fig. 6)

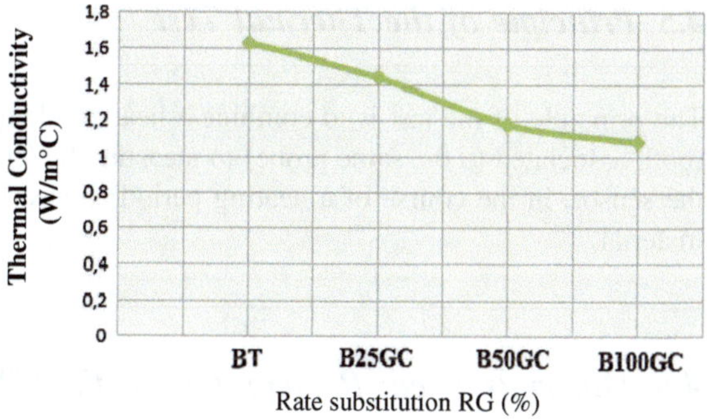

Fig. 6 Effect of adding the RG on the thermal conductivity of different compositions of concrete

4.4 Thermal Conductivity "λ"

Thermal conductivity is one of the thermo-physical parameter which is most important for the description of heat transport properties of materials, since it has a key to the process of heat transfer and is mainly used for the prediction of the temperature profile and heat flow (Photograph 3 and Table 3).

Given the extensive use of concrete, the thermal conductivity is a very important property, especially in the case of large area applications (cement concrete roads for example), this property can affect the temperature gradient between the free surface exposed to solar radiation and the lower surface in contact with the ground. Non-uniform thermal expansion together with a curvature of the member (upper surface tension) can trigger premature cracking as damaging as the dimensional variations due to withdrawal [9].

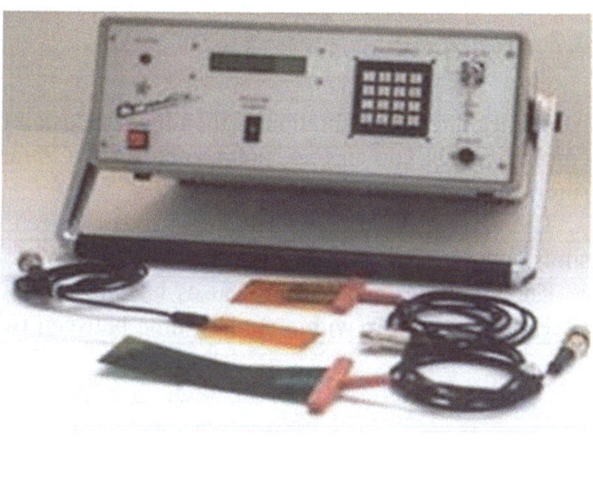

Photograph 3 The CT meter

4.5 Principle of the Thermal Test

The principle of the test is to combine a heating element and a temperature sensor (both associated to the same probe) to measure the temperature rise experienced by the sensor, in the course of a heating period chosen by the user depending on the material.

4.6 Materials Used: Presentation of the CT Meter

A *CT* meter is a device easily transportable developed to allow an accurate assessment of the thermal characteristics of a number of materials such as brick, rocks, soil, concrete, asphalt, substances powders, liquids, resins and complex products.

A *CT* meter consists of two elements:

- The control: responsible for generating the heat output and interpret the curve of temperature rise induced in the test material.
- Probe: load transmitting the heating power and temperature induced collect.

Both types of probes are available:

- The sensor ring: for insertion between two planes of the sample pieces of circular section.
- The probe wire (or single rod) designed to be inserted between two pieces of the sample shots of rectangular section.

4.7 Procedure

After checking the calibration of the *CT* meter using other materials whose thermal conductivity is known (e.g. polystyrene), the *CT* meter is adjusted and the measurement time set to 120 s.

The placing the probe wire (single rod) between the two pieces of the sample shots and perform the measurement while waiting for temperature stabilization.

Knowing that the measure causes heating of the material was measured, it is necessary to comply with a cooling time between two measurements.

5 Calculation Method

The *CT* meter directly displays the value of the thermal conductivity λ with an accuracy determined by the manufacturer (±5 %).

Thermal conductivity is one of the thermo-physical parameters which is most important for the description of heat transport properties of materials. It has a key to the process of heat transfer and is mainly used for the prediction of the temperature profile and heat flow.

Given the extensive use of concrete, the thermal conductivity is a very important property, especially in the case of large area applications (cement concrete roads, for example), this property can affect the temperature gradient between the free surface exposed to solar radiation and the lower surface in contact with the ground. Non-uniform thermal expansion together with a curvature of the member (upper surface tension) can trigger premature cracking as damaging as the dimensional variations due to withdrawal [9].

6 Conclusion

One can justify the fall in the modulus of rigidity by the low RG. Indeed, according to Hobbs [10] the modulus of elasticity of concrete depends on the modulus of elasticity of the aggregates and their volume fraction in within the material.

The second reason is the inherent defects of adhesion between the rubber and the cement matrix or the poor quality of the transition zone between these two phases.

The last reason is related to the porosity of concrete incorporating rubber aggregates that is deemed higher than the reference concrete.

From what has been said before, we find that the incorporation of RG is very damaging in the evaluation of the Young's modulus. The latter decreases with an increase in the percentage of RG. This trend means that the concrete incorporating RG is a material that attenuates the ultrasonic velocities, meaning that it has potential in terms of vibration damping capacity. In other words, the rubber granules reduce the kinetics of ultrasonic pulses in the material. This reduction is due to the decreasing density of RG, with respect to gravel.

We can say that the thermal conductivity test tells us about the impact of the incorporation of RG in concrete, the higher the RG increases the more thermal conductivity decreases. This effect can be justified by the amorphous structure of RG with a thermal conductivity of 0.4 W/m °C. Another factor that comes into play is the porosity of the concrete, which is greater with the addition of RG air with a much lower thermal conductivity (0.023 W/m °C) and fills the pores of the concrete and limits the conduction of heat.

Therefore the heat transfer is directly related to the components, to the morphology of the medium (solid matrix and porous network).

References

1. Abdalouahab F, Djidjeli Z (2009) Valorisation des pneumatiques usagées dans la protection de l'environnement. 19ème Congrès Français de mécanique
2. Bonnet S (2004) Matériaux cimentaires à haute déformabilité par incorporation de granulats issus du broyage de pneus usagés. PhD thesis, Université Paul Sabatier, Toulouse. France
3. Chettah A (2008) Comportement vibroacoustique des structures élaborées à partir de poudrettes de pneus recyclés, Ph.D. thesis, Université de Reims, France
4. Cuong HA (2010) Optimisation de la composition et caractérisation d'un béton incorporant des granulats issus du broyage de pneus usagés: application aux éléments de grande surface. Ph.D. thesis, Université de Toulouse, France
5. Eldin NN and Senouci AB (1993) Observations on rubberized concrete behavior. J Cem Concr Aggr
6. Ganjian E, Khorami E, Maghsoudi A (2009) Scrap-tyre-rubber replacement for aggregate and filler in concrete. Constr Build Mat 23(5):1828–1836
7. Garros M (2007) Thèse de doctorat : composites cimentaires incorporant des granulats caoutchouc issus du broyage de pneus usagés: optimisation de la composition et caractérisation. Ph.D. thesis, Université Paul Sabatier de Toulouse, France
8. Güneyisi E, Gesoğlu M, Özturan T (2004) Properties of rubberized concretes containing silica fume. Cem Concr Res 34(12):2309–2317
9. Hadjab H (2007) Zone de processus de rupture dans les matériaux quasi-fragile: étude expérimentale et modélisation numérique. Ph.D. thesis, USTHB, Algeria
10. Hobbs DW (1971) The dependence of the bulk modulus, Young's modulus, creep, shrinkage and thermal expansion of concrete upon aggregate volume concentration

The Use of PSC Technique to Estimate the Damage Extension During Three Point Bending Test

Charalampos Stergiopoulos, Ilias Stavrakas, Dimos Triantis, George Hloupis and Filippos Vallianatos

Abstract It is already known that when a mechanical loading is applied to cement-based specimens weak electrical currents are generated. Their existence is attributed to the creation of cracks and the eventual evolution of the cracks' network in the bulk of the specimen. This work introduces the simultaneous recording of electrical signal emissions at both the tension and the compression region of cement mortar beams of rectangular cross-section that were subjected to mechanical loading using the Three-Point-Bending technique. During the experiments the behavior of the electrical signal was studied during four sequential load stages: (1) Abrupt load increase up to the vicinity of 3 PB strength, (2) maintaining the high load level for relatively long time, (3) abrupt load decrease to a low load level (i.e. 50 % of the 3 PB strength approximately), (4) maintaining the low load level for relatively long time. The electrical signal analysis was conducted using non extensive statistical physics (NESP) and specifically the Tsallis entropy model studying the values of its q-parameter. The aim of this work was to study the electrical signal relaxation

C. Stergiopoulos · I. Stavrakas (✉) · D. Triantis · G. Hloupis
Laboratory of Electrical Characterization of Materials and Electronic Devices, Department of Electronics Engineering, Technological Educational Institute of Athens, Agiou Spiridonos, 12210 Athens, Greece
e-mail: ilias@ee.teiath.gr

C. Stergiopoulos
e-mail: cstergio@gmail.com

D. Triantis
e-mail: triantis@teiath.gr

G. Hloupis
e-mail: hloupis@teiath.gr

C. Stergiopoulos
School of Engineering and Design, Brunel University, West London, Middlesex, UK

F. Vallianatos
Technological Educational Institute of Crete, Chania, Greece
e-mail: fvallian@chania.teicrete.gr

A. Öchsner and H. Altenbach (eds.), *Mechanical and Materials Engineering of Modern Structure and Component Design*, Advanced Structured Materials 70,
DOI 10.1007/978-3-319-19443-1_29

process that follows the change of the mechanical load and the law that describes this relaxation with respect to the mechanical status of the specimen using statistical physics analysis.

Keywords Failure diagnostics · Non-destructive testing · Cement based materials · Electrical current emissions · Non-extensive statistical modelling · Tsallis entropy

1 Introduction

The wide use of cement-based construction materials has created the need for developing techniques able to evaluate their mechanical strength, health status and condition. Several Non-Destructive Techniques (NDT) have been applied using in situ and laboratory measurements regarding mechanical status monitoring and damage assessment [1, 2]. A novel experimental method known as Pressure Stimulated Currents (PSC) technique has been introduced that monitors the fracture process and contributes to the health status evaluation of quasi-brittle materials. This technique is based on the detection of electrical signals generated during the formation and growth of microcracks [3]. PSC technique has already been applied to marble [4–6], amphibolite [7] and cement based specimens [8–10].

It is already known that transient electric phenomena appear when a material is subjected to mechanical loading [3, 11–13]. The generation of weak electrical currents termed as Pressure Stimulated Currents (PSC) [3] can be detected and studied in terms of their behavior and characteristics. The motion of charged dislocations (MCD) [13, 14] is one of the basic mechanisms that can best describe such phenomena of electrical signal generation and detection. The first aim of this work was to perform a systematic study in order to examine in detail the recording of PSCs during three-point-bending (3 PB) tests and conclude to the PSC behavior introduced to previous works [8].

2 Non-extensive Statistical Physics (NESP) Approach

The microfracture mechanism is based on disorder and long range interactions. The generalization of the Boltzmann-Gibbs (BG) statistical physics is a proper theoretical frame for the analysis of signals produced by this mechanism as PSCs are. This generalization is known as non-extensive statistical physics (NESP) [15–17].

As PSCs are produced by the microfracture mechanism that is based on disorder and long range interactions, a generalization of the Boltzmann-Gibbs (BG) statistical physics is a proper theoretical frame for the analysis of parameters related to the evolution of cracks as PSC are. This generalization is known as non-extensive statistical physics (NESP). The non-additive entropy S_q that NESP introduces is not proportional to the number of the system's elements, as in the Boltzmann-Gibbs entropy S_{BG}. The Tsallis entropy S_q reads as:

$$S_{\mathrm{q}} = k_{\mathrm{B}} \frac{1 - \sum_{i=1}^{W} p_i^q}{q - 1}, \quad \mathrm{q} \in R, \tag{1}$$

where k_B is Boltzmann's constant, p_i is a set of probabilities, W is the total number of microscopic configurations, and q the entropic index. This index is a measure of the non-additivity of the system, and for the particular case $q = 1$, the Boltzmann-Gibbs entropy S_{BG} is obtained;

$$S_{BG} = -k_B \sum_{i=1}^{W} p_i \ln p_i \tag{2}$$

We note that for q = 1, we obtain the well-known exponential distribution [14]. The cases $q > 1$ and $q < 1$ correspond to sub-additivity and super-additivity, respectively.

In previous works [8] following a stepwise process of abruptly increased bending load increased the PSC signal and presented a peak during the loading change. Moreover while the loading remained constant after such a change, the PSC signal presented a restoration process trying to reach its original low background. There was an attempt to describe the PSC restoration process using an empirical equation described by two exponential decays. In this work, we focused on trying to describe the PSC relaxation phase using a uniform way. This phase is discussed under the frame of statistical physics analysis and specifically the Tsallis entropy model [18–22].

3 Materials and Experimental Setup

In order to produce identical specimens for conducting the experiments a cement mortar mixture that contained Ordinary Portland Cement (OPC) was used. The OPC was mixed with sand of fine aggregates and water and its weight ratio was 1:3:0.5 respectively. A refinement process was performed to the aggregates and the size of the ones that were selected ranged from 0.6 up to 3.0 mm approximately. Low speed was maintained during the mixing process for achieving the best moisturizing of the cement powder. The mixture stayed in wooden molds for 24 h and the porosity of the mixture was estimated to be 8 % approximately. The beams were 200 mm long with a square cross-section of 50 mm × 50 mm. After their extraction from the molds, the beams were stored in a special chamber that offered constant ambient conditions (22 °C and 75–80 % humidity) for their proper curing. The specimens remained in the chamber for 90 days in order to obtain 95 % of their total strength [23].

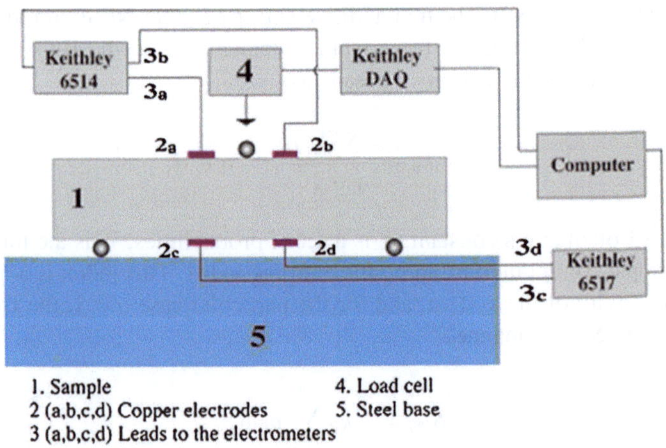

Fig. 1 The experimental setup for measuring PSC and mechanical load during 3 PB tests

A first set of the beams was used for conducting preliminary 3 PB strength tests for investigating their ultimate 3 PB strength (L_f). It was found that for the selected experimental topology their ultimate 3 PB strength was 4.8 ± 0.2 kN.

The basic experimental setup for measuring PSC is shown in Fig. 1. For conducting the 3 PB tests three identical metallic cylindrical rollers were used. The two of them supported the beam and the distance of their contact point from the centre of the specimen was 85 mm. The third roller was placed at the loading point. Teflon underlays of 2 mm thickness supported the rollers to electrically isolate the measuring system and the specimen. Two pairs of electrodes made of thin copper foils were used for capturing the electric signals. Their shape were orthogonal having dimensions of 30 mm × 10 mm. They were placed at the tensile and compression zone. Their distance from the centre was 20 mm. Before their attachment to the beams silver paste intermediate conductive layer was applied in order to ensure the best electrical path and minimize the external noise effects favored by specimen movement during the tests. Two sensitive electrometers were used (Keithley models 6514 and 6517) for measuring the electric signals.

4 Results and Discussion

A mechanical process of 3 PB loading/unloading was applied on the specimens (see Fig. 2a). The temporal behavior of the electrical signals were concurrently recorded (see Fig. 2b).

The loading protocol included five distinct stages as described below. The experimental recordings are discussed for each of the sequential stages. Specifically:

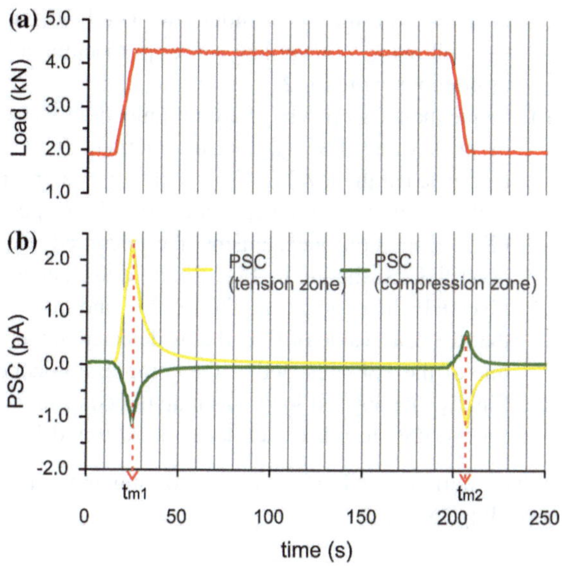

Fig. 2 **a** The temporal behavior mechanical loading and **b** the temporal behavior of the corresponding PSC at the tension (*yellow line*) and the compression (*green line*) zones

(a) A mechanical load (L_l) that was equal to the 50 % of the ultimate 3 PB strength (L_f) was initially applied.

(b) Sequentially, an abrupt mechanical load increase of high constant rate was applied up to a high load level (L_h). The load value (L_h) was equal to the 90 % of the L_f. As expected from previous works the PSC demonstrates a spike-like shape with respect to time. The PSC obtained its peak value, PSC(t_{m1}), at the moment (t_{m1}) when the applied mechanical load reached the level $L_{h.}$ This behavior was observed with both signals detected at the tension (lower side of the beam) and compression (upper side of the beam) zones of the specimen. The recorded electric signals had opposite polarities as shown in Fig. 2b. It was also observed that the peak value of PSC detected at the tension region was higher than the one of the PSC detected at the compression region. This is due to the fact that the cement mortar specimens obtain significantly lower tensional than compressional strength, thus, crack initiation and formation processes that activate the charge separation phenomena and the corresponding PSC emissions initiate from the lower side of the specimen (tensional zone).

(c) This high load value (L_h) is maintained on the cement mortar beam for long time in order to record a long PSC relaxation process both at the tensile and the compression region as shown on Fig. 2a, b. The duration of this stage was maintained until the PSC relaxed back to a background level $PSC(t \gg t_{m1})$. The behavior of the PSC seemed to be a typical one as it has already been reported on experiments that included compression stress tests and/or bending loading tests on cement based materials. The PSC also exhibited a similar behavior during experiments performed on geomaterials (i.e. marble, amphibolite) [4, 7]. For first time it was observed that the PSC relaxation deviates from the exponential law. The PSC recorded during this time frame was following a power law while approaching a large time range as also stated in previous works [22].

(d) Sequentially, a process of abrupt unloading was conducted. The PSC obtained its peak value, PSC(t_{m2}), at the moment (t_{m2}) when the applied mechanical load reached the level L_l. This behavior was observed for both signals detected at the tensile and compression region of the specimen. As expected due to the PSC generation mechanisms the recorded electric signals show opposite polarities as shown in Fig. 2b. It must be noted that the PSC polarity is reversed when comparing the loading and the unloading stages. It is also clear that the peak value of PSC that is detected at the tensile zone is higher than the one of the PSC detected at the compression zone.

(e) Finally, the applied load was maintained at L_l for relatively long time in order for the PSC to relax back to its background level $PSC(t \gg t_{m2})$.

The normalized PSC relaxation during the stages c and e can be calculated according to Eq. 3:

$$\xi_{ij}(t) = \frac{PSC_{ij}(t - t_{mj}) - PSC_{ij}(t \gg t_{mj})}{PSC_i(t_{mj}) - PSC_{ij}(t \gg t_{mj})} \tag{3}$$

where i notes the zone that the PSC emissions is recorded from (i.e. t: for the compression zone and b: for the tension zone), j corresponds to the applied load level (i.e. 1: represents the stage c, high load level and 2 the stage e, low load level), $PSC_{ij}(t \gg t_{mj})$ is the value of the PSC background level and $PSC_i(t_{mj})$ is the peak value of the PSC.

The function $\xi(t)$ was calculated while the mechanical load was maintained constant (stages c and e). In other words the above mentioned equation was applied during the time frame starting from the moment t_{mj} until the PSC relaxed back to its background level ($PSC_{ij}(t \gg t_{mj})$). Specifically, using the basic non-extensive statistical mechanics formulation for modeling the PSC relaxation processes at the:

(a) compression region while maintaining at high load value (4.2 kN)—ξt1
(b) compression region while maintaining at low load value (2 kN)—ξt2
(c) tension region while maintaining at high load value (4.2 kN)—ξb1
(d) tension region while maintaining at low load value (2 kN)—ξb2

In order to further analyze the experimental results non-extensive statistical physics will be used that are based on Tsallis entropy. The following equation was used to perform fitting to the results [20, 22].

$$\xi(t) = \left[1 + (q-1) \cdot \beta_q \cdot t\right]^{1/1-q} = \exp_q(-\beta_q \cdot t) \tag{4}$$

Figure 3 shows the values of $\xi_{ij}(t)$ and the corresponding fitting curves during the PSC relaxation at the high and low load levels, at both the compression and tension zones. In order to evaluate the use of the NESP theoretical background and

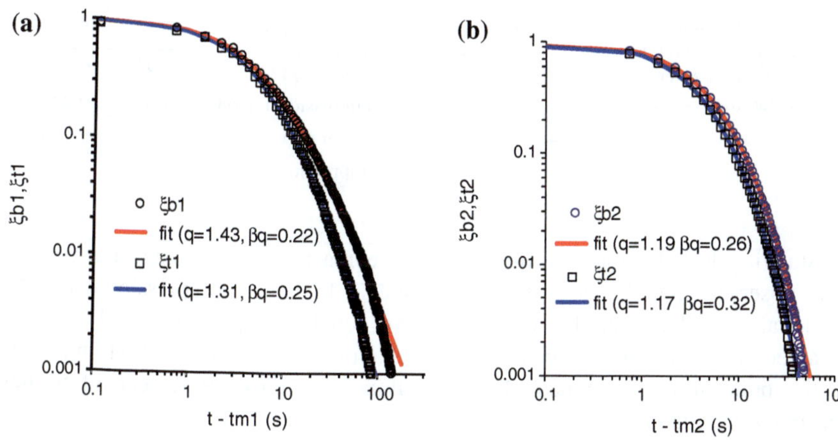

Fig. 3 The values of $\xi_{ij}(t)$ and the corresponding fitting curves during the PSC relaxation at: **a** the high load level L_h and **b** the low load level L_l, at the compression (*blue lines*) and tension (*red lines*) zones

the Tsalli's model applicability for the presented experiments, comparative plots are presented in Fig. 3a for the PSC relaxation when the applied mechanical load was at the level L_h. (i.e. 90 % of the 3 PB strength approximately, stage c). The fitting and the modeling was conducted for both the tension (red line) and compression (blue line) zones. The corresponding curves when the applied mechanical load was maintained at relatively low value (i.e. L_L = 50 % of the 3 PB strength approximately, stage e) are presented in Fig. 3b.

The fitting shows that during the stage c, the value of q parameter is greater than 1.3 regarding the compression zone and greater than 1.4 regarding the tensile zone. These values are in agreement with results exhibited in other works where similar experiments were performed on geomaterials and cement based materials [20–22].

The fact that the value of the q parameter is equal to 1.43 regarding the tension region when the applied mechanical load is high, demonstrates the disorganization of the mechanical tensions' distribution in the bulk of the specimen. This phenomenon signifies the beginning of dynamic processes related to the evolution of the inner damages that inevitably lead to the ultimate fracture of the specimen.

On the other hand, the value of the q parameter regarding the compression zone remains lower than 1.4 (i.e. 1.3). The dynamic process is not yet triggered and sufficiently evolved. The observations described above are supported by the behavior of the electrical signal that is significantly higher at the tension zone than the one detected at the compression region.

During stage d the PSC emission that is attributed to the crack and local discontinuities movements is significantly lower than stage b. This observation is consistent with the existing literature and it is attributed to the significantly lower number of the activated electrical signal sources (i.e. pre-existing damages, crack opening and propagation processes) according to the MCD model.

Table 1 Relaxation of PSC under constant mechanical loading and non-extensive entropy parameters q and β_q

Load, L (kN)	PSC relaxation	q	β_q (s^{-1})
2–4.2	Tension region	1.43	0.222
	Compression region	1.31	0.251
4.2–2	Tension region	1.19	0.260
	Compression zone	1.17	0.316

Studying the behavior of the q-parameter after reaching the low load value (stage e) it is observed that the corresponding value at both the compressional and the tensional zone have similar values that are equal to approximately 1.2. The β parameter also has a low value as the PSC is attributed to local polarization process in the bulk of the specimen. These local polarizations are dependent to new specimen bulk damage.

The fitting results as calculated according to the Tsalli's entropy model (Eq. 4) are summarized in Table 1.

The high value of the q parameter (see Table 1) denoted that the specimen is near to fracture which was expected as the high level of the mechanical load is approximately at 90 % of the 3 PB strength. Lower values of the q parameter recorded when the value of the mechanical load was low (see Table 1) makes clear that its level was not critical for the specimens' health status. The study of the q-parameter may constitute a strength Non-Destructive Testing evaluation index. On the contrary the β_q parameter exhibited a gradual decrease as the mechanical load increased and an increase for low level loads.

5 Conclusions

Several experiments were conducted including cement mortar beams of rectangular cross sections. During the experimental process mechanical loading was applied on the specimens using the Three-Point Bending (3 PB) technique. After reaching a high value (L_h) that is 90 % approximately of the ultimate fracture value (L_f) the loading remained constant with respect to time. The unloading took place during the next phase and the mechanical level returned to its initial low level. The recorded electrical signals exhibited a behavior that were identical to the one observed during previous experiments and other works regarding cement based material and geomaterials. The behavior of the PSC signal is due to the crack initiation and formation processes that activate the charge separation phenomena. The process of electrical current generation can be attributed to the motion of charged dislocations (MCD) which is one of the basic mechanisms involved in such phenomena. A processing of the electrical signal during its relaxation was performed using non-extensive statistical physics (NESP) based on Tsallis entropy equation. The values of the Tsallis q-parameter as well as the β_q parameter showed that they have physical plausibility and can be further investigated for this kind of experiments if they could be used as indices for monitoring the crack formation and propagation mechanism and predicting the upcoming fracture of a cement mortar beam. To summarizing, it seems that the q-parameter value of 1.4 is critical and indicates the massive bulk damages in the specimen provided that the applied mechanical load is in the vicinity of the 3 PB strength.

Acknowledgments This research has been co-financed by the European Union (European Social Fund—ESF) and Greek national funds through the Operational Program "Education and Lifelong Learning" of the National Strategic Reference Framework (NSRF)—Research Funding Program: ARCHIMEDES III. Investing in knowledge society through the European Social Fund.

References

1. Balayssac JP, Laurens S Breysse D et al (2013) Evaluation of concrete properties by combining NDT methods. In: Güneş O, Akkaya Y (eds) Nondestructive Testing of Materials and Structures, vol 6. Springer, Netherlands, pp 187–192
2. Kumar Mehta P, Monteiro PJM (2005) Concrete: microstructure, properties, and materials. McGraw-Hill Prof Med/Tech ISBN 978-0-07-179787-0
3. Stavrakas I, Triantis D, Agioutantis Z et al (2004) Pressure stimulated currents in rocks and their correlation with mechanical properties. Nat Hazards Earth Syst Sci 4:563–567
4. Anastasiadis C, Triantis D, Stavrakas I et al (2004) Pressure stimulated currents (PSC) in marble samples after the application of various stress modes before fracture. Ann Geophys-Italy 47:21–28
5. Triantis D, Stavrakas I, Anastasiadis C et al (2006) An analysis of pressure stimulated currents (PSC), in marble samples under mechanical stress. Phys Chem Earth 31:234–239
6. Anastasiadis C, Triantis D, Hogarth CA (2007) Comments on the phenomena underlying pressure stimulated currents (PSC) in dielectric rock materials. J Mater Sci 42:2538–2542
7. Triantis D, Anastasiadis C, Vallianatos F et al (2007) Electric signal emissions during repeated abrupt uniaxial compressional stress steps in amphibolite from KTB drilling. Nat Hazards Earth Syst Sci 7:149–154
8. Kyriazopoulos A, Anastasiadis C, Triantis D et al (2011) Non-destructive evaluation of cement-based materials from pressure-stimulated electrical emission—preliminary results. Constr Build Mater 25:1980–1990
9. Stergiopoulos C, Stavrakas I, Hloupis G et al (2013) Nondestructive testing electrical methods for sensing damages in cement mortar beams. Open J Appl Sci 3:50–55
10. Stergiopoulos C, Stavrakas I, Hloupis G et al (2012) Monitoring acoustic emissions and electrical signals during three-point bending tests performed on cement mortar specimens. In: Proceedings of 8th international conference fracture mechanics of concrete and concrete structures, FraMCoS-8, abs no. 591, Toledo, Spain, 11–14 March
11. Enomoto J, Hashimoto H (1990) Emission of charged particles from indentation fracture of rocks. Nature 346:641–643
12. O'Keefee SG, Thiel DV (1995) A mechanism for the production of electromagnetic radiation during fracture of brittle materials. Phys Earth Planet In 89:127–135
13. Vallianatos F, Triantis D, Tzanis A et al (2004) Electric earthquake precursors: from laboratory results to field observations. Phys Chem Earth 29:339–351
14. Vallianatos F, Tzanis A (1998) Electric current generation associated with the deformation rate of a solid: preseismic and coseismic signals. Phys Chem Earth 23(9):933–938
15. Tsallis C (2009) Introduction to nonextensive statistical mechanics: approaching a complex world. Springer, Berlin

16. Tsallis C (1999) Nonextensive statistics: theoretical, experimental and computational evidences and connections. Braz J Phys 29:1–35
17. Vallianatos F (2013) On the statistical physics of rockfalls: a non-extensive view. Europhys Lett. doi:10.1209/0295-5075/101/10007
18. Vallianatos F (2012) On the non-extensive nature of the isothermal depolarization relaxation currents in cement mortars. J Phys Chem Solids 73:550–553
19. Triantis D, Vallianatos F, Stavrakas I et al (2012) Relaxation phenomena of electrical signal emissions from rock following application of abrupt mechanical stress. Ann Geophys-Italy. doi:10.4401/ag-5316,207-212
20. Vallianatos F, Triantis D (2012) Is pressure stimulated current relaxation in amphibolite a case of non-extensivity? Europhys Lett (EPL). doi:10.1209/0295-5075/99/18006
21. Stergiopoulos C, Stavrakas I, Hloupis G et al (2013) Electrical and acoustic emissions in cement mortar beams subjected to mechanical loading up to fracture. Eng Fail Anal 35:454–461
22. Vallianatos F, Triantis D (2013). A non-extensive view of the pressure stimulated current relaxation during repeated abrupt uniaxial load-unload in rock samples. Europhys Lett. doi: 10.1209/0295-5075/104/68002
23. Kosmatka S, Kerkhoff B, Panarese W (2003) Design and control of concrete mixtures, 14th edn. Portland Cement Association, Skokie Illinois USA

Numerical Modelling of Young's Modulus of Single-Layered Cubic Zirconia Nanosheets

Ibrahim Dauda Muhammad, Mokhtar Awang and Lee Kain Seng

Abstract It has been established that zirconia nanosheets have many potential applications when compared with other materials possessing similar properties; however, the utilization of most of their potentials is constrained due to minimal data currently available on its mechanical properties. In this paper, the Young's modulus of single-layered zirconia nanosheets is predicted based on the concept of the finite element analysis. The nanosheet was modelled structurally as a hexagonal network of bonds connected by zirconium and oxygen atoms. Zirconia nanosheets with different dimensions and chirality were simulated with bonds between the atoms regarded as beam elements. The Young's modulus of the nanosheet was determined based on the combination of molecular mechanics and structural mechanics. The results obtained from the modeling indicates that the technique used is a viable tool for predicting mechanical properties of cubic zirconia nanosheets at a lower computational cost when compared to complex ab initio molecular dynamics and sophisticated experimental techniques.

Keywords Zirconia nanosheets · Material modelling · Young's modulus · Finite element analysis (FEA)

1 Introduction

Graphene is one of the most researched two-dimensional (2D) nanomaterials due to its unique mechanical, and other, properties [1]; but it has some defects which led to additional attention to other inorganic 2D materials. Zirconium oxide or zirconia

---

I.D. Muhammad (✉) · M. Awang · L.K. Seng
Department of Mechanical Engineering, Universiti Teknologi PETRONAS, Seri Iskandar, Perak, Malaysia
e-mail: ibrahimuhd@gmail.com

M. Awang
e-mail: mokhtar_awang@petronas.com

L.K. Seng
e-mail: kianseng.lee001@gmail.com

A. Öchsner and H. Altenbach (eds.), *Mechanical and Materials Engineering of Modern Structure and Component Design*, Advanced Structured Materials 70,
DOI 10.1007/978-3-319-19443-1_30

(ZrO_2) at nanoscale has some useful properties and exists in three phases: monoclinic, tetragonal and cubic with possibility for transformation at various temperatures [2, 3]. Cubic zirconia nanosheets (CZNS) are a two-dimensional nanostructure with thickness in a scale ranging from 1 to 100 nm and can be single or multi layered with different lattices. CZNS having 111 miller indices have hexagonal lattices, similar to that of graphene [3]. Potential areas of applications of CZNS include implementation as a catalyst and/or catalyst support, solid electrolyte, component of capacitors and structural ceramics [4–7]. As a confirmation of the potentials of zirconia nanosheets (ZNS), synthesis in various phases has been achieved using several techniques [8–12].

For effective application of CZNS in various areas, details of the mechanical properties are required as most of the potential uses are structurally based [2]. But mechanical properties of synthesized CZNS are not readily available as delicate and expensive equipment are required and methods to be used are still under development [10]. This lack of detailed understanding of the mechanical behavior of CZNS will limit the design and optimization of related composites or materials.

As an alternative, computational methods with emphasis on molecular dynamics (MD), have been used to simulate some properties of zirconia at the nanoscale [9]. In most of the MD simulations, studies are mainly conducted on nanotubes and the emphasis is on specific properties related to electronic, bulk, surface, optical and thermal [11–14]. Even where mechanical simulations were conducted [9, 15, 16], the computational expense of MD simulations limited the size of ZNS studied [10].

In the present work, the finite element analysis (FEA) approach used for graphene sheets is adopted for CZNS and it is based on molecular mechanics theory [17]. It is achieved by relating the potential energy of the atoms in the bonds of the structure with the mechanical properties. From the simulation results , the Young's modulus of CZNS was predicted.

2 Modeling and Simulation

2.1 Geometry

Cubic zirconia nanosheets can be described geometrically as the lattice planes of zirconia. An earlier study [12] indicated that the plane obtained on the (111) miller indices have a hexagonal orientation similar to that of graphene. The *Surface Builder* tool of *Material Studio* [18] was used to generate CZNS with different chirality and sizes as illustrated in Fig. 1, where a and b are the length and width respectively.

Thereafter, each generated CZNS structure was saved as a text file in Protein Data Bank (PDB) format and converted into APDL code for input into FEA software (*ANSYS*) using *PYTHON*. The output indicates the xyz coordinates all the atoms and their connectivity in the CZNS structure.

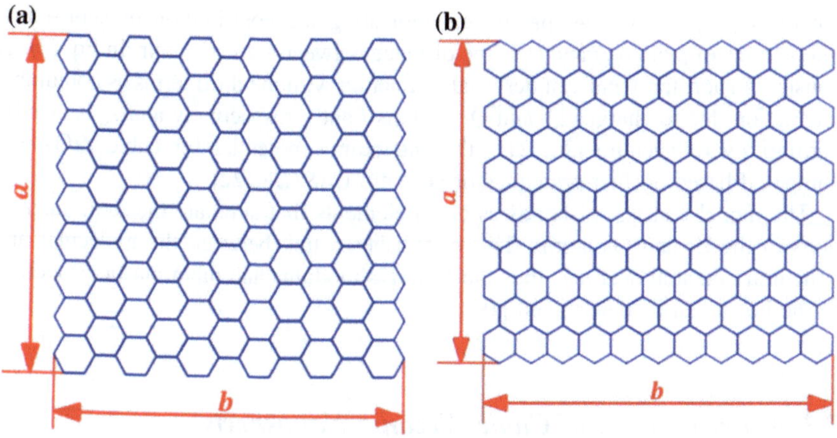

Fig. 1 Schematic diagram of the armchair (n, n) and zigzag (n, 0) CZNS

2.2 Interatomic Potential Energy of Zr-O Bond

Based on the concept of molecular mechanics, the total potential energy gives the summation of specific potential components caused by constraint and unconstrained interactions of atoms and is expressed [19] as

$$U = U_{\rho} + U_{\theta} + U_{\omega} + U_{\tau} + U_{vdw} + U_{elect} \tag{1}$$

In which U_{ρ}, U_{θ}, U_{ω} and U_{τ} are energies associated with stretching, bond-angle variation, bond inversion and torsion respectively; while U_{vdw} and U_{elect} represents van der Waals and electrostatic interaction respectively as expressed in Eq. (1). Figure 2 illustrates the basic structural unit of CZNS (hexagon).

For Zr-O bonds, interactions between the constituent atoms are mainly ionic and described using Buckingham (or Born-Meyer-Huggins) model [20]. The potential energy of ZrO_2 is a sum of two-body interactions represented as

$$U(r_{ij}) = A_{ij}e^{\left(-\frac{r_{ij}}{\rho_{ij}}\right)} - \frac{C_{ij}}{r_{ij}^{6}} + \frac{Q_iQ_j}{4\pi\varepsilon_o r_{ij}} = \emptyset_s r_{ij} + \frac{Q_iQ_j}{4\pi\varepsilon_o r_{ij}} \tag{2}$$

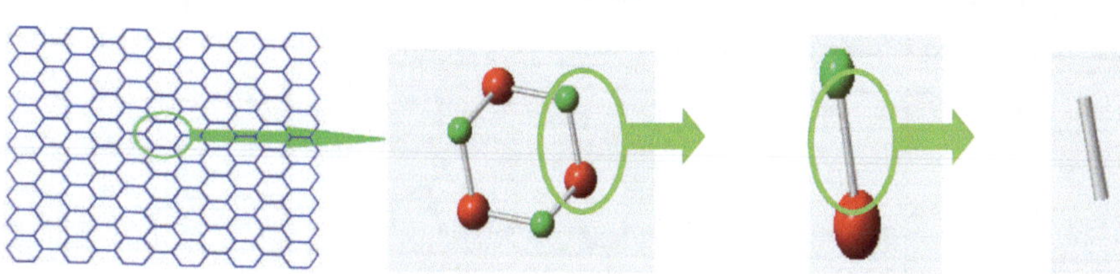

Fig. 2 Space-frame modelling for CZNS

where A_{ij}, ρ_{ij} and C_{ij} are parameters indicating the contribution of short-range interactions of pairs (ϕ_s) and r_{ij} is the distance between ions of a pair. In Eq. (2), ϕ_s consist of short term and van der waals attraction with the third terms as coulumbic interaction due to charges Q_i and Q_j for ions *i* and *j* respectively and ε_o is electric permittivity of vacuum [20, 21]. The interatomic potential for cubic ZrO_2 was computed based on first-principles or ab initio *CASTEP* [22].

The model consists of bonds as beam elements that simulate the ionic interactions in the Zr and O_2 atoms. This establishes a link between the molecular and structural potential energies, thus the Young's modulus and other elastic properties of the beam can be determined [19].

2.3 Atomistic FEA of Cubic Zirconia Nanosheets

The Zr and O_2 atoms in CZNS are bonded with and held together mainly by ionic bonds [20, 21] forming a hexagonal lattice. The bonds constrain movement or displacement and rotation of atoms due to external forces, leading to partial or total deformation. As applied to graphene [22], CZNS was simulated as space-frame structures by assuming the bonds as connected load-carrying elements and the atoms as joints or nodes. To simulate the uniaxial load, one CZNS side was retrained in all the six degrees of freedom and a displacement was applied in the opposite end as showed in Fig. 3. For the bond model, Beam188 in *ANSYS* was used as the element with 1.28 Å as diameter [22]; other properties of the element such as cross-sectional area, area moment of inertia ($I_{yy} = I_{zz}$) and torsional moment of inertia (I_{xx}) were computed to be 1.28579 Å^2, 0.131496 Å^4 and 0.26299 Å^4. Based on the aforementioned data, armchair (n, n) and zigzag (n, 0) CZNS with different chirality were simulated with the same boundary conditions.

After simulation by using displacement as applied load, the Young's modulus can be determined based on the elasticity law [23]

$$E = \frac{\sigma}{\varepsilon} = \frac{F/A}{\Delta L/L} = \frac{FL}{\Delta LWt} \tag{3}$$

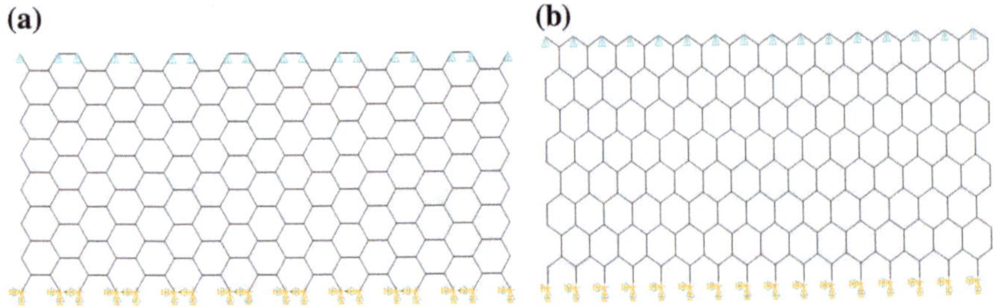

Fig. 3 Finite element model for armchair (**a**) and zigzag (**b**) CZNS

where F is the resultant force applied to the nodes at one end, $A = Wt$ is the cross sectional area of the sheet with width W and thickness t, L and ΔL are the initial length and extension respectively. The thickness (t) is taken as the interlayer spacing of cubic ZrO_2 of the plane with (111) miller indices and was found out to be 0.29387 nm using *CrystalMaker* [22].

3 Results and Discussion

As stated earlier, ANSYS code was used to model the nanosheets with their geometries described by their length and width. CZNS with different lengths and widths were studied. The geometries of the sheets for armchair and zigzag are not precisely the same with each other, so the closest possible widths were chosen and the length for all fixed at 25 Å.

The plot of the deformed shape is illustrated in Fig. 4. This shows the structural pattern of displacement of the nanosheet when load in form of displacement was applied. The pattern is uniform and in the direction of the applied loads, with maximum displacement at the nodes where the loads were applied (free end) and minimum displacement at the fixed ends where all the six degrees of freedom were restricted.

To sum the displacement vector, the contour plot was generated as displayed in Fig. 5. The plot shows how the applied load varies over the model as indicated by contour lines that are continuous across the entire model. The plot was generated from reaction forces, which are typically discontinuous from element to element. Thus forces are averaged at the nodes so that continuous contour lines can be displayed.

After obtaining the contour plot for the nodal solution, the nodes were selected and the reaction solution as the constrained nodes listed. The list contains the forces on each of the selected nodes and the resultant force; in this simulation, all forces

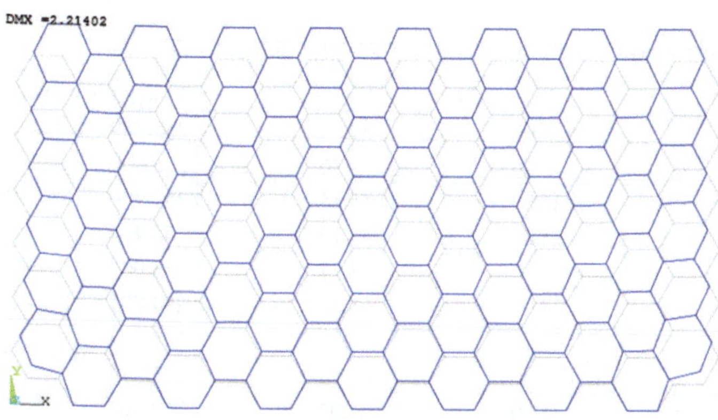

Fig. 4 Plot of deformed (*darker*) and undeformed (*lighter*) displacement for CZNS

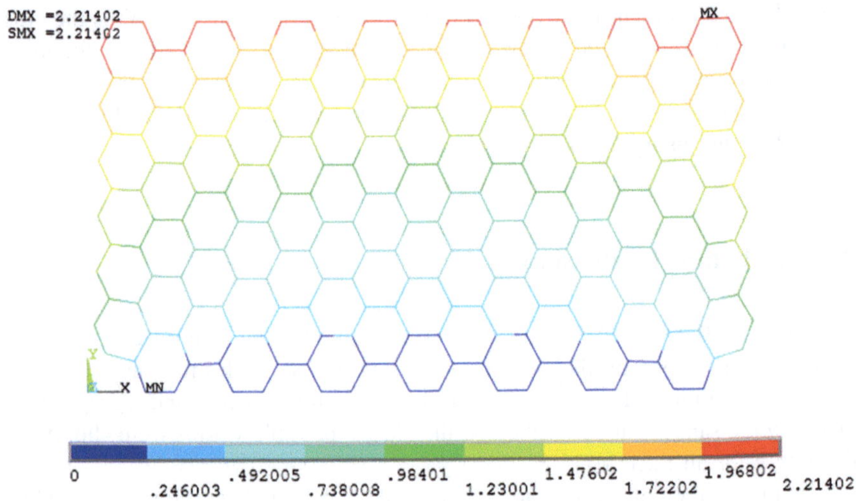

Fig. 5 Contour plot for nodal solution for displacement of (8 × 8) CZNS

given for all nodes are in one direction as the applied load is uniaxial. The nodal solution was used as the average for body while the selected nodes solution is the contributions of all elements connected to the node. The procedure was repeated for all the nanosheets having different chirality and sizes in other to obtain their respective resultant reaction forces and displacements.

Table 1 enumerates the computed Young's modulus (*E*) of the CZNS using the resultant force from simulations by ANSYS code for different model sizes. It can be observed that the Young's moduli of the CZNS obtained are almost independent of sizes as also concluded for graphene sheets [24]. The value of *E* for CZNS is 253.26 ± 1.2560 GPa compared to 1.1079 TPa for graphene [25], 330 GPa for MoS_2 nanosheets [26] and 323 GPa for h-BN nanosheets [27].

Table 1 Computed Young's moduli of the CZNS for different model sizes

Chirality (n × m)	Length (Å)	Width (Å)	Young's modulus (GPa)
6 × 6	25.00	39.58	255.780
10 × 0	25.00	39.76	253.900
7 × 7	25.00	46.18	251.760
12 × 0	25.00	45.70	253.274
8 × 8	25.00	52.77	253.067
14 × 0	25.00	53.32	254.887
9 × 9	25.00	59.37	252.567
16 × 0	25.00	60.94	251.777
10 × 10	25.00	65.97	252.567
17 × 0	25.00	64.75	252.684

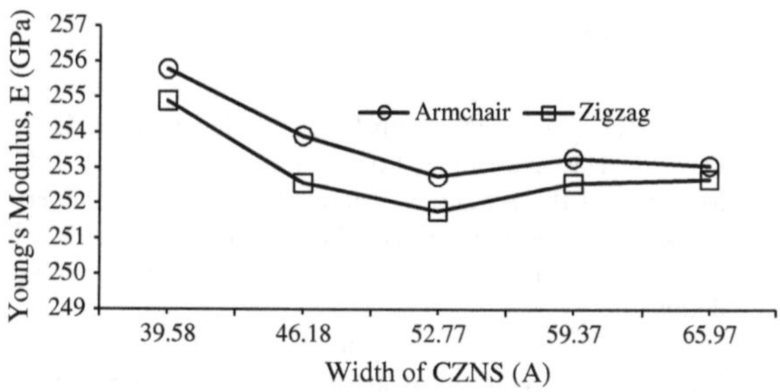

Fig. 6 Effect of width on the Young's modulus of CZNS

From Fig. 6, it may be concluded that the trend in the variation of E is similar for armchair and zigzag CZNS, and thus the effect of sheet chirality is minimal. It is also observed that the effect of the width on E of both armchair and zigzag CZNS is evident. With increasing width the Young's modulus of both armchair and zigzag decreases, which may be due to reduction in load distribution per node. It should be noted that all sheets used for the simulation have rectangular shape and when sheets with square shapes (7×0 and 5×0) were simulated, the values obtained for E were higher.

By computing the Young's modulus of CZNS, important quantitative data is obtained in relation to its level of elasticity or stiffness. The value is necessary for realistic comparison with other nanosheets and will also assist in designing nanocomposites for identified and other applications.

4 Conclusion

The Young's moduli of armchair and zigzag single-layered cubic zirconia nanosheets were simulated using ANSYS. Under uniaxial load, the average Young's modulus of cubic zirconia nanosheets was found out to be 253.26 ± 1.25 GPa. Additionally, the Young's moduli are similar for armchair and zigzag nanosheets but with variation according in relation to width of the structure. The technique used provides a means of obtaining the Young's modulus of the nanosheet at minimal computational cost. Thus the approach is a valuable tool for further detail studying of other mechanical properties of the zirconia nanosheet.

Acknowledgments The authors are grateful for the supports provided by Universiti Teknologi PETRONAS and Malaysian Ministry of Higher Education (MOHE) through the Long Term Research Grant Scheme (LRGS) for One Baja Research Programme (Project 6).

References

1. Ansari R, Ajori S, Motevalli B (2012) Super Micro 51(2):274–289
2. Davar F, Hassankhani A, Loghman-Estarki MR (2013) Ceram Int 39(3):2933–2941
3. Muhammad ID, Awang M, Mamat O (2014) Adv Mater Res. 845:387–391
4. Cook J, Hamburg D, Kaiser W, Logothetis E (1983) Engine dynamometer study of the transient response of ZrO_2 and TiO_2 exhaust gas oxygen sensors. SAE Technical Paper 830985
5. Möbius HH (1997) J solid state electron 1(1):2–16
6. Yamamoto N, Sato S, Takahashi R, Inui K (2006) J Mole Catal A Chem 243(1):52–59
7. Ahmed A, El-Hakam S, Samra S, El-Khouly A, Khder A (2008) Colloids Surf A Physicochem Enzi Aspects 317(1):62–70
8. Takenaka S, Uwai S, Ida S, Matsune H, Kishida M (2013) Chem Lett 42(10):1188–1190
9. Bandura AV, Evarestov RA (2012) Comp Mater Sci 65:395–405
10. Mohammadpour E, Awang M (2012) Appl Phys A 106(3):581–588
11. Caravaca M, Mino J, Pérez V, Casali R, Ponce C (2009) J Phys Condens Matter 21(1):015501
12. Xia X, Oldman R, Catlow R (2009) Chem Mater 21(15):3576–3585
13. Gennard S, Cora F, Catlow CRA (1999) J Phys Chem B 103(46):10158–10170
14. Nadeem M, Akhtar M, Shaheen R, Haque M, Khan A (2001) J Mater Sci Technol 17 (6):638–642
15. Cai J, Anastassakis E (1995) Phys Rev B 51(11):6821
16. Stapper G, Bernasconi M, Nicoloso N, Parrinello M (1999) Phys Rev B 59(2):797
17. Meo M, Rossi M (2006) Comp Sci Technol 66(11):1597–1605
18. Accelrys Software Inc., (2012) Materials studio. Accelrys Software Inc., San Diego
19. Ansari R, Rouhi S, Mirnezhad M, Sadeghiyeh F (2013) Appl Phys A 112(3):767–774
20. Lewis G, Catlow C (1985) J Phys C Solid State Phys 18(6):1149–1155
21. Wang C (2009) Multiscale modeling and simulation of Nanocrystalline zirconium oxide. PhD Thesis, Mechanical Engineering Department, University of Nebraska
22. Muhammad ID, Awang M (2014) Appl Mech Mater 446:151–157
23. Ansari R, Rouhi S (2010) Physica E 43(1):58–69
24. Chang Y, Wang H, Zhu Q, Luo P, Dong S (2013) J Adv Ceram 2(1):21–25
25. Lee J, Yoon D, Cheong H (2012) Nano Lett 12(9):4444–4448
26. Castellanos-Gomez A, Poot M, Steele GA, van der Zant HSJ, Agraït N, Rubio-Bollinger G (2012) Adv Mater 24(6):772–775
27. Boldrin L, Scarpa F, Chowdhury R, Adhikari S (2011) Nanotechnology 22(50):505702

Closed Form of a Transverse Tapered Cantilever Beam Fundamental Frequency with a Linear Cross-Area Variation

Farid Chalah, Lila Chalah-Rezgui, Salah Eddine Djellab and Abderrahim Bali

Abstract The present work is based on the Rayleigh quotient formula to express the transverse vibration frequency of a tapered cantilever beam by a closed form equation. In this investigation the tapered case considered is that of a linear cross-section variation. Thus, a shape equation is needed. It can be obtained through an exact solution of the fourth order differential equation, with non constant coefficients governing the equilibrium of the tapered beam element. The shape form of a uniform beam is considered as being the first mode shape in the present investigation. By applying the Rayleigh quotient, a simple closed form equation of the circular frequency as a function of the taper degree is suggested for practical use. A validation of the numerical results with the extreme case corresponding to the uniform beam is done. By comparing the curves of the two shapes corresponding to the tapered and uniform cantilever beams, they are found to be in good compliance.

Keywords Closed form · Rayleigh quotient · Tapered beam · Transverse vibration · Fundamental frequency

F. Chalah (✉) · L. Chalah-Rezgui · S.E. Djellab
Faculty of Civil Engineering, USTHB, 16111 Algiers, Algeria
e-mail: f_chalah@yahoo.fr

L. Chalah-Rezgui
e-mail: lrezguichalah@gmail.com

S.E. Djellab
e-mail: sedjellab@gmail.com

A. Bali
Ecole Nationale Polytechnique, URIE, Algiers, Algeria
e-mail: balianl@yahoo.fr

A. Öchsner and H. Altenbach (eds.), *Mechanical and Materials Engineering of Modern Structure and Component Design*, Advanced Structured Materials 70,
DOI 10.1007/978-3-319-19443-1_31

1 Introduction

The vibration analysis of tapered beams is an attractive research objective. Many researchers have been interested in this field of investigation and because of that; several studies are still being carried out. They direct their research in various directions, regarding the cross area variations and the boundary conditions. Different approaches are used to improve the result accuracy.

In structural analysis, beams with distributed characteristics are frequently used. Adding to this, a variation of the cross section of the beam is introduced to take more efficiently of the real structure forms. These beam types are non uniform or tapered. The aim of this investigation is to suggest contracted formulas describing the frequency variations of a tapered beam for the two cases of linear and cubic inertia variations. In a previous work, frequency variation graphs were proposed. It therefore constitutes a first step for this purpose and will be followed by a mathematical development based on the energetic method of Rayleigh. On the other hand, to apply the FEM approach to analyzing a vibrating tapered beam, a discretization process is used which consists of dividing it in a number of parts. The stiffness and mass matrices are obtained by assembling these elementary beams to finally construct a model which is considered in place of the real structure. Although, the matrices are correctly evaluated, the resulting constructed matrix remains not fully exact. However, the different beam parts maintain their sections uniformly along their lengths with their corresponding geometrical characteristics in terms of inertia and cross section. The analysis of non-uniform beams was largely reported in the literature. Several approaches were used for the free vibration problem resolution. Different methods were developed and improved to conduct the stiffness matrix evaluation for a non uniform beam element. They are relative to the direct integration of the differential equation governing the equilibrium in static and dynamic analysis [1–3], the flexibility matrix construction followed by its inversion [4–8] and the use of transfer matrices [9, 10].

The exact or approximate solution process, for the study of free vibration of a tapered beam with different restraint conditions, allowed to solve problems of multistep non uniform beams [11, 12]. Additional research consisting of transforming or reducing the equation of motion to that of uniform beam was carried out [13, 14]. On the other hand, the free vibrations of tapered beams analysis were investigated considering flexible ends [15] and associated spring-mass systems [16, 17].

In the present work, the fundamental vibration natural frequencies of a tapered cantilever beam for linear and cubic inertia variations are given by applying the Rayleigh formula. It is evident that the mass variation is linear for the two situations. The used shape is that corresponding to the normalized displacement resulting of a uniform beam loaded by its distributed uniform mass. This is introduced in place of the exact solution of the differential equation of fourth order, with no constant coefficients governing the equilibrium [18].

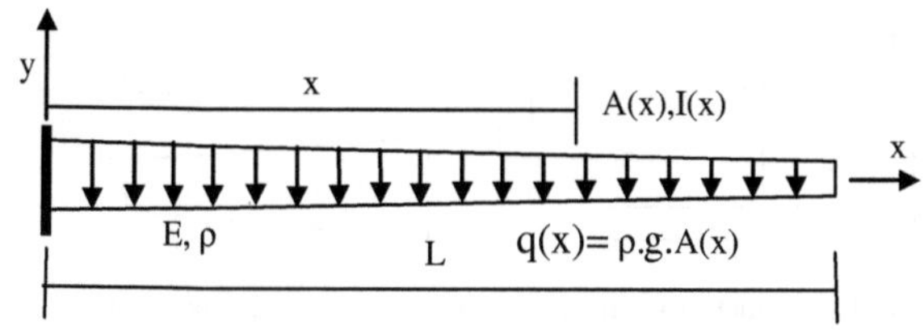

Fig. 1 Example of a tapered cantilever beam loaded by a distributed mass

2 Problem Formulation

A tapered cantilever beam, as seen in Fig. 1, is considered to be a material having an elastic modulus E, a material density ρ. The inertia at the x abscissa is $I(x)$ inertia and the cross section $A(x)$. For all the treated cases, the loading is that corresponding to its weight.

The differential equation of the fourth order governing the equilibrium of a non uniform beam is given by the general expression:

$$\frac{\partial^2}{\partial x^2}\left(E \cdot I(x) \cdot \frac{\partial^2 v(x)}{\partial x^2}\right) = q(x) \tag{1}$$

which is developed in:

$$E \cdot \left(I(x) \cdot \frac{\partial^4 v(x)}{\partial x^4} + 2 \cdot \frac{\partial I(x)}{\partial x} \cdot \frac{\partial^3 v(x)}{\partial x^3} + \frac{\partial^2 I(x)}{\partial x^3} \cdot \frac{\partial^2 v(x)}{\partial x^2}\right) = q(x) \tag{2}$$

It can be seen that the derivatives of the inertia appears. Many possible variations of the inertia can be considered depending on the geometry of the section (rectangular, circular, …). The linear and cubic inertia variations correspond to the two situations of Fig. 2.

By identifying the mean plane position, the following projections of the tapered beam are illustrated in Fig. 3.

The expressions describing the possible inertia variations are deduced from Fig. 4. They are obtained by extending the two symmetrical lines corresponding to the sides of the trapeze to their cross point until forming a triangle. Then, a line is

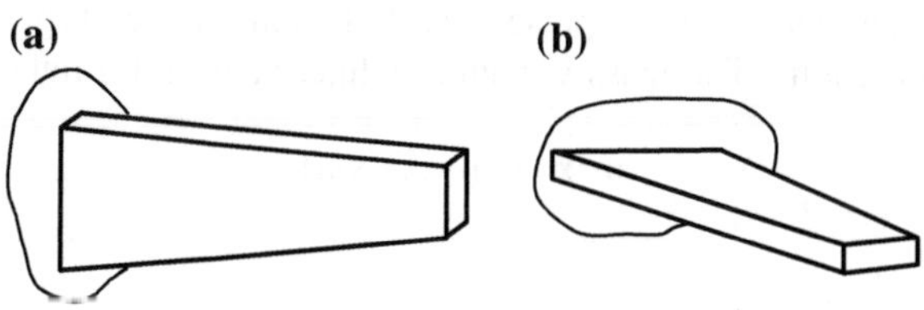

Fig. 2 Spatial views of tapered cantilever beams with linear cross section variation

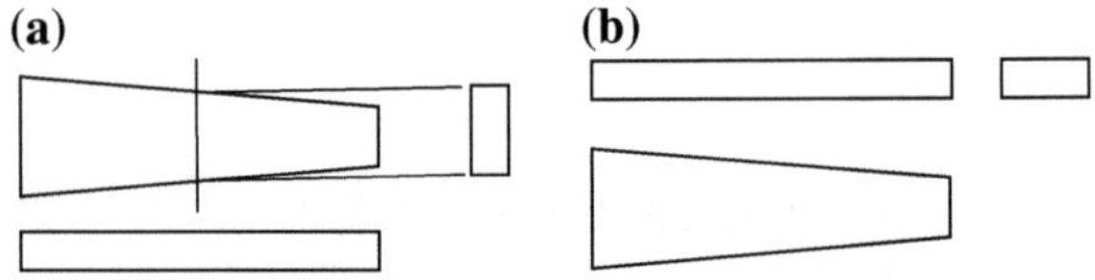

Fig. 3 Projections of the tapered beam depending on the spatial position

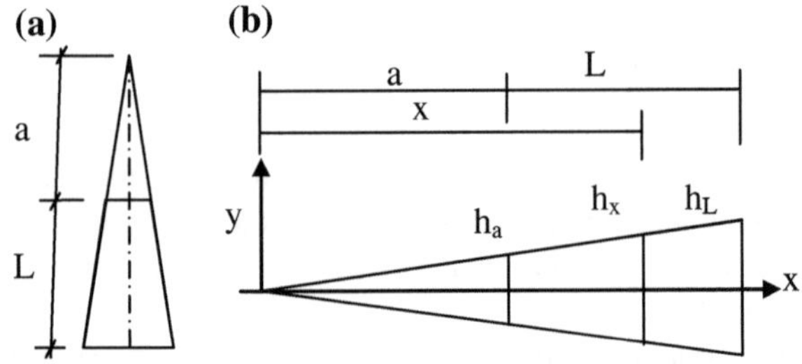

Fig. 4 Dimensions and lengths of the reference triangle of the tapered beam

drawn along the symmetry axis of the formed triangle from the top to the upper base of the triangle. Its length is designated by the letter a. Let us consider the following dimensions b, h_a, h_L and L respectively the depth, widths and length of a tapered beam, as shown in Fig. 4.

The variable of width h_x is easily obtained by applying simple geometric ratios. It is expressed by a function of the reference triangle lengths of Fig. 4:

$$h_x = h_a \cdot (x/a) \tag{3}$$

Thus, the inertia variation of Fig. 2a, is:

$$I_x = I_a(x/a)^3 \tag{4}$$

And that of Fig. 2b, is:

$$I_x = I_a(x/a) \tag{5}$$

It should be noted that some research works focused on various shape variations using different geometries. In the present work the only treated types are linear and cubic inertia variations. The mass variation is linear and is formulated as follows:

$$m(x) = \rho \cdot A_a(x/a) \tag{6}$$

3 Numerical Validation

The Rayleigh's formula is a useful tool generally used in structural analysis to evaluate approximately the fundamental vibration frequency. It uses the transverse deflection of the studied structure. It results from principle of total energy conservation of a system vibrating harmonically. This relationship is expressed as:

$$\omega_1 = \sqrt{\frac{\int_L E \cdot I(x) \cdot [\psi''(x)]^2 \, dx}{\int_L \rho \cdot A(x) \cdot [\psi(x)]^2 \, dx}} \tag{7}$$

For each case study, the deflection is first evaluated. The shape of the uniformed beam loaded by its own distributed mass is used rather than using that of the normalized homogeneous solution of the fourth order differential equation. Based on the curvature/moment relations for a uniform beam, the deflection of a cantilever beam subjected to uniform loading is:

$$\mathrm{v(x)} = \left(-\mathrm{q} \cdot \mathrm{L}^2 \cdot \mathrm{x}^2/4 + q \cdot \mathrm{L} \cdot \mathrm{x}^3/6 - \mathrm{q} \cdot \mathrm{x}^4/24\right)/\mathrm{E} \cdot \mathrm{I} \tag{8}$$

A normalization of this expression, gives:

$$\Psi_1(\mathrm{x}) = \mathrm{x}^2\left(6 \cdot \mathrm{L}^2 - 4 \cdot \mathrm{L} \cdot \mathrm{x} + \mathrm{x}^2\right)/\left(3 \cdot \mathrm{L}^4\right) \tag{9}$$

In the case where a concentrated force is applied at the end of the cantilever beam, the shape to be taken into account is:

$$\Psi_2(\mathrm{x}) = \left(3\mathrm{Lx} - \mathrm{x}^3\right)/2\mathrm{L}^3 \tag{10}$$

Using the Rayleigh quotient in the case of a cantilever uniform beam, the following results are given:

$$\text{Shape } \Psi_1(\mathrm{x}) : \omega_1 = 1.88879^2 \cdot \left(\mathrm{EI}/\rho \mathrm{AL}^4\right)^{1/2} (0.41\,\%)$$

$$\text{Shape } \Psi_2(\mathrm{x}) : \omega_1 = 1.87885^2 \cdot \left(\mathrm{EI}/\rho \mathrm{AL}^4\right)^{1/2} (1.48\,\%)$$

The ω_1 value was compared, using the $\Psi_1(x)$ and $\Psi_2(x)$ shapes, relatively to the theoretical $\omega_1 = 1.875^2 \, (EI/A\rho L^4)^{½}$ value.

For all the proposed formulas, the given expressions are of the form:

$$\omega_1(\alpha) = f(\alpha) \cdot \sqrt{\frac{E \cdot I_a}{A_a \cdot \rho \cdot L^4}} \tag{11}$$

with

$$\alpha = \mathrm{a}/\mathrm{L}.$$

4 Case Where the Tapered Cantilever Beam Is Fixed at Its Large Base

Many research works on the tapered beam vibrations in specialized literature, provided various approaches to address these typical analysis. A proposal of closed form formulas is relevant and as useful as possible is an interesting task. Some of the existing approaches concern particular case studies, whereas others are based on beam shapes not often encountered. Some others have tackled for fixed geometries and their exploitation is not extensible easily.

Linear inertia variation

An explicit integration of the Rayleigh formula with an assumption on the shape corresponding to that of a uniform cantilever beam, leads to two expressions of the circular frequency in closed forms.

$$\omega_1(\alpha) = 3\sqrt{21}\sqrt{\frac{5+6\cdot\alpha}{18+91\cdot\alpha}}\cdot\sqrt{\frac{E\cdot I_a}{A_a\cdot\rho\cdot L^4}} \tag{12}$$

$$\omega_1(\alpha) = 3\sqrt{210}\sqrt{\frac{3+4\cdot\alpha}{49+264\cdot\alpha}}\cdot\sqrt{\frac{E\cdot I_a}{A_a\cdot\rho\cdot L^4}} \tag{13}$$

As known, the Rayleigh quotient results are higher than that provided by the theory and this concerns all the considered shapes. In this comparison conditions, the restraints are fixed-free. The application of the closed form equations in the case of a linear variation in the inertia for great values of a/L, give:

$$\omega_1 = 1.906^2\left(\mathrm{EI}/\mathrm{A}\rho\mathrm{L}^4\right)^{1/2} \quad \text{closed form } \Psi_1(x)(3.34\,\%)$$

$$\omega_1 = 1.898^2\left(\mathrm{EI}/\mathrm{A}\rho\mathrm{L}^4\right)^{1/2} \quad \text{closed form } \Psi_2(x)(2.49\,\%)$$

At this level of comparison, it appears that the proposed formulas give quite good results in view of the difficulty of the problem formulation. The error appreciation level generally admitted for the Rayleigh quotient, in analyzing a uniform cantilever beam, is about 1.5 %. A graphical reading difficulty appears in the curve of Fig. 5a, and the accuracy would give:

$$\omega_1 = 1.90^2\cdot\left(\mathrm{EI}/\mathrm{A}\rho\mathrm{L}^4\right)^{1/2} \quad \text{with an error of } 2.7\,\%$$

Against an evaluation with the permanent error risk with a handle calculation:

$$\omega_1 = 1.897^2\left(\mathrm{EI}/\mathrm{A}\rho\mathrm{L}^4\right)^{1/2}(2.41\,\%)$$

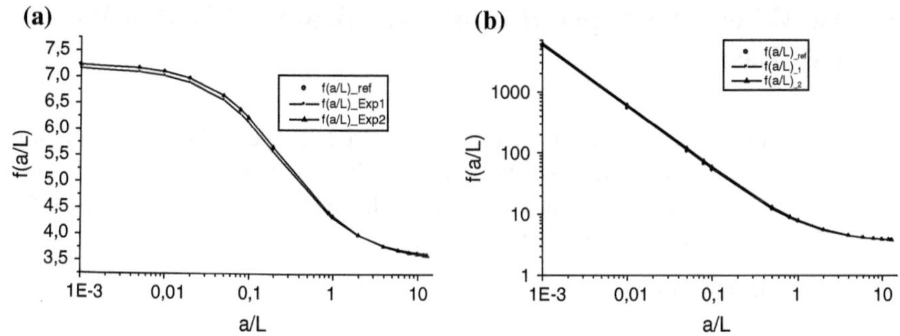

Fig. 5 Graphs for determining ω_1 in function of the α ratio for the tapered cantilever beam fixed at its large base. **a** Linear inertia variation. **b** Cubic inertia variation

Cubic inertia variation

The two shapes introduced in the Rayleigh formula, give the formulas:

$$\omega_1(\alpha) = 2 \cdot \sqrt{42}\sqrt{\frac{(10+\alpha\cdot(36+5\cdot\alpha\cdot(9+4\cdot\alpha)))}{\alpha^2\cdot(49+264\cdot\alpha)}}\cdot\sqrt{\frac{E\cdot I_a}{A_a\cdot\rho\cdot L^4}}, \tag{14}$$

for $\alpha > 0.5$,

$$\omega_1(\alpha) = \frac{9}{2}\sqrt{\frac{(35+4\alpha\cdot(30+7\cdot\alpha\cdot(5+2\cdot\alpha)))}{\alpha^2\cdot(18+91\cdot\alpha)}}\cdot\sqrt{\frac{E\cdot I_a}{A_a\cdot\rho\cdot L^4}} \quad \text{for } \alpha > 0.5, \tag{15}$$

For the case a cubic variation in the inertia, for great values of α, the results are:

$$\omega_1 = 1.931^2\left(\mathrm{EI}/\mathrm{A}\rho\mathrm{L}^4\right)^{1/2} \quad \text{curve } (6.06\,\%)(\text{with the normalized homogenous solution})$$
$$\omega_1 = 1.933^2\left(\mathrm{EI}/\mathrm{A}\rho\mathrm{L}^4\right)^{1/2} \quad \text{closed form } \Psi_2(x)(6.28\,\%)$$
$$\omega_1 = 1.937^2\left(\mathrm{EI}/\mathrm{A}\rho\mathrm{L}^4\right)^{1/2} \quad \text{closed form } \Psi_2(x)(6.72\,\%)$$

Note: It should not be hoped to obtain a uniform beam by forcing a beam with a very low taper. Numerically, perturbations will appear.

The closed form equations proposed in the fixed-free cases do not suffer from the taper degree. It is easy to note that the uniform beam shape generally used without preliminary confirmation and verification is adapted for analyzing tapered cantilever beams fixed at their large base.

5 Case Where the Tapered Cantilever Beam Is Fixed at Its Small Base

The applications made in the case where the tapered cantilever beam is fixed at its large base are reused for analyzing the situation when the tapered cantilever beam is fixed at its small base and free at its second end. Then, the proposed closed forms for the linear inertia variation, are:

Linear inertia variation

$$\omega_1(\alpha) = 3\sqrt{210}\sqrt{\frac{1+4\alpha}{215+264\cdot\alpha}}\cdot\sqrt{\frac{E\cdot I_a}{A_a\cdot\rho\cdot L^4}} \quad \text{for } \alpha > 0.5$$

In this interval, it varies similarly but still maintains a constant distance. The following one is more accurate:

$$\omega_1(\alpha) = 3\sqrt{21}\sqrt{\frac{1+6\cdot\alpha}{73+91\cdot\alpha}}\cdot\sqrt{\frac{E\cdot I_a}{A_a\cdot\rho\cdot L^4}} \quad \text{for } \alpha > 0.3$$

$$\omega_1 = 1.861^2\left(EI/A\rho L^4\right)^{1/2} \quad \text{curve } (1.5\,\%)$$

$$\omega_1 = 1.864^2\left(EI/A\rho L^4\right)^{1/2} \quad \text{closed form } \Psi_1(x)(1.17\,\%)$$

$$\omega_1 = 1.876^2\left(EI/A\rho L^4\right)^{1/2} \quad \text{closed form } \Psi_2(x)(1.07\,\%)$$

These two suggested formulas are valid for limited α values ($\alpha > 0.3$).

Cubic inertia variation

The suggested closed forms for the cubic inertia variation, are:

$$\omega_1(\alpha) = 2\sqrt{42}\cdot\sqrt{\frac{(1+\alpha\cdot(6+5\cdot\alpha\cdot(3+4\cdot\alpha)))}{\alpha^2\cdot(215+264\cdot\alpha)}}\cdot\sqrt{\frac{E\cdot I_a}{A_a\cdot\rho\cdot L^4}} \quad \text{for } \alpha > 2$$

$$\omega_1(\alpha) = \frac{9}{2}\cdot\sqrt{\frac{(1+4\alpha\cdot(2+7\cdot\alpha\cdot(1+2\cdot\alpha)))}{\alpha^2\cdot(73+91\cdot\alpha)}}\cdot\sqrt{\frac{E\cdot I_a}{A_a\cdot\rho\cdot L^4}} \quad \text{for } \alpha > 2$$

$$\omega_1 = 1.873^2\left(EI/A\rho L^4\right)^{1/2} \quad \text{curve } (0.21\,\%)$$

$$\omega_1 = 1.872^2\left(EI/A\rho L^4\right)^{1/2} \quad \text{closed form } \Psi_1(x)(0.32\,\%)$$

$$\omega_1 = 1.888^2\left(EI/A\rho L^4\right)^{1/2} \quad \text{closed form } \Psi_2(x)(1.39\,\%)$$

These two suggested formulas are valid for limited α values ($\alpha > 2$).

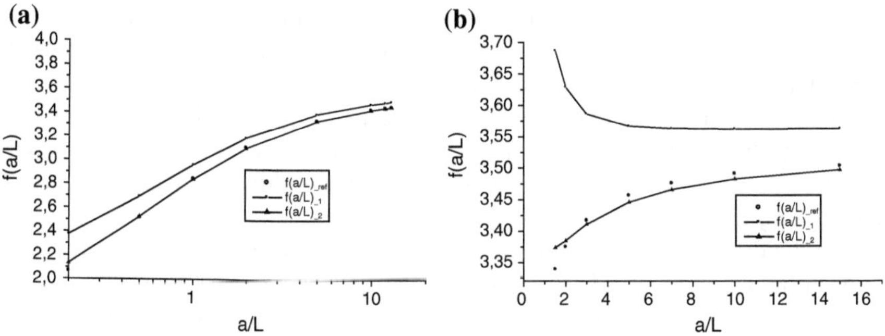

Fig. 6 Graphs for determining ω_1 in function of the (*a*/*l*) ratio for the tapered cantilever beam fixed at its small base. **a** Linear inertia variation. **b** Cubic inertia variation

Comments for the cubic inertia variation

It was observed that for great values of a/L, the tapered cantilever beam gives more accurate values than that given by analyzing the vibration of a uniform beam. The validation consists of comparing the results when the a/L ratio takes a great value so that the studied cantilever tapered beam can be considered a uniform beam. A value of $a/L = 20$ seems to be reasonable. At this time, it is comparable to the theoretical value.

The closed form function as suggested remains valid only for the intervals [0.3–10] and [3–10] respectively for linear and cubic inertia variations. Otherwise, the expected accuracy of a reading operation on the curve, remains surely less than that which will be induced by the tentative to propose an approximate formula. It is noted that the variations on 2.3–3.4 in Fig. 6a and 3.38–3.48 in Fig. 6b, are not large. The main conclusion is that the closed form equation maintains the accuracy order of the Rayleigh quotient in the proposed intervals. The concerned interval where a proposal of a closed form formula is possible, is located beyond a/L = 0.3 in Fig. 6a. Some difficulties are encountered to adjust correctly the formula because the variation order is low and for the second curve of about 0.2 in Fig. 6b, (It is noted that the variation of f(a/L) is 3.3–3.56) while the concerned interval of (a/L) variation is still significant from the geometric point of view but not from physical meaning. This is due to the fact that the variations of both the stiffness and the mass are linked to each other and exert influence differently on the period rendering a prediction of its evolution impossible. It does not require extensive research to improve the solution which should not bring in a significant precision.

6 Conclusion

The closed forms are in compliance with the normalized solution of the forth order differential equation, which led to complete and complex formulations of the expression. This allows using the shapes corresponding to that of a uniform beam, independently of the loading; end force or uniform load. For the case where the inertia variation is cubic, the mass effect is not sufficient to cause any influence against the inertia importance, especially for small values of a/L ratios. These remarks are relative to the Fixed-Free situations (large base fixed). It should be noted that the uniform beam shape generally used in analyzing any given tapered cantilever beam fixed is interesting to reduce the problem difficulty induced by considering its inertia variation.

For the case where the small base is fixed, the proposed expressions are related only to large values of a/L ratios, i.e. weak taper degrees. By evidence, the recommendation resulting from this investigation consists of the fact that it is not appropriate to consider the usual shapes of a cantilever uniform beam to analyze the free vibrations of a tapered cantilever beam (Free-fixed: small base fixed) having a strong taper degree.

References

1. Karabalis DL, Beskos DE (1983) Static, dynamic and stability analysis of structures composed of tapered beams. Comput Struct 16(6):731–748
2. Biondi B, Caddemi S (2007) Euler Bernoulli beams with multiple singularities in the flexural stiffness. Eur J Mech A Solids 26:789–809
3. Djellab SE, Chalah F, Falek K (2011) Modeling and analysis of buildings with irregular shear walls. In: 1st National civil engineering seminar on materials and environmental protection. Mostaganem
4. Eisenberger M (1985) Explicit stiffness matrices for non-prismatic members. Comput Struct 20(4):715–720
5. Vu-Quoc L, Léger P (1992) Efficient evaluation of the flexibility of tapered I-beams accounting for shear deformations. Int J Numer Methods Eng 33:553–566
6. Frieman Z, Kosmatka JB (1992) Exact stiffness matrix of a nonuniform beam extension, torsion and bending of a Bernoulli-Euler beam. Comput Struct 42(5):671–682
7. Frieman Z, Kosmatka JB (1993) Exact stiffness matrix of a nonuniform beam bending of a Timoshenko beam. Comput Struct 49(3):545–555
8. Tena-Colunga A (1996) Stiffness formulation for nonprismatic beam elements. J Struct Eng ASCE 122(12):1484–1489
9. Luo Y, Wu F, Xu X (2006) Element stiffness matrix and modified coefficients for circular tube with tapered ends. J Constr Steel Res 62:856–862
10. Luo Y, Xu X, Wu F (2007) Accurate stiffness matrix for non prismatic members. J Struct Eng ASCE 133(8):1168–1175
11. Attarnejad R, Manavi N, Farsad A (2006) Exact solution for the free vibration of a tapered beam with elastic end rotational restraints. In: Computational methods. Springer, Berlin, pp 1993–2003
12. Li QS (2000) An exact approach for free flexural vibrations of multistep nonuniform beams. J Vib Control 6:963–983
13. Abrate Serge (1995) Vibration of non-uniform rods and beams. J Sound Vib 185(4):703–716
14. Ece MC, Aydogdu M, Taskin V (2007) Vibration of a variable cross-section beam. Mech Res Commun 34:78–84
15. De Rosa MA, Auciello NM (1996) Free vibrations of tapered beams with flexible ends. Comput Struct 60:197–202

16. Qiao H, Li QS, Li GQ (2002) Vibratory characteristics of flexural non-uniform Euler Bernoulli beams carrying an arbitrary number of spring-mass systems. Int J Mech Sci 44(4):725–743
17. Wu JJ (2003) Use of effective stiffness matrix for the free vibration analyses of a non-uniform cantilever beam carrying multiple two degree-of-freedom spring-damper-mass systems. Comput Struct 81(24–25):2319–2330
18. Chalah F, Djellab SE, Falek K, Chalah-Rezgui L, Bali A (2013) Tapered beam fundamental natural frequency, based on Rayleigh quotient. Appl Mech Mater 330:526–530. doi:10.4028/www.scientific.net/AMM.330.526 (Trans Tech Publications, Switzerland)

Free Vibration of a Beam Having a Rotational Restraint at One Pinned End and a Support of Variable Abscissa

Lila Chalah-Rezgui, Farid Chalah, Salah Eddine Djellab, Ammar Nechnech and Abderrahim Bali

Abstract To carry out a dynamic analysis of vibrating systems, different methods exist and numerical methods take a large place. In this study, a beam with one overhang, having a rotational restraint at one pinned end and a support of variable abscissa is investigated. Many situations are studied using the finite element method based on the Euler-Bernoulli assumptions to analyze beams vibration, without elastic restraints. The purpose of this investigation consists of treating similar beam cases with and without a rotational restraint at one node. The first step of validation is relative to the extreme situations where either the overhang or the intermediate span length is zero (span length approx. zero). In these two studied cases, the rotational restraint is not considered. It is compared to theoretical, energetic Rayleigh method results and those due to the finite element method for the simply supported beam, pinned-clamped and free-fixed beam (cantilever beam). When the rotational restraint value increases the beam is considered first pinned-pinned and then pinned-fixed behavior is obtained. The results as found are in agreement with analytical ones and after introducing the rotational restraint at the right-hand end, an intermediate behavior is initiated by varying the s/L (span length/beam length) ratio, allowing a simplified extraction of the fundamental vibration frequency for various values of K_r (rotational restraint).

L. Chalah-Rezgui (✉) · F. Chalah · S.E. Djellab · A. Nechnech
Faculty of Civil Engineering, USTHB, 16111 Algiers, Algeria
e-mail: lrezguichalah@gmail.com

F. Chalah
e-mail: f_chalah@yahoo.fr

S.E. Djellab
e-mail: sedjellab@gmail.com

A. Nechnech
e-mail: nechnech_a@yahoo.fr

A. Bali
URIE, Ecole Nationale Polytechnique, Algiers, Algeria
e-mail: balianl@yahoo.fr

A. Öchsner and H. Altenbach (eds.), *Mechanical and Materials Engineering of Modern Structure and Component Design*, Advanced Structured Materials 70,
DOI 10.1007/978-3-319-19443-1_32

Keywords Free vibration · Supported beam · FEM · Overhang · Rotational restraint

1 Introduction

Many research works have been devoted to the analysis of the transversal vibration of uniform beams. The finite element method (FEM) presented in literature [1–3] allows solving such problems where the structure is divided into a number of elements constituted of the same material with constant geometric characteristics. These parts which are assembled and reconnected to each other, constitute an idealization of the initial structure. The followed path for this situation consists of writing a Fortran program in which, different subroutines are implemented for solving various and complex problems of structures. A free vibration beam analysis is achieved by an eigenvalue problem solving.

Based on the Euler-Bernoulli theory, as reported in the literature, it is admitted to neglect the shear deformation and the rotary (gyrating) inertia, in the study of an oscillating uniform beam. For basic study cases, solving the differential equation of fourth order differential equation with a null second member that governs the oscillatory motion allows the fundamental vibration frequency determination.

Details are given concerning the analysis of beam lateral vibrations in [4–6]. Other approaches consider the use of the inverse mode shape problem techniques for finding the shape of structures and are reported in [7–10]. An interesting work made on a simply supported beam with symmetric overhangs was presented in [11]. This was treated by an algebraic solution. It was studied again using the FEM [12] and followed by extended study cases [13].

Several research works included springs differently located along the beams. They act as supports or are simply linked to discrete masses [14–17]. Also, another work deals with beams having end spring-hinged with translational restraint at another end [18].

Different approaches are still developed and improved [19–21]. In this investigation, the powerful FEM has been used mainly for its analysis possibilities.

2 FEM Approach

To apply in this study the FEM method, the beam finite element is used. It is constituted by a prismatic beam of length L. At its two nodes, shear forces and bending moments act. The unknown quantities are the vertical deflections and the slopes at the two ends.

2.1 Beam Stiffness Matrix

In a 2D analysis the acting forces at the two nodes of an elementary beam are as shown in Fig. 1.

The nodes displacements are: v_1, θ_1, v_1 and θ_2.

The stiffness matrix is:

$$[K] = E.I. \begin{bmatrix} \frac{12}{L^3} & \frac{6}{L^2} & \frac{-12}{L^3} & \frac{6}{L^2} \\ \frac{6}{L^2} & \frac{4}{L} & \frac{-6}{L^2} & \frac{2}{L} \\ \frac{-12}{L^3} & \frac{-6}{L^2} & \frac{12}{L^3} & \frac{-6}{L^2} \\ \frac{6}{L^2} & \frac{2}{L} & \frac{-6}{L^2} & \frac{4}{L} \end{bmatrix} \quad (1)$$

where

E.I is the flexural rigidity of the beam
L Beam length

The rotational spring value is introduced in the diagonal position of the stiffness matrix for the concerned rotational degree of freedom of the beam.

2.2 Mass Matrix

Based on FEM classical mathematical developments, the following mass matrix is obtained. It is constituted by 4 × 4 coefficients. It is said a consistent mass type. The beam elementary mass representation is given below:

$$[M] = \frac{\rho AL}{420} \begin{bmatrix} 156 & 22L & 54 & -13L \\ 22L & 4L^2 & 13L & -3L^2 \\ 54 & 13L & 156 & -22L \\ -13L & -3L^2 & -22L & 4L^2 \end{bmatrix} \quad (2)$$

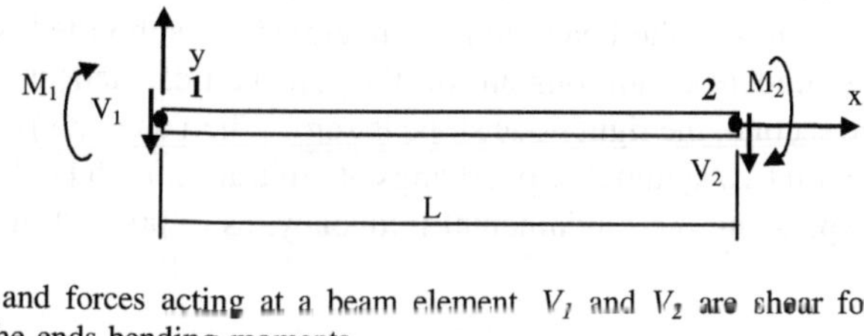

Fig. 1 Forces and forces acting at a beam element V_1 and V_2 are shear forces, M_1 and M_2 correspond to the ends bending moments

where

A cross-section area
ρ density of material
L beam length

The problem to be solved is:

$$\left|[K] - \omega^2[M]\right| = 0 \tag{3}$$

where

ω circular frequency of the vibrating system.

3 Results

The Jacobi method is used for the eigenfrequencies determination. This is an iterative process to solve a general eigenvalue problem. It consists of reaching diagonal forms for the stiffness and mass matrices.

3.1 Validation of the Results Concerning the Two Extreme Cases for $K_r = 0$

The fundamental free vibration of a uniform beam on a simple support of variable abscissa and a rotational elastic constraint at the right-hand end is investigated. The only considered mass is that of its material density. This analysis was carried out using the FEM and followed by plotted curves for a better interpretation of results. The sought values are the eigenvalues resulting from the free vibration of a beam differently supported. Only transverse oscillations were considered. In this study, the approach focuses on the s/L (span length/beam length) ratio effect. In the first two studied cases, the rotational restraint is not considered. It is compared to theoretical results and those obtained by the FEM method for the simply supported, pinned-fixed and a cantilever beam. The two beams represented in Fig. 2 were studied in [13] without elastic restraint. The boundary conditions for each situation depend on the existing constraints at the beam ends or intermediate supported nodes. For the first beam, the boundary conditions are on the zero vertical deflections at the two ends. But for the second, the right node slope is zero. When analyzing a transversally vibrating beam with two symmetric overhangs of arbitrary length [11], a K_1 constant is defined for expressing the fundamental frequency, as expressed in Eq. (4):

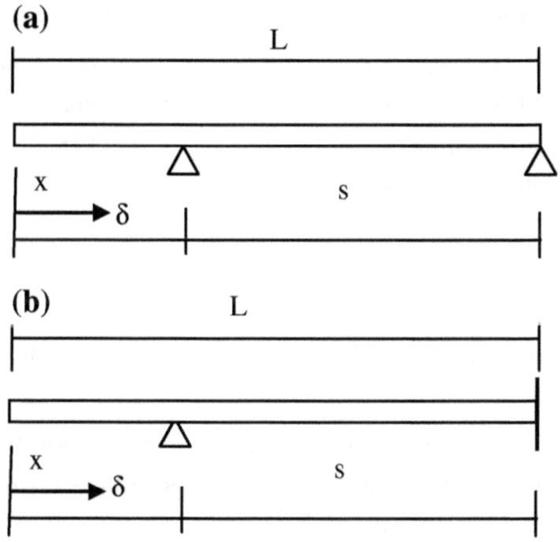

Fig. 2 Representations of a beam with one overhang. **Case a** Pinned-pinned. **Case b** Pinned-clamped

$$f^2 = K_1(EI/\rho AL^4) \tag{4}$$

It should be noted that an overhang of variable length and an intermediate span result from these boundary conditions. By varying the intermediate span length, different K_1 values are deduced and they are plotted against the span length/total beam length. The curves of Fig. 3, show the studied cases.

For comparison, an analysis of the results relatively to theoretical and the energetic Rayleigh method is made. The later is an approximate commonly used formula to find the fundamental vibration circular natural frequency of an oscillating system. The results are compared with the pinned-pinned, pinned-clamped and free-clamped cases where an agreement is reached. Table 1 shows numerical values of the ω_1 frequencies and the K_1 values for the various boundary conditions.

With

$$\omega_1 = \alpha \cdot \left(EI/\rho AL^4\right)^{1/2} \tag{5}$$

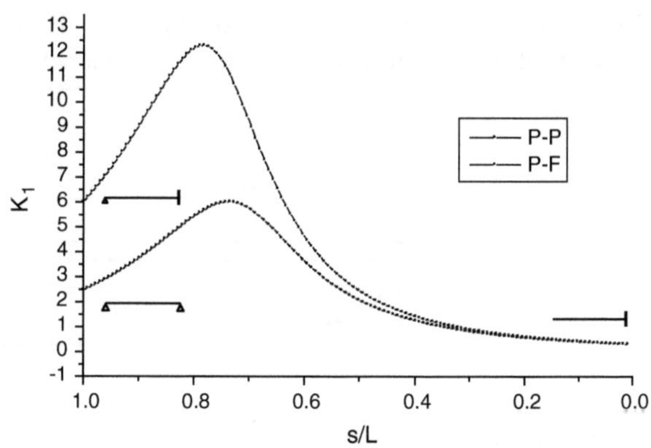

Fig. 3 Variations of K_1 in function of the S/L ratio for both the P–P and P–F cases

Table 1 α^2 and K_1 values for various ends restraints

Beam case	K_1	α^2 based on K_1	α^2 Theory	α^2 Rayleigh
Pinned-pinned	2.467	9.8688	9.8696	9.8767
Free-clamped	0.313	3.5152	3.5160	3.5675
Pinned-clamped	6.021	15.4175	15.4182	15.4511

where

ω_1 is the fundamental vibration circular frequency of the beam

The values of the K_1 factor corresponding to the 1st case was found with a good agreement relatively to an algebraic solution largely discussed in [11] and reported in [12] using an FEM based analysis. For the 2nd case, it is evident that the presented beam length is twice with regards to [11], thus at that time a correction is to be made: $0.31314/0.5^4 = 5.010$ (as given in [11, 12] and mentioned in [13]). The 3rd case study first introduced in [13] easily verifiable was confirmed by both, i.e., the theory and the Rayleigh method.

3.2 *Simply Supported Beam with a Variable Overhang Having a Rotational Restraint at One End*

This investigation type was first reported in [11] to analyze a simply supported beam with two symmetric overhangs. This numerical approach is used to determine the elastic modulus of a material. A program based on the FEM approach was developed [12] for this aim and the obtained results by the two approaches concerning this problem were found in compliance. Then, it was extended to analyze the cases of an only one overhang relatively to two ends conditions, namely pinned and fixed [13]. Also, it becomes then attractive to investigate the effect of disposing a varying rotational restraint at the right-hand end. The final idea is to regroup together all these close situations and after gather all the possible interpretations. The variable is still the S/L ratio while the K_r rotational elastic constraint is introduced, as shown in Fig. 4, after as a parameter.

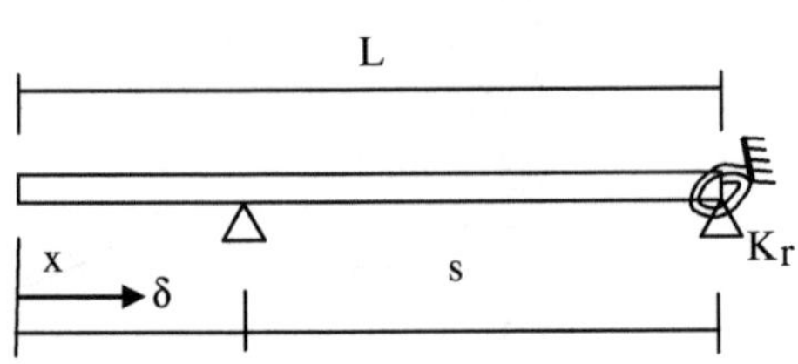

Fig. 4 Beam simply supported having a rotational restraint at one end with one overhang of arbitrary length

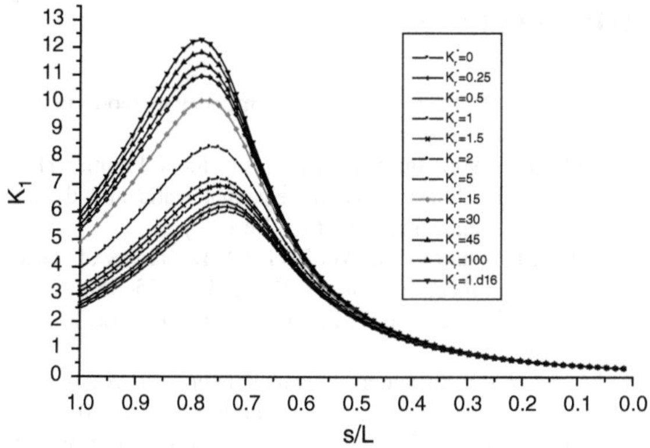

Fig. 5 Variation of K_1 as a function of the S/L ratio and the K_r restraint at the right-hand end with $K_r^* = K_r \cdot L/E.I.$

Figure 5 shows the K_1 variations in function of the S/L ratio and the K_r rotational restraint at the beam right-hand end as a parameter.

When the rotational restraint is important the beam is considered first pinned-fixed and after a cantilever behavior is obtained. The obtained results are in agreement with analytical ones and after introducing the rotational restraint at the right-hand end an intermediate behavior is shown allowing a simplified extraction of the fundamental vibration frequency for various values of the K_r (rotational restraint).

4 Conclusion

The K_1 values describing the transverse free vibration frequencies of a beam having a rotational restraint at one pinned end with a support of variable abscissa are investigated. They are obtained by applying the FEM method. The extreme situations corresponding to a simply supported beam and the case where the right-hand pinned end becomes clamped (cantilever beam) with one overhang of an arbitrary length studied in a previous work served as reference to validate the first investigation stage. At this step, the value K_r which represents the rotational restraint at the right-hand extreme end is zero.

The results show the K_1 variation in function of the S/L ratio, taking into account a varying parameter K_r. It is noted that these intermediate curves are delimited by two extreme situations. It should be noted that it is interesting to investigate other ends restraints conditions by considering a rotational end restraint showing an end constraint passing from a pinned case to a clamped situation.

The main remark consists of the fact that when the S/L tends to zero, the K_r influence may be ignored and at that time K_1 takes the value of 0.31314 corresponding to a cantilever beam. But, when the K_r value increases the K_1 factor varies from 2.467 to 6.021 representing pinned-pinned and pinned-clamped situations, respectively and a vertical translation of the curves is observed to tend to a pinned-clamped beam behavior.

References

1. Bathe KJ, Wilson EL (1976) Numerical methods in finite element analysis. Prentice-Hall, Englewood Cliffs
2. Bathe KJ (1996) Finite element procedures. Prentice Hall, Englewood Cliffs
3. Wilson EL (2008) Three dimensional static and dynamic analysis of structures, 3rd edn. Computers and Structures, Inc., Berkeley
4. Han SM, Benaroya H, Wei T (1999) Dynamics of transversely vibrating beams: using four engineering theories. J Sound Vib 225(5):935–988
5. Clough RW, Penzien J (2003) Dynamics of structures, 3rd edn. Computers and Structures, Inc., Berkeley
6. Paz M (2000) Structural dynamics theory and computation. Chapman & Hall, London
7. Meenakshi Sundaram M, Ananthasuresh GK (2012) A note on the inverse mode shape problem for bars, beams, and plates. Inverse Prob Sci Eng 21:1–16
8. Gladwell GML (2004) Inverse problems in vibrations, 2nd edn. Kluwer Academic Publications, Dordrecht
9. Chu MT (1992) Inverse eigenvalue problems. SIAM Rev 40:1–39
10. Chu MT, Gene HG (2005) Inverse eigenvalue problems theory, algorithms, and applications. Oxford University Press, Oxford
11. Murphy JF (1997) Transverse vibration of a simply supported beam with symmetric overhang of arbitrary length. J Test Eval JTEVA 25(5):522–524
12. Falek K, Rezgui L, Chalah F, Bali A, Nechnech A (2013) Structural element vibration analysis. In: ICA2013, vol 133, no. 5, Part 2 of 2 May 2013, Montreal, Canada
13. Chalah-Rezgui L, Chalah F, Falek K, Bali A, Nechnech A (2013) Transverse vibration analysis of uniform beams under various ends restraints. 2013 2nd international conference on civil engineering (ICCEN 2013), Stockholm, Sweden
14. Wu J-S, Hsu T-F (2007) Free vibration analyses of simply supported beams carrying multiple point masses and spring-mass systems with mass of each helical spring considered. Int J Mech Sci 49:834–852
15. Hamdan MN, Jubran BA (1991) Free and forced vibrations of a restrained cantilever beam carrying a concentrated mass. J KAU Eng Sci 3:71–83 (1411 A.H./1991 A.D.)
16. Lin H-Y, Tsai Y-C (2007) Free vibration analysis of a uniform multi-span beam carrying multiple spring–mass systems. J Sound Vib 302:442–456
17. Darabi MA, Kazemirad S, Ghayesh MH (2012) Free vibrations of beam–mass–spring systems: analytical analysis with numerical confirmation. Acta Mech Sin 28(2):468–481. doi:10.1007/s10409-012-0010-1
18. Maurizi MJ, Rossi RE, Reyes JA (1976) Vibration frequencies for a uniform beam with one end spring hinged and subjected to a translational restraint at the other end. J Sound Vib 48 (4):565–568
19. Banerjee JR (2012) Free vibration of beams carrying spring-mass systems—a dynamic stiffness approach. Comput Struct 104–105:21–26
20. Liu Y, Gurram CS (2009) The use of He's variational iteration method for obtaining the free vibration of an Euler-Bernoulli beam. Math Comput Model 50:1545–1552
21. Lai H-Y, Hsu J-C, Chen C-K (2008) An innovative eigenvalue problem solver for free vibration of Euler–Bernoulli beam by using the Adomian decomposition method. Comput Math Appl 56:3204–3220

On the Buckling Behavior of Curved Carbon Nanotubes

Sadegh Imani Yengejeh, Seyedeh Alieh Kazemi and Andreas Öchsner

Abstract In this study, numerous armchair and zigzag single-walled carbon nanotubes (SWCNTs) were simulated by a commercial finite element package and their buckling behavior was investigated through performing several computational tests with cantilevered boundary conditions and different bending angles. Both computational and analytical results were compared in the case of straight tubes. It was pointed out that the computational results are in good agreement with the analytical calculations. It was also concluded that the first critical buckling load of both straight armchair and zigzag CNTs increases by increasing the chiral number. In addition, it was indicated that the first critical buckling load of straight CNTs decreases by introducing the bending angle to the structure of CNTs. However, this decrease is more noticeable in the case of armchair and zigzag CNTs with higher number of chirality and it is almost negligible for CNTs with lower number of chirality.

Keywords Finite element method · Buckling behavior · Carbon nanotubes · Curved shape

S. Imani Yengejeh (✉)
Department of Solid Mechanics and Design, Faculty of Mechanical Engineering, Universiti Teknologi Malaysia—UTM, 81310 Skudai, Johor, Malaysia
e-mail: Imani.sd@gmail.com

S.A. Kazemi
Department of Mechanical Engineering, The University of Birjand, Birjand, Iran
e-mail: Sa.kazemi83@gmail.com

A. Öchsner
School of Engineering, Griffith University, Gold Coast Campus, Southport 4222, Australia
e-mail: Andreas.Oechsner@gmail.com

A. Öchsner
School of Engineering, The University of Newcastle, Callaghan, NSW 2308, Australia

A. Öchsner and H. Altenbach (eds.), *Mechanical and Materials Engineering of Modern Structure and Component Design*, Advanced Structured Materials 70,
DOI 10.1007/978-3-319-19443-1_33

1 Introduction

After the discovery of carbon nanotubes (CNTs) by Iijima [1], numerous opportunities and possibilities have been opened to produce an entire generation of new materials and structures that possesses unique physical properties [2]. Study of the CNTs is one of the most promising domains in the area of physics, mechanics, chemistry, and material science [3]. The application of CNTs lies within a wide range, including nanocomposites, nanodevices, and nanoelectronics [4–9]. The studies indicate that CNTs possess significant material properties that distinguish them from any known material. For example, the elastic modulus of CNTs is reported to be more than 1 TPa. In addition, they are also particularly flexible in bending and can undergo large elastic deformation without breaking [10]. The investigation of the behavior of CNTs are mainly divided into two groups of computational and experimental approaches. Molecular dynamics and continuum mechanics approaches (e.g. the finite element method (FEM)) have been the most common computational techniques to investigate the behavior of CNTs [11]. Due to numerous applications of CNTs in mechanical aspects, the investigation of their buckling behavior is crucially important. The following paragraph summarizes several related studies where the buckling behavior of CNTs was investigated.

In 2003, Pantano et al. [12] presented a nonlinear structural mechanics based approach for simulating the deformation of single- and multi-walled carbon nanaotubes (MWCNTs). They applied shell finite elements for the modeling the individual tubes, where a particular pairing of elastic properties and mechanical thickness of the tube wall was identified to enable successful modeling based on the shell theory. They also simulated the effects of van der Waals forces with special interaction elements. They validated their molecular dynamics simulations with high resolution micrographs available in literature [13]. They also investigated the mechanics of wrinkling of MWCNTs, demonstrating the role of the multi-walled shell structure and interwall van der Waals interactions in governing buckling and post-buckling behavior. Later in 2010, Guo and Zhang [14] investigated the bending stiffness of SWCNTs by using the molecular mechanics model and the deformation mapping technique. They proposed an analytical expression for the bending stiffness of SWCNTs. They finally showed that the bending stiffness of a SWCNT is approximately proportional to the cube of its radius.. The effect of defects on the buckling behavior of CNTs was investigated by Ghavamian and Öchsner [15]. Their study was based on the FEM. In detail, they modeled two basic CNTs in their perfect form. Then the buckling behavior of CNTs was evaluated by comparing their critical loads obtained from the simulation and analytical calculations. They concluded that the existence of any curvature in the structure of nanotubes decreases their buckling strength.

The main objective of the actual work is to continue and broaden the previous studies and investigate the buckling behavior of CNTs in the case of different bending angles.

2 Methodology

2.1 Geometric Definition

CNTs are some kind of hollow cylinder shaped structure based on the major similarity between a CNT and graphene atomic configuration, as illustrated in Fig. 1. Having the diameters ranging from 1 to 50 nm and length over 10 μm, these nanostructures can be imagined by rolling a graphene sheet into a cylinder. The geometry of a CNT is defined by the chiral vector $\vec{C}_h$ and the chiral angle θ. The chiral vector is presented by two unit vectors $\vec{a}_1$ and $\vec{a}_2$ and two integers m and n as it is presented by the following equation [16]:

$$\vec{C}_h = m\vec{a}_1 + n\vec{a}_2, \tag{1}$$

The structure of CNTs is defined based on the chiral vector or angle by which the sheet is rolled into a cylinder, in three different configurations including zigzag, armchair and chiral. In the case of ($\theta = 0°$) or ($m = 0$) the zigzag CNT is constructed. An armchair CNT is obtained if, in terms of chiral vector ($m = n$) or in terms of chiral angle ($\theta = 30°$), and finally a chiral CNT is shaped if ($0° < \theta < 30°$) or ($m \neq n \neq 0$) [16].

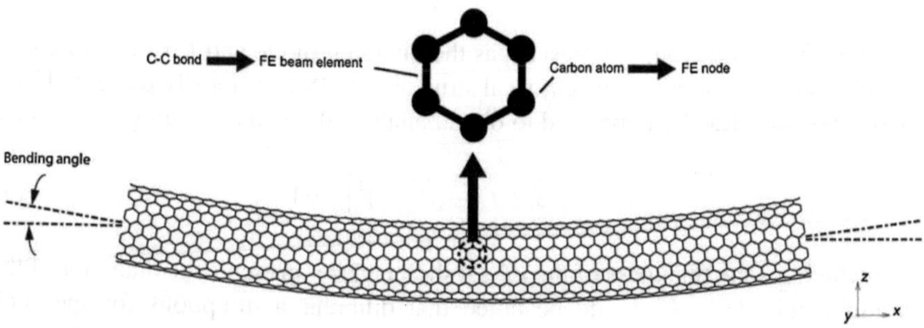

Fig. 1 Side view of the (10,10) CNT as a space-frame structure

Based on the following equation, the diameter of the CNT can be calculated:

$$d_{\mathrm{CNT}} = a_0\sqrt{m^2 + mn + n^2}/\pi, \tag{2}$$

where $a_0 = \sqrt{3}b$ and $b = 0.142$ nm is the length of the C–C bond [16].

Our modeling method follows the idea first suggested in [17] where the theory of classical structural mechanics was extended into the modeling of carbon nanotubes: In a CNT, carbon atoms are bonded together by covalent bonds which have their characteristic lengths and angles in a three-dimensional space. Then, it was suggested that CNTs, when subjected to loading, act as space-frame configurations. Therefore, the bonds between carbon atoms are considered as connecting load-carrying generalized beam members, while the carbon atoms behave as joints of the members. This idea is illustrated in Fig. 1.

In this paper, the models of armchair and zigzag SWCNTs were simulated by the CoNTub software [18]. Defining the chirality and the length of the tubes, the spatial coordinates of the C-atoms and the corresponding connectivities (i.e. the primary bonds between two nearest-neighboring atoms) were calculated. Then the gathered data was transferred to a commercial finite element package, where C–C bonds were modeled as circular beam elements [19]. Afterwards, the FE analyses were conducted and buckling behavior of different zigzag and armchair CNTs was evaluated.

The phenomenon of buckling is in its simplest form a specific kind of elastic instability in a slender structure that occurs under certain compressive loads. In the basic theory of elasticity, the critical buckling load of a straight elastic beam is presented in Eq. 3 [20]:

$$P_{\mathrm{cr}} = \frac{n^2\pi^2 EI}{(KL)^2} \tag{3}$$

In the above equation, n represents as the buckling mode and I is the structure's second moment of area. As the classical structure of CNTs is mostly presented by a hollow cylinder, Eq. 4 can be used to obtain analytical results for straight CNTs as:

$$I = \pi\left[(d+t)^4 - (d-t)^4\right]/64 \tag{4}$$

In which t is the thickness of the tube's shell and d represents the diameter of the pertaining tube [17]. It should be noted that different assumptions for the shell thickness t can be found in literature, see for example [21, 22]. We assign here the same value as in [21, 22] where the thickness of a single-walled carbon nanotube was assumed to be the same as the interlayer spacing of graphite (0.34 nm).

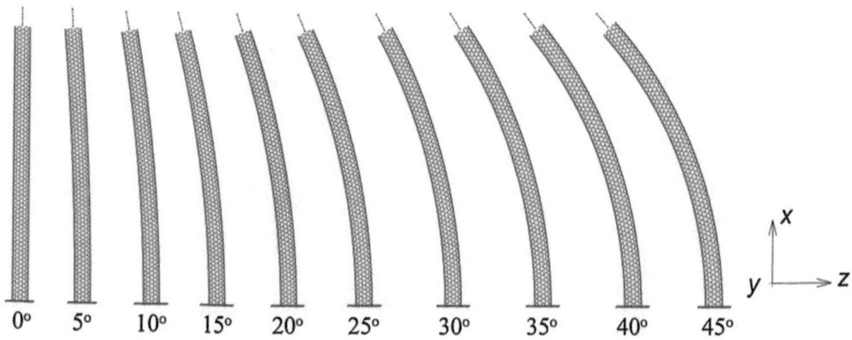

Fig. 2 Armchair CNT with cantilevered boundary conditions and different forms from 0° to 45° bending angle

2.2 Material Parameters and Boundary Conditions

In this study, the buckling behavior of numerous types of SWCNTs under cantilevered boundary conditions is investigated, in which one end is fully fixed and the other end is completely free. Figure 2 illustrates an armchair CNT under cantilevered boundary condition. The models of CNTs were simulated from straight CNTs to curved CNTs with 45° bending angle.

In order to define the properties of the equivalent beam elements for the CNT bonds, the same values for the equivalent beam elements are assumed as in the approach proposed in [2, 17, 22]. These effective material and geometrical properties were obtained in the mentioned references based on a molecular mechanics approach.

3 Results and Discussion

The buckling properties of different zigzag and armchair SWCNTs were investigated and the effect of bending angle was taken into consideration. Their critical loads (to be more precise, the first positive buckling mode since this is in practical applications the most critical one) of CNTs were obtained analytically and computationally, applying an FE approach. Afterwards, both methods were compared in the case of the straight CNTs in order to have some kind of validation of the FE approach. For the FE method, zigzag and armchair CNTs with lengths of 15 and 15.4 nm, respectively, were simulated by a commercial finite element package (MSC. Marc). Then, as illustrated in Fig. 3, by introducing arbitrary compressive point loads to one of the CNT's end, the buckling behavior of the CNTs was

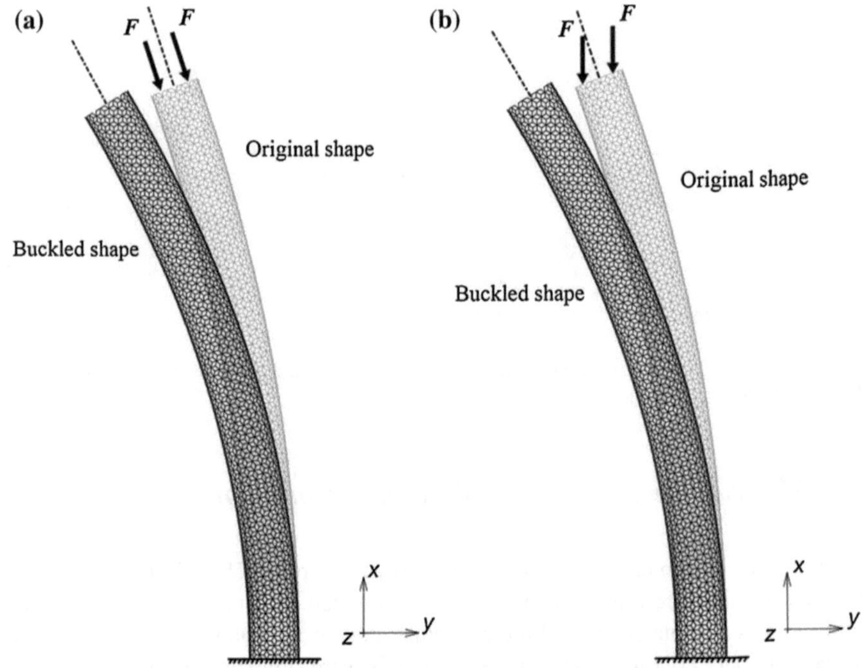

Fig. 3 (10,10) zigzag CNT's first mode under buckling load in original and buckled shape with **a** the first case and **b** the second case of cantilevered boundary conditions

investigated for cantilevered boundary condition with different curvatures from 0° to 45° bending angle (one end fixed and the other end free). Finally, the computations yielded the critical buckling loads of the CNTs. Two different boundary conditions were considered at the free end. For the first case the applied load was perpendicular to the end surface, i.e. there is an angle between the *X*-axis and the line of action, and for the second case the applied load was in direction to the *X*-axis as shown in Fig. 3. The CNTs were assumed to be hollow cylinders, so that by using Eq. 3 their critical loads were evaluated as an analytical solution. In this equation we considered $K = 2$ for the cantilevered case.

The simulated CNTs and their characteristics are presented in Table 1.

The relative difference between the analytical solution and FEM result is defined by the following equation:

$$\text{Relative difference in } \% = \left|\frac{\text{FEM result} - \text{analytical solution}}{\text{analytical solution}}\right| \times 100 \quad (5)$$

Table 1 Characteristic of simulated CNTs

CNT Type	Chirality (n,m)	Length (nm)	Diameter (nm)	Young's modulus (TPa)	Critical load P_{cr} (nN)		Relative difference in %
					Analytical solution	FEM result	
Armchair	(3,3)	15	0.407	1.039	0.172	0.122	29.07
Armchair	(5,5)	15	0.678	1.039	0.594	0.549	7.52
Armchair	(7,7)	15	0.949	1.039	1.468	1.492	1.64
Armchair	(9,9)	15	1.220	1.039	2.981	3.152	5.73
Armchair	(10,10)	15	1.356	1.039	4.032	4.312	6.95
Armchair	(11,11)	15	1.492	1.039	5.312	5.682	6.97
Armchair	(12,12)	15	1.612	1.039	7.471	7.63	2.13
Armchair	(13,13)	15	1.754	1.039	9.519	9.54	0.22
Armchair	(14,14)	15	1.898	1.039	10.743	11.68	8.72
Armchair	(16,16)	15	2.171	1.039	15.92	17.3	8.67
Zigzag	(8,0)	15.4	0.616	1.024	0.405	0.41	1.23
Zigzag	(9,0)	15.4	0.705	1.025	0.643	0.588	8.59
Zigzag	(10,0)	15.4	0.786	1.028	0.782	0.809	3.45
Zigzag	(11,0)	15.4	0.861	1.031	1.107	1.081	2.35
Zigzag	(12,0)	15.4	0.939	1.032	1.408	1.406	0.14
Zigzag	(14,0)	15.4	1.088	1.034	2.58	2.72	5.43
Zigzag	(16,0)	15.4	1.244	1.037	3.182	3.34	4.96
Zigzag	(18,0)	15.4	1.392	1.038	4.49	4.77	6.24
Zigzag	(20,0)	15.4	1.556	1.039	6.376	6.503	1.99
Zigzag	(22,0)	15.4	1.721	1.040	8.875	9.12	2.76
Zigzag	(24,0)	15.4	1.875	1.041	12.42	12.6	1.45
Zigzag	(26,0)	15.4	2.021	1.042	13.87	14.1	1.66

This difference is listed in Table 1 for each CNT. Based on the results, it can be concluded that the computational solutions are in good agreement with the analytical calculations. It was revealed that the critical buckling load of both straight armchair and zigzag CNTs increases by increasing the chiral number as shown in Fig. 4.

Having a closer look on the results, it can be also concluded that the buckling strength of armchair and zigzag CNTs decreases by increasing the bending angle for both cases of boundary condition, as illustrated in Fig. 5. However, it should be noted that this trend is more obvious for the CNTs with higher chirality, as the decrease of critical buckling load for both armchair and zigzag CNTs with lower

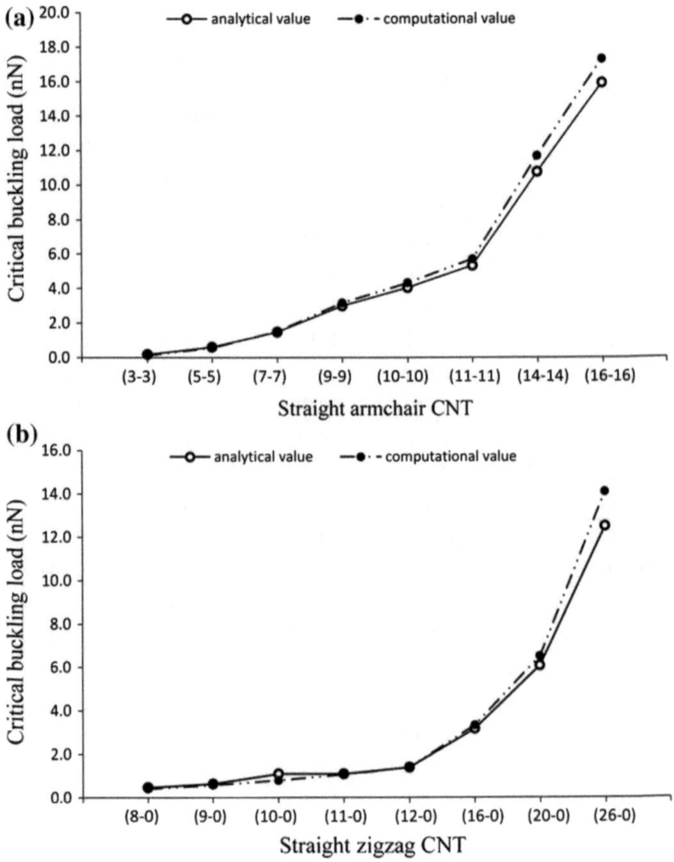

Fig. 4 (10,10) zigzag CNT's first mode under buckling load in original and buckled shape with **a** the first case and **b** the second case of cantilevered boundary conditions

chirality is almost negligible. It is also indicated that the gradual decrease for the first case of boundary condition is more significant comparing to the second case at higher curvatures.

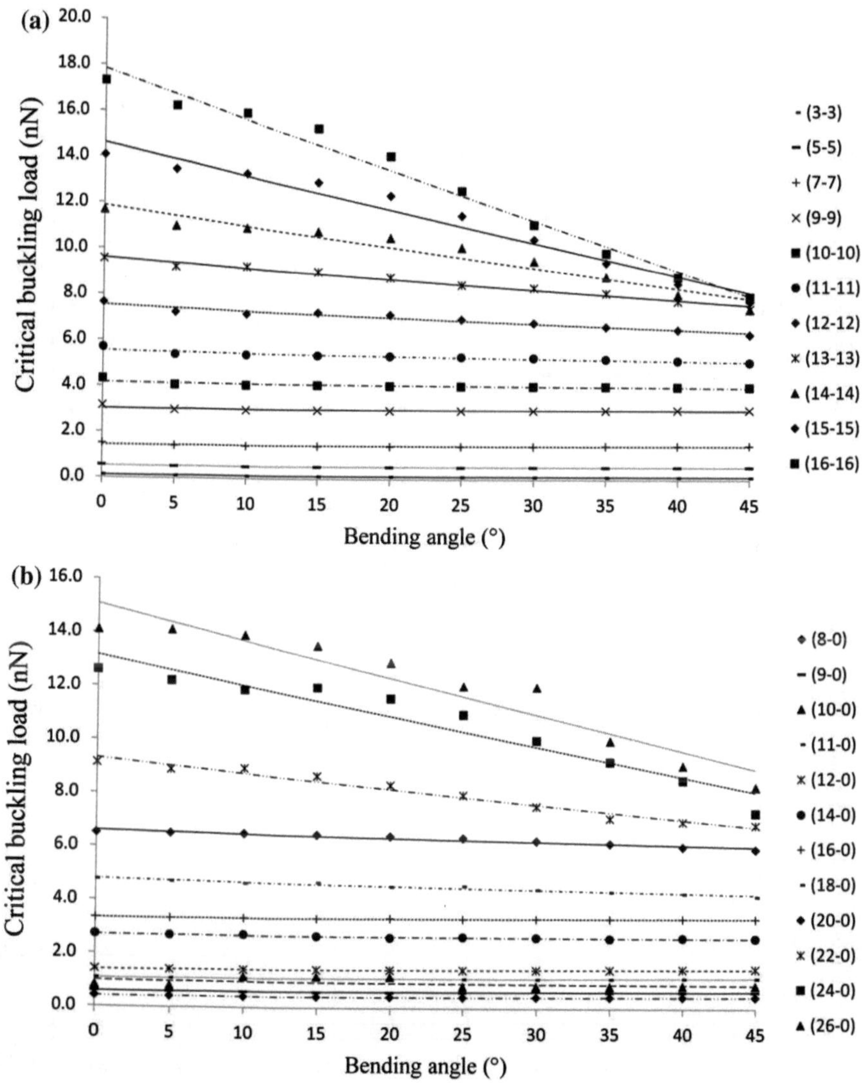

Fig. 5 Change in critical buckling load (P_{cr}) with **a** armchair and **b** zigzag CNTs for the first case and with **c** armchair and **d** zigzag CNTs for the second case

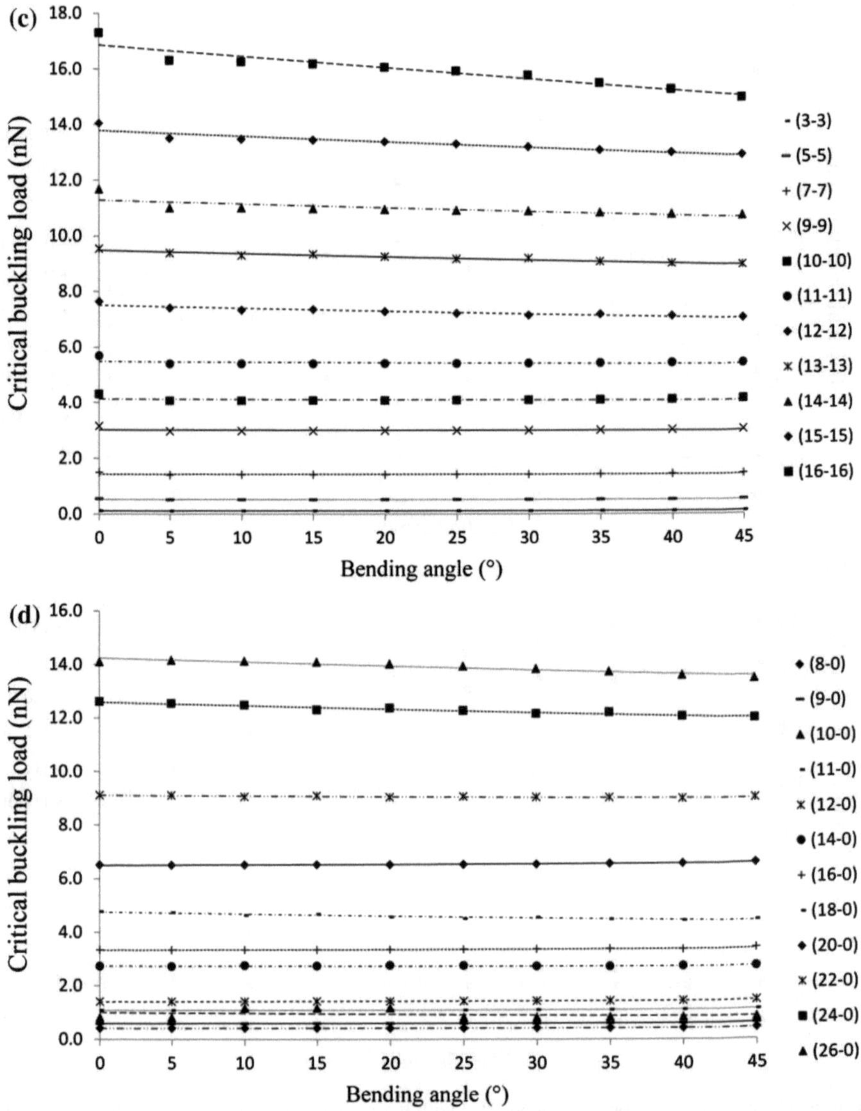

Fig. 5 (continued)

4 Conclusions

In this research, several SWCNTs (zigzag and armchair) were simulated by an FE approach and their buckling behavior was investigated through performing several computational tests with cantilevered boundary condition and different variable bending angles. Towards achieving the most accurate results the buckling behavior of these CNTs were evaluated analytically and computationally and were compared together. It was concluded that the critical buckling load of both straight armchair and zigzag CNTs increases by increasing the chiral number. It was also shown that the buckling strength of armchair and zigzag CNTs with higher chirality decreases by introducing bending angles for both cases of boundary condition. Nevertheless, the change of increasing the bending angle on the critical buckling load of armchair and zigzag CNTs with lower number of chirality is almost negligible. The finding of this study may have useful effects on further investigations of armchair and zigzag CNTs.

References

1. Iijima S (1991) Helical microtubules of graphitic carbon. Nature 354:56–58
2. To CWS (2006) Bending and shear moduli of single-walled carbon nanotubes. Finite Elem Anal Des 42:404–413
3. Mehdipour I, Barari A, Kimiaeifar A, Domairry G (2012) Vibrational analysis of curved single-walled carbon nanotube on a Pasternak elastic foundation. Adv Eng Softw 48:1–5
4. Hai-Yang S, Xin-Wei Z (2010) Mechanical properties of nickel-coated single-walled carbon nanotubes and their embedded gold matrix composites. Phys Lett A 374:1068–1072
5. Lai PL, Chen SC, Lin MF (2008) Electronic properties of single-walled carbon nanotubes under electric and magnetic fields. Physica E 40:2056–2058
6. Deretzis I, La Magna A (2008) Electronic transport in carbon nanotube based nano-devices. Physica E 40:2333–2338
7. Chowdhury R, Adhikari S, Mitchell J (2009) Vibrating carbon nano-tube based bio-sensors. Physica E 42:104–109
8. Hornbostel B, Pötschke P, Kotz J, Roth S (2008) Mechanical properties of triple composites of polycarbonate, single-walled carbon nano-tubes and carbon fibres. Physica E 40:2434–2439
9. Hwang CC, Wang YC, Kuo QY, Lu JM (2010) Molecular dynamics study of multi-walled carbon nanotubes under uniaxial loading. Physica E 42:775–778
10. Wang X, Zhang YC, Xia XH, Huang CH (2004) Effective bending modulus of carbon nanotubes with rippling deformation. Int J Solids Struct 41:6429–6439
11. Imani Yengejeh S, Akbar Zadeh M, Öchsner A (2014) On the buckling behavior of connected carbon nanotubes with parallel longitudinal axes. Appl Phys A 115:1335–1344
12. Pantano A, Boyce MC, Parks DM (2003) Nonlinear structural mechanics based modeling of carbon nanotube deformation. Phys Rev Lett 91:145504
13. Poncharal P, Wang ZL, Ugarte D, de Heer WA (1999) Electrostatic deflections and electromechanical resonances of carbon nanotubes. Science 283:1513–1516
14. Guo X, Zhang T (2010) A study on the bending stiffness of single-walled carbon nanotubes and related issues. J Mech Phys Solids 58:428–443
15. Ghavamian A, Öchsner A (2012) Numerical investigation on the influence of defects on the buckling behavior of single-and multi-walled carbon nanotubes. Physica E 46:241–249
16. Dresselhaus MS, Dresselhaus G, Saito R (1995) Physics of carbon nanotubes. Carbon 33:883–891
17. Li C, Chou TW (2003) A structural mechanics approach for the analy-sis of carbon nanotubes. Int J Solids Struct 40:2487–2499
18. Melchor S, Martin-Martinez FJ, Dobado JA (2011) Dobado JA. CoNTub v2.0-algorithms for constructing C3-symmetric models of three-nanotube junctions. J Chem Inf Model 51: 1492–1505
19. Kang Z, Li M, Tang Q (2010) Buckling behavior of carbon nanotube-based intramolecular junctions under compression: Molecular dynamics simulation and finite element analysis. Comput Mater Sci 50:253–259

20. Lui EM, Chen WF (1987) Steel frame analysis with flexible joints. J Construct Steel Res 8:161–202
21. Tserpes KI, Papanikos P (2005) Finite element modeling of single-walled carbon nanotubes. Compos B 36:468–477
22. Kalamkarov AL, Georgiades AV, Rokkam SK, Veedu VP, Ghasemi-Nejhad MN (2006) Analytical and numerical techniques to pre-dict carbon nanotubes properties. Int J Solids Struct 43:6832–6854

Metrology by Image: Discussing the Accuracy of the Results

Fabiana Rodrigues Leta, Juliana F.S. Gomes, Pedro B. Costa and Felipe de O. Baldner

Abstract The development of new inspection techniques have been studied due to their important role in ensuring the quality of products and services. Achieving technical specifications of products considering acceptance criteria requires knowledge of assured inspection methods. Inspection techniques have evolved expressively, lacking, however, in some cases, of certified standardization. The past ten years have experienced considerable development of inspection methods. In part, this is due to the need to meet the security and reliability requirements for consumer protection. Moreover, the extraordinary advances in electronics and computers were also important factors for this advance. In this context visual inspection by images is being applied in many applications, allowing the identification of innumerous product features, like cracks, corrosion, distortion, misalignments, porosity, among others. But, inspections and measurements require calibration of instruments, devices and systems. In order to obtain confident measurements it is important to study and specify the suitable lighting system, camera, lens and principally the best computer measurement algorithms. Thus the aim of this paper is to discuss the relevance of the main parameters that influence the measurement of images using computer vision techniques.

Keywords Computer vision · Image · Metrology · Measurement

F.R. Leta (✉) · P.B. Costa · F. de O. Baldner
Universidade Federal Fluminense -UFF, Mechanical Engineering Department,
Computational and Dimensional Metrology Laboratory, Niterói, RJ 24.210-240, Brazil
e-mail: fabianaleta@id.uff.br

P.B. Costa
e-mail: pedrobcostarj@gmail.com

F. de O. Baldner
e-mail: fbaldner@gmail.com

J.F.S. Gomes · P.B. Costa · F. de O. Baldner
National Institute of Metrology, Standardization and Industrial Quality—Inmetro,
Duque de Caxias, RJ 25.250-020, Brazil
e-mail: jfgomes@inmetro.gov.br

A. Öchsner and H. Altenbach (eds.), *Mechanical and Materials Engineering of Modern Structure and Component Design*, Advanced Structured Materials 70,
DOI 10.1007/978-3-319-19443-1_34

1 Introduction

Visual inspection is one of the oldest techniques for evaluating products. It is considered to be a non-destructive test (NDT) used largely to evaluate materials and products due to its ease of execution and low cost. However, its effectiveness relies directly on the standardization of the observation environment and the experience of the technician in recognizing characteristics and defects.

With an ease of access to new technologies such as software and digital imaging equipment, the computer vision area emerges as a versatile tool for the development of innovations in NDTs for many applications. The usage of automated image inspections in the industry is becoming more and more an interesting solution, especially for identifying defects and the analyzing a product's quality.

Such systems offer a greater level of repeatability and reproducibility than human performed visual inspections, mostly due to the elimination of inherently human factors like subjectivity, fatigue and sluggishness. Image driven inspection and measurement systems allow for the reduction of errors intrinsic to the subjectivity of the human being, justifying even more its employment in industry and trade. These systems not only comprise dimensional and geometrical aspects, but also characteristics of appearance, such as the case would be in food, fabric, cosmetics, paints, etc., encompassing the most varied industrial sectors.

Image inspection processes that required a high level of accuracy must account for key aspects in metrology, such as measurement standards, device calibrations, measurement uncertainty, among others, bringing forth the importance of the theme of image metrology, which binds the concepts of computer vision [1], with its algorithms for image processing and analysis, to the rigors of the measurement science.

1.1 The Importance of Inspection in the Industry

All fields of Engineering, especially those related to industrial production are intimately connected to product inspection (including materials), whose properties and characteristics must be verified by the use of tests, with the objective of complying with some project specification. The fulfillment of products' technical specifications and acceptance criteria requires a knowledge of inspection methods, in order to enable the choice of that which will be more appropriate for the application [2, 3].

Society has been demanding more and more products and services that are both of quality and safe, bringing great challenges to the industry. Meeting such challenges imply the development of new materials, devices and technologies, therefore requiring an availability of new methods to perfect or replace usual inspection techniques in the industry, in companies and research facilities. Thereby, to improve the efficiency of inspection systems brings about the minimization of error and time spent in such processes, allowing a consequential improvement of product quality in response to consumer demand.

The development of new inspection techniques has been the study focus regarding its important role in ensuring quality for both products and services. By using tests in order to prevent accidents and in preventive and predictive maintenance techniques contributes for cost reduction in maintenance, repairs and equipment replacement. If in the past tests were treated as a means of knowing the properties of a material, allowing the use in different applications, they are, nowadays, strongly bound to the quality of goods and services, aiming at cost reduction and preservation of life and the environment, becoming thus a competitive factor in companies that employ them.

What is noteworthy in discussions and studies in several countries is the deepening of the relation between inspection, quality and competitiveness. Inspection techniques have been evolving in the world market systematically but lacking, in some cases, of standardization in what regards the certification of the personnel involved and execution procedures.

The last ten years have presented a considerable development in testing methods. In part, this is due to the necessity of satisfying requisites for safety and reliability demanded for consumer protection. Besides, the advances in electronics and computers have also been important factors to this growth.

The ever increasing need for automation in industry processes was one of the boosters in the development of new techniques and testing equipment to be used in serial production. Its application in product control, in companies of any size, is motivated by the growing needs of quality and economy in current society.

By working to protect the consumer and improve quality in products traded worldwide, several standards and technical regulations were created by World Trade Organization as a way to standardize products' acceptance criteria.

Standards and regulations are documents that establish characteristics for a product, such as its function, performance, packaging and labeling, or methods and processes in related products. However, standards are voluntary while regulations are mandatory.

With the purpose of increasing exportation and complying with international demands, a notable effort has been observed in implementing quality systems in both private and government sectors. Standardization defines the kinds of inspections, considering quantitative and qualitative methods for analysis, in order to provide supervision and to guarantee quality of the marketed products. Metrology has the decisive role in results' reliability and the inherent improvement of the final product (Fig. 1).

1.2 The Evolution of Visual Inspection in the Industry

As a central focus in sector discussions, the increase in industrial productivity is considered as one of the greatest objectives of the world economy and, in the latest

Fig. 1 Parameters of influence in a product's final quality

years, has assumed an outstanding importance, demanding more reliable tests in product inspection.

Traditional testing techniques are performed with the naked eye (sometimes associated with magnifying glasses) and usually when the surface to be tested is of easy access. When access is not so easy, a solution to perform such tests is the usage of cameras connected to computers, eventually with equipment to magnifying the field of observation. Therefore surfaces may be visually inspected by using endoscopes with fiber optic cables, allowing access to the insides of equipment, machines, pipes, among others, which are also used in the medical field and surgical interventions.

Recent research in the field has been producing technologies using computer vision systems for recognition and inspection of objects and complex surfaces, sometimes in real time. Automation has promoted a development of testing equipment in production lines and the evolution of sensor technology has allowed the establishment of new measurement techniques in products, by means of testing, allowing for a permanent monitoring during the process, as will be discussed later.

An inspection process, in its broadest form, may encompass the steps (Fig. 2).

After the definition of what needs to be evaluated, the most appropriate inspection process must be selected. In dimensional inspections, the tolerance field of the quantity being verified must be taken into consideration, the resolution of the instrument and the means of interaction with the measurand. It is recommended that the instrument has a reading that is at least one tenth of the tolerance field of the part, or in the worst case scenario, equal to the fifth part.

In inspection systems that rely on quantitative evaluation of the characteristics of products or materials, it is recommended for the used instruments to be calibrated and verified by accredited laboratories.

Fig. 2 Steps of an inspection process

1.3 The Use of Metrology in Industrial Inspections

The world needs of the commerce require inspections and measurements with guaranteed worldwide reliability and many of those transactions demand calibration of devices or measurement systems. In this way, recognition of measurement equivalence and its acceptance by national authorities reduce the effects of technical barriers to commerce, thus allowing exporting [4].

It can be said that scientific knowledge becomes concrete when it is possible to express it numerically. However, a number on its own just represents an intensity. To comprehend and transmit technical information, the quantification of a property is not enough. The practice of metrology comprehends the importance of attaching a unit of measurement to such a numerical quantity. In the industrial field, metrology is fundamental to determine conformity of products and services in Engineering with its specifications.

Metrology is the science of measurement and its applications, dealing with its basic concepts, methods, errors and error propagation, units, standards for quantifying physical quantities as well as static and dynamic behavior of measurement systems [5, 6]. Metrology encompasses all theoretical and practical aspects of measurements for any measurement uncertainty and field of application.

Metrology encompasses:

- Units of measurement, allowing for conversion of abstractions or quantities (length, mass, etc.) in a way that it can be quantified (meter, kilogram, etc.);
- Instruments, that are calibrated in terms of such standardized units of measurement;
- Measurements, which consists in the use of such instruments by means of achieving quantification of a given characteristic being studied of a product or process.

Inspections and measurements are not only about retain or reject manufactured products that do not comply with standards, but also to guide manufacturing and the assembly of parts as a way to ensure quality. Measuring or simply verifying to prevent failure is an important task not only in the industry but also in the daily lives of the citizens. It can be noted that metrology is present in every moment of the daily lives of citizens such as when they are shopping for food, as they fuel their cars, take a cab or even when they monitor their blood pressure, among others. The universe of needs in the usage of metrology is not exhausted by these examples. Much of the scientific and technological advance of men has occurred by experiments and measurements.

When looking at the importance of measurements in the many applications mentioned, it is possible to say that guaranteeing exact results that will represent the characteristic in question is of the utmost importance to. Small differences in measurement results may lead to enormous differences in any conclusion leading to a decision. Being able to believe in the results of a measurement is paramount to making safe conclusions.

1.4 Image Metrology

With the evolution of software, digital cameras and computers with greater resources and interfaces, all of those with declining cost, image processing and analysis algorithms have become tools of interest in the development of new methodologies for non-destructive testing with big applications in industry.

The usage of automated inspections in many sectors is increasingly becoming a solution, especially in the analysis of compliance for parts, search of manufacturing defects, final quality analysis of a product, object tracking, among others. In Fig. 3 there is a simple schematic of a transporting treadmill with products to be tracked or inspected in an automated production line shown, which uses a computer system for identification and image analysis.

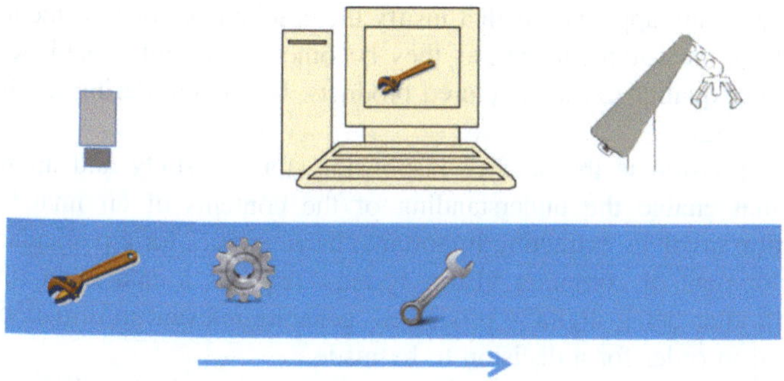

Fig. 3 Schematic of an automated object recognition system

As previously commented before, unlike problems that may arise in an human-driven visual inspection process, systems such as these may offer accuracy and repeatability in contactless measurement.

However, the obtainment of characteristics from digital images may have many diverse objectives, thus indicating that different degrees of accuracy from the measurement results may be accepted, in accordance with the application.

In applications where the purpose is to identify or recognize objects, for example, generally a set of characteristics has to be obtained and analyzed through a decision system, therefore tolerating small variations in the measurements of such characteristics, which will be defined with an acceptance criteria without compromising the final result. The same happens in qualitative analysis, when the objective of the system is a simply identification of presence of absence of elements in an object. However, high levels of accuracy and low measurement uncertainties and tolerances are needed in many applications of measurement for calibration, as illustrated in Fig. 4.

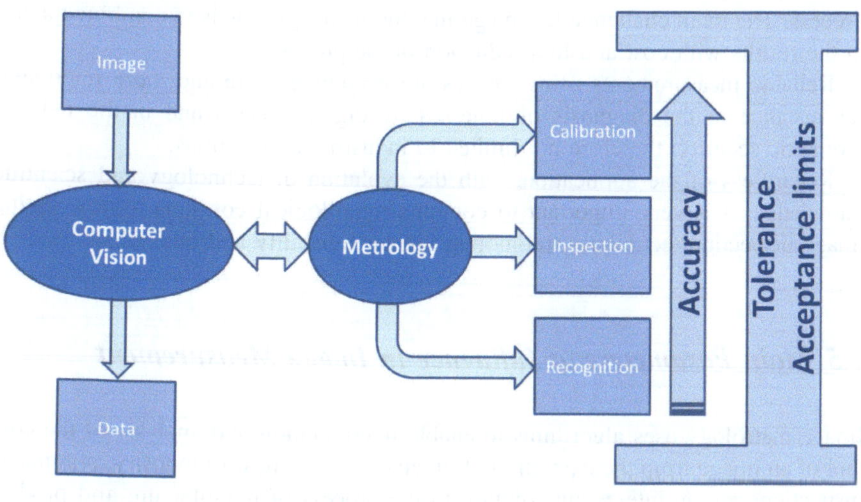

Fig. 4 Automatic measuring system

There are many applications that justify the employment of new methodologies of contactless measurement that, as they become increasingly sophisticated, may assure greater quality in manufactured products, when considering the fundamentals of metrology.

Computer vision is the science responsible for the study and application of methods that enable the understanding of the contents of an image, and this interpretation involves extracting important characteristics for a particular purpose. The development of computer vision systems requires a data input (the image) which will, after being digitally processed, generate relevant information that will be analyzed in order for a decision to be made.

In many applications, mathematical operations are applied to digital images in order to obtain measurements or to conduct inspections in the target object of the analysis. This process can be automated and requires high accuracy, to result in reliable information. The quantities or evaluated characteristics may involve not only dimensional aspects, but also color and texture characteristics and general shape, as is the case with food, cosmetics, paint, etc. Currently, computer vision has been used in many areas of industry, significantly contributing in many steps of the productive cycle of a product, for example, in the guidance of robots, task automation, quality control, etc.

Even though a computer vision system needs to be organized in accordance with a given application, there are typical steps that one follows, such as image acquisition, pre-processing, segmentation, characteristic extraction and analysis.

An image metrology system (IMS) is equivalent to a computer vision system due to the same basic steps but prioritizing the metrological reliability in the final result. An IMS has the purpose of assessing the compliance of a product, in the case of recognition or inspection, or simply a measurand, in the case of calibration of standards and instruments. Automated inspection systems are commonly used in production lines when human activity is repetitive or when product manufacture is fast, requiring fast and accurate measurements for decision-making during the process. The main challenge for image measurement systems is to combine quality in the results with cost and loss reduction in the process.

Reliable measurements using images are becoming more and more important, for instance in the diagnosis of illnesses by digital images and in the field of biometric security, therefore not limited to industrial applications.

Regardless of the application, with the evolution of technology and scientific knowledge, it is very important to combine metrological concepts to new digital image inspection techniques, promoting greater reliability and quality.

1.5 Main Parameters of Influence in Image Measurement

Image metrology uses algorithms to enable interpretation and analysis of the content of an image from the extraction of given characteristics. In industry, automated inspections are an interesting solution to the process of manufacture and product assembly, as well as the final quality analysis of these, because they enable the evaluation of characteristics of color, texture and shape, which are criteria usually used by the end user when acquiring products.

To develop an IMS, it is necessary to obtain input images, usually acquired by digital cameras, which will then be processed and transformed into output data, in response to a measurement or inspection need. These systems are basically composed of a light source, an image acquisition device and a computer system to analyze the image, as show in Fig. 5.

Some parameters influence the quality of the acquired image, such as the device used in acquiring the image, the lighting system used, and others.

The image acquisition device is the one responsible for creating the representation of an object using the light reflected by it. Such a device is, by itself, a complex system where each of its components may influence the final result of a measurement, that is, in the digital image generated. The schematic of Fig. 6 presents all the basic components of an image acquisition device, such as a camera.

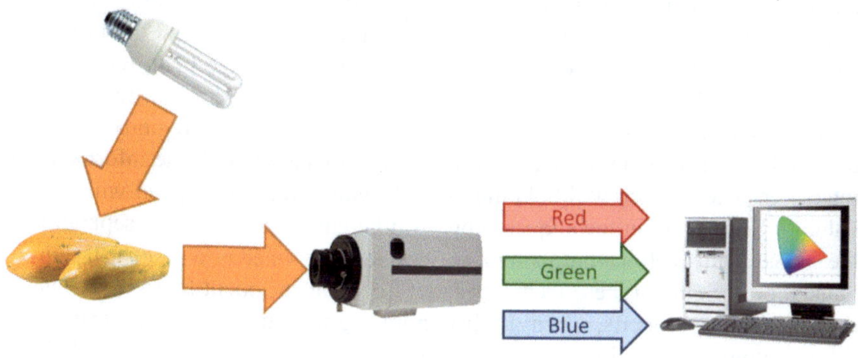

Fig. 5 Components of an image metrology system [7]

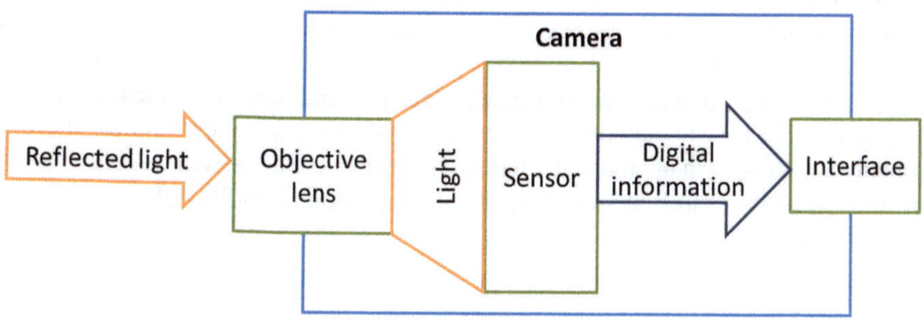

Fig. 6 Basic components of an image acquisition device [8]

There are many devices for image acquisition that may be selected in accordance with the needed physical characteristics, such as: the type of objective lens (standard, high resolution, macro or telecentric); the type of image acquisition device (digital cameras, optical microscopes, profile projector, vision measuring machine); the type of sensor (CMOS or CCD) and the interface with the computer.

The digital cameras and objective lens presents some factors that need to be adjusted in accordance with the application, like focal distance, aperture and exposure time.

For lighting, many types have been used in measurement and classification systems and the final choice of the best light source for a given application requires experimental tests. A system that is successfully used in an industrial process of a given product will not be effective in all possible situations. Each product has its own specificity, size, shape, texture and color, requiring specific detailing when designing a lighting system for higher accuracy.

After an image is captured, it needs to be processed in a way that it becomes easily understandable by the computer. An image contains the data of intensity for all its area, and the pre-processing step has the purpose of isolating regions or objects of interest in the image, or at least making it easier. Several pre-processing techniques can be used such as spatial domain filters (average filter, median filter, mode filter, order filter, Gaussian filter) and also frequency domain filters.

The segmentation step consists in separating an image into regions or objects, usually by transforming an image into a binary image, with all the information of interest in pixels which will hold the value 1, while every other pixel will hold the value 0. Segmentation techniques propose different ways to do this separation on the image, with each having its own way of defining whether a pixel will hold the value 0 or 1. Some of these techniques are: Edge enhancement filters segmentation (Sobel, Prewitt, Roberts, Canny); Histogram segmentation (such as Otsu's algorithm, among others); Image subtraction segmentation. Besides these techniques, there are many morphological operations that are based on the geometric transformation of an image, the most common ones being dilation and erosion, which are used in binary images.

As time passes, with the dissemination of techniques of image processing and analysis, computer vision has begun being used by many areas of science. Before these numerous applications, it was observed the necessity of obtaining values with greater accuracy in edge identification. This demand has motivated studies to develop techniques to determine edges with subpixel accuracy. Unlike the presented techniques, mathematical models were created as a way to identify the position of a border in fractions of a pixel, opposed to the traditional representation by integers.

2 Case Studies

2.1 Dimensional Metrology

Line scales are one of the main standards in dimensional metrology used for image measurements due to its application versatility and high accuracy achieved. Line scales are material standards used mostly to verify and optically calibrate measurement systems, in other words, contactless measurement. These measurement systems require line scales calibrated with high accuracy.

The calibration of a line scale may be achieved in many different ways, depending on the accuracy needed, which can vary from tenths of a micrometer in manual measurements or even a few nanometers in methodologies that use interferometric lasers, etc.

In this case study, a methodology is presented to measure line scales fully based on image processing and analysis, which is the measurement using image registration, where the objective of using this technique is to increase the number of measurement points in a calibration. Usually in a measurement using conventional methods ten points are measured along the scale. The image registration technique enables the measurement of a complete line scale without a significant increase of time.

The calibration of a line scale consists in the determination of the distances between the center of the marks along its length, usually in reference to the first mark of the scale, as shown in Fig. 7.

2.1.1 Image Registration

The image registration technique arises from the need of comparing images obtained in different times or by different sensors. In short, the main objective of image registration is to unite those images. This procedure is done by using

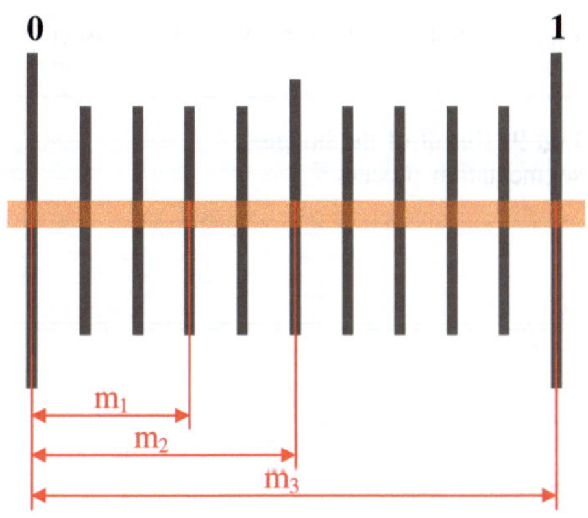

Fig. 7 Example of the determination of distances between marks in a line scale [9]

redundant points in different images [10]. In other words, points that can be seen in two different images must occupy the same location in a new image, causing the reallocation of all the other points of the images [11].

The application of the image registration technique in the calibration of a line scale consists in acquiring several images from its beginning to its end, keeping, in every image redundant parts in order for the application of the technique. With this, it is possible to rebuild the line scale, with good definition, in only one image, enabling the measurement of all the distances.

2.1.2 Technique Application

For the image acquisition a CCD camera coupled to a measuring microscope was used. The camera has a resolution of 768 × 576 pixels and a firewire interface. During image acquisition care must be taken so that in every image there are redundant points with the previous image, as shown in Fig. 8. For a line scale of 6 mm a total of three images were acquired.

In the pre-processing step the images were segmented with a 50 % cut, with the purpose of highlighting only the scale marks [12], resulting in the image shown in Fig. 9.

In the process of applying image registration, the extraction of characteristics is of the utmost importance. In this step, the algorithm must identify the redundant points in sequential images. In this case the geometric center of repeating marks

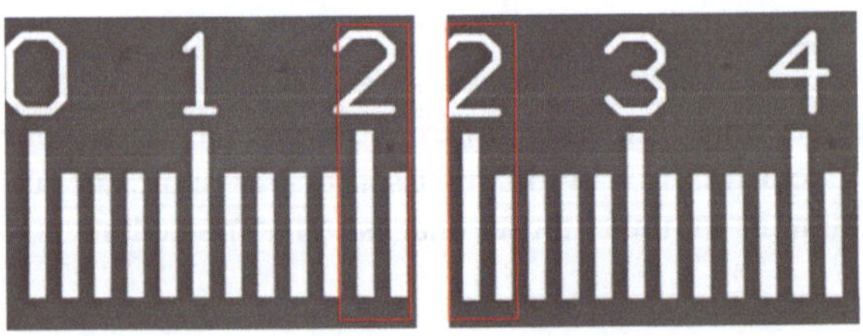

Fig. 8 Common region between two consecutive images

Fig. 9 Result of the image segmentation process

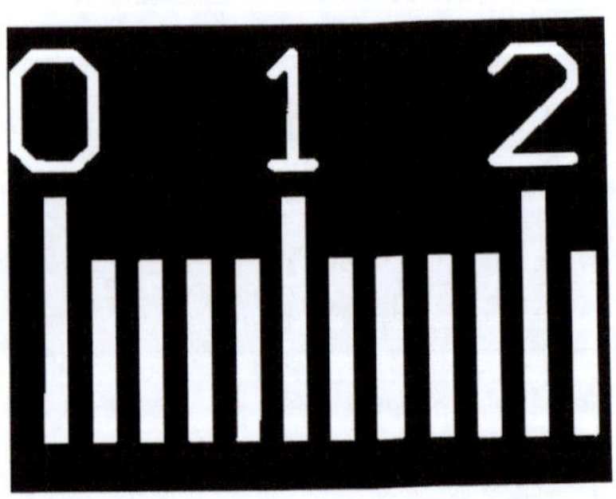

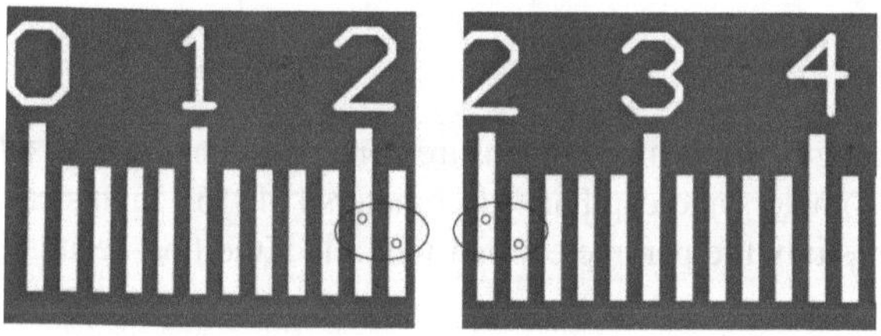

Fig. 10 Points used for image registration

were chosen, as indicated in Fig. 10. The algorithm then sweeps the entire image looking for the marks' center in both images and then storing the x and y coordinates for each center. With those coordinates, it is possible to apply the image registration algorithm [13].

The steps for image registration are as follows:

- In the first image, the final two marks' centers coordinates are stored: x_1, y_1 and x_2, y_2.
- In the second image, the first two marks' (the same marks from the first image) centers coordinates are stored: X_1, Y_1 e X_2, Y_2.
- With the linear system of Eq. (1), both sets of coordinates are mathematically related.

$$\begin{cases} X_1 = a_{11} + a_{21}x_1 \\ Y_1 = a_{31} + a_{41}y_1 \\ X_2 = a_{11} + a_{21}x_2 \\ Y_2 = a_{31} + a_{41}y_2 \end{cases} \tag{1}$$

The results of this system will supply values of the constants a_{ij} for the polynomial that will then position the pixels of the second image on the first image. After doing this procedure for every image, a final image is obtained merging images of 0–2 mm, of 2–4 mm and of 4–6 mm (Fig. 11).

After image registration, the distances between all the marks along the scale can be obtained much faster than conventionally, in which all the distances would be measured separately.

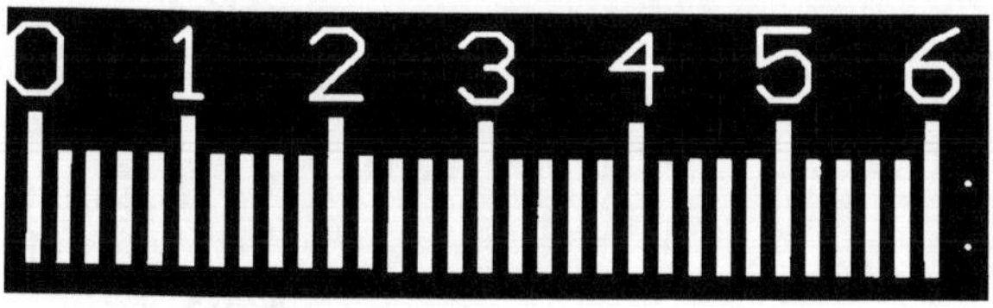

Fig. 11 A 6 mm line scale after image registration

2.2 Fluid Metrology

In fluid metrology, dimensional measurements influence the results of the calibration of hydrometers and capillary viscometers [14–16]. In this case study an IMS will be used to study the parameters that will affect the final results for this specific application.

2.2.1 Lighting System

The depiction of an object in an image is dependent on the type of material in which it is built, being different from the way in which it is seen due to how the light reflects on it. Glassware and glass-made instruments in general are objects that, due to its high reflection index, are difficult to be photographed. Proper lighting will enhance the glass object without generating reflections which can hinder image processing, as shown in Fig. 12 [17].

2.2.2 Image Acquisition

For the image acquisition, a DSLR photographic camera was used. By using this type of camera, it becomes important to properly select the available parameters, the aperture and the exposure time being those of greater influence [8, 18].

While in photography an image is said to be "properly exposed" when its histogram is well divided without peaks on its extremities, this aspect is entirely aesthetic [19], thus being necessary to determine what characterizes a properly exposed image in the metrological sense [18]. The exposure value (EV) guarantees

Fig. 12 Lighting system to enhance glass objects

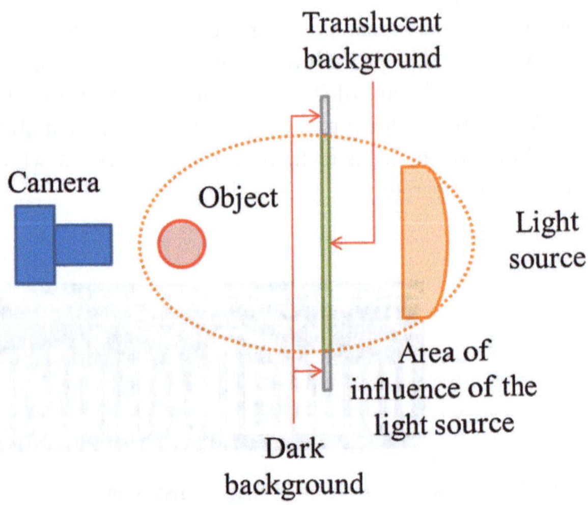

that, from an initial image with a given exposure time and aperture, the change of these parameters in a certain way will generate an image that is apparently equal. However, by varying the aperture, the depth of field will also be changed and, therefore, the sharpness of the object in the image will also change. In this way, the aperture must then be set in the lowest possible value and then the exposure time must be set in a way that will generate a properly exposed image. After acquiring the image, the aperture is increased while diminishing the exposure time, therefore keeping the EV and acquiring a new image. This process is then repeated until the entire set of aperture values are exhausted, generating a set of seemingly identical images. By analyzing this set of images, it is possible to find an aperture value that is optimal for the system in question. As seen in the graph of Fig. 13 it is possible to notice a variation in the measurements, the lowest measure being the one in which the aperture is the greatest, indicating that a higher aperture and lowest depth of field generate a smaller result.

A similar approach is used to evaluate the exposure time. However, instead of using the concept of a properly exposed image, the exposure time varies from a high value in which the object is still observable due to an excess of light to a low value in which the object is still observable due to a lack of light. Any time outside this range will provide the camera an image in which the object is not observable. Inside this exposure time range, images are acquired for each value, and all of them will be analyzed in the same way as the aperture, that is, to find an exposure time that is shown to be the optimum for this system. The graph of Fig. 14 shows the results of exposure time and its influence in the measurement. Considering only the linear region highlighted, a smaller error between each point can be observed. Thus, the optimal exposure time range for this IMS is between 1/80 and 1/160 s.

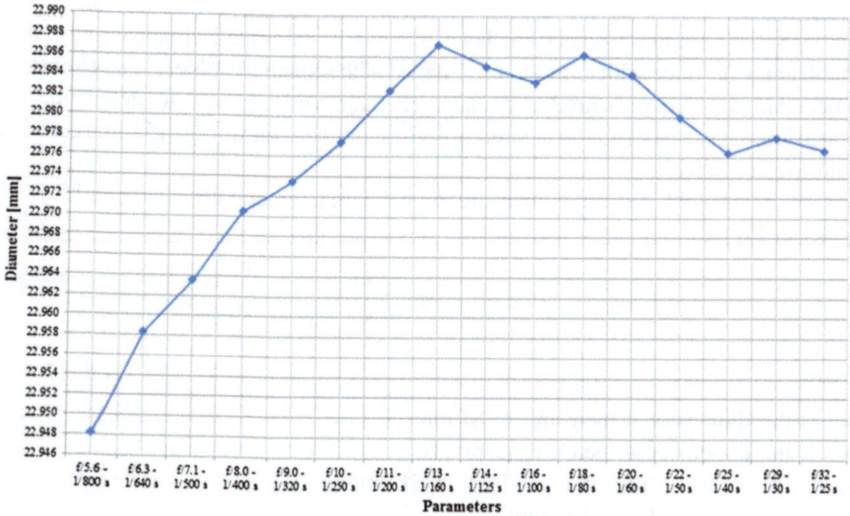

Fig. 13 Diameter variation per each image with the same EV

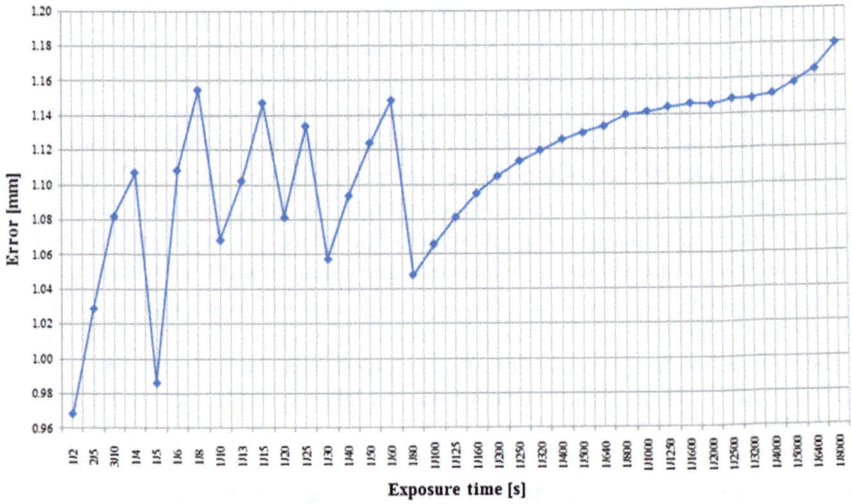

Fig. 14 Measurement error for each exposure time, and region of interest

2.2.3 Traceability by Comparison

After determining parameters that are optimal for image acquisition in this IMS, an acquired image will supply the measurement of any dimension in the object that needs to be used in the mathematical model, where this measure will have the pixel as a unit of measurement. However, for the proper usage of this measurement in the mathematical model, it must have a unit of the SI (International System of Units) attached to it, that being the meter and its multiples and submultiples [6].

To this end, it is necessary to propose that an object is a dimensional standard for this IMS. This standard object has been previously calibrated dimensionally, thus establishing its traceability. From the image of the standard object, the calibrated dimension must be measured and its pixel value will be used to determine the dimension measured from the desired object as can be seen in Fig. 15, guaranteeing metrological reliability.

That being said, the final measure may be mathematically described by the model in Eq. (2).

$$x_D = x_P \cdot \frac{p_D}{p_P} \tag{2}$$

where:

x_D is the dimensional measure, in SI units;
x_P is the dimensional measure, in pixels;
p_D is the measure from the standard, in SI units;
p_P is the measure from the standard, in pixels.

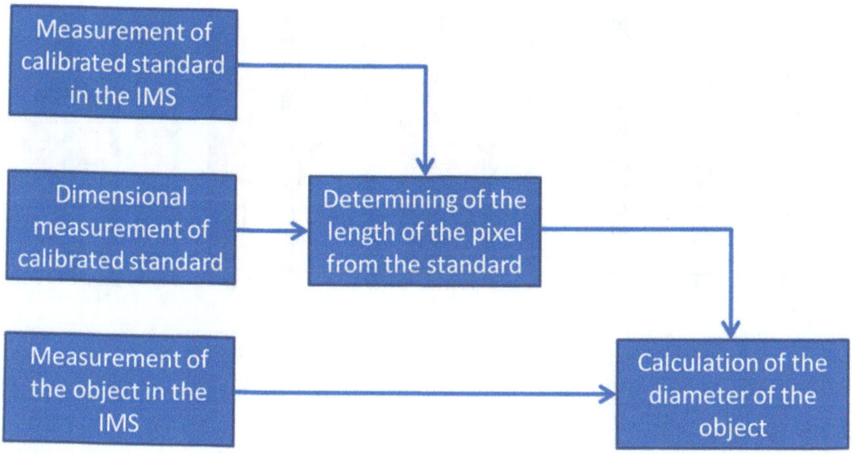

Fig. 15 Steps to determine a traceable diameter by comparison

2.2.4 Measurement Uncertainty

The measurement uncertainty of the measurand represents the doubt in the final result from the measuring process by analyzing the influences described in its mathematical model [20]. Then, to determine the expanded uncertainty from the dimensional measurements from this IMS, it is necessary to calculate the uncertainties for each measurement, from the object and from the standard, both in the IMS and its calibration certificate.

First, it is necessary to determine the estimates for each quantity of influence. This is done by assuming a uniform distribution for the results from the image measurement. The estimates are, in pixel, half of the amplitude of their respective distribution. In other words, it is half of the difference between the highest and lowest values. The estimate for the standard measure in SI units comes from the calibration certificate itself. Another factor to be taken into account is the system repeatability, where its estimate is the standard deviation from the N measurements.

With the estimates available, the standard uncertainties are calculated for each quantity, which will enable, together with the sensibility coefficient to calculate the combined standard uncertainty. The sensibility coefficient determines how much the variation of a quantity influences the final result. The combined standard uncertainty is given by the square root of the quadratic sum of the products of the sensibility coefficient and the standard uncertainties.

To express the uncertainty in terms of the measurement results, the expanded uncertainty is needed, being the product between the combined standard uncertainty and the effective coverage factor, given from a student t distribution.

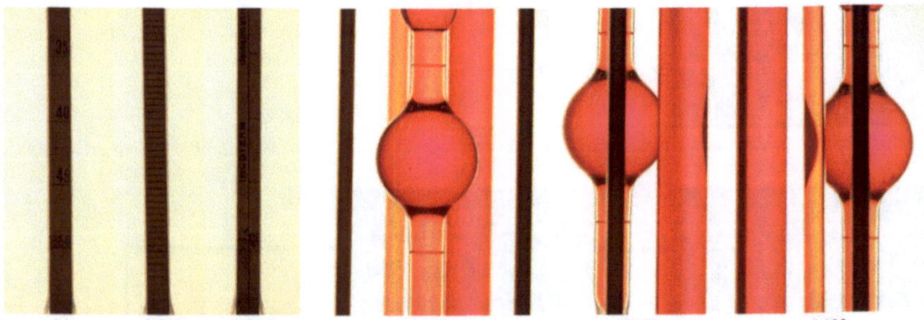

Fig. 16 A hydrometer and a capillary viscometer in three different angles

2.2.5 Results

By applying this IMS to instruments of fluids metrology, the measurements of the diameter of the hydrometer's rod and the diameter of the upper and lower menisci can be found with metrological reliability. For each instrument, 15 measurements were taken, in order to evaluate its precision. An example of hydrometer and viscometer are show in Fig. 16, in three different positions, with the purpose of also evaluating circularity.

The results for each instrument are show, respectively, in Tables 1 and 2, with the measurements for each quantity and the results for the measurement uncertainty.

Thus, the final result of the diameters can be reported with their respective uncertainties in the following way:

- Diameter of the stem of the hydrometer: 4.90 ± 0.14 mm
- Diameter of the upper meniscus of the viscometer: 5.99 ± 0.30 mm
- Diameter of the lower meniscus of the viscometer: 6.00 ± 0.33 mm

Table 1 Results and measurement uncertainty for the hydrometer

	Symbol	Value	Unit
Diameter of the object	x_P	265.62	pixel
Measurement of the standard	p_D	22.98	mm
Measurement of the standard from image	p_P	1244.94	pixel
Calculated diameter	x_D	4.90	mm
Combined uncertainty	$u_C(x_D)$	0.07	mm
Effective degrees of freedom	ν_{eff}	2129	–
Effect coverage factor	k_{eff}	2.00	–
Expanded uncertainty	U	0.14	mm

Table 2 Results and measurement uncertainty for the capillary viscometer

	Symbol	Upper meniscus	Lower meniscus	Unit
Diameter of the object	x_P	345.30	345.39	pixel
Measurement of the standard	p_D	22.98	22.98	mm
Measurement of the standard from image	p_P	1323.44	1323.44	pixel
Calculated diameter	x_D	5.99	6.00	mm
Combined uncertainty	$u_C(x_D)$	0.15	0.16	mm
Effective degrees of freedom	ν_{eff}	34607	22807	–
Effect coverage factor	k_{eff}	2.00	2.00	–
Expanded uncertainty	U	0.30	0.33	mm

3 Conclusion

Reliable measurements using images are becoming more and more important, for example, in the diagnostic of diseases using digital image and in the biometric security field, not restricting itself to industrial applications that aim for increase of productivity and reduction of production cost.

In inspection systems that consider quantitative evaluations of the characteristics of a product or material, it is recommended that the used instruments are properly calibrated and verified by accredited institutions. It is also of the utmost importance that the used procedures generate reproducible results with assured metrological reliability.

There are many applications that justify the employment of new methodologies for contactless measurement. As the applications get more and more sophisticated, they can assure greater quality to manufactured products when considering fundamental metrological questions. In this context, image metrology is presented with the purpose of evaluating a product's conformity, in the case of recognition of inspection, or simply of evaluating mensurand's value, in the case of calibration of standards or instruments. Obtaining reliable results is the main challenge for image measuring systems.

Image metrology systems encompasses all steps of a regular computer vision systems but also allied with all the fundamental care spent in measuring systems, using concepts of repeatability, reproducibility, minimization of errors and uncertainties.

For each step, from image acquisition to the extraction of characteristics, it is analyzed considering the system which will present the least possible errors. It can be observed that the choice of equipment for image acquisition and lighting consists in one of the more relevant steps of the system, followed by image segmentation, in which the dimensional information of objects and standards is extracted.

Regardless of the application, with the evolution of technology and scientific knowledge, it becomes essential that metrological concepts be allied with new digital image inspection techniques, thus promoting results with greater reliability and quality.

References

1. Conci A, Azevedo E, Leta FR (2008) Computação Gráfica: Teoria e Prática, vol 2. Elsevier Editora Ltda, Rio de Janeiro
2. Gomes JFS, Leta FR (2012) Applications of computer vision techniques in the agriculture and food industry: a review. Eur Food Res Technol. doi:10.1007/s00217-012-1844-2
3. Costa PB, Leta FR (2010) Measurement of the aperture area: an edge enhancement algorithms comparison. In: 17th international conference system signals image process, Rio de Janeiro
4. Instituto Nacional De Metrologia Qualidade e Tecnologia (2005) Manual de Barreiras Técnicas às Exportações: o que são e como superá-las. http://www.inmetro.gov.br/barreirastecnicas/Manual_BarrTecnicas.pdf. Accessed 02 Jan 2015
5. BIPM, IEC, IFCC et al (2008) The international vocabulary of metrology—basic and general concepts and associated terms (VIM), 3rd edn. JCGM 200:2012
6. Bureau International des Poids et Mesures (2006) SI Brochure, 8th edn. http://www.bipm.org/utils/common/pdf/sibrochure8.pdf. Accessed 02 Jan 2015
7. Gomes JFS, Leta FR, Costa PB, Baldner F de O (2014) Important parameters for image color analysis: an overview. In: Leta FR (ed) Visual computing: scientific visualization and imaging systems, 1st edn. Springer, Berlin
8. Baldner F de O (2014) Metodologia para Utilização de Sistemas de Visão Computacional em Calibrações da Metrologia de Fluidos. Rio de Janeiro
9. Acko B (2012) Final report on EUROMET key comparison EUROMET.L-K7: calibration of line scales. Metrologia 49:04006. doi:10.1088/0026-1394/49/1A/04006
10. Borges RO, Lima CA, Nascimento ERP, Machado EGB, Lopes GO, Françoso RD, Pereira FJ, Ramos VM, Galvão WS (2007) Métodos e Técnicas de Registro de Imagens SAR/SIPAM a Partir de Dados Orbitais Ópticos e Sensores SAR/SIPAM. 13° Simp. Bras. de Senso. Remoto. Florianópolis
11. Goshtasby AA (2005) 2-D and 3-D image registration. Wiley, New Jersey
12. Pratt WK (2007) Digital image processing, 4th edn. PIKS Scientific Inside, New Jersey
13. Costa PB, Marques A, Baldner F de O, Leta FR (2013) Line scale measurement using image registration. Int J Metrol Qual Eng 4:121–125. doi:10.1051/ijmqe/2013047
14. International Organization for Standardization (1994) ISO 3105: glass capillary kinematic viscometers—specifications and operating instructions. ISO, Geneva
15. ASTM International (2007) ASTM D446: standard specifications and operating instructions for glass capillary kinematic viscometers. West Conshohocken
16. Cuckow FW (1949) A new method of high accuracy for the calibration of reference standard hydrometers. J Soc Chem Ind 68:44–49. doi:10.1002/jctb.5000680203
17. Hunter F, Biver S, Fuqua P (2007) Light science and magic, 3rd edn. Focal Press, Oxford
18. Baldner F de O, Costa PB, Gomes JFS, et al (2013) Determining camera parameters for round glassware measurements. 7° Congr. Bras. Metrol. Ouro Preto
19. Hedgecoe J (2005) O novo manual de Fotografia, 4th edn. SENAC, São Paulo
20. BIPM, IFCC, IUPAC (2008) OIML. Evaluation of measurement data Guide to the expression of uncertainty in measurement JCGM 100: 2008 (GUM 1995 with minor corrections)

Experimental and Numerical Studies of Fiber Metal Laminate (FML) Thin-Walled Tubes Under Impact Loading

Zaini Ahmad, Muhammad Ruslan Abdullah and Mohd Nasir Tamin

Abstract Fiber metal laminate (FML) in form of tubular structures is a modern light-weight structure fabricated by incorporating metallic and composite materials. This present study deals with the impact characteristics and energy absorption performances of fiber metal laminate (FML) thin-walled tubes subjected to impact loading. Dynamic computer simulation techniques validated by experimental testing are used to perform a series of parametric studies of such devices. The study aims at quantifying the crush response and energy absorption capacity of FML thin-walled tubes under axial impact loading conditions. A comparison has been done in terms of crush behaviour and energy absorption characteristics of FML tubes with that of pure aluminium and composite tubes. It is evident that FML tubes are preferable as impact energy absorbers due to their ability to withstand greater impact loads, thus absorbing higher energy. Furthermore, it is found that the loading capacity of such tubes is better maintained as the crush length increases. The primary outcome of this study is design information for the use of FML tubes as energy absorbers where impact loading is expected particularly in crashworthiness applications.

Keywords Fiber metal laminate · Impact · Finite element · Energy absorption · Thin-walled

Z. Ahmad (✉) · M.R. Abdullah · M.N. Tamin
Department of Solid Mechanics and Design, Faculty of Mechanical Engineering, Universiti Teknologi Malaysia, 81310 Johor Bahru, Malaysia
e-mail: azaini@fkm.utm.my

Z. Ahmad · M.N. Tamin
Computational Solid Mechanics, Faculty of Mechanical Engineering, Universiti Teknologi Malaysia, Johor Bahru, Malaysia

A. Öchsner and H. Altenbach (eds.), *Mechanical and Materials Engineering of Modern Structure and Component Design*, Advanced Structured Materials 70,
DOI 10.1007/978-3-319-19443-1_35

1 Introduction

For many years, a number of manufactured lightweight vehicles have become increasingly attractive due to their efficient fuel consumption. Most importantly, the lightweight components incorporated into the vehicle design have to be compromised because of the required level of safety. Thin-walled structures have been categorized as a passive safety measure. This structure is designed to deform permanently to mitigate the crash energy and forces transmitted to the vehicles. A lot of extension research has been carried out to produce light weight thin-walled structures in the form of tubes and plates. Various types of thin-walled tubular structures, namely metallic, composite and polymers tubes, have been used as lightweight energy absorbers to mitigate the adverse effect of impact and hence protect the vehicle or structure in a well-controlled manner [1–5].

Composite (GFRP) tubes often exhibit a great energy absorption capacity under impact loading, only to a certain extent since they tend to eventually collapse under catastropic or brittle failure. It is well-known that fiber delamination or breakage (catastropic failure) may cause a rapid decrease in energy absorption capacity [6, 7]. From an energy absorption point of view, a thin-walled tube should collapse progressively with a stable crush response in order to effectively mitigate the kinetic energy. A FML thin-walled tube is a combination of suitable properties of metals and fiber reinforced composite to form a hybrid material. Such a tube has a capability of providing ductile, stable plastic collapse mechanisms [8]. Therefore, it is anticipated that FML tubes may also offer a progressive crushing and confine the brittle failure of the composite layer due to the presence of aluminium layers on the tube configuration. There have been numerous studies on the crushing and energy absorption response of FML materials, however only in the form of sandwich panels under impact loading conditions [9–11]. For instance, glass fiber laminate (GLARE) plates have previously been found that could sustain impact load effectively in aviation applications [12]. However, the application of FML materials in the forms of thin-walled tubular structures has not yet been available and may be sparse in open literature.

It is therefore of particular interest to explore and examine the use of FML tubes as energy absorbing devices. It is anticipated that the crush response of the device may differ to that of pure metallic or composite device. The presence of metallic and composite layer stacks seems to improve crush stability and enhance the energy absorption capacity of the tubular structure [13–15]. The present study discusses the energy absorption performance of FML thin-walled circular tubes under axial impact loadings using validated computer models. The relative effect of material parameters and the number of layers on the energy absorption responses was examined. Overall the results demonstrate the advantages of using FML tubes as energy absorbing devices with a noticeable improvement in energy absorption capacity.

2 Fabrication of FML Tubes and Experimental Tests

For experimental testing, a few numbers of samples (2×1 ($m \times n$) and 3×2 stacking layers) with a diameter of 55 mm and length of 100 mm was fabricated using a conventional technique. In particular, m and n signify the number of layer for metallic and composite layers, respectively. The fabrication process involved is quite bold for producing the best tube samples. Anodized aluminium sheet with a thickness of 0.3 mm has been used as the metallic layer while glass reinforced composites in the form of chop strand mats (CSM) and woven roving with also a thickness of 0.3 mm has been used for composite layers. Al sheet has the following mechanical properties: initial yield stress, σ_Y = 160.4 MPa; Young's modulus, E = 71 GPa; Poisson ratio, ν = 0.31 and density, ρ = 2700 kg/m^3, while the glass reinforced composite has a Young's modulus, E_a = 30.9 GPa, E_b = 8.3 GPa, E_c = 0 GPa; Poisson ratio, ν = 0.087 and density, ρ = 1800 kg/m^3.

Figure 1 illustrates the fabrication process of 2×1 FML thin-walled tubes. A steel mandrel with the same diameter to the tube was used to form the FML tube. Firstly, poly-ease is applied on the mandrel. The Al sheet is then spun on the mandrel to form the first layer of the tube (inner layer). CSM or woven roving prepreg 0.3 mm is manually wrapped on the first Al layer to form the second layer. Next, the second Al sheet is spun on the second layer to form the third layer (outer layer). A similar process will be repeated to fabricate 3×2 samples. The later step is to inject the resin into the composite layer. Both end tubes are covered with fiber tissue and a spiral tube is firmly inserted at both ends. The spiral tubes are connected to inlet and outlet hoses. The whole specimen is covered with plastic and sealed with a tar by using the vacuum infusion technique. In this project, epoxy resin has been used as matrix constituent. The resin flows through the inlet hose and spread evenly along the tube for 20–30 min. The tube sample is cured for nearly 24 h before being removed from the mandrel. Lastly, a cured FML tube is removed from the mandrel by using a special designed rig. Surface finishing has to be done prior to the testing.

To ensure repeatability with minimal experimental errors, five FML tubes were considered for each case. However, only the best representative test results are presented in this paper. The compression test set-up is shown schematically in Fig. 2. Loading was achieved by lowering the crosshead such that the specimen was crushed at its free end onto the load-bearing surface of the testing machine. The specimen was also free to slide over the load-bearing surface. All tests were conducted at a crosshead speed of 5 mm/min.

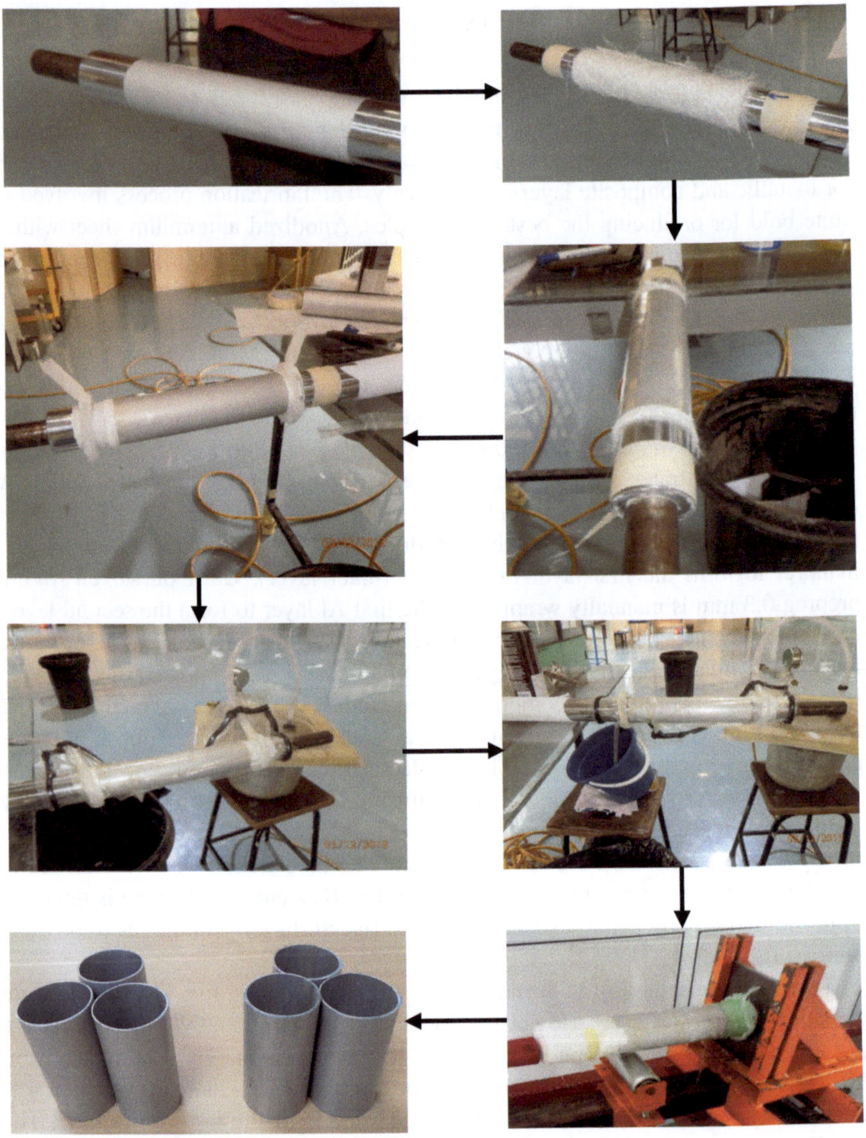

Fig. 1 Fabrication process of 2 × 1 FML thin-walled tubes

Fig. 2 Quasi static compression test set-up for axial loading

3 Development and Validation of the Finite Element Model

All computers modelling in the present study was conducted using the nonlinear finite element (FE) code LS-DYNA 971 [16]. A model of the FML tubes was developed using quadrilateral Belytschko-T say shell elements with 5 integration points through the thickness as shown in Fig. 3. From the convergence study,

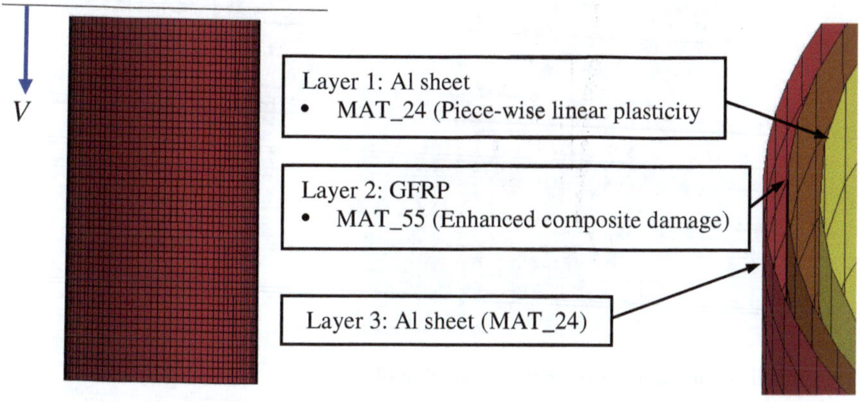

Fig. 3 Finite element discretization of FML thin-walled tube

element sizes of 2.5, 2.8 and 3.0 mm were chosen for the inner, second layer and outer tubes, respectively. A piece-wise linear plasticity and enhanced composite damage material model were used to model the Al tube and GPRP respectively. The moving mass in the experiments was modelled as the analytical rigid body since its properties do not contribute significantly to the analysis. Loading was simulated by prescribing a ramp velocity-time history to the moving rigid body over a crushing duration in order to speed up the analyses and to ensure an accurate and efficient quasi static analysis motion of this moving mass. The moving mass was constrained to translate vertically only along the *z* axis.

A reliable contact algorithm for the interaction between the tube layers is rigorous and needs to be established precisely. Self-contact interaction was simulated using an "automatic single surface" to the part of each tube in order to avoid interpenetration of the tube wall. Since the FML tubes consist of two different materials and their stiffness, the contact condition between those parts has to be duly simulated. Thereby, "automatic node to surface" and "automatic surface to surface" contacts were defined for the contact interfaces of the top rigid body with the top end points of the whole tube and the contact interfaces between each tube layer, respectively. Initial imperfections were also considered in the present study to represent the tube wall warping of the fabricated sample. The collapse initiator is modelled to reduce the initial peak load and hence provide a relatively stable load response as it is crushed (Fig. 4).

Figure 5 shows the comparison of the experimental and numerical static load-deflection responses for the FML tubes. In this figure, the average value of all five tests is compared with the numerical result. The load in general fluctuates about a

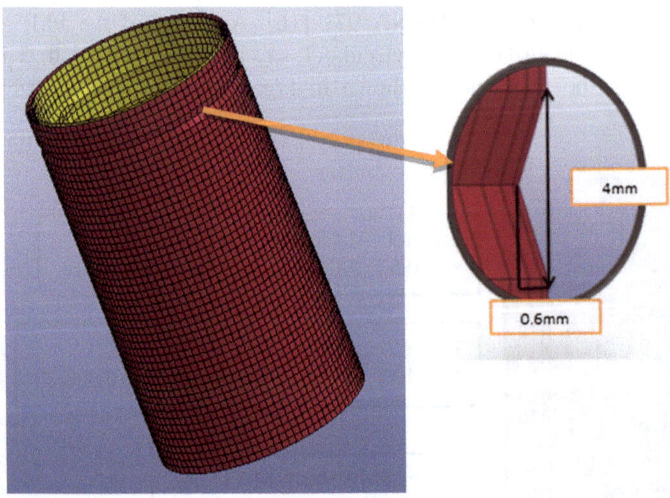

Fig. 4 Collapse initiator representing imperfection of tube wall

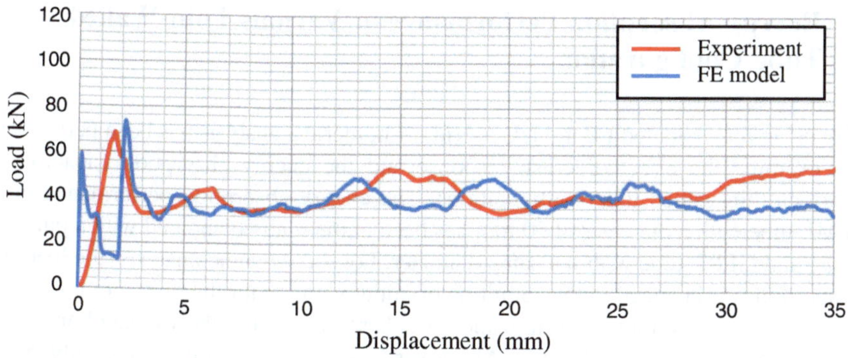

Fig. 5 Load displacement response of FML thin-walled tube under axial loading

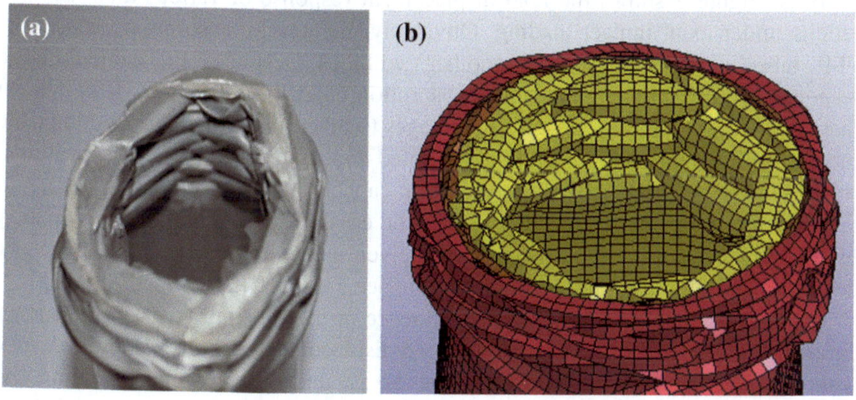

Fig. 6 Deformation profiles of FML thin-walled tube **a** experiment, **b** FE model

mean value due to the formation of lobes as the tube under goes progressive crushing. The main trends in the experimental results for the tube are well captured by the numerical simulation. Also of interest for validating the numerical model is the deformation mode simulated by the FE model. The deformation mode predicted by the FE model was hence compared with that of the experimental results as depicted in Fig. 6. All the specimens in the set of experiments underwent progressive crushing initiated at the loaded end. It can be seen that the predicted lobe formation and the number of lobes for the tube differ slightly from the experimental result. This is mainly due to a slight rupture occurring in all the experimental samples which influenced the lobe formation very slightly. In general, the lobe formation and deformation mode predicted by the FE model agree favorably with the experimental results, and thus the FE model is capable of simulating the physical response of FML tubes undergoing axial loading.

4 Energy Absorption Performance of Various Thin-Walled Tube Configuration

By using the validated FE model, a series of numerical analysis has been performed to examine the effect of crush response and the energy absorption capacity of various thin-walled tube configurations. The energy absorption performance of a FML tube with a diameter of 60 mm has been numerically compared with a pure Al tube and GFRP tube with the same diameter. For the dynamic analysis, the loading condition was changed to the definition of initial velocity for the rigid moving mass, and the rest of the model discretization was kept identical to the validated one. The initial velocity used is 10 m/s. Moreover, the strain rate effect has only been included in the composite material model. Since the aluminium material is strain rate independent, no strain rate parameter has been employed in the numerical analysis. Figure 7 shows the load displacement response of various tube configurations under axial impact loading. Obviously, the energy absorption capacity of FML tubes is greater than that of another two tubes. Despite of the stable collapse response of GFRP tubes, the energy absorption capacity is the least. It is evident that the FML tube may offer some advantages to other purely metallic or composite tubes. More importantly, it should be noted that a FML tube shows profound responses and a desirable energy absorption capacity when compared to a pure Al tube and a GFRP tube. In addition to this, the crush response and absorbed energy by a FML tube is more prominent up to a given deflection and such a tube can be deemed as an effective energy absorbing device.

It is also of interest to evaluate the influence of the stacking layer on the energy absorption capacity. Figure 8 shows the load displacement response of an FML tube with a configuration layer of 2 × 1 and 3 × 2 under axial impact loading. As anticipated, the 3 × 2 FML tube displays the higher energy absorption capacity due to the increasing amount of material deformation and greater interaction effect between the layers. It is also noteworthy that the energy absorption capacity increases with increasing deflection and numbers of tube layers. However, the

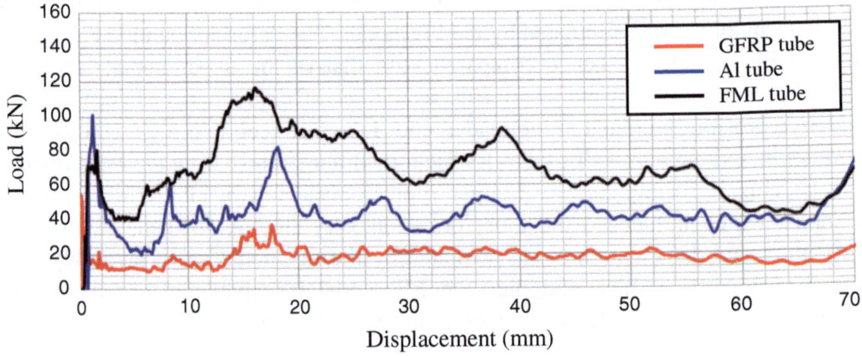

Fig. 7 Load displacament response of various tube configuration

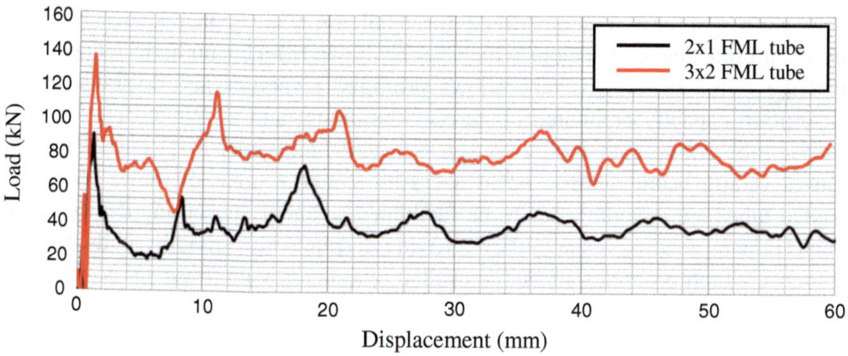

Fig. 8 Load displacament response of FML tube with two stacking layer configurations

Table 1 Energy absorption indicator of various tube configuration

Tube configuration	P_m (kN)	P_P (kN)	SEA (kJ/kg)	η_C
GFRP	18.5	57	8.75	0.32
Al	37.5	102	10.26	0.36
FML	61.3	82	17.16	0.74

presence of additional stacking layers on FML tubes does not greatly influence the load-deflection response of the energy absorbers.

From an energy absorption point of view, the energy absorption performance does not solely depend on the absorbed energy. There is another energy absorption indicator to evaluate this performance, namely specific energy absorption, mean load, initial peak load and crush force efficiency. Table 1 tabulates the numerical results of energy absorption indicator for the aforementioned tube configuration.

Owing to the highest energy absorption capacity of the FML tube, most of the energy absorption indicators are more desirable by using such tube configurations, particularly the mean load, specific energy absorption and crush force efficiency. From Table 1, it is obvious that the FML tube may offer superior energy absorption performance with a stable progressive crushing as evident by crush force efficiency (ratio of mean load and initial peak load). From a crashworthiness point of view, minimizing the initial peak load is indispensible in a crash event to prevent volatile impact load transmitted to the survival room. However, the initial peak load of the FML tube is yet greater than that of GFRP tube.

5 Conclusion and Remark

The present paper has investigated the crush and energy absorption response of a FML tube in comparison with other tube configurations. A finite element model of thin-walled tubes was validated using experimental techniques, and the energy absorption response was quantified for axial dynamic impact loads. The most important practical conclusions which can be drawn from the numerical analysis are that the thin-walled FML tube shows profound responses and desirable energy absorption capacity compared to purely Aluminium and GFRP tubes. Moreover, such a tube configuration offers progressive crushing under axial impact loading, thus avoiding brittle failure. From a crashworthiness point of view, FML thin-walled tubes are advantageous to be facilitated as energy absorbing devices under axial impact loadings since they provide a relatively smooth load-deflection response which promotes stable, controlled retardation with lower initial peak load as required during an impact event.

Acknowledgments This project is supported by the Ministry of Higher Education (MOHE) Malaysia under Fundamental Research Grant Scheme (FRGS) Vote No. R.J130000.7824.4F248. Sincere appreciation and acknowledgement also goes to Universiti Teknologi Malaysia (UTM) for the continuos support in completing this project.

References

1. Nagel GM, Thambiratnam DP (2004) Dynamic simulation and energy absorption of tapered tubes under impact loading. Int J Crashworthines 9(4):389–399
2. Hanssen AG, Langseth M, Hopperstad OS (2000) Static and dynamic crushing of circular aluminium extrusions with aluminium foam filler. Int J Impact Eng 24(5):475–507
3. Ahmad Z, Thambiratnam DP (2009) Dynamic computer simulation and energy absorption of foam-filled conical tubes under axial impact loading. Comput Struct 87(3–4):186–197
4. Alghamdi AAA (2001) Collapsible impact energy absorbers: an overview. Thin-walled Struct 39(2):189–213
5. Santosa SP, Wierzbicki T, Hanssen AG et al (2000) Experimental and numerical studies of foam-filled sections. Int J Impact Eng 24(5):509–534
6. Kathiresan MK, Manikandan V (2014) Crashworthiness analysis of glass fibre/epoxy laminated thin walled composite conical frusta under axial compression. Compos Struct 108:584–599
7. Lau STW, Said MR, Yaakob MY (2012) On the effect of geometrical designs and failure modes in composite axial crushing: A literature review. Compos Struct 94(3):803–812
8. Vlot A (1996) Impact loading on fibre metal laminates. Int J Impact Eng 18(3):291–307
9. Sadighi M, Alderliesten RC, Benedictus R (2012) Impact resistance of fiber-metal laminates: a review. Int J Impact Eng 49:77–90
10. Chai GB, Manikandan P (2014) Low velocity impact response of fibre-metal laminates—a review. Compos Struct 107:363–381
11. Morinière FD, Alderliesten RC, Benedictus R (2014) Modelling of impact damage and dynamics in fibre-metal laminates—a review. Int J Impact Eng 67:27–38
12. Kathiresan M, Manisekar K, Manikandan V (2012) Performance analysis of fibre metal laminated thin conical frusta under axial compression. Compos Struct 94(12):3510–3519
13. Lin JS, Wang X, Lu G (2014) Crushing characteristics of fiber reinforced conical tubes with foam-filler. Compos Struct 116:18–28
14. Tarlochan F, Samer F, Hamouda AMS et al (2013) Design of thin wall structures for energy absorption applications: enhancement of crashworthiness due to axial and oblique impact forces. Thin-walled Struct 71:7–17

15. Kim HC, Shin DK, Lee JJ et al (2014) Crashworthiness of aluminum/CFRP square hollow section beam under axial impact loading for crash box application. Compos Struct 112:1–10
16. Hallquist JO (2006) LS-DYNA 3D theoretical manual. Livermore Software Technology Corporation, Livermore

Dynamic Calibration Methods of Accelerometer in Vibration-Temperature Combined Environment

Chun zhi Li, Ying Chen and Tong Zhou

Abstract The ground simulation experiment of vibration-temperature combined environments is an important means of aircraft testing for environmental adaptability research and structural response analysis. An accelerometer is used to measure structure mechanics parameters with vibration loading, while the sensitivity of the accelerometer drifts as temperature starts to rise, something which could affect the precision of vibration experiment as well as the data obtained from the test. Based on the acceleration calibration method and standards, this paper expounds the key points of dynamical calibration technique with vibration-temperature combined loading, and also introduces some latest experimental methods and calibration devices, which provide an accurate and controlled means of sensitivity calibration of the accelerometers in accordance with back-to-back comparative calibration standards. At last, we combined accelerometer's calibrations and the obtained data, to establish the sensitivity correction of the temperature drift model to improve precision of the combined load during the ground simulation experiment.

Keywords Combined environment · Accelerometer · Dynamic calibration · Temperature response

C.z. Li (✉) · Y. Chen · T. Zhou
Institute of System Engineering, China Academy of Engineering Physics,
P.O. Box 919-402, 621900 Mianyang, Sichuan, People's Republic of China
e-mail: liczz@aliyun.com

Y. Chen
e-mail: Chenying0315@163.com

T. Zhou
e-mail: zhoutong@163.com

A. Öchsner and H. Altenbach (eds.), *Mechanical and Materials Engineering of Modern Structure and Component Design*, Advanced Structured Materials 70,
DOI 10.1007/978-3-319-19443-1_36

1 Introduction

The ground simulation experiment of vibration-temperature combined environments is an important means of aircraft testing for environmental adaptability research and structural response analysis. MIL standard STD-810-F and GJB150A also recommend that experiments of vibration-temperature combined environments should be studied in case of requirements [1, 2]. Thus the combined environmental research is now achieved more and more attention in many studying areas.

Dynamic calibration of combined environments is one of the key factors for a successful experiment. In the beginning of the last century, the United States National Institute of Standards and Technology has made criterions of dynamic testing calibration methods for high temperature-vibration accelerometer specifications from different aspects, which is international standard ISO16063-11, Methods for the calibration of vibration and shock transducer-Part21: Vibration calibration by comparison to a reference. The manufacturer Endevco and B&K company have established their own vibration-temperature combined environment of dynamic calibration devices for accelerometers, which are also suitable for the production of their own sensor calibration and other test products. Though the technology of the combined calibration device is difficult, and patent design is protected, less literature about combined devices and relevant methods of calibration data is published so far. At present there are few domestic research works carrying out combined dynamic calibration techniques mainly including the criterions of the acceleration calibration method. The National standard GB/T20485 specifies vibration and shock sensor calibration experiments and its testing method [3], yet the dynamic calibration research in combined environments is still in the initial research stage, whose application is not popularized with less dynamic calibration standards.

An accelerometer is used for measuring the structure dynamic response in vibration environment or response to a vibration environment, while the thermal sensitivity drift of accelerometer is an indicator of the sensitivity deviation changing with the temperature shift, and also is one of the most important temperature stability indexes of the sensor. Currently there exist unsolved problems such as vibration control and response bias caused by accelerometer thermal sensitivity drift under combined environment. Research shows that accelerometers used in environments of high temperatures (higher than 250 °C), due to its internal combination of quartz, ceramic cutting technology or microelectronics circuit, the output response at 120 °C of its sensitivity deviation is almost at 8 % with the temperature rising [1, 2]. So the control precision of simulation environment testing and analysis of data have been greatly influenced by temperature sensitivity deviation. Therefore it has been greatly significance for obtaining accuracy data to develop dynamic calibration technology in combined environment.

2 Calibration Method

The dynamic calibration method for accelerometers using comparison method to measure the sensitivity of a calibrated sensor on the vibration device given by a certain g-value and a specific frequency in different temperature conditions [3, 5], and then according to Eq. 1 the sensitivity is calculated which is the basic calculation of thermal sensitivity drift as described as Eq. 2 below.

$$S2 = (X2/X1) * S1 \tag{1}$$

where:

S1 standard accelerometer sensitivity, mV/g;
S2 calibrated accelerometer sensitivity, mV/g;
X1 standard accelerometer, mV;
X2 is calibrated accelerometer outputs are, mV.

The thermal sensitivity drift as Eq. 2 follows:

$$\alpha = (STi - S)/(Ti - T) \tag{2}$$

where

α thermal sensitivity drift, %/FS/°C;
STi sensitivity under different temperature, mV/g;
S standard temperature sensitivity, mV/g;
(Ti − T) difference between the test temperature and standard temperature, °C.

The sine fixed frequency method [6–9] is adopted by combined dynamic calibration given by the excitation frequency and amplitude according to the different temperatures, and the amplitude linearity relative to the reference output is measured and the relative deviation is calculated as well. The least square is recommended for calculating the amplitude linearity of sensitivity in the case of a higher acceleration response. The continuous scanning method (or point by point) can also be used for dynamic calibration, within the scope of the working frequency by 1/3 octave frequency, respectively to different temperature environments.

A combined calibration device is schematically drawn in Fig. 1. First the calibration device and the selected equipment should meet the requirements of GB/T13823.3 comparison test method of basic index on temperature response. Second thermal isolation measures should be taken between the sensor and temperature box to ensure the vibration generator working well while the temperature changes. Additionally according to the GB/T13823.3 regulation, the join stinger made by ceramic should have a coaxial rigid connection between the standard and calibrated sensors with less influence on the transfer function of the join assembly.

As Fig. 1 shows, the main factors influencing calibration results have the following aspects: first the thermal sensitivity drift of the standard accelerometer, for

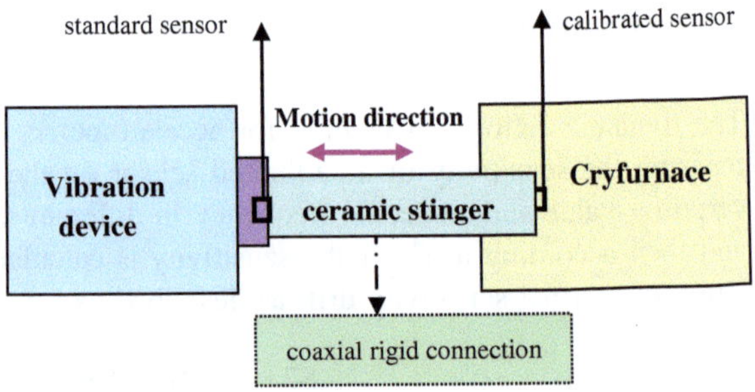

Fig. 1 Schematic of combined calibration device

most piezoelectric sensors, its thermal sensitivity drift index is not given by generally applies sheet. In fact, thermal drift is certainly existing the drift would directly affect the corresponding temperature sensitivity. So the standard accelerometer should be installed on the outside of the cryofurnace, as to try to have the standard sensor working in constant temperature environment. Second is the ceramic stinger whose stiffness should be tested in a high and low temperature cryofurnace. It is requested that the rigidity of the connecting stinger should be designed large enough in order to transfer the vibration without distortion, as well as to ensure two accelerometers having the same value of vibration response.

One of the ceramic stinger ends is fixed to the vibration device, the other end into a cryofurnace. According to the theory of mechanics of materials [6], bending stiffness kw and tensile stiffness kl are calculated respectively:

$$\mathrm{kw} = 3\mathrm{EJ}/\mathrm{l3} \tag{3}$$

$$\mathrm{kl} = \mathrm{SE}/\mathrm{l} \tag{4}$$

For the circular cross section, the Eq. 3 as follows:

$$\mathrm{kw} = \pi \mathrm{Ed4}/(21.3\mathrm{l3}) \tag{5}$$

The connecting stinger natural frequency is given by Eq. 6:

$$\omega \mathrm{n} = \mathrm{k}/(\mathrm{m} + 0.23\mathrm{M})1/2 \tag{6}$$

where

M total mass of the accelerometer and installation, in kg;

m mass of connecting stinger quality, in kg.

To have the sensor working without distortion on the vibration table, Eq. 7 must meet the natural frequency/ωn five times greater than the testing frequency of ω.

$$\omega n \geq 5\omega \tag{7}$$

Thus the ceramic stinger's size of Eq. 8 can be determined by the Eqs. 6 and 7.

$$d4/l3 \geq 532.5\omega 2(M + 0.23m)/\pi E \tag{8}$$

Through Eq. 8, the measured frequency can be improved by changing the ceramic stinger geometry size to make the thermal sensitivity drift of the accelerometer calibration more accuracy.

3 Combined Calibration Experiment

3.1 Calibration Device

The calibration device adopts combined assembly as shown in Fig. 2 which is developed by PCB, the rigid connection rod is designed with high pressure strength and heat-resistant ceramic connections. The calibrated accelerometer is installed inside the cryofurnace, which is of the type is B&K 4375, working temperature range is from −50 to +250 °C. The steady sine excitation comparison method is adopted to improve the calibration accuracy, while the device's main characteristics are from −74 to +650 °C with a frequency response from 5 Hz to 2 kHz.

Testing process: combining a sine function wish a 1 g amplitude and a vibration loading with a 160 Hz frequency, the temperature range was from room temperature to 250 °C with interval of 10 °C, therefore, the calibration experiment will be carried out in a combined assembly with a cryofurnace.

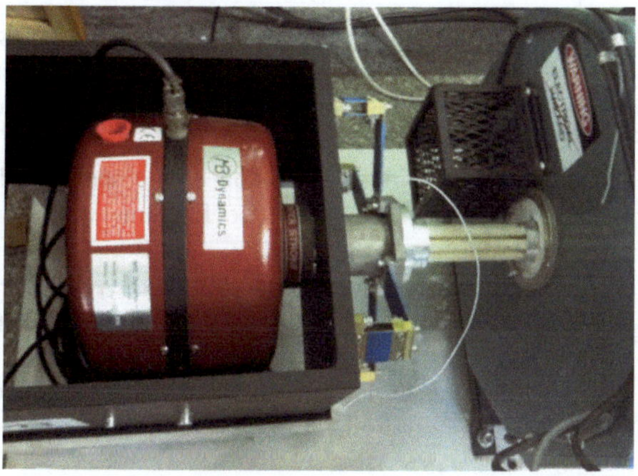

Fig. 2 Combined assembly with cryofurnace

Table 1 Calibration data

Temperature (°C)	Sensitivity (pC/g)	Sensitivity deviation (%)
25	3.109	0.00
60	3.219	3.54
70	3.231	3.92
80	3.249	4.50
100	3.287	5.73
110	3.320	6.79
120	3.317	6.69
130	3.340	7.43
140	3.348	7.69
150	3.359	8.04
160	3.362	8.14
170	3.384	8.85
180	3.393	9.13
190	3.433	9.44
200	3.441	9.69
220	3.537	12.75
240	3.566	13.68
250	3.577	14.03

3.2 Calibration Data

Calibrated data of the accelerometer sensitivity and its temperature sensitivity deviation are shown in Table 1.

3.3 Analysis Results

A combined calibration device can provide sine excitation vibrations with frequencies of 160 Hz. From Table 1, the sensitivity increases with temperature drift, the higher the temperature is, the greater the thermal sensitivity drift is, as shown in Fig. 3. Sensitivity deviation is close to 10 % when the temperature is about 200 °C. Generally the sensitivity deviation of the sensor is linearly increasing with temperature increases, also it can be fundamentally correction used as temperature compensation.

Through with the combined calibration experiment, the limitation is that the calibration temperature load is not precisely as designed to heat up from inside the cryfurnace, and thus it can affect the calibration results due to the internal air control. It is suggested that the characters of the calibration device should be verified compared to other similarly assembly.

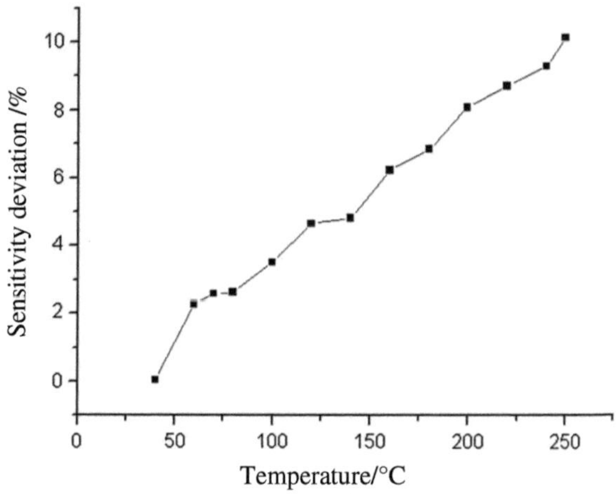

Fig. 3 Temperature—sensitivity curve

4 Conclusion

Dynamic calibration methods for accelerometers in vibration-temperature combined environments were explored in this project. The combined calibration device and testing methods were also introduced including its key technique of connecting rod design. It is applied for accelerometer calibration by a single assembly. Accuracy data was also provided. With the calibration results the sensitivity correction model of temperature drift was established which is effective in improving precision of the combined load during the ground simulation experiment. Furthermore it was proven that a reasonable design of the combined calibration device has a high importance in developing the calibration technique.

References

1. MIL-STD-810F (2000) Environmental engineering considerations and lab test, USA
2. GJB150A-2009 (2009) Military equipment laboratory test method, Beijing, China
3. The national standard GB/T20485 (2008)
4. Endevco vibration&shock acceleration handbook. V2010, USA
5. B&K vibration&shoch acceleration manual. V2011, Danmark
6. Random Vibration (2003) Mechanical vibration and shock series, vol III. Applied mechanics reviews, USA
7. International standard ISO16063-11 (2003) Methods for the calibration of vibration and shock transducer-Part 21: Vibration calibration by comparison to a reference transducer
8. Piezoelectric accelerometer calibration procedures (2008) National metrological verification procedures JJG233
9. Xiguang Y (1989) Sensor technology handbook. Beijing, China
10. Chan CC Analytical method validation and instrument calibration. Wiley, New York
11. S.M etc (2002) Material mechanics. Beijing, China

Erratum to: Effect of Simultaneous Plasma Nitriding and Aging Treatment on the Microstructure and Hardness of Maraging 300 Steel

Adriano Gonçalves dos Reis, Danieli Aparecida Pereira Reis, Antônio Jorge Abdalla, Jorge Otubo, Susana Zepka, Antônio Augusto Couto and Vladimir Henrique Baggio Scheid

Erratum to: Chapter 'Effect of Simultaneous Plasma Nitriding and Aging Treatment on the Microstructure and Hardness of Maraging 300 Steel' in: A. Öchsner and H. Altenbach (eds.), *Mechanical and Materials Engineering of Modern Structure and Component Design*, Advanced Structured Materials 70, DOI 10.1007/978-3-319-19443-1_22

The original version of the chapter 'Effect of Simultaneous Plasma Nitriding and Aging Treatment on the Microstructure and Hardness of Maraging 300 Steel' Fig. 1 was inadvertently published without the part Fig. d.

The online version of the original chapter can be found under
DOI 10.1007/978-3-319-19443-1_22

A.G. dos Reis (✉) · D.A.P. Reis · J. Otubo · S. Zepka
Instituto Tecnológico de Aeronáutica, 12228-900 São José dos Campos, Brazil
e-mail: areis@ita.br

D.A.P. Reis
e-mail: danieli.reis@unifesp.br

D.A.P. Reis
Universidade Federal de São Paulo, 12231-280 São José dos Campos, Brazil

A.J. Abdalla · V.H.B. Scheid
Instituto de Estudos Avançados, 12228-001 São José dos Campos, Brazil
e-mail: abdalla@ieav.cta.br

A. Öchsner and H. Altenbach (eds.), *Mechanical and Materials Engineering of Modern Structure and Component Design*, Advanced Structured Materials 70,
DOI 10.1007/978-3-319-19443-1_37

The correct Fig. 1 is appear here.

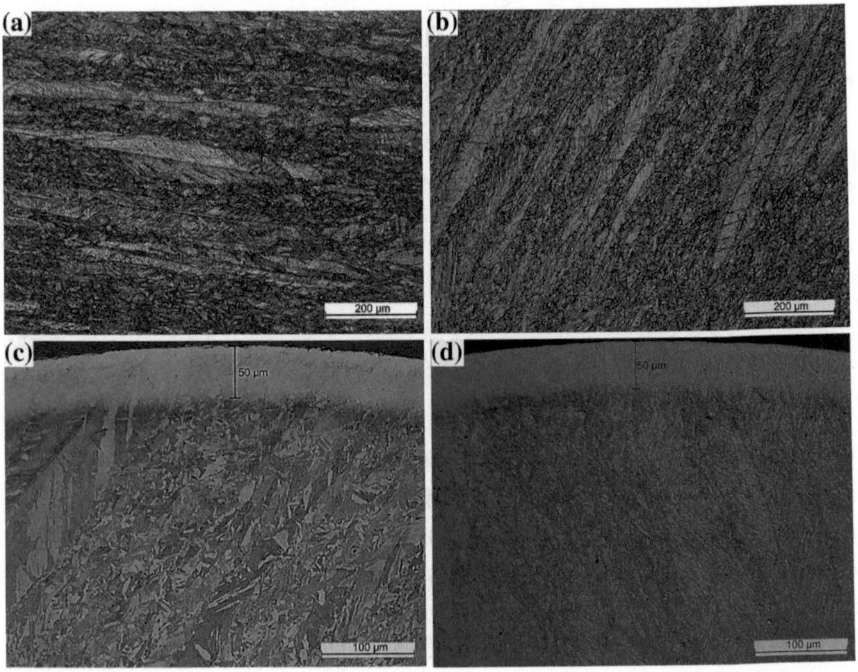

Fig. 1 Optical micrographs of maraging 300 **a** solution annealed, **b** solution annealed and aged, **c** solution annealed and plasma nitrided and **d** solution annealed, aged and plasma nitrided

A.A. Couto
Instituto de Pesquisas Energéticas e Nucleares and Mackenzie, 05508-900 São Paulo, Brazil
e-mail: acouto@ipen.br

A.G. dos Reis
Rua Matias Peres, 364, Bairro Floradas de São José, São José dos Campos SP, 12230-082, Brazil

MIX
Papier aus verantwortungsvollen Quellen
Paper from responsible sources
FSC® C105338

If you have any concerns about our products,
you can contact us on
ProductSafety@springernature.com

In case Publisher is established outside the EU,
the EU authorized representative is:
Springer Nature Customer Service Center GmbH
Europaplatz 3, 69115 Heidelberg, Germany

Printed by Libri Plureos GmbH
in Hamburg, Germany